Office Hours with a Geometric Group Theorist

Office Hours with a Geometric Group Theorist

Edited by Matt Clay and Dan Margalit

PRINCETON UNIVERSITY PRESS

PRINCETON AND OXFORD

Published by Princeton University Press,
41 William Street, Princeton, New Jersey 08540

In the United Kingdom: Princeton University Press,
6 Oxford Street, Woodstock, Oxfordshire, OX20 1TR

press.princeton.edu

Cover art: Illustration of door courtesy of Shutterstock.
Other illustrations by Anne Karetnikov

Photo of Tara Brendle by Jonathan Tickner

ISBN 978-0-691-15866-2
Library of Congress Control Number: 2016960820

British Library Cataloging-in-Publication Data is available

This book has been composed in Times

Printed on acid-free paper. ∞

press.princeton.edu

Typeset by Nova Techset Pvt Ltd, Bangalore, India

Printed in the United States of America

10 9 8 7 6 5 4 3 2

To Julianna and Magdalena, Lily and Simon

Contents

Preface

This book grew out of an invited paper session, *An Invitation to Geometric Group Theory*, at the 2010 MAA MathFest in Pittsburgh. Geometric group theory is an exciting, relatively new field of mathematics that explores groups using tools and techniques from geometry and topology. A common theme throughout the field is:

> *Algebraic properties of a group are reflected in the geometry of spaces on which the group acts.*

While the notion of a group forms a key component of an undergraduate education in mathematics, the above viewpoint is rarely taken. The typical undergraduate algebra curriculum focuses mainly on finite groups and students can graduate without being exposed to the vast and interesting world of infinite groups. The purpose of this book is to invite undergraduates, beginning graduates students, and curious researchers to explore this world by means of considering examples central to geometric group theory.

Each chapter—rather, office hour—of this book, driven by examples, covers one topic from geometric group theory. The authors were charged with the task of writing their answer to an undergraduate who had posed one of the following questions:

- *Will you tell me about your research?*
- *Will you help me find a topic for a senior project?*

As such, the office hours have an intentionally informal tone, each conveying the individual voice of the author. Ideally, the student reading this book should feel as if he or she is visiting a professor during office hours. When a student comes to your office curious about advanced topics in mathematics, you don't start the conversation by: "An amalgamated free product decomposition is." Rather, you draw a picture of SL(2, $\mathbb{Z}$) acting on the hyperbolic plane and explain how this action naturally decomposes the group. We hope that this style conveys the authors' enthusiasm for the subject and helps to inspire the next generation of researchers.

To facilitate the introduction to mathematical research, each office hour ends with suggestions for further reading and a selection of advanced exercises and research projects.

We do not think of this book as a replacement for John Meier's excellent text, *Groups, Graphs and Trees*, but instead we see it more as an undergraduate version of Pierre de la Harpe's book, *Topics in Geometric Group Theory*. In choosing topics for this book, we focused on the concepts that appear in several areas of geometric group theory—such as quasi-isometry—or on the classes of groups that prevail throughout geometric group theory—such as Coxeter groups and right-angled Artin groups. We also tried to choose topics that we felt were accessible to an undergraduate who had taken a first course in abstract algebra, at least with the help of a patient adviser. Thus, we omitted some key geometric group theory concepts such as finiteness properties. In the rare office hours that deal with a specialized topic, such as asymptotic dimension, we felt that a student could find some interesting concept to latch onto, such as that of finding appropriate arrangements of bricks.

With the exception of the two introductory office hours, the office hours do not depend on one another, although connections within the material are highlighted. The topics covered in introductory office hours are:

- Groups
- Free groups and free abelian groups
- Group actions
- Cayley graphs
- Metric spaces

After reading the introductory office hours, an undergraduate should be able to browse the remaining office hours independently and focus on the topics that grab his or her interest.

We would be remiss not to list some of the other excellent general reference books on geometric group theory:

- *A Course in Geometric Group Theory*, Brian H. Bowditch
- *Metric Spaces of Non-positive Curvature*, Martin Bridson & André Haefliger
- *Topics in Geometric Group Theory*, Pierre de la Harpe
- *Groups, Graphs and Trees*, John Meier
- *Classical Topology and Combinatorial Group Theory*, John Stillwell

It has been a great pleasure working with the 15 other authors to compile some of our favorite gems of geometric group theory. We would like to thank them all for their hard work, persistence, patience, and enthusiasm. Most of all we would like to thank them for sharing their mathematical knowledge and their unique voices, and for believing in this project when we started with little more than an idea six years ago. The process has been invigorating and inspiring for us. We hope that you enjoy your journey through geometric group theory as much as we enjoyed laying down the path.

Matt Clay and Dan Margalit

Fayetteville and Atlanta, November 2016

Acknowledgments

Matt Clay and Dan Margalit would like to thank Mladen Bestvina and Benson Farb for first teaching us about geometric group theory and for being constant sources of support and inspiration. We would also like to thank Linda Chen, Yen Duong, Sudipta Kolay, and Caglar Uyanik for sending several comments and corrections. We are grateful to Ryan Dickmann for help with the typesetting of the book. We would also like to thank Justin Lanier for many insightful suggestions. We thank Bhisham Bherwani for his careful copyediting. Finally, this project would not have been possible without Vickie Kearn at Princeton University Press, whose unwavering support of our idea got us started and kept us going.

Tara Brendle, Leah Childers, and Dan Margalit are grateful to Joan Birman and Saul Schleimer for helpful conversations. They would also like to thank Mante Zelvyte for comments on an earlier draft of their office hour.

Moon Duchin would like to thank the participants in the 2012 Undergraduate Faculty Program at the Park City Mathematics Institute, especially Thomas Mattman and David Peifer, for their comments. She is also grateful to Sarah Bray and Brendan Foley for attempting several of the exercises.

Johanna Mangahas would like to express gratitude for comments and corrections offered by the participants of the 2012 PCMI Undergraduate Summer School taught by Jennifer Taback and the Undergraduate Faculty Program led by Moon Duchin, especially Allison Miller and Joe Ochiltree and David Peifer.

Timothy Riley thanks Margarita Amchislavska, Costandino Moraites, and Kristen Pueschel for careful readings of drafts of this chapter; Sasha Ushakov for conversations on the difficulty of describing "puzzle solutions"; and Robert Kleinberg and Jonathan Robbins for conversations on Project 4 of Office Hour 8.

Jennifer Taback thanks Daniel Studenmund, Peter Davids, Michael Ben-Zvi, Kevin Buckles, and all the students in Math 2602 and Math 3602 at Bowdoin College for helpful comments on her office hour.

Matt Clay, Sean Cleary, Johanna Mangahas, Dan Margalit, Timothy Riley and Jennifer Taback gratefully acknowledge support from NSF Grant Nos. DMS-1006898, DMS-1417820, DMS-1204592, DMS-1057874, DMS-1101651, and DMS-1105407, respectively.

The non-editing office hour authors thank Matt Clay and Dan Margalit for conceiving this book, inviting our participation, providing encouragement and vision along the way, and skillfully and patiently seeing the project to completion.

PART 1

Groups and Spaces

Office Hour One

Groups

Matt Clay and Dan Margalit

Symmetry, as wide or as narrow as you define its meaning, is one idea by which man through the ages has tried to comprehend and create order, beauty and perfection.

Hermann Weyl

An important problem in mathematics is to determine the symmetries of objects. The objects we are interested in might be concrete things such as planar shapes, higher-dimensional solids, or the universe. Or they might be more abstract, such as groups, spaces of functions, or electric fields.

We'll begin in this office hour with a general discussion of groups; the focus will be on interpreting every group as a group of symmetries of some object. In the second office hour we will make precise what it means for a group to be a group of symmetries of an object. These ideas are at the very heart of geometric group theory, the study of groups, spaces, and the interactions between them.

Symmetry. By understanding the symmetries of an object we come closer to fully understanding the object itself. For instance, consider a cube in Euclidean 3–space:

A symmetry of the cube is a rigid transformation of the cube (for now, no reflections allowed). In other words, a symmetry of the cube is what you get if you pick the cube up, rotate it with your hands, and then put in back in the same spot where it started. One way to obtain a symmetry is to skewer the cube through the centers of two opposite faces and rotate by any multiple of $\pi/2$. We could also skewer the cube through the midpoints of two opposite edges, or through two opposite corners.

Here is a question:

What are all of the symmetries of the cube?

If we don't know the answer to this question, then we don't know the cube!

There is a fairly straightforward analysis which tells us that there are exactly 24 symmetries of the cube: there are eight places to send any given corner, and once a corner is fixed, there are three choices for how to turn the cube at that corner.

This is a start, but leaves a lot to be desired: What is a good way to list the 24 symmetries? If we do one symmetry followed by another symmetry, we get a third symmetry; where is this third symmetry on our list?

Here is a simple idea that completely illuminates the problem. Draw the four long diagonals of the cube:

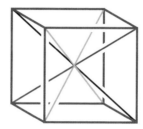

We can check the following two facts:

- Any symmetry of the cube gives a permutation of the four long diagonals.
- Any permutation of the four long diagonals gives a unique symmetry of the cube.

The validity of the first statement is not too hard to see. The second one takes a little thought. First notice that if we take the permutation that swaps two long diagonals, then we obtain a symmetry of the cube, namely, a symmetry obtained by skewering the cube through the midpoints of the two edges that connect the endpoints of the two long diagonals in question. But an arbitrary permutation of the four long diagonals can be obtained by swapping two long diagonals at a time, so we are done.

It follows that the set of symmetries of the cube is in bijection with the set of permutations of its four long diagonals. There are $4! = 24$ permutations of the four long diagonals, agreeing with our calculation that there are 24 symmetries of the cube. Moreover, since a permutation is just a bijective function from a set to itself, we know what happens when we do one permutation, then another—we get the composition of the two functions. We'll say more about permutations in a few minutes, when we talk about symmetric groups.

That was satisfying—we certainly understand the cube better now than we did at the start. In geometric group theory we aim to understand symmetries of much more complicated (and beautiful!) objects.

Here is the plan for this first office hour. We'll start by revisiting the basic examples of groups from any undergraduate course in abstract algebra. But we will look at these groups from the point of view that *every* group should be the collection of symmetries of some geometric object, just as the symmetric group on four letters is the collection of symmetries of a three-dimensional cube. To some extent, a lot of the material in this first office hour is a review of some of the important ideas from a first course in abstract algebra; you may want to skim through it if you are comfortable with those concepts.

In the next office hour, we will make sense of and prove the following theorem (Theorem 2.2 below):

> *Every group is naturally identified with the collection of symmetries of some geometric object, namely, the Cayley graph of the group.*

The goal of this book is to convince you that this theorem—along with the many other theorems in this book—is the beginning of a beautiful and fruitful dictionary relating the algebraic and geometric structures arising in the wild and fascinating world of infinite groups.

1.1 GROUPS

Let's recall the definition of a group. Then we'll explain why you should think of a group in terms of symmetries.

First, a *multiplication* on a set G is a function

$$G \times G \to G.$$

The image of (g, h) is usually written gh. In other words, a multiplication is a way of combining two elements of G in order to get a third one.

Then, a *group* is a set G together with a multiplication that satisfies the following three properties:

- *Identity.* There is an element 1 of G such that $1g = g1 = g$ for all $g \in G$. The element 1 is called the *identity element* of G.
- *Inverses.* For each $g \in G$, there is an element $h \in G$ such that $gh = hg = 1$. The element h is often denoted by g^{-1} and is called the *inverse* of g.
- *Associativity.* The multiplication on G is associative, that is:

$$(fg)h = f(gh)$$

for all $f, g, h \in G$.

What does this abstract, formal definition have to do with symmetries? Let's first think about this in the case of the cube. We defined a symmetry of the cube as a rigid transformation of the cube. The way we multiply two symmetries g and h is:

"do h, then do g." This is just like composing functions: in the composition $g \circ h$ we apply h first.

We now see that the set of symmetries of the cube satisfies the definition of a group. The identity symmetry 1 is the rigid transformation of the cube that leaves it alone. Indeed, doing one rigid transformation g and then doing 1 is the same as doing g. The inverse of a symmetry g is the symmetry that undoes g. Associativity follows from the associativity of composition of functions: doing h, then g and f is the same as doing h and g, then f.

As we already said, we'd like to eventually convince you that every group— that is, every set with a multiplication as above—is the set of symmetries of some object. First, let's look at a few examples of groups and explain how they each can be thought of as a set of symmetries.

The dihedral group.   The dihedral group D_n is the set of symmetries (rigid transformations) of a regular n–gon in the plane. The multiplication is the same as in the example of the cube. Two symmetries of the regular pentagon are the reflections s and t shown here:

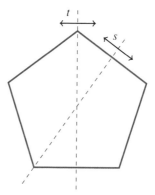

For general n we can take s to be the reflection through any line that passes through the center of the n–gon and a vertex of the n–gon, and t to be the reflection about the line that differs from the first line by a π/n rotation. The elements s and t *generate* the group D_n, by which we mean that every symmetry of D_n is obtained by multiplying, s, t, s^{-1}, and t^{-1} (actually, in this example $s = s^{-1}$ and $t = t^{-1}$). For example, the clockwise rotation by m "clicks" is $(st)^m$.

Adam Piggott's Office Hour 13 on Coxeter groups discusses examples of groups that are all based on the idea of the dihedral groups.

The symmetric group.   The symmetric group S_n is the set of permutations of the set $\{1, \ldots, n\}$, with multiplication given by composition of functions. This makes sense since a permutation of $\{1, \ldots, n\}$ is really just a bijective function $\{1, \ldots, n\} \to \{1, \ldots, n\}$. If we want to think of S_n as the set of symmetries of some object, we can think of it as the symmetries of a set of n points.

A *cycle* in S_n is a permutation that can be described as $i_i \mapsto i_2 \mapsto \cdots \mapsto i_k \mapsto i_1$ (and all other elements of $\{1, \ldots, n\}$ stay where they are). We write this cycle as $(i_1\ i_2\ \cdots\ i_{k-1}\ i_k)$. For instance, the cycle (1 2 4) in S_6 sends 1 to 2, 2 to 4, and 4 to 1, while sending 3, 5, and 6 to themselves.

Every element of S_n is a product of disjoint cycles (disjoint means that each element of $\{1, \ldots, n\}$ appears in at most one cycle). For instance, in S_6 the element

$$(1\ 2\ 4)(3\ 5)$$

is the product of the disjoint cycles (1 2 4) and (3 5). This is the permutation that sends 1 to 2, 2 to 4, and 4 to 1, and also 3 to 5 and 5 to 3, and finally 6 to itself.

As we said, the multiplication is function composition. Say, for example, we want to multiply (1 2 4)(3 5) by (2 6):

$$(2\ 6) \cdot (1\ 2\ 4)(3\ 5).$$

Where does this product send 1? Well, the first permutation (on the right) sends 1 to 2, and the second permutation sends 2 to 6, so the product (= composition) sends 1 to 6. You can use the same procedure to determine where this product sends 2, 3, 4, 5, and 6. You will find that $(2\ 6) \cdot (1\ 2\ 4)(3\ 5) = (1\ 6\ 2\ 4)(3\ 5)$.

A *transposition* is a cycle of length 2, for instance, (3 5). It is an important fact that S_n is generated by transpositions of the form $(i\ \ i+1)$, where $1 \leq i \leq n-1$ (we actually used this fact implicitly in our discussion of the cube at the beginning). Let us check that we can write (1 2 4)(3 5) as a product of such elements in S_6. We can draw a diagram of this permutation as follows:

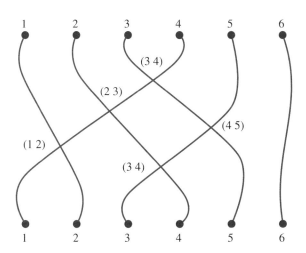

Each crossing in the diagram corresponds to an element of S_6 of the form $(i\ i+1)$. Reading top to bottom (and writing right to left), we find:

$$(1\ 2\ 4)(3\ 5) = (3\ 4)(1\ 2)(4\ 5)(2\ 3)(3\ 4).$$

The point is that multiplication in S_n can be realized by stacking diagrams. And the diagram for $(1\ 2\ 4)(3\ 5)$ can be obtained by stacking the diagrams for the five permutations on the right-hand side of the equation. Since we can always make a diagram where the crossings occur at different heights, this argument can be used to show that every element of S_n is equal to a product of transpositions $(i\ i+1)$.

Exercise 1. Use the idea of stacking diagrams to prove that S_n is generated by $\{(i\ i+1) \mid 1 \leq i \leq n-1\}$.

The picture we drew of the permutation $(1\ 2\ 4)(3\ 5)$ can be called a braid diagram. See Aaron Abrams' Office Hour 18 on braid groups for more on this idea.

Exercise 2. We showed that S_4 is the collection of symmetries of a three-dimensional cube. Can any of the other symmetric groups be thought of as the symmetries of higher-dimensional cubes? Or other shapes?

The integers modulo n. Let n be an integer greater than 1. The *integers modulo n* is the set

$$\mathbb{Z}/n\mathbb{Z} = \{0, 1, \ldots, n-1\}$$

with the multiplication

$$(a, b) \mapsto \begin{cases} a+b & a+b \leq n-1 \\ a+b-n & a+b \geq n. \end{cases}$$

The identity for $\mathbb{Z}/n\mathbb{Z}$ is 0. The inverse of 0 is 0, and the inverse of any other m is $n-m$. Associativity is a little trickier: to check that $(a+b)+c = a+(b+c)$, there are a few cases, depending on which sums are greater than n. We'll leave this as an exercise.

Exercise 3. Show that if m is any element of $\{0, \ldots, n-1\}$ with $\gcd(m, n) = 1$, then $\{m\}$ is a generating set for $\mathbb{Z}/n\mathbb{Z}$.

You are most familiar with $\mathbb{Z}/n\mathbb{Z}$ in three cases:

- When $n = 12$, the multiplication we defined is clock arithmetic. For example, 3:00 plus 5:00 is 8:00, and 9:00 plus 4:00 is 1:00 (although perhaps we should think of 12:00 as 0:00 instead).
- When $n = 10$, the multiplication we defined is an operation you use when balancing your checkbook. By just keeping track of the last digit, you have a quick first check for whether the sum of many big numbers is equal to what your bank says it is.
- When $n = 2$, the multiplication is light switch arithmetic. Identify 1 with flipping the switch and 0 with do nothing. Then flip plus flip is the same as doing nothing, flip plus do nothing is the same as flip, etc.

The idea of light switch arithmetic inspires a whole class of groups called lamplighter groups; see Jen Taback's Office Hour 15.

For what object is $\mathbb{Z}/n\mathbb{Z}$ the group of symmetries? Using the idea of the clock we might try to use a regular n–gon. We can certainly see $\mathbb{Z}/n\mathbb{Z}$ as a set of symmetries of the n–gon, namely, the rotational symmetries. Thus, $\mathbb{Z}/n\mathbb{Z}$ can be regarded as a *subgroup* of the full symmetry group, D_n. This means that we have realized $\mathbb{Z}/n\mathbb{Z}$ as a subset of D_n in such a way that the multiplication in $\mathbb{Z}/n\mathbb{Z}$ agrees with the multiplication in D_n. We can see that $\mathbb{Z}/n\mathbb{Z}$ is a proper subgroup of D_n (i.e., it is not the whole thing), since D_n also has reflectional symmetries.

That's pretty good. But we can modify the n–gon so that $\mathbb{Z}/n\mathbb{Z}$ is exactly the group of symmetries. One way to do this is to draw a little arrow in the middle of each edge:

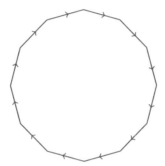

We think of the arrow as defining a preferred direction on each edge. Then we define a symmetry of this modified n–gon to be a rigid transformation that preserves the directions of the arrows. This is still a perfectly fine notion of symmetry! The group $\mathbb{Z}/n\mathbb{Z}$ is the full group of symmetries of the n–gon in this modified sense.

1.2 INFINITE GROUPS

Many undergraduate courses in abstract algebra give short shrift to infinite groups. As we will see in this book, this is where all of the fun is! Let's start with the simplest infinite group.

The integers.   You are probably familiar with the set of *integers*, or whole numbers:

$$\mathbb{Z} = \{\ldots, -3, -2, -1, 0, 1, 2, 3, \ldots\}.$$

The multiplication $\mathbb{Z} \times \mathbb{Z} \to \mathbb{Z}$ is the usual addition of integers:

$$(m, n) \mapsto m + n.$$

The group $\mathbb{Z}$ is generated by $\{1\}$. But it is also generated by $\{2, 3\}$, for example.

Of what object is $\mathbb{Z}$ the set of symmetries? A first guess is that $\mathbb{Z}$ is the symmetries of the number line L, by which we mean the real line $\mathbb{R}$ with a distinguished point at each integer:

Here, by a symmetry of L, we just mean a rigid transformation of L that preserves the set of distinguished points. Examples of symmetries of L are: translate L to the left or right by n units, where n is a positive integer; reflect it about an integer point; and reflect it about a half-integer point.

The group $\mathbb{Z}$ gives a set of symmetries of L as follows: the integer n is the symmetry that translates L by n units to the right if $n \geq 0$ and translates L by $-n$ units to the left if $n < 0$. If we think of L as an ∞–gon, this is analogous to what we did for $\mathbb{Z}/n\mathbb{Z}$. As in the case of $\mathbb{Z}/n\mathbb{Z}$, this identifies $\mathbb{Z}$ as a proper subgroup of the set of symmetries of L, since we don't get any of the reflections.

We can use the same trick as in the $\mathbb{Z}/n\mathbb{Z}$ case to modify L in such a way that $\mathbb{Z}$ is the full symmetry group: on each edge we draw a little arrow that points to the right:

The integers squared.   Our next group is

$$\mathbb{Z}^2 = \{(m, n) : m, n \in \mathbb{Z}\}$$

with the multiplication

$$((m, n), (m', n')) \mapsto (m + m', n + n').$$

If we think of $\mathbb{Z}^2$ as the set of vectors in $\mathbb{R}^2$ with integer coordinates, then this is simply vector addition. One generating set for $\mathbb{Z}^2$ is $\{(1, 0), (0, 1)\}$.

We can use the same idea as above to build an object whose symmetries are exactly described by $\mathbb{Z}^2$:

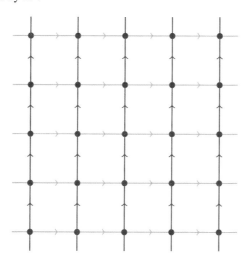

In order for this to work, we need to again refine what we mean by symmetry: this time we insist that colors are preserved as well as the red points and the arrows. The symmetry corresponding to $(m, n) \in \mathbb{Z}^2$ is the translation by m units to the right and n units up. (If m or n are negative, then translate left and down accordingly.)

Multiplicative groups of numbers. We would be remiss to not mention two important groups, $\mathbb{R}^*$ and $\mathbb{C}^*$. These are the groups of nonzero real numbers and nonzero complex numbers. The group multiplication in each group is the usual multiplication in $\mathbb{R}$ and $\mathbb{C}$. In both cases the identity is 1. An important distinction between these groups and our previous examples is that $\mathbb{R}^*$ and $\mathbb{C}^*$ are not finitely generated, that is, there does not exist a finite generating set.

If you've already seen these groups, you might want to learn about the quaternions, a generalization of the complex numbers which is very important in physics. Instead of just i, the quaternions have i, j, and k! There is a corresponding multiplicative group, consisting again of the nonzero elements.

Matrix groups. There is another set of examples of groups that you learned about in linear algebra (although in such a class you might not have used the word "group"). Denote by $\mathrm{GL}(n, \mathbb{R})$ the set of real $n \times n$ matrices with nonzero determinant. Why is this a group? Well, given two matrices, we already know how to multiply them. Note that this multiplication is closed (that is, the product of two matrices with nonzero determinant is another matrix with nonzero determinant). Indeed, this follows immediately from the following basic fact you learned in linear algebra:

$$\det(MN) = \det(M)\det(N).$$

Two other things you learned in linear algebra is that every square matrix with nonzero determinant is invertible and that multiplication is associative. The identity matrix has determinant 1, which is not zero, so we have checked that $\mathrm{GL}(n, \mathbb{R})$ is indeed a group. It is called the *general linear group*. There are really lots of general linear groups, since n can be any nonnegative integer and $\mathbb{R}$ can be replaced with any ring (see below), but when the context is clear we still call it *the* general linear group.

There are many variations. The group $\mathrm{SL}(n, \mathbb{R})$ is the subgroup of $\mathrm{GL}(n, \mathbb{R})$ consisting of matrices with determinant 1 (this is the *special linear group*). The group $\mathrm{GL}(n, \mathbb{Z})$ is the subgroup of $\mathrm{GL}(n, \mathbb{R})$ consisting of all invertible integral matrices (note: these are exactly the integral matrices with determinant ± 1). Then $\mathrm{SL}(n, \mathbb{Z})$ is the subgroup of $\mathrm{GL}(n, \mathbb{Z})$ consisting of integral matrices with determinant 1. There is also $\mathrm{GL}(n, \mathbb{C})$, the group of complex $n \times n$ matrices with nonzero determinant, and $\mathrm{SL}(n, \mathbb{C})$, the subgroup of $\mathrm{GL}(n, \mathbb{C})$ consisting of $n \times n$ matrices with determinant 1. All of the above examples are subgroups of $\mathrm{GL}(n, \mathbb{C})$, and there are many other examples of interesting subgroups of $\mathrm{GL}(n, \mathbb{C})$ (subgroups of $\mathrm{GL}(n, \mathbb{C})$ are called *linear groups*).

We will study $\mathrm{SL}(2, \mathbb{Z})$ in detail in Office Hour 3, giving a finite presentation for this group as well as many of its relatives.

Exercise 4. Construct a matrix group whose entries lie in a finite field.

The free group of rank 2. Our next example is the most complicated (and interesting!) one yet. And it is most probably not a group that you saw in your undergraduate algebra course.

First we define a *word* in the letters a and b to be an arbitrary finite string made up of the symbols a, b, a^{-1}, and b^{-1}. Some examples are:

$$ababab a, aaaaaaa, aba^{-1}b^{-1}, \text{ and } aa^{-1}.$$

We allow the empty word as well; this is the string containing no letters. We can multiply two words by concatenating them:

$$(ab, b^{-1}a) \mapsto abb^{-1}a.$$

Next, we define a *reduced word* to be a word with the property that we never see an a followed by an a^{-1} (or vice versa) or a b followed by a b^{-1} (or vice versa). If we have a word that is not reduced, we can make it shorter by deleting an offending pair of symbols:

$$aa^{-1}abb^{-1}a \rightsquigarrow abb^{-1}a.$$

Performing this procedure inductively will eventually lead to a reduced word. You should check that this reduction procedure always results in the same reduced word. In the above example, there are three places where we could start reducing. In this case, though, no matter what we do we will end up with the reduced word aa.

Exercise 5. Show that every word in a, b, a^{-1}, and b^{-1} can be reduced to a unique reduced word.

We are finally ready to define the *free group of rank 2*. It is the set

$$F_2 = \{\text{reduced words in } a \text{ and } b\}$$

with the multiplication:

$$\text{concatenate, then reduce.}$$

Let's check that this is a group. The identity is the empty word. The inverse of a word is obtained by reversing the word, and then replacing each a by a^{-1}, a^{-1} by a, b by b^{-1}, and b^{-1} by b. For example, the inverse of

$$aba^{-1}b$$

is

$$b^{-1}ab^{-1}a^{-1}$$

Indeed, if we concatenate these two (in either order!) and reduce, we obtain the empty word. Associativity follows immediately from the fact that the reduction procedure always results in the same reduced word (but this fact needs to be proved!). By the very definition of F_2, we see that this group is generated by $\{a, b\}$.

Exercise 6. Check the details of the construction of the free group of rank 2, in particular the assertion that the multiplication is associative. *Hint: Show by induction on the length of (unreduced) words that if there are two choices for a reduction, then either choice leads to the same reduced word. There are two cases, according to whether the two reductions are adjacent or not.*

Free groups are important examples of groups in geometric group theory. Office Hours 3, 4, 5, and 6 explore these groups in more detail.

How can we construct an object whose symmetry group is F_2? This is not so easy! In Office Hour 2 we will explain a way to do this for an arbitrary group.

Other free groups.   Given a set S, we can define a free group $F(S)$ as follows. First, we artificially create inverses for each element of S. For $s \in S$, we can write this inverse as s^{-1} (let's not allow s and s^{-1} to both be elements of S). Then the elements of $F(S)$ are reduced words in the elements of S, and the multiplication is defined in the same way as for F_2. As such, when S is a finite set with n elements, we can denote $F(S)$ by F_n (different choices of S with $|S| = n$ technically give different groups, but they are all isomorphic in the sense defined below). Convince yourself that $F(S)$ still is a well-defined group when S is infinite, or even uncountable. In general, the cardinality of S is the *rank* of the corresponding free group.

As we will explain in more detail in the next section, free groups are important because every group is the quotient of a free group.

And many more.   One aspect of geometric group theory is to explore the zoo of infinite groups that are out there. Some examples of interesting classes of groups in this book are the Coxeter groups already mentioned (Adam Piggott's Office Hour 13), the right-angled Artin groups (Bob Bell and Matt Clay's Office Hour 14), the lamplighter groups (Jen Taback's Office Hour 15), Thompson's groups (Sean Cleary's Office Hour 16), the mapping class groups (Tara Brendle and Leah Childers' Office Hour 17), and the braid groups (Aaron Abrams' Office Hour 18).

1.3 HOMOMORPHISMS AND NORMAL SUBGROUPS

We'll now take a few minutes to give a quick overview of the most basic and important ideas from a first course in group theory. If you already know all about the first isomorphism theorem and about group presentations, then you might want to skip ahead to the next office hour.

Homomorphisms.   We have talked about groups and we now have some examples. Next we would like to discuss how groups relate to each other. A *homomorphism* from a group G to a group H is a function

$$f : G \to H$$

so that

$$f(ab) = f(a) f(b)$$

for all $a, b \in G$. We can say this succinctly as: f is a function that preserves the multiplication in G. A homomorphism is an *isomorphism* if it is bijective, and two groups are *isomorphic* if there is an isomorphism from one to the other. Convince yourself that isomorphic groups are essentially the same group.

To illustrate the idea of an isomorphism, let's look at a pair of groups. Consider the group $\mathbb{Z}$ on one hand, and the free group F_1 on one generator a on the other hand. Elements of F_1 are finite, freely reduced strings in a and a^{-1}, and the multiplication is given as usual by concatenation and free reduction. What is an isomorphism $\mathbb{Z} \to F_1$? (There is more than one!)

Besides isomorphisms, we might also be interested in injective homomorphisms and surjective homomorphisms. A homomorphism is injective if and only if the preimage of the identity in H is the identity in G. (Why?)

Note that any homomorphism $f \colon G \to H$ can be turned into a surjective homomorphism by replacing H with $f(G)$. For that reason, the surjectivity of a homomorphism is not nearly as crucial as the injectivity. As such we will focus on injective and non-injective homomorphisms.

Injective homomorphisms.   If we have an injective homomorphism $f \colon G \to H$, then we can think of f as realizing G as a subgroup of H. Here are a few examples:

1. There is an injective homomorphism $\mathbb{Z}/n\mathbb{Z} \to D_n$ where m maps to rotation by $2\pi m/n$.
2. There is an injective homomorphism $\mathbb{Z}/2\mathbb{Z} \to D_n$ where the nontrivial element of $\mathbb{Z}/2\mathbb{Z}$ maps to any reflection.
3. There is an injective homomorphism $\mathbb{Z}/2\mathbb{Z} \to S_n$ where the nontrivial element of $\mathbb{Z}/2\mathbb{Z}$ maps to any product of disjoint transpositions.
4. For any integer $m \neq 0$, there is an injective homomorphism $\mathbb{Z} \to \mathbb{Z}$ given by multiplication by m.
5. For any choice of $(a, b) \neq (0, 0)$, there is an injective homomorphism $\mathbb{Z} \to \mathbb{Z}^2$, where n maps to (na, nb).
6. For any choice of nontrivial element $w \in F_2$, there is an injective homomorphism $\mathbb{Z} \to F_2$ given by $n \mapsto w^n$.

Exercise 7. How many other injective homomorphisms can you find between the groups we have already discussed?

Non-injective homomorphisms.   Non-injective homomorphisms can be just as interesting and just as useful. Since we have multiple elements of G mapping to the same element of H, we are definitely losing some information. But the fact that a homomorphism preserves the multiplication of G means we are still remembering some important data. Often we can set up a non-injective homomorphism to remember exactly the data we care about and throw away everything else. Here are some examples:

1. The most basic example is the homomorphism $\mathbb{Z} \to \mathbb{Z}/2\mathbb{Z}$ where 1 maps to 1. A light switch is a great example of this: it does not remember how many times it has been flipped, just whether it has been flipped an even or an odd number of times.

2. There is a homomorphism $\mathrm{GL}(n, \mathbb{R}) \to \mathbb{R}^\star$ given by the determinant. So if you want to know the determinant of a product of matrices, you do not need to remember the actual matrices, just their determinants.

3. There is a homomorphism $\mathbb{R}^\star \to \mathbb{Z}/2\mathbb{Z}$ where positive numbers map to 0 and negative numbers map to 1. The fact that this is a homomorphism can be rephrased as: in order to tell if a product of nonzero numbers is positive or negative, you do not need to remember what the numbers are, just whether they are positive or negative.

4. The composition $\mathrm{GL}(n, \mathbb{R}) \to \mathbb{R}^\star \to \mathbb{Z}/2\mathbb{Z}$ is a homomorphism that remembers whether an element of $\mathrm{GL}(n, \mathbb{R})$ has positive or negative determinant.

5. There is a homomorphism $S_n \to \mathbb{Z}/2\mathbb{Z}$ that records whether the diagram for a permutation (as above) has an even or an odd number of crossings. In other words, this homomorphism records the parity of the number of transpositions needed to write a given element of S_n.

6. There is a homomorphism $D_n \to \mathbb{Z}/2\mathbb{Z}$ where rotations map to 0 and reflections map to 1. This homomorphism remembers whether the polygon has been flipped over or not.

7. There is a homomorphism $F_2 \to \mathbb{Z}^2$ where the generators a and b for F_2 map to $(1, 0)$ and $(0, 1)$, respectively. This homomorphism remembers the sum of the exponents on a and b in any element of F_2.

8. Any linear map $\mathbb{R}^n \to \mathbb{R}^m$ is a homomorphism. For example, the map $\mathbb{R}^2 \to \mathbb{R}$ that projects to the x–axis is a homomorphism that remembers just the first coordinate. This is a fancy way to state the obvious fact that in order to know the first coordinate of a sum of vectors, you need to know just the first coordinate of each vector.

Exercise 8. How many other non-injective homomorphisms can you find between the groups we have already discussed?

Normal subgroups.   Next we will present a definition that will seem unrelated to homomorphisms; and then we will make a deep connection. A *normal subgroup* of a group G is a group N where N is a subset of G and where gng^{-1} lies in N for all $n \in N$ and all $g \in G$.

Why are normal subgroups important? Well, one way to make a normal subgroup of G is to find a group homomorphism f from G to another group H. The *kernel* of f is the set of $g \in G$ so that $f(g)$ is the identity (we use the same terminology in linear algebra). We have the following basic fact: the kernel of a homomorphism $f : G \to H$ is a normal subgroup of G.

Here is why: if n is in the kernel of f, then

$$f(gng^{-1}) = f(g)f(n)f(g^{-1}) = f(g)\, 1\, f(g^{-1})$$
$$= f(g)f(g^{-1}) = f(gg^{-1}) = f(1) = 1,$$

and so gng^{-1} is also in the kernel of f. Thus the kernel of f is normal.

The above examples of non-injective homomorphisms have kernels as follows:

1. The kernel of our $\mathbb{Z} \to \mathbb{Z}/2\mathbb{Z}$ is $2\mathbb{Z}$.
2. The kernel of our $GL(n, \mathbb{R}) \to \mathbb{R}^\star$ is $SL(n, \mathbb{R})$.
3. The kernel of our $\mathbb{R}^\star \to \mathbb{Z}/2\mathbb{Z}$ is $\mathbb{R}_+$.
4. The kernel of our $GL(n, \mathbb{R}) \to \mathbb{R}^\star \to \mathbb{Z}/2\mathbb{Z}$ is the subgroup $GL^+(n, \mathbb{R})$ consisting of matrices with positive determinant.
5. The kernel of our $S_n \to \mathbb{Z}/2\mathbb{Z}$ is the *alternating group* A_n; this consists of the elements that can be written as the product of an even number of transpositions.
6. The kernel of our $D_n \to \mathbb{Z}/2\mathbb{Z}$ is the subgroup consisting of rotations; this subgroup is isomorphic to $\mathbb{Z}/n\mathbb{Z}$.
7. The kernel of our $F_2 \to \mathbb{Z}^2$ is the set of elements whose a–exponents and b–exponents both add to zero, e.g., $a^7 b^{-5} a^{-10} b^5 a^3$. This group is also known as the commutator subgroup of F_2 and is denoted by $[F_2, F_2]$.
8. The kernel of the projection $\mathbb{R}^2 \to \mathbb{R}$ is the same as the kernel from linear algebra; in this case it is isomorphic to $\mathbb{R}$.

So non-injective homomorphisms give us nontrivial normal subgroups. Can we go the other way? That is, if N is a normal subgroup of G, is there a homomorphism $G \to H$ so that N is the kernel? The answer is yes! Let's explain this. The first step is to define the *quotient group* G/N. Informally, this is the group obtained from G by declaring every element of N to be trivial.

Let us make this more precise. Declare two elements g_1, g_2 of G to be equivalent if $g_1 g_2^{-1} \in N$, and write $[g]$ for the equivalence class of g. Then the elements of G/N are the equivalence classes of the elements of G (notice that the equivalence class of the identity is precisely N). Now we need to say how to multiply two equivalence classes. We declare that the product of the equivalence class $[g_1]$ with the equivalence class $[g_2]$ is the equivalence class $[g_1 g_2]$. This is precisely the place where the definition of a normal subgroup comes from—it is exactly what is needed so that this multiplication is well defined.

Exercise 9. Check that the multiplication in G/N is well defined if and only if N is normal.

There is an obvious homomorphism $G \to G/N$, where $g \in G$ maps to $[g] \in G/N$. It is not hard to see that the kernel of this homomorphism is precisely N. So we have completed the loop: normal subgroups and surjective homomorphisms are the same thing!

The first isomorphism theorem.   The so-called first isomorphism theorem succinctly summarizes the relationship between homomorphisms and normal subgroups that we have just described.

THEOREM 1.1 (First isomorphism theorem). *If $G \to H$ is a surjective homomorphism with kernel K, then H is isomorphic to G/K.*

Again using our above examples, we have the following isomorphisms:

1. $(\mathbb{Z})/(2\mathbb{Z}) \cong \mathbb{Z}/2\mathbb{Z}$ (this explains the notation $\mathbb{Z}/2\mathbb{Z}$)
2. $\mathrm{GL}(n, \mathbb{R})/\mathrm{SL}(n, \mathbb{R}) \cong \mathbb{R}^\star$
3. $\mathbb{R}^\star/\mathbb{R}_+ \cong \mathbb{Z}/2\mathbb{Z}$
4. $\mathrm{GL}(n, \mathbb{R})/\mathrm{GL}^+(n, \mathbb{R}) \cong \mathbb{Z}/2\mathbb{Z}$
5. $S_n/A_n \cong \mathbb{Z}/2\mathbb{Z}$
6. $D_n/(\mathbb{Z}/n\mathbb{Z}) \cong \mathbb{Z}/2\mathbb{Z}$
7. $F_2/[F_2, F_2] \cong \mathbb{Z}^2$
8. $\mathbb{R}^2/\mathbb{R} \cong \mathbb{R}$

Exercise 10. Use the first isomorphism theorem to come up with your own isomorphisms.

1.4 GROUP PRESENTATIONS

You may not have covered group presentations in your first course in abstract algebra, but they are closely tied to the other topics in this section and will loom large in this book.

One naive way to completely describe a group is to list all of the elements and write down the multiplication table (that is, explicitly give the product of any two elements). For a very small finite group, this is a reasonable way of understanding the group. But once you start considering larger—or even infinite—groups this approach becomes unwieldy.

We already have a more economical way of listing the elements of a group: we can list a set of generators. Often the infinite groups we care about have finite sets of generators. This is good, but it certainly doesn't tell us enough about a group: the free group F_2 and the free abelian group $\mathbb{Z}^2$ both have generating sets with two elements and we know that these are different groups since one is abelian and the other is not. We need a way of describing the multiplication table. But obviously we don't want to just list all of the entries.

The key observation is that some entries in the multiplication table imply other entries. In other words, equations between group elements imply other equations. For instance, if $gh = k$ and $h = pq$, then (substituting pq for h in the first equation) we know that $gpq = k$. So what we can hope for is that we don't need to write down the entire multiplication table, but rather a small set of equalities that imply all the others. So now we have exactly hit upon the idea of what a group presentation is: it is firstly a list of generators and secondly a list of defining equalities so that all equalities follow from the listed ones.

Here is a first example, the group $\mathbb{Z}/n\mathbb{Z}$. Let's denote the elements of $\mathbb{Z}/n\mathbb{Z}$ by $a^0, \ldots, a^{n-1}$ instead of $0, \ldots, n - 1$. Naturally, the multiplication is $a^j a^k = a^{j+k}$, where the exponents are all taken modulo n. With this notation, $\mathbb{Z}/n\mathbb{Z}$ has the presentation

$$\mathbb{Z}/n\mathbb{Z} \cong \langle a \mid a^n = 1 \rangle.$$

So the generating set is a and the only defining equality is $a^n = 1$. Every entry in the multiplication table for $\mathbb{Z}/n\mathbb{Z}$ follows from this one equality. For instance, one entry in the multiplication table is $a^{n-3}a^4 = a$. But we can easily derive this using the defining equality $a^n = 1$ as follows:

$$a^{n-3}a^4 = a^{n+1} = a^n a^1 = 1 \cdot a = a.$$

A second important example of a group presentation is the following:

$$\mathbb{Z}^2 = \langle x, y \mid xy = yx \rangle.$$

Can you see why this is a presentation of $\mathbb{Z}^2$? We'll come back to this in a little bit.

Exercise 11. Convince yourself that $\mathbb{Z}^2$ indeed has the purported presentation.

Hopefully these examples give you an idea of what a group presentation is. The way we make things formal—and the way we really think about group presentations—is through the use of free groups. At first our definition of a presentation won't look exactly like the examples we just gave, but we'll eventually explain why it is really the same thing.

A *group presentation* is a pair (S, R), where S is a set and R is a set of words in S. If we freely reduce the elements of R, then we can regard them as elements of the free group $F(S)$, and indeed it will be important to take this point of view. The presentation is finite when both sets S and R are finite.

We say that a group G has the presentation $\langle S \mid R \rangle$ if G is isomorphic to the quotient of $F(S)$ by the normal closure of R, the smallest normal subgroup of $F(S)$ containing R (equivalently, the subgroup of $F(S)$ generated by all conjugates of all elements of R). We write:

$$G \cong \langle S \mid R \rangle.$$

Here is how we think of this: the group G consists of reduced words in $S \cup S^{-1}$ but with the added caveat that words in R are the same as the empty word (this is where the normal closure comes in: if some element of R is supposed to represent the identity, then any conjugate of that element in $F(S)$ should also represent the identity, and any products of conjugates of such elements ...). The elements of S are called *generators* for G and the elements of the normal closure of R are called *relators* for G, and the elements of R are called defining relators. If we have a presentation (as in our above discussion) with a relation like $ab = ba$ or $aba = bab$, then we turn this into a relator by moving everything to one side: $aba^{-1}b^{-1}$ or $abab^{-1}a^{-1}a^{-1}$.

How does this formal definition match up with what we said before? Let's look more closely at $\mathbb{Z}^2$. This group has the presentation

$$\mathbb{Z}^2 \cong \langle a, b \mid aba^{-1}b^{-1} \rangle.$$

(Notice that this is different from the presentation we gave a few minutes ago, but bear with us!) What are the elements of the group $\langle a, b \mid aba^{-1}b^{-1} \rangle$? Well, we start with the free group on a and b, which consists of all freely reduced words in a and b. Then we need to take the quotient by the normal closure of $aba^{-1}b^{-1}$. What is the quotient as a set? Well, in the quotient we can change a word w into an equivalent word by inserting $aba^{-1}b^{-1}$ anywhere in w, or removing it from anywhere in w.

Why is that the same as our first presentation of $\mathbb{Z}^2$ as $\langle x, y \mid xy = yx \rangle$? Well, we claim that two elements of the free group on $\{a, b\}$ are equivalent in the quotient (by the normal quotient by $aba^{-1}b^{-1}$) if and only if they differ by a sequence of swaps, where we swap ab for ba or vice versa. Let's convince ourselves of this. Start with a word $wbaw'$, where w and w' are both words in a and b. We said we can insert $aba^{-1}b^{-1}$ anywhere we want; so in the quotient,

$$wbaw' = w(aba^{-1}b^{-1})baw' = wab(a^{-1}b^{-1}ba)w' = wabw'.$$

By applying swaps like this we see that every element of the quotient is equivalent to $a^m b^n$ for some m and n. But this gives us an obvious isomorphism between our quotient and $\mathbb{Z}^2$; the isomorphism is given by $a^m b^n \leftrightarrow (m, n)$.

We can also see from the above discussion how our new presentation of $\mathbb{Z}^2$ (with generators a and b) matches up with our old one (with generators x and y). Both presentations are really saying the same thing. So which one you use just depends on your preference. When we have an equality between generators like $xy = yx$, then the equality is called a *relation*. And when we have an equality like $aba^{-1}b^{-1} = 1$ we say that $aba^{-1}b^{-1}$ is a *relator*, as in our formal definition. We can always change a relation to a relator by moving all of the generators to one side (as in our $\mathbb{Z}^2$ example), and the net effect is the same. In this book we will tend to use relators more than relations, but both will come up.

Every group can be given by a group presentation. In particular:

> *Every group is a quotient of a free group.*

This is one reason why free groups are so important. In many ways, geometric group theory grew out of the study of combinatorial group theory, a subject largely concerned with studying groups via their presentations.

Exercise 12. Show that every group has a presentation. *Hint: Your presentation does not have to be efficient.*

Exercise 13. Find (nice) group presentations for F_n, D_n, and S_n.

Armed with the notion of a group presentation, you can start to list lots of different groups. Which presentations represent isomorphic groups? It turns out that this is an unsolvable problem in general; there is no algorithm that will take two presentations and decide in finite time if they present the same group. It is good to keep this in mind: group presentations are very helpful, but they do have limitations.

That's it for the first office hour—we introduced a number of examples of groups that will play key roles in the rest of the book. We want to understand these groups from as many points of view as possible. At first glance you might wonder what there is to learn about a free abelian group or a free group. The answer: more than you might imagine!

Hopefully we have convinced you that the concept of a group is deeply inter-twined with the idea of symmetries: groups lead to symmetries and symmetries lead to groups. In order to study infinite groups, we will want to take advantage of this and think of groups geometrically. A key step on this path is to show that every group really is the group of symmetries of some geometric object. That's the goal of the next office hour—see you there!

Office Hour Two

... and Spaces

Matt Clay and Dan Margalit

As we said at the outset of Office Hour 1, our goal is to exhibit every group as a collection of symmetries of some geometric object. We have given some examples of finite and infinite groups and in some cases we have given an appropriate geometric object with that group as its symmetries.

Our first task in this office hour is to say precisely what it means for a group to be a group of symmetries of a geometric object (answer: group actions). We will then define, for any group G, the Cayley graph of G and show that the group of symmetries of this graph is precisely the group G.

We will then introduce metric spaces, which formalize the notion of a geometric object. We'll see lots of metric spaces that groups can act on. Furthermore, we will see that groups themselves *are* metric spaces; in other words, groups themselves can be thought of as geometric objects. At the end we will use these ideas to frame the motivating questions of geometric group theory.

Group actions. In Office Hour 1 we got the idea that groups can *do* things to geometric objects (or sets) in the sense that elements of the group give symmetries of the object. For instance, each element of the dihedral group D_n does something to the n-gon, each element of the symmetric group S_n does something to the set $\{1, \ldots, n\}$, and each element of $\mathbb{Z}$ does something to the real line. In other words, we can say that these groups *act* on those objects. This is precisely the idea of a group action.

Formally, an *action* of a group G on a set X is a function

$$G \times X \to X$$

where the image of (g, x) is written $g \cdot x$ and where

- $1 \cdot x = x$ for all $x \in X$ and
- $g \cdot (h \cdot x) = (gh) \cdot x$ for all $g, h \in G$ and $x \in X$.

If A is a subset of X then $g \cdot A$ is the set $\{g \cdot a \mid a \in A\}$. When there is no confusion we sometimes write gx and gA instead of $g \cdot x$ and $g \cdot A$.

Notice that for any group G and any set X we always have the trivial action, defined by $g \cdot x = x$ for all $g \in G$ and $x \in X$.

Let S_X denote the symmetric group of X, that is, the group of permutations of X (this makes sense even if X is infinite). So not only does each element of G give rise to a permutation of X, but it does so in such a way as to respect the multiplications in the two groups.

The group S_X is the group of symmetries of X, thought of as a set (hence the name). An action of G on X is the same thing as a homomorphism $G \to S_X$. So an action of G on X is the formal way to realize G as a group of symmetries of the set X.

Exercise 1. Check that actions of G on X are the same as homomorphisms $G \to S_X$.

Here are some examples of group actions on sets:

1. S_n acts on $\{1, \ldots, n\}$
2. D_n acts on the set of vertices (or edges, or diagonals ...) of an n–gon
3. $\mathrm{GL}(n, \mathbb{R})$ acts on the set of vectors (or planes, or bases ...) in $\mathbb{R}^n$
4. $\mathbb{Z}/2\mathbb{Z}$ acts on $\mathbb{Z}$ via multiplication by ± 1
5. $\mathbb{Z}$ acts on $\mathbb{R}^n$ via $m \cdot v = A^m v$, where A is some fixed $n \times n$ matrix
6. $\mathbb{Z}$ acts on the unit circle in $\mathbb{C}$ via the formula $m \cdot z = e^{im\theta} z$, where θ is any real number (so m rotates the circle by $m\theta$)

Exercise 2. List some other group actions.

If we have an action of a group G on a set X, then the *orbit* of $x \in X$ under G is the set

$$G \cdot x = \{g \cdot x \mid g \in G\}.$$

So, for example, the orbit of a vertex of an n–gon under the action of D_n on the set of vertices is the full set of all vertices; in such a situation we say that the action is *transitive*. At the other extreme, the orbit of x under the trivial action is simply $\{x\}$. If no nontrivial element of G fixes any $x \in X$, we say that the action of G on X is *free*

Exercise 3. List all of the orbits for your favorite group actions (for instance, the ones listed above).

Groups acting on themselves. If G is any group, then G acts on itself in two ways:

1. G acts on itself by left multiplication: $g \cdot h = gh$
2. G acts on itself by conjugation: $g \cdot h = ghg^{-1}$

The first action is transitive and free, but the second usually is not (the orbits in the second case are called *conjugacy classes*).

Exercise 4. Check that the purported actions of G on itself are really actions.

Exercise 5. Describe the conjugacy classes in S_n, D_n, F_n, and $\mathbb{Z}^n$.

The action of G on itself by left multiplication gives a homomorphism $G \to S_G$. This homomorphism is injective (why?). Therefore, we have the following theorem.

THEOREM 2.1 (Cayley's theorem). *Every group is isomorphic to a subgroup of some symmetric group.*

Cayley's theorem already tells us that every group can be thought of as a set of symmetries (of the group itself). However, there is no geometry here: we are just thinking of the group as a discrete set of points. The real goal is to find group actions on geometric objects so that we can use geometry to study the group. In this book the geometric objects we will think about are graphs and metric spaces, the subjects of the next two sections.

2.1 GRAPHS

We'll now describe our first class of geometric objects—graphs. We'll say what it means for a group to act on a graph, and finally we will show that every group is the group of symmetries of some graph, its Cayley graph.

Graphs. A *graph* Γ is a combinatorial object defined as follows: there is a nonempty set V (the set of vertices) and a set E (the set of edges), and a function from E to the set of unordered pairs of (not necessarily distinct) elements of V; we'll call this function the *endpoint function*. We sometimes say that an edge *connects* its two endpoints.

We can draw a picture of Γ by drawing one dot for each element of V and connecting dots with arcs—one for each edge—according to the endpoint function; see, for instance, Figure 2.1.

Exercise 6. Write down a formal description of the graph in Figure 2.1, using sets V and E and the endpoint function.

The *valence* of a vertex in a graph is the number of times that vertex appears as the endpoint of an edge. In other words, this is the number of half-edges incident to the vertex. In Figure 2.1 every vertex has valence 4.

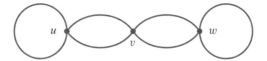

Figure 2.1 A graph.

Graphs usually are described not by explicitly listing the sets V and E, but rather by giving some rule for listing the vertices and edges. Here are some examples:

1. The *complete graph* on n–vertices K_n has vertex set $\{1, \ldots, n\}$ and an edge for every pair of distinct vertices.

    Notice that in our picture of K_4 there is an intersection right in the middle. This is not a vertex! It is simply an artifact of the picture. Can you find a picture of K_4 without any extra crossings like this? What about K_5?

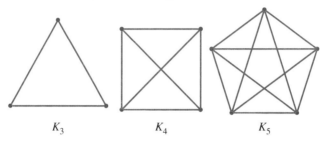

K_3 $\qquad\qquad$ K_4 $\qquad\qquad$ K_5

2. The *complete bipartite graph* on $K_{m,n}$ has vertex set $V = V_1 \bigsqcup V_2$ where V_1 and V_2 have m and n elements, respectively. There is an edge for each pair $\{v, w\}$, where $v \in V_1$ and $w \in V_2$. If we think of the elements of V_1 as "red vertices" and the elements of V_2 as "green vertices," then there are edges connecting every red vertex to every green vertex, and no edges connecting vertices of the same color. (More generally, a *bipartite graph* is one where there are "red" and "green" vertices and all edges connect vertices of different colors.)

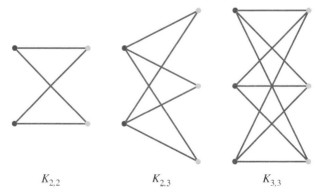

$K_{2,2}$ $\qquad\qquad$ $K_{2,3}$ $\qquad\qquad$ $K_{3,3}$

3. Here's an interesting example: consider the graph where the vertices are the two-element subsets of $\{1, 2, 3, 4, 5\}$ and where there is an edge between two

subsets if they are disjoint. This graph is known as the Petersen graph; it is a famous and important example in graph theory. Each vertex of the Petersen graph has valence 3. For example, the vertex $\{1, 2\}$ is adjacent to $\{3, 4\}$, $\{3, 5\}$, and $\{4, 5\}$.

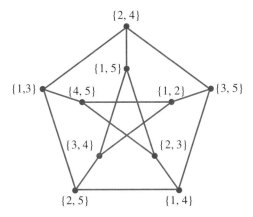

4. A *tree* is a connected graph without a cycle. Finite trees are very important when studying searching and sorting algorithms in computer science. They also make an appearance in Sean Cleary's Office Hour 16 on Thompson's group. Infinite trees play an essential role in geometric group theory; see Office Hour 3. Below are two pictures of trees, the one on the right is a portion of T_3, the tree in which every vertex has valence 3. Sometimes T_3 is called a regular 3–valent tree.

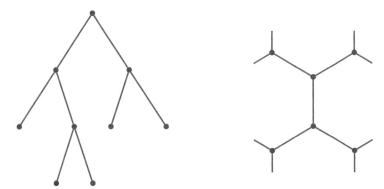

5. Graphs are commonly used to describe networks or to show complex relationships. For instance, we could create a graph whose vertices are the members of a social network (such as Facebook) and whose edges connect two members that are friends. (In some social networks, such as Google Plus, it is possible for friendships to be one-directional; to describe such a social network we need directed graphs, which we'll get to in a little bit.)

Here is a similar example: consider the graph that has as vertex set the set of movie actors and has edges between actors who have appeared in the

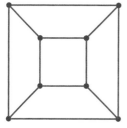 

Figure 2.2 Two isomorphic graphs.

same movie. Extensive knowledge of this graph would be useful in playing the parlor game Six Degrees of Kevin Bacon.

Variations. In certain contexts, we might be interested in variations on the definition of a graph. For instance, in some cases we will care about *loopless graphs* (no edges connecting a vertex to itself) or *simple graphs* (at most one edge between two given vertices).

We also might be interested in *directed graphs*, where the edges have preferred directions, and in *labeled graphs*, where the vertices and/or edges come with a label (e.g., a letter or a color). When we draw a picture of a directed graph, we usually draw little arrows on the edges to indicate the preferred direction. We already encountered directed graphs in Office Hour 1 with edge labels when we constructed the geometric object that has $\mathbb{Z}^2$ as its group of symmetries.

Exercise 7. Give the formal definitions of simple graphs, loopless graphs, directed graphs, and labeled graphs.

Graph isomorphisms. You should have some intuitive idea of when two graphs are the same, for example, the two graphs in Figure 2.2. Formally, if Γ and Γ' are graphs with vertex sets V and V' and edge sets E and E', then an *isomorphism* $\Gamma \to \Gamma'$ is a pair of bijections $V \to V'$ and $E \to E'$ that are compatible with the endpoint function in the sense that the endpoints of the image of an edge are the images of the endpoints of the edge. So just as an isomorphism of groups is a bijection that preserves all of the group structure, an isomorphism of graphs is a bijection (or rather a pair of bijections) that preserves all of the graph structure. For either groups or graphs, an isomorphism is a certificate that two a priori different objects are really the same object.

Sometimes the isomorphism is not clear, as in Figure 2.2. You might be surprised to learn that both graphs in Figure 2.3 are isomorphic to the Petersen graph!

Exercise 8. Find an explicit isomorphism between the graphs in Figure 2.2.

Exercise 9. Show that $K_{m,n}$ is isomorphic to $K_{n,m}$.

Exercise 10. Show that both graphs in Figure 2.3 are isomorphic to the Petersen graph.

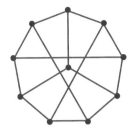

Figure 2.3 Two graphs isomorphic to the Petersen graph.

If we are working with graphs that have extra structure (i.e. directed and/or labeled edges) then an isomorphism must preserve this extra structure. For instance, if we have two graphs with blue and green edges, then an isomorphism should take blue edges to blue edges and green edges to green edges. Similarly, if we have directed graphs, then a graph isomorphism should respect the arrows on the edges.

Graph automorphisms. An isomorphism from a graph to itself is called an *automorphism*. The automorphisms of a graph are exactly what you should think of as its symmetries. The set of automorphisms of a graph Γ forms a group denoted by $\mathrm{Aut}(\Gamma)$.

The graph on the left-hand side of Figure 2.2 has lots of obvious automorphisms, for instance, rotation by $\pi/2$ and reflection about the horizontal. It also has less-obvious automorphisms, for instance, ones that interchange the inner and outer squares.

Since the graph on the right-hand side of Figure 2.2 is isomorphic to the one on the left-hand side, it has the same group of automorphisms. What do the less-obvious automorphisms from the left-hand side look like on the right-hand side?

Exercise 11. Determine the automorphism group of the graph in Figure 2.2. *Hint: the discussion at the beginning of Office Hour 1 should help.*

We already have a number of examples of graphs. Here are some of their automorphisms:

1. Since each vertex in K_n is connected to every other vertex, all vertices are interchangeable; thus $\mathrm{Aut}(K_n) \cong S_n$.
2. Similarly, we find that if $m \neq n$, then $\mathrm{Aut}(K_{m,n}) \cong S_m \times S_n$. What about $\mathrm{Aut}(K_{n,n})$?
3. From our original definition of the Petersen graph (using subsets of $\{1, 2, 3, 4, 5\}$) we see that S_5 is a subgroup of the automorphism group.
4. The regular 3–valent tree T_3 has lots of automorphisms. For instance, we can rotate by $2\pi/3$ at any vertex or swap two branches at a vertex (of course we need to be careful about how to define the automorphism on the rest of the tree). There are infinite order automorphisms as well. It turns out that $\mathrm{Aut}(T_3)$ is uncountable!

5. If two people have the same friends on Facebook, then there is an automorphism of the Facebook graph that interchanges these two people and fixes every other vertex (whether or not the two people are friends with each other!).

Exercise 12. Is S_5 the full group of automorphisms of the Petersen graph?

Exercise 13. Describe an infinite order element in $\mathrm{Aut}(T_3)$.

Exercise 14. Besides the one mentioned above, can you think of other scenarios that lead to graph automorphisms of the Facebook graph?

If a graph Γ has extra structure (i.e., directed and/or labeled edges), then automorphisms of the graph are required to preserve this extra structure (after all, a graph automorphism is a special case of a graph isomorphism). The result is still a group and is still denoted by $\mathrm{Aut}(\Gamma)$. For instance, in Office Hour 1 we considered automorphisms of a regular n–gon, thought of as a directed graph where the edges are all directed in the counterclockwise direction. As we saw, the group of automorphisms of this directed graph is $\mathbb{Z}/n\mathbb{Z}$.

Group actions on graphs. An *action* of a group G on a (possibly directed, possibly labeled) graph Γ is a homomorphism $G \to \mathrm{Aut}(\Gamma)$. So it is based on the same idea as a group action, but an action on a graph needs to preserve the structure of the graph.

If we unravel the definition of a group action of G on an (unlabeled, undirected) graph Γ with vertex set V and edge set E, we have:

- for each $g \in G$ and $v \in V$ a choice of element of V, denoted by $g \cdot v$,
- for each $g \in G$ and $e \in E$ a choice of element of E, denoted by $g \cdot e$,
- for x in either V or E, we have $1 \cdot x = x$,
- for $g, h \in G$ and x in V or E, we have $g \cdot (h \cdot x) = (gh) \cdot x$, and
- if the endpoints of $e \in E$ are v and w, then the endpoints of $g \cdot e$ are $g \cdot v$ and $g \cdot w$.

In (yet) other words, a group action of G on Γ is a pair of group actions of G on V and E that are compatible with the endpoint function.

If we take Γ to be a labeled graph, then the second condition above would further stipulate that the labels on e and $g \cdot e$ need to be the same.

If we take Γ to be a directed graph, then the last condition above would say: if the initial and terminal endpoints of $e \in E$ are v and w, respectively, then the initial and terminal endpoints of $g \cdot e$ are $g \cdot v$ and $g \cdot w$, respectively.

We can rephrase our list of automorphisms of graphs in terms of group actions:

1. S_n acts on K_n by permuting the vertices
2. $S_m \times S_n$ acts on $K_{m,n}$ by permuting the two sets of vertices
3. S_5 acts on the Petersen graph via its action on $\{1, 2, 3, 4, 5\}$
4. $\mathbb{Z}/3\mathbb{Z}$ acts on the regular 3–valent tree T_3 (as do $\mathbb{Z}/2\mathbb{Z}$ and $\mathbb{Z}$)
5. $\mathbb{Z}/2\mathbb{Z}$ sometimes acts on the Facebook graph (for instance, when there are two people with the same set of friends, as above)

Cayley graphs.   Finally, we come to the whole reason we have been studying graphs. Let G be an arbitrary group and let S be a generating set for G. For simplicity we assume that S does not contain the identity. The *Cayley graph* for G with respect to S is a directed, labeled graph $\Gamma(G, S)$ given as follows: the vertex set is G, there is a directed edge from g to gs for every $g \in G$ and $s \in S$, and we label the edge from g to gs by the element s of G. (If S is small we can sometimes use colors for the edges instead of labels.)

You might wonder what happens if you take two different generating sets for the same group. How are the two Cayley graphs related? They might not be isomorphic (for instance, if the cardinalities of the generating sets are different). However, there is a sense in which they are the same; see Office Hour 7 on quasi-isometries for an explanation.

The Cayley graphs for $\mathbb{Z}$ with respect to $\{1\}$ and for $\mathbb{Z}^2$ with respect to $\{(1, 0), (0, 1)\}$ are familiar. Indeed, we already discovered them in Office Hour 1! Each blue edge in the graph for $\mathbb{Z}$ corresponds to the generator $\{1\}$. The green and blue edges in the graph for $\mathbb{Z}^2$ correspond to the generators $(1, 0)$ and $(0, 1)$, respectively. (Of course, 1 here is the number 1, not the identity.)

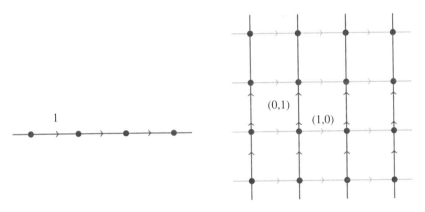

Here is the Cayley graph for S_3 with respect to the generating set $\{(1\ 2), (2\ 3)\}$. The blue edges correspond to the generator $(1\ 2)$ and the green edges correspond to the generator $(2\ 3)$.

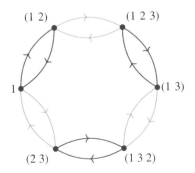

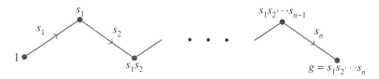

Figure 2.4 Part of a Cayley graph.

Exercise 15. Draw the Cayley graph for $\mathbb{Z}$ with the generating set $\{2, 3\}$, for D_4 with the generating set (s, t), for S_3 with the generating set $\{(1\ 2), (1\ 2\ 3)\}$, and for S_4 with the generating set $\{(1\ 2), (2\ 3), (3\ 4)\}$.

Here's a useful observation about the Cayley graph made evident by Figure 2.4:

> *If s_1, s_2, ..., s_n are the labels (in order) on an edge path in $\Gamma(G, S)$ from 1 to g, then $g = s_1 s_2 \cdots s_n$. And vice versa, if $g = s_1 s_2 \cdots s_n$, then there is an edge path from 1 to g in $\Gamma(G, S)$ whose labels are s_1, s_2, ..., s_n (in that order).*

(We allow s_i to be the inverse of an element s of S, meaning that our path crosses an s–edge against the direction of the edge.) So there is a correspondence between paths in the Cayley graph from 1 to g on one hand and words in the generators representing g on the other hand.

Exercise 16. What do the labels on an edge path from g to h in $\Gamma(G, S)$ tell you about g and h?

Automorphisms of the Cayley graph. Each element of a group G gives rise to an automorphism of the Cayley graph $\Gamma(G, S)$, as follows. The automorphism Φ_g associated to $g \in G$ is given on the vertices by:

$$\Phi_g(v) = gv.$$

This is just the action of G on itself by left multiplication. Since there is at most one directed edge connecting any two vertices with a specified direction, and since no edges connect a vertex to itself, there is only one way to extend this to an automorphism of the entire directed, labeled graph: Φ_g sends the edge $v \rightarrowtail vs$ to the edge $gv \rightarrowtail gvs$. This works since $\Phi_g(v) = gv$ and $\Phi_g(vs) = gvs$.

THEOREM 2.2. *Let S be a generating set for G. Then the map $G \to \mathrm{Aut}(\Gamma(G, S))$ is an isomorphism.*

The proof requires one new idea. Let S^{-1} denote the set of inverses of elements of S. The *word length* of $g \in G$ with respect to S is the length of the shortest word in $S \cup S^{-1}$ that is equal to g in G. The identity $1 \in G$ has word length 0. By the definition of a generating set, every element of G has finite word length with respect to S.

Our earlier observation about edge paths in Cayley graphs shows that we can also think of word length geometrically: the word length of g is the fewest number of edges in an edge path from 1 to g.

Let's prove Theorem 2.2. To simplify notation, let $\Gamma = \Gamma(G, S)$. It is not hard to see that the map $G \to \text{Aut}(\Gamma)$ we have already defined is a homomorphism. It is injective since different elements of G send the identity vertex of Γ to different vertices. The main thing is to show that it is surjective, that is, every element of $\text{Aut}(\Gamma)$ is equal to some Φ_g.

Let $\Phi \in \text{Aut}(\Gamma)$, that is, Φ is an isomorphism $\Gamma \to \Gamma$, where Γ is regarded as a directed graph with edge labels. Suppose that $\Phi(1) = g$ (we are using the same symbol for a vertex and the corresponding element of Γ). We will show that $\Phi = \Phi_g$. As above, since directed edges are determined by their endpoints, it is enough to show that Φ agrees with Φ_g on the vertices of Γ.

We use induction on word length in G with respect to S. The base case is word length 0, which is just the statement $\Phi(1) = g = \Phi_g(1)$. Assume Φ agrees with Φ_g on all elements of G of word length n with respect to S. Suppose $v \in G$ has word length $n + 1$. This means that $v = ws$, where the word length of w is n and $s \in S \cup S^{-1}$. To simplify matters, assume $s \in S$; the other case is similar.

By assumption, $\Phi(w) = \Phi_g(w)$. There is a unique edge labeled s emanating from w, namely, the edge from w to $ws = v$. Similarly, there is a unique edge labeled s emanating from $\Phi(w)$, with ending point $\Phi(w)s$. Since Φ and Φ_g both respect edge labels, it must be that $\Phi(v) = \Phi_g(v) = \Phi(w)s$. In particular, $\Phi(v) = \Phi_g(v)$, which is what we wanted to show!

The Cayley graph for the free group.   By Theorem 2.2, we know there is a geometric object with F_2 as its group of symmetries, namely, the Cayley graph for F_2 (with respect to any generating set). But what does such a Cayley graph look like?

Let a and b be the standard generators for F_2. We can draw a picture of the Cayley graph $\Gamma = \Gamma(F_2, \{a, b\})$ as follows. We start by drawing the vertex for the identity element 1 of F_2 (the empty word):

The four edges containing 1 connect 1 to the vertices a, b, a^{-1}, and b^{-1}, the four elements of F_2 with word length 1:

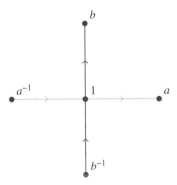

What are the edges incident to these newly drawn vertices? In general, the number of edges incident to a vertex is twice the cardinality of S, so there will be three new edges at each vertex:

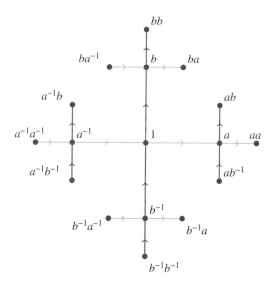

At the nth stage of this process, we will have $4(3^{n-1})$ new vertices, and we will have to draw $4(3^n)$ new edges. Since the label of each new vertex is a reduced word in a and b, and since distinct reduced words give distinct elements of F_2, we will never see a repeated vertex. Thus, the Cayley graph of F_2 is a tree. In fact it is T_4, the tree in which every vertex has valence 4.

Let's think about how elements of F_2 give symmetries of this Cayley graph. The first key thing to remember is that all edges look the same from the point of view of an abstract graph even though we have drawn them with different lengths in our picture.

To present a concrete example, we'll consider the element a, and its symmetry Φ_a of the Cayley graph. First let's think about what Φ_a does to the vertices, i.e., reduced words in a and b. Multiplying a word that starts with a on the left by a results in a word that starts aa. Likewise, multiplying a word that starts with b or b^{-1} on the left by a results in a word that starts ab or ab^{-1}, respectively. What about words that start with a^{-1}? Well, it depends on the second letter; this will be the new first letter. So the options are a^{-1}, b, and b^{-1}. We cannot have a as the new first letter since that would have meant that our initial reduced word started $a^{-1}a$, which means it was not reduced.

To see the action on Γ it is useful to divide the vertices of the Cayley graph Γ into four regions, corresponding the first letter of the corresponding words. There is the a region Γ_a, the b region Γ_b, the a^{-1} region $\Gamma_{a^{-1}}$, and the b^{-1} region $\Gamma_{b^{-1}}$ (we can take 1 to be a fifth region on its own):

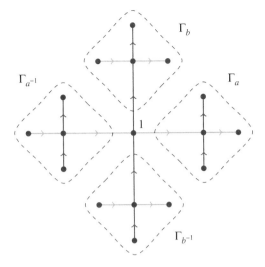

Using these regions, we can give the following rough description of what Φ_a does to the vertices of Γ: it takes $\Gamma_{a^{-1}}$ onto $\Gamma_{a^{-1}} \cup \Gamma_{b^{-1}} \cup \Gamma_b \cup \{1\}$. It also takes 1 to a, and $\Gamma_a \cup \Gamma_b \cup \Gamma_{b^{-1}} \cup \{1\}$ onto Γ_a.

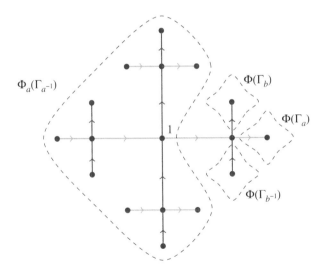

It might look like $\Phi_a(\Gamma_{a^{-1}})$ is three times the size of $\Gamma_{a^{-1}}$. However, you should convince yourself that these two (infinite) subgraphs of the Cayley graph are isomorphic as graphs. The distortion is just an artifact of our picture.

Try to describe the picture of Φ_w for more complicated w. And read Office Hour 3 for a much deeper discussion of what these symmetries can look like.

2.2 METRIC SPACES

We have solved the problem of finding a geometric object whose set of symmetries is a given group G. If we really want to understand a group G, we should try to understand an even bigger variety of geometric objects for which G is a set of symmetries. We'll now introduce metric spaces—a fairly general notion of a geometric object—and then we'll say what it means for G to be the collection of symmetries of a metric space. The idea is to mimic (rather, generalize) the notion of a group action on a graph. At the end we will state an observation that forms the cornerstone of geometric group theory: the group G can itself be thought of as a metric space, that is, G is itself a geometric object.

Metric spaces. In short, a metric space is a set with a way of measuring the distance between two points. You know lots of examples of these. The most basic example is $\mathbb{R}$, with the distance between x and y being $|x - y|$ (or more generally, the Euclidean metric on $\mathbb{R}^n$; see below). But there are other, more abstract, examples. For instance, the set could be all of the subway stations in a city and the "distance" could be the average transit time between those stations. We'll give more examples after we state the definition.

A *metric space* is a set X together with a function

$$d \colon X \times X \to \mathbb{R}$$

satisfying the following properties:

- *positive definiteness*: $d(x, y) \geq 0$ for all $x, y \in X$ and
$$d(x, y) = 0 \text{ if and only if } x = y$$
- *symmetry*: $d(x, y) = d(y, x)$ for all $x, y \in X$
- *triangle inequality*: $d(x, z) \leq d(x, y) + d(y, z)$ for all $x, y, z \in X$

The function d is called a *metric*, and is also called a *distance function*.
Let's consider some examples.

1. *The Euclidean metric.* A metric space you are very familiar with is Euclidean n–space, that is, $\mathbb{R}^n$ with the Euclidean distance:

$$d((a_1, \ldots, a_n), (b_1, \ldots, b_n)) = \sqrt{(a_1 - b_1)^2 + \cdots + (a_n - b_n)^2}.$$

2. *The taxicab metric.* Consider $\mathbb{R}^2$ with the following metric:

$$d((x_1, y_1), (x_2, y_2)) = |x_1 - x_2| + |y_1 - y_2|.$$

This distance is the (Euclidean) length of any shortest path from (x_1, y_1) to (x_2, y_2) that uses only horizontal and vertical line segments. The taxicab metric models the distance that a taxi must travel in a city whose streets are laid out in a rectangular grid.

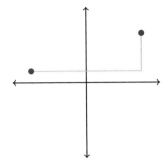

The taxicab metric makes sense in $\mathbb{R}^n$ as well:

$$d((a_1, \ldots, a_n), (b_1, \ldots, b_n)) = |a_1 - b_1| + \cdots + |a_n - b_n|.$$

3. *The hub metric.* Consider $\mathbb{R}^2$ in polar coordinates (r, θ). Assume $r_1, r_2 \geq 0$ and $\theta_1, \theta_2 \in [0, 2\pi)$. We define a metric by:

$$d((r_1, \theta_1), (r_2, \theta_2)) = \begin{cases} |r_1 - r_2| & \text{if} \quad \theta_1 = \theta_2 \\ r_1 + r_2 & \text{otherwise.} \end{cases}$$

In other words, the distance between two points is the length of the shortest path, where all paths must travel along rays from the origin. This is reminiscent of a railway system where all rail lines pass through a single hub.

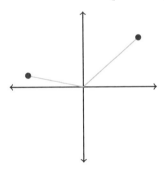

4. *The sup metric.* Consider $\mathbb{R}^2$ with the following metric:

$$d((x_1, y_1), (x_2, y_2)) = \max\{|x_1 - x_2|, |y_1 - y_2|\}.$$

A form of this metric is used by the U.S. Postal Service, which defines the length of a rectangular package to be the largest of its three dimensions.

5. *The L^1 norm.* Here is an example that is more abstract in nature. Let $C([0, 1])$ denote the set of continuous functions from $[0, 1]$ to $\mathbb{R}$. We define a metric on $C([0, 1])$ by:

$$d(f, g) = \int_0^1 |f(x) - g(x)| \, dx.$$

6. *The Hamming distance.* Here is an example that arises in coding theory. Let X denote the set of binary strings with n digits, that is, strings of 0's or 1's

that have length n. Given two strings s_1 and s_2 we define $d(s_1, s_2)$ to be the number of digits in which s_1 and s_2 differ. For example, if $n = 7$ we have

$$d(0110101, 1110000) = 3$$

as the two given strings differ in the first, fifth, and seventh digits.

7. *The path metric on a graph.* Now, for perhaps the most important example on this list (at least for us), let Γ be a *connected* graph, that is, a graph where there is an edge path between any two vertices. We can impose a metric on the set of vertices of Γ: the distance between two vertices is the length of the shortest edge path (that is, sequence of edges) connecting the vertices.

We'll see this metric a lot in the book. We'll refer to it as the *path metric* on a graph. We saw this idea when we defined the word length as the length of the shortest edge path from the identity to a given group element in the Cayley graph.

8. *The geometric realization of a graph.* The path metric on a graph is really a metric on the vertices of a graph. In particular, all distances are integers. But when you think about a graph you probably think of it as a continuous space, where it makes sense to move a small distance ϵ away from a vertex. Let's turn this intuition into reality.

The *geometric realization* X of a graph $\Gamma = \Gamma(V, E)$ is the space formed as follows. We start with one point for each element of V and one copy of the unit interval $[0, 1]$ for each element of E. Then we form X by identifying the endpoints of each interval with the V–points according to the endpoint function; in other words, as a set, X is the quotient of a disjoint union of intervals and points. The images of elements of V and E in the quotient are called the vertices and edges of X.

There is now a natural metric on X: for instance, if x and y are two points on the same edge of X (and assuming that the endpoints of the edge are not identified), then since we identified the edge with $[0, 1]$, the expression $|x - y|$ makes sense, and that is the distance between x and y. We'll leave it to you to define the distance between two points that do not lie on the same edge. Project 6 in Office Hour 3 asks you to fill in the details of this construction.

On an informal level you should think of Γ and X as being the same, but in practice there are situations where it is important to distinguish the one you are talking about from the other. Intuitively, we should think of Γ as a list of instructions for building a metric space and X as the finished product.

Exercise 17. Check that all of the above distance functions indeed define metrics on their respective spaces.

Exercise 18. Draw the unit sphere around the origin (that is, the set of points that have distance 1 from the origin) in the Euclidean metric, the taxicab metric, the hub metric, and the sup metric on $\mathbb{R}^2$.

Exercise 19. In which of the above spaces is there a unique shortest path between two points?

A subset of a metric space is also a metric space using the same distance functions. So, for instance, we can consider a regular n–gon in $\mathbb{R}^2$ as a metric space with the Euclidean distance.

Groups as metric spaces. We now come to the most important examples of metric spaces in this book: groups!

Let G be a group and S a generating set. We declare the distance between g, $h \in G$ to be the word length of $g^{-1}h$ with respect to S. In particular, the distance between g and 1 is the word length of g. It is straightforward to see that this rule does indeed define a metric on G. We call it the *word metric*.

The word metric is the same as the path metric on the Cayley graph of G (thought of as an unlabeled, undirected graph): the distance between two group elements is the length of the shortest edge path between the corresponding vertices of the Cayley graph.

Here are two key examples: the word metric on $\mathbb{Z}$ with the generating set $\{1\}$ is simply the usual Euclidean metric that $\mathbb{Z}$ inherits as a subset of $\mathbb{R}$; the word metric on $\mathbb{Z}^2$ with the generating set $\{(1, 0), (0, 1)\}$ is the taxicab metric.

In summary, the word metric gives us:

A group together with a generating set is a metric space.

This is one of the important themes of geometric group theory. Some people might even say that the study of groups as metric spaces *is* geometric group theory.

Once we have turned a group into a geometric object (that is, a metric space), we can start to ask interesting questions about what a group looks like. There are many ways to interpret this: Timothy Riley's Office Hour 8 discusses isoperimetric problems for a group (what is the largest area enclosed by a loop of length ℓ?), Moon Duchin's Office Hour 9 discusses groups that look roughly like hyperbolic spaces, Nic Koban and John Meier's Office Hour 10 talks about what a group can look like at infinity, Greg Bell's Office Hour 11 discusses the dimension of a group, and Eric Freden's Office Hour 12 studies the number of group elements in a ball of radius r around the identity.

Exercise 20. Write down the word metrics for your favorite finite groups (with your favorite generating sets) by listing all distances explicitly.

Isometries between metric spaces. Let X and Y be metric spaces with distance functions d_X and d_Y. An *isometry* from X to Y is a bijective function $f \colon X \to Y$ so that

$$d_Y(f(x), f(y)) = d_X(x, y)$$

for all pairs $x, y \in X$. In other words, an isometry is a bijective function that preserves distances. So an isometry is to metric spaces what an isomorphism is to groups or graphs.

Here is an example. Let X be the set of vertices of the cube graph from Figure 2.2 and let d_X be the path metric on the graph. Let Y be the set of binary strings of length 3 and let d_Y be the Hamming distance. We claim that X and Y

are isometric. To see this we put the cube graph in $\mathbb{R}^3$ so that it exactly corresponds to the vertices and edges of the standard unit cube in $\mathbb{R}^3$, with all coordinates equal to 0 or 1. This allows us to label each vertex of the graph by its coordinates in $\mathbb{R}^3$. Then the isomorphism $X \to Y$ is given by $(x, y, z) \to xyz$; so, for instance, the vertex $(1, 0, 1)$ gets sent to the string 101. Edges of the cube graph correspond to pairs of vertices that differ in one coordinate—exactly the situation where the corresponding binary strings have distance 1.

The previous discussion suggests a third metric space isometric to both X and Y. Let Z be the vertices of the unit cube in $\mathbb{R}^3$, as above, and endow this set with the taxicab metric in $\mathbb{R}^3$.

Exercise 21. Complete the proof that X, Y, and Z are isometric. Generalize to higher dimensions.

Exercise 22. Let X be $\mathbb{R}$ with the usual metric, and let Y be any line in $\mathbb{R}^n$ with the Euclidean metric on $\mathbb{R}^n$. Show that X and Y are isometric.

Isometries of a metric space. Next, an *isometry* of a metric space X is an isometry $X \to X$ (as above, isometries are to metric spaces what automorphisms are to groups and graphs). The set of isometries of a metric space forms a group, which we denote by $\text{Isom}(X)$. This is the group of symmetries of the metric space X. Here are some examples:

1. Rotations about a point in $\mathbb{R}^n$ and translations by some vector in $\mathbb{R}^n$ are both examples of isometries of $\mathbb{R}^n$ with the Euclidean metric.
2. Reflection across a hyperplane (that is, an $(n-1)$–dimensional plane) in Euclidean n–space is an isometry. Piggott's Office Hour 13 on Coxeter groups studies groups generated by reflections.
3. In the taxicab metric on $\mathbb{R}^2$, translations are isometries.
4. In the hub metric on $\mathbb{R}^2$, rotations about the origin and reflections about lines through the origin are isometries.
5. Here are two isometries of $C([0, 1])$:

$$\rho_1(f)(x) = -f(x) \quad \text{and} \quad \rho_2(f)(x) = f(1 - x).$$

6. Let's consider now the Hamming distance on the set of binary strings of length n. First, there is an isometry that swaps $0 \leftrightarrow 1$ in each string. Permuting the digits of all strings (by a fixed permutation) defines an isometry too; thus S_n is a subgroup of the group of isometries.

Exercise 23. Check that each of the above functions is an isometry. In each case, try to to find other isometries. Or even better: compute the entire isometry group of the given metric space.

Exercise 24. Let X be the subspace of functions $f \in C([0, 1])$ such that $f(0) = f(1)$. Every isometry of $C([0, 1])$ that preserves X is an isometry of X. Can you find additional isometries of X?

Group actions on metric spaces.   We said that we should regard Isom(X) as the group of symmetries of a metric space X. So what does it mean for a group G to act by symmetries on X? Well, as in the case of graphs, it should mean we have a homomorphism

$$G \to \text{Isom}(X).$$

If we have such an action $G \to \text{Isom}(X)$, then each $g \in G$ gives an isometry $X \to X$. As is usual with group actions, we will write the image of x under this isometry as $g \cdot x$ or gx.

Again unraveling the definitions, a group action of G on the metric space X is a map

$$G \times X \to X$$

where the image of (g, x) is written $g \cdot x$ and where the following properties are satisfied:

- $1 \cdot x = x$ for all $x \in X$,
- $g \cdot (h \cdot x) = (gh) \cdot x$ for all $g, h \in G$ and $x \in X$, and
- $d(g \cdot x, g \cdot y) = d(x, y)$ for all $g \in G$ and $x, y \in X$.

The first two items form the definition of a group action of G on X and the third item makes it an action by isometries.

Exercise 25. Verify that our two definitions of a group action by isometries are equivalent.

What are some examples of group actions by isometries?

- The dihedral group D_n acts by isometries on a regular n–gon in the plane.
- The symmetric group S_n acts by isometries on a regular n–simplex by permuting the vertices.[1]
- Every group acts by isometries on itself with the word metric by left multiplication.
- $\mathbb{Z}^2$ acts by isometries on $\mathbb{R}^2$ in the Euclidean metric or the taxicab metric by translation.
- $\mathbb{Z}/n\mathbb{Z}$ acts by isometries on $\mathbb{R}^2$ with the Euclidean metric or the hub metric (via rotation).
- Every group acts by isometries on every metric space by the *trivial action*: $g \cdot x = x$ for all $g \in G$ and all $x \in X$.

Of course, the last example is not so interesting. It tells us nothing about G or X, as every element of G maps to the identity element of Isom(X). Recall that the kernel of a homomorphism is the preimage of the identity. In general, we can expect an action of G on X to yield more information if the kernel of the corresponding homomorphism $G \to \text{Isom}(X)$ is small. In other words, a group action of G on X is most useful when most elements of G do something to X.

---

[1] A regular n–simplex is the smallest convex shape containing the points in $\mathbb{R}^{n+1}$ that lie on the positive part of a coordinate axis and have distance 1 from the origin; so, for instance, a regular 2–simplex is an equilateral triangle, and a regular 3–simplex is a regular tetrahedron.

From the geometric to the algebraic. The next theorem shows the power of applying geometry to the study of groups. In the statement, a *torsion* element of a group is an element of finite order, and a group is *torsion free* if its only torsion element is the identity.

THEOREM 2.3. *Suppose that G is a group that has a free action by isometries on $\mathbb{R}^n$ with the Euclidean metric. Then G is torsion free.*

This theorem really typifies the theme of this book, as it uses the interplay between the G and the metric space $\mathbb{R}^n$ in order to gain purely algebraic information about G.

Proof. Suppose that $g \in G$ has finite order m and let $\langle g \rangle$ denote the (cyclic) subgroup of G generated by g. We would like to show that $g = 1$. Let v be an element of $\mathbb{R}^n$ and consider the orbit of v under $\langle g \rangle$:

$$\mathcal{O} = \{v, g \cdot v, \ldots, g^{m-1} \cdot v\}.$$

Every finite set in $\mathbb{R}^n$ has a unique *centroid*, that is, there is a unique point w that minimizes the sum of the distances to the points of $\mathcal{O}$:

$$\sum_{i=1}^{m} d(w, g^i \cdot v).$$

Notice that

$$g \cdot \mathcal{O} = \{g \cdot v, g^2 \cdot v, \ldots, g^m \cdot v\} = \mathcal{O}.$$

In other words, the orbit of v under $\langle g \rangle$ is preserved by g (this is true for any group action). It follows that the centroid of $g \cdot \mathcal{O}$ is again w. But since G acts by isometries the centroid of $g \cdot \mathcal{O}$ is $g \cdot w$, and so this means $g \cdot w = w$. Since we assumed that the action of G is free, it must be that $g = 1$, as desired. $\qquad\square$

Exercise 26. Prove that any finite set of points in $\mathbb{R}^n$ has a unique centroid. *Hint: Convex functions have unique minima and the sum of convex functions is convex.*

2.3 GEOMETRIC GROUP THEORY: GROUPS AND THEIR SPACES

In the first two office hours we have discussed two types of objects: groups and spaces. The overarching question in geometric group theory is:

> *Which groups act by isometries on which spaces? If a group does act on a space, what does it say about the group? What does it say about the space? If we think of a group as a metric space, what does it look like?*

We already saw a first example of this in Theorem 2.3. Another theorem in this vein will be discussed in Office Hour 3:

> *If a group acts freely on a tree, then the group is a free group.*

Like Theorem 2.3, this theorem typifies geometric group theory: from just the information that our group acts on a certain kind of space in a certain way, we deduce concrete algebraic information about the group. There are similar theorems in Office Hour 5: if a group acts in a prescribed way on certain sets, graphs, or metric spaces, we can deduce that the group is free.

Here is a theorem from Office Hour 7, regarding the geometry of a group thought of as a metric space via the word metric.

> *If a group is quasi-isometric to $\mathbb{Z}$, then it has a subgroup of finite index that is isomorphic to $\mathbb{Z}$.*

We'll define quasi-isometric in Office Hour 7, but for now you can read it as "roughly isometric." The point of the theorem is that rough geometric information about a group can tell us fine algebraic information about the group. We'd like to emphasize that we could not have even formulated this theorem without our new perspective on groups as metric spaces.

Geometric group theory is full of beautiful examples of groups and spaces and interactions between the two. But don't take our word for it—get going!

PART 2

Free Groups

Office Hour Three

Groups Acting on Trees

Dan Margalit

In the introduction we discussed the following line of questioning:

Which groups act on which spaces? If a group does act on a space, what does it say about the group?

The purpose of this office hour is to give a first, but compelling, answer to this question. The spaces we are going to look at are trees, that is, connected graphs without cycles. We say that a group action on a tree is *free* if no nontrivial element of the group preserves any vertex or any edge of the tree. In this office hour we will prove the following striking characterization.

THEOREM 3.1. *If a group G acts freely on a tree, then G is a free group.*

We already discussed the converse to this statement in Office Hour 2. So the condition that G is free is equivalent to the condition that G acts freely on a tree.

If we have a subgroup H of a free group G, then H acts freely on a tree as well. Indeed, take any free action of G on a tree and restrict the action to H. So Theorem 3.1 has the following consequence, known as the Nielsen–Schreier theorem:

THEOREM 3.2. *Any subgroup of a free group is a free group.*

Before you scoff at this seemingly obvious theorem, try to prove it from scratch. Not as obvious as you thought!

It is nice to have theory, but can we find any interesting, real-life examples to which Theorem 3.1 applies? We will give infinitely many! Consider the group of 2×2 integral matrices of determinant 1, namely, SL$(2, \mathbb{Z})$. This group is not isomorphic to a free group since it has nontrivial elements of finite order (can you find them?). However, it has many subgroups of finite index that are isomorphic to free groups. Let me explain.

For any natural number m, denote by SL$(2, \mathbb{Z})[m]$ the kernel of the homomorphism

$$\text{SL}(2, \mathbb{Z}) \rightarrow \text{SL}(2, \mathbb{Z}/m\mathbb{Z})$$

that reduces entries modulo m. In other words, SL$(2, \mathbb{Z})[m]$ is the group of 2×2 integral matrices that have determinant 1 and that are equal to the identity matrix modulo m. This group is called the *level m congruence subgroup* of SL$(2, \mathbb{Z})$. This group has finite index in SL$(2, \mathbb{Z})$ because SL$(2, \mathbb{Z}/m\mathbb{Z})$ is finite (there are finitely many choices for each entry!).

We will prove the following theorem:

THEOREM 3.3. *For $m \geq 3$, the group* SL$(2, \mathbb{Z})[m]$ *is isomorphic to a free group.*

We will prove this theorem by examining the actions of these groups on a beautiful tree, the Farey tree, which is surprisingly ubiquitous in mathematics. Flip ahead to Figure 3.2 for a preview.

But what about SL$(2, \mathbb{Z})$ itself? It is not a free group; but what is it? It still acts on the Farey tree, but not freely. By analyzing our theory of group actions on trees a little more closely, we will be able to give explicit presentations for groups acting on trees without the freeness condition. As one example, we will show that SL$(2, \mathbb{Z})$ has the following presentation:

$$\text{SL}(2, \mathbb{Z}) \cong \langle a, b \mid a^4 = b^6 = 1, a^2 = b^3 \rangle.$$

At the end of the office hour, we will discuss SL$(3, \mathbb{Z})$, the group of 3×3 integral matrices with determinant 1. Does this group have a presentation as simple as the one for SL$(2, \mathbb{Z})$? We will see.

Okay, we have a lot to do. Let's get cracking!

3.1 THE FAREY TREE

Before we get about proving that a group acting freely on a tree is free, let me convince you that there are interesting group actions on trees that pop up naturally in mathematics.

My favorite example of a tree is the Farey tree. The easiest way to define it is to first define two auxiliary objects.

The Farey graph.   Say that $(m, n) \in \mathbb{Z}^2$ is *primitive* if it is not equal to $k(m', n')$ with $k > 1$ an integer and $(m', n') \in \mathbb{Z}^2$ (note that (m, n) is primitive if and only if $\gcd(|m|, |n|) = 1$, and also remember that $\gcd(m, 0) = m$). Define an equivalence relation $\sim$ on the primitive elements of $\mathbb{Z}^2$ by declaring (m, n) to be equivalent to $-(m, n)$.

The *Farey graph* is the graph whose vertex set is the set

$$\{(m,n) \in \mathbb{Z}^2 \mid (m,n) \text{ primitive}\}/\sim .$$

We can denote a vertex by $\pm(m,n)$. Two vertices $\pm(p,q)$ and $\pm(r,s)$ are connected by an edge if the determinant of the matrix

$$\begin{pmatrix} p & r \\ q & s \end{pmatrix}$$

is ± 1.

Before we even draw a picture of this graph, let's prove that $\mathrm{SL}(2,\mathbb{Z})$ acts on it. But how? By matrix multiplication, of course! If we take $A \in \mathrm{SL}(2,\mathbb{Z})$ and a vertex $\pm(p,q)$ of the Farey graph, then $A \cdot (\pm(p,q))$ is defined as

$$A \cdot (\pm(p,q)) = \pm A \begin{pmatrix} p \\ q \end{pmatrix}.$$

To check that this is really an action, you should first use the definition for a group action on a set to show that this gives a well-defined action on the set of vertices, and then check that the action on vertices preserves adjacency and non-adjacency of vertices.

Exercise 1. Check that the purported action of $\mathrm{SL}(2,\mathbb{Z})$ on the Farey graph really is an action.

Now, how can we draw the Farey graph? Well, let's start with the vertices $\pm(1,0)$ and $\pm(0,1)$. These are obviously connected by an edge. Are there any vertices connected to both endpoints? A quick calculation shows that there are exactly two such vertices: $\pm(1,1)$ and $\pm(-1,1)$. So we already have two triangles in our Farey graph.

Now look at any of the four new edges, say the one connecting $(1,0)$ and $(1,1)$. What vertices are connected to both of these? Again, a quick calculation tells us that there are exactly two: $\pm(0,1)$, which we already had, and $\pm(2,1)$, which is new.

We can now see the pattern: every time we draw a new edge, we can obtain one new vertex connected to both endpoints by adding the appropriate representatives of the two endpoints (above, we had to add $+(1,0)$ to $+(1,1)$ to get the new vertex). Can you see from linear algebra that this operation produces a vertex connected to the two being added? (Hint: How do column operations affect the determinant?)

This procedure tells us how to draw the Farey graph inductively. First we draw the two big triangles in Figure 3.1. Then we draw the four next-largest triangles, ad infinitum. Figure 3.1 shows the fifth stage of this process. In the figure we left out the $\pm$ symbols so as not to clutter the picture. Note that each vertex is connected to infinitely many other vertices.

How do we know that this procedure really gives the entire Farey graph? There are two steps. First check that this process eventually draws all edges emanating from $\pm(1,0)$ (which ones are they?), and then check that there is an element of $\mathrm{SL}(2,\mathbb{Z})$ that takes an arbitrary vertex to $\pm(1,0)$. That does it!

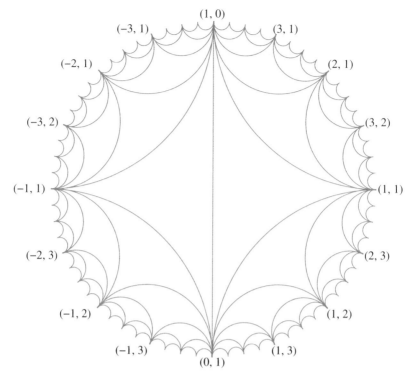

Figure 3.1 The Farey graph.

Exercise 2. Fill in the details of the proof that the given inductive procedure really draws the whole Farey tree.

The Farey complex. Once we have the Farey graph, we can construct the Farey complex. In our picture of the Farey graph, we can see lots of boundaries of triangles: triples of vertices that are pairwise connected by an edge. We can imagine gluing in (two-dimensional) triangles in all of those places (this can be done formally using the notion of a quotient space). The *Farey complex* is the space obtained by gluing in all possible triangles to the Farey graph. We already argued that every edge of the Farey graph is contained in exactly two triangles, and so near any point of the Farey complex, it looks like the plane. It shouldn't be too hard to see that the action of SL(2, ℤ) on the Farey graph induces an action on the Farey complex, meaning that the action on the graph takes the boundary of a triangle to the boundary of a triangle.

The Farey tree. Finally, we can define the *Farey tree*. The set of vertices is the union of the set of edges of the Farey complex and the set of triangles of the Farey complex. We connect two vertices when there is a containment relation. In other words, when an edge of the Farey complex is contained in a triangle of the Farey

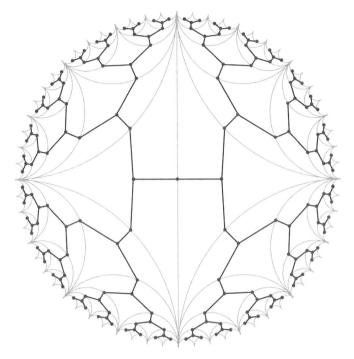

Figure 3.2 The Farey tree superimposed on the Farey complex.

complex, we connect the corresponding vertices of the Farey tree.[1] Because of the way we defined the Farey tree, we can visualize it as being superimposed on the Farey complex; see Figure 3.2. Just to emphasize: there are two types of vertices of the Farey tree—corresponding to edges and to triangles of the Farey complex—and adjacent vertices have different types.

Exercise 3. Prove that the Farey tree is indeed a tree. *Hint: Show that there is a unique (non-backtracking) path from each vertex to the vertex corresponding to the edge connecting $(0, 1)$ to $(1, 0)$.*

Again, the action of $\mathrm{SL}(2, \mathbb{Z})$ on the Farey complex carries over to an action on the Farey tree. The action is not free (the negative of the identity matrix fixes the whole tree!), but we will see later that there are many subgroups of $\mathrm{SL}(2, \mathbb{Z})$ that do act freely.

Who was Farey?   The Farey graph is named after John Farey, Sr. Farey made his living as a geologist, but he was a polymath, having written over 250 papers on subjects ranging from horticulture to canals and surveying to geology and meteorology to music and mathematics, to pacifism. His most famous contribution came in the form of a note he wrote to the *Philosophical Magazine* in 1816 entitled

---

[1] The construction of the Farey tree from the Farey complex is a special case of a general construction called a *dual* to a simplicial complex.

"On a Curious Property of Vulgar Fractions" ("vulgar fraction" is an old-fashioned term for "simple fraction"). The letter fits on one page; here is the most salient part:

> *If all the possible vulgar fractions of different values, whose greatest denominator (when in their lowest terms) does not exceed any given number, be arranged in the order of their values, or quotients; then if both the numerator and the denominator of any fraction therein, be added to the numerator and the denominator, respectively, of the fraction next but one to it (on either side), the sums will give the fraction next to it; although, perhaps, not in its lowest terms.*

If we concentrate on the lower-right-hand quadrant of our Farey graph, and replace $\pm(m, n)$ with the fraction m/n, we can see immediately how this is related. We started with the only simple fractions with denominator at most 1: $0/1$ and $1/1$. Then when we allow the denominator to be at most 2, there is one new fraction $1/2$, obtained by the so-called Farey addition of $0/1$ and $1/1$: $(1 + 0)/(1 + 1) = 1/2$, etc. (the algebra student's dream!).

In other words, the fractions between 0 and 1 with denominator at most n are the ones created in the lower-right quadrant of the nth stage of the Farey graph. And moreover, by constructing the Farey graph in this way, the simple fractions appear *in order* around the circle into which we inscribed the Farey graph.

The attribution of this observation to Farey is a little bit of a historical accident. As it happens, this fact was stated and proved by Charles Haros in 1802. The famous mathematician Augustin Cauchy saw Farey's note, proved the fact that Farey observed, and attributed the result to Farey; for more on this, see the entertaining article by Bruckheimer and Arcavi [64].

Two centuries later, the Farey graph has become an important example in many areas of mathematics, including group theory, graph theory, number theory, geometry, and dynamics; see Allen Hatcher's undergraduate text [163] as well as the surveys by Caroline Series [234, 235] for more about this amazing graph.

3.2 FREE ACTIONS ON TREES

As in our discussion of the action of $\mathrm{SL}(2, \mathbb{Z})$ on the Farey tree, a group action on a tree is a group action on the sets of vertices and edges that respects the edge relations. As in the beginning of the office hour, a group action on a tree is free if no vertex of the tree is fixed by a nontrivial group element and no edge of the tree is preserved by a nontrivial group element. Just to emphasize: if an element of the group *inverts* an edge—that is, preserves it by flipping it over—then the action is not free.

Our main theorem. The first big goal in this office hour is to prove Theorem 3.1, which states:

> *If a group G acts freely on a tree, then G is isomorphic to a free group.*

Our proof of this theorem has three steps. First we find a tiling of the tree that is consistent with the action of G, then we use the tiling to find a generating set for G, and finally we show that the generating set is a free generating set.

Step 1: Tiling the tree.   Again, the key to the proof is a certain tiling of our tree T. By a tile, we mean a subtree T_0 of the barycentric subdivision T' of T (the *barycentric subdivision* of a graph is the graph obtained by subdividing each edge; that is, we place a new vertex at the center of each edge of the original graph). And a tiling of T is a collection of tiles with the following properties:

1. No two tiles share an edge, so two tiles can only intersect at one vertex.
2. The union of the tiles is the entire tree T'.

Of course we want our tiling to have something to do with the action of G on T (and the induced action on T'), so we will impose one more restriction:

3. There is a single tile T_0 so that the set of tiles is equal to $\{gT_0 \mid g \in G\}$.

A tiling of T with all three properties could be called a G–tiling of T. Notice that the last condition is really two conditions rolled into one: first we need that each gT_0 is a tile, and second we need that every tile is of this form.

Where can we find a G–tiling of T? Here is an idea. Choose an arbitrary vertex v of T, and consider the orbit of v under G. Since G acts freely on T, it follows that the points of the orbit are in bijection with the elements of G.

Intuitively, each tile will be the set of points of T' that are closest to some vertex gv. This doesn't quite make sense because the path metric on T' is only defined on the vertices (if you replace T' with its geometric realization as in Project 6, then this intuitive idea can be made precise).

For each element g of G, we take T_g to be the subtree of the barycentric subdivision T' whose vertex set is the set of vertices w of T' so that $d(w, gv) \leq d(w, g'v)$ for all $g' \in G$ and whose edge set is the set of edges e of T' so that both vertices of e lie in T_g (the distance here is the path metric on T'). We now need to check that the collection $\{T_g\}$ forms a tiling of T.

Claim. Each T_g is a tile.

In other words, we want to check that each T_g is a subtree of the subdivision T'. To do this, we need to show that T_g is a connected subgraph of T', as any connected subgraph of a tree is a tree (since T' has no cycles, no subgraph has a cycle).

Let w be a vertex of T_g. We will show that every vertex of the (unique!) edge path in T' from w to gv lies in T_g. It follows from this that T_g is connected.

Say that $d(w, gv) = n$. Now let u be the first vertex after w on the edge path from w to gv in T'. Note that $d(u, gv) = n - 1$ (convince yourself of this!). If u were not in T_g, that would mean that there was some g' with $d(u, g'v) = m < n - 1$. But then $d(w, g'v) \leq m + 1 < n$, a contradiction. We have thus shown that each T_g is a tile.

Claim. The union of the T_g is all of T'.

It is obvious that every vertex of T' lies in some T_g, as every vertex must be closest to some gv. So it remains to show that each edge of T' lies in some T_g.

The key observation is that each edge e of T' has one vertex u that comes from T and one vertex w that does not. Thus any edge path alternates between these two types of vertices. In particular, the distance from u to the G–orbit of v is even and the distance from w to the G–orbit of v is odd (in general the distance from a point x to a set A is the infimum of $d(x, a)$, where a is in A). These two distances are not equal.

Suppose that the first distance is the smaller one and that u lies in T_g. We have assumed that the distance from w to the G–orbit of v is greater than $d(u, gv)$ and by the triangle inequality we have that $d(w, gv) \leq d(u, gv) + 1$. Since the distance from w to the G–orbit of v is an integer, it must be that $d(w, gv)$ equals the distance between w and the G–orbit of v; in other words, w lies in T_g. As u and w both lie in T_g, it follows that e lies in T_g as well, and so we are done.

Claim. For each $g, h \in G$, we have $gT_h = T_{gh}$.

By taking h to be the identity element, we obtain the tile T_0 as in the third condition for $\{T_g\}$ to be a G–tiling. So let's prove this claim.

Since T_h and T_{gh} are completely determined by their vertex sets, it is enough to show that g takes the vertex set of T_h to the vertex set of T_{gh}. Say that u is a vertex of T_h. This means that for any $k \in G$, we have

$$d(u, hv) \leq d(u, kv).$$

We would like to show that gu is a vertex of T_{gh}. This is the same as saying that

$$d(gu, (gh)v) \leq d(gu, kv)$$

for all $k \in G$. But g acts by isometries on T', and so applying g^{-1} to T' we see that this is equivalent to the statement that:

$$d(u, hv) \leq d(u, (g^{-1}k)v)$$

for all $k \in G$. But since multiplication by g^{-1} is a bijection $G \to G$, this is the same as saying that

$$d(u, hv) \leq d(u, kv)$$

for all $k \in G$. This is equivalent to the assumption that $u \in T_h$, and so we are done.

Step 2: Finding a generating set.

We have our action of the group G on the tree T, and we have our G–tiling of T. The next step in the proof is to use the tiling in order to find a symmetric generating set S for G. We take

$$S = \{g \in G \mid (gT_0) \cap T_0 \neq \emptyset\}.$$

Remember that two tiles can only intersect in a single vertex of T'. Therefore, we could replace the condition $(gT_0) \cap T_0 \neq \emptyset$ with the condition that $(gT_0) \cap T_0$ is a single vertex of T'.

We now need to show that our set S really is a symmetric generating set for G. First we will show that S is symmetric. So let $s \in S$. This means that

$$(sT_0) \cap T_0 = \{w\}$$

for some vertex w of T'. Applying s^{-1}, we immediately conclude that

$$T_0 \cap (s^{-1}T_0) = \{s^{-1}(w)\}.$$

But this means that $s^{-1} \in S$, as desired.

To finish Step 2, we need to show that S generates G. To this end, let g be an arbitrary element of G. We want to write g as a product of elements of S. We would like to use our group action, so we look at the vertex gv. We can draw the unique path from gv back to v. We can keep track of the tiles encountered along this path:

$$T_{g_n} \ , \ T_{g_{n-1}} \ , \ \cdots \ , \ T_{g_1} \ , \ T_0,$$

where $g_n = g$ and $g_0 = e$.

Claim. Each $g_{i-1}^{-1}g_i$ is equal to some $s_i \in S$.

Once we prove the claim, it follows easily by induction that

$$g = g_n = s_1 s_2 \cdots s_n,$$

and so we will be done.

To prove the claim, notice that if a path travels through tiles $T_{g_{i+1}}$ and T_{g_i} without traveling through any tiles in between, then $T_{g_{i+1}} \cap T_{g_i}$ must be nonempty (in fact, we know that the intersection is a single vertex). But then applying g_i^{-1}, we see that

$$(g_i^{-1}T_{g_{i+1}}) \cap (g_i^{-1}T_{g_i}) = T_{g_i^{-1}g_{i+1}} \cap T_0$$

is nonempty. But this means exactly that $g_i^{-1}g_{i+1}$ is in S, which is what we claimed.

Step 3: Free generation.   We started with an action of our group G on a tree T. We then found a G–tiling of T, and used this to construct a symmetric generating set S for G. It remains to show that S is a free generating set for G. In other words, if we have an element g of G, then there is only one way to write g as a freely reduced product of elements of S.

Here is how we will do this. Suppose that g is a product of elements of S, say

$$g = s_1 s_2 \cdots s_k,$$

that is freely reduced (that is, s_i is never equal to s_{i+1}^{-1}). We will construct a path from gv to v that passes through the following tiles (and no others), in order:

$$T_g = T_{s_1 \cdots s_k} \ , \ T_{s_1 \cdots s_{k-1}} \ , \ \cdots \ , \ T_{s_1} \ , \ T_0.$$

If we can do this, we will be done. Why? Because there is a unique (non-backtracking) path from gv to v in T, and this argument will show that unique path from gv to v completely determines the word $s_1 s_2 \cdots s_k$ representing G.

So how do we find the path associated to the product $s_1 s_2 \cdots s_k$? Well, we just reverse the process from before. First we find a path from v to $s_1 v$. By definition of S, the tiles T_0 and $s_1 T_0$ intersect in a single vertex. It follows that the union $T_0 \cup s_1 T_0$ is a tree! And this means there is a path—unique, by the way—from v to $s_1 v$ contained in $T_0 \cup s_1 T_0$.

To get to $s_1 s_2 v$ we continue reversing the process from before. Applying s_1^{-1} to $T_{s_1 s_2}$ and T_{s_1}, and using our rule that $h T_k = T_{hk}$, we see that $T_{s_1 s_2} \cap T_{s_1}$ is a vertex, and so $T_{s_1 s_2} \cup T_{s_1}$ is a tree, and so we can find a path from $s_1 v$ to $s_1 s_2 v$ contained in $T_{s_1 s_2} \cup T_{s_1}$. Continuing inductively, we obtain the desired path. This completes the proof of Theorem 3.1!

Exercise 4. We showed that if G acts freely on a tree T, then it is free. But what is the rank of the free group (i.e., the number of free generators)? *Hint: Consider the quotient T/G, which will in general be a graph, but not a tree.*

Example: Congruence subgroups of SL(2, $\mathbb{Z}$). We will use our newfound theory of group actions on trees to give an infinite list of natural finite-index subgroups of SL(2, $\mathbb{Z}$), each of which is isomorphic to a free group. Specifically, as advertised in the beginning, we will prove the following theorem.

THEOREM 3.4. *For $m \geq 3$, the group $\mathrm{SL}(2, \mathbb{Z})[m]$ is isomorphic to a free group.*

To prove the theorem, it suffices to show that $\mathrm{SL}(2, \mathbb{Z})[m]$ acts freely on the Farey tree. We will do something better: we will compute the stabilizer of a vertex or of an edge in the entire group $\mathrm{SL}(2, \mathbb{Z})$, and then we will check that the only stabilizer that lies in $\mathrm{SL}(2, \mathbb{Z})[m]$ is the identity.

No inversions. Since the action of SL(2, $\mathbb{Z}$) on the Farey complex cannot interchange an edge and a triangle, the action of SL(2, $\mathbb{Z}$) on the Farey tree cannot interchange vertices of two different types. In particular, an element of SL(2, $\mathbb{Z}$) cannot interchange the two endpoints of a single edge. In other words, SL(2, $\mathbb{Z}$) acts on the Farey tree without inversions. The remaining step is to understand the stabilizer in SL(2, $\mathbb{Z}$) of each vertex of the Farey tree.

Stabilizers of vertices corresponding to edges of the Farey complex. Let us first deal with the vertex v of the Farey tree corresponding to the edge of the Farey complex connecting the vertices $\pm(1, 0)$ and $\pm(0, 1)$. For an element of SL(2, $\mathbb{Z}$) to stabilize v, it simply must preserve the set

$$\{(1, 0), (-1, 0), (0, 1), (0, -1)\}.$$

Now, the columns of a matrix are just the images of the standard basis vectors under the action of that matrix. Therefore, the columns of our stabilizer must lie in this list of four elements. That gives exactly $\binom{4}{2} = 6$ matrices to think about. But we cannot choose a vector and its negative, for then the determinant will be 0. It turns out that the other four matrices all have determinant 1:

$$\begin{pmatrix} 1 & 0 \\ 0 & 1 \end{pmatrix}, \begin{pmatrix} 0 & 1 \\ -1 & 0 \end{pmatrix}, \begin{pmatrix} -1 & 0 \\ 0 & -1 \end{pmatrix}, \begin{pmatrix} 0 & -1 \\ 1 & 0 \end{pmatrix}$$

These are exactly the elements of the cyclic group generated by the second matrix on the list. So the stabilizer of v is a cyclic group of order 4.

Notice that the only two of these four matrices that lie in $SL(2, \mathbb{Z})[m]$ for $m \geq 2$ are the identity and its negative. However, since $-1 \not\equiv 1 \mod m$ for $m \geq 3$, only the identity lies in $SL(2, \mathbb{Z})[m]$ for $m \geq 3$. This is a good start, but that was just one vertex!

Claim. The group $SL(2, \mathbb{Z})$ acts transitively on the vertices of the Farey tree corresponding to edges of the Farey complex.

How does the claim help? Well, let w be any vertex of the Farey tree corresponding to an edge of the Farey complex. By the claim, there is some $M \in SL(2, \mathbb{Z})$ with $Mv = w$ (where v is the vertex discussed above). We can then use the following basic formula concerning group actions:

$$\mathrm{Stab}(Mv) = M \, \mathrm{Stab}(v) M^{-1}.$$

Now, since the stabilizer of v in $SL(2, \mathbb{Z})[m]$ is trivial, and since $SL(2, \mathbb{Z})[m]$ is normal in $SL(2, \mathbb{Z})$ (it is a kernel!), it follows that the stabilizer of $Mv = w$ in $SL(2, \mathbb{Z})[m]$ is trivial. To say it another way, if there were a nontrivial element N of the stabilizer of w in $SL(2, \mathbb{Z})[m]$, then $M^{-1}NM$ would be a nontrivial element of $SL(2, \mathbb{Z})[m]$ in the stabilizer of v. That does it.

Exercise 5. Check the formula $\mathrm{Stab}(Mv) = M \, \mathrm{Stab}(v) M^{-1}$.

OK, so let's check the claim. Take the vertex v of the Farey tree corresponding to $\{\pm(p, q), \pm(r, s)\}$. We will show there is an element of $SL(2, \mathbb{Z})$ taking the standard vertex v_0 given by $\{\pm(1, 0), \pm(0, 1)\}$ to v. By the definition of the Farey tree, the determinant of

$$\begin{pmatrix} p & r \\ q & s \end{pmatrix}$$

is 1 (after possibly replacing (r, s) with $-(r, s)$). But then this is the matrix we were looking for!

Stabilizers of vertices corresponding to triangles of the Farey complex. As in the last case, it suffices to consider the case of the vertex corresponding to the triple

$$\pm(1, 0) \quad \pm(0, 1) \quad \pm(1, 1)$$

(prove the analogous claim). If an element of $SL(2, \mathbb{Z})$ fixes this vertex of the Farey tree, it must take $(1, 0)$ to one of the six vectors listed and it must take $(0, 1)$ to another one of the six vectors. But the images of these vectors are the columns of the given element of $SL(2, \mathbb{Z})$, and so there are only a few possibilities for the stabilizer of this vertex in $SL(2, \mathbb{Z})$, namely, the matrices

$$\begin{pmatrix} 1 & 0 \\ 0 & 1 \end{pmatrix}, \begin{pmatrix} 0 & 1 \\ -1 & 1 \end{pmatrix}, \begin{pmatrix} -1 & 1 \\ -1 & 0 \end{pmatrix}$$

and their negatives. Simply using the facts that $-1 \not\equiv 1 \mod m$ and $\pm 1 \not\equiv 0 \mod m$, we can easily see that—aside from the identity—none of these 12 matrices is congruent to the identity modulo m when $m \geq 3$, and hence none lie in $SL(2, \mathbb{Z})[m]$. This completes the proof that $SL(2, \mathbb{Z})[m]$ is free!

Exercise 6. Compute the ranks of some of the congruence subgroups $SL(2, \mathbb{Z})[m]$. To do this you should compute the quotient of the Farey complex by $SL(2, \mathbb{Z})[m]$. The first few are objects that you are familiar with!

3.3 NON-FREE ACTIONS ON TREES

We would really like a presentation for our favorite group, $SL(2, \mathbb{Z})$. First, we will attack a slightly simpler group, the quotient of $SL(2, \mathbb{Z})$ by the order 2 subgroup generated by the negative of the identity. This quotient group is denoted by $PSL(2, \mathbb{Z})$ (the P stands for "projective," the standard term for taking quotients by scalar multiplication). Our next goal is to show that $PSL(2, \mathbb{Z})$ has the following presentation:

$$PSL(2, \mathbb{Z}) \cong \langle a, b \mid a^2 = b^3 = 1 \rangle.$$

This presentation looks almost like a free group—there are no relations between a and b except for the relations which say that a and b have finite order. That brings us to our next definition.

Free products. The groups $SL(2, \mathbb{Z})$ and $PSL(2, \mathbb{Z})$ are not free, but they do act on a tree, the Farey tree. Moreover, the action is very close to being free in that the stabilizer of each vertex and edge is a finite group.

This inspires us to define a type of group that is *almost* free, a free product. We will see that $PSL(2, \mathbb{Z})$ can be described as a free product, and later we will see that $SL(2, \mathbb{Z})$ can be described as a further generalization of a free product.

Let G and H be two groups. The free product of G and H, written $G * H$, is the group of alternating words in the disjoint union of the elements of G and H. In other words, the elements are of one of four forms:

$$g_1 h_1 g_2 \cdots g_k h_k, \quad g_1 h_1 g_2 \cdots h_{k-1} g_k, \quad h_1 g_1 h_2 \cdots h_k g_k, \quad h_1 g_1 h_2 \cdots g_{k-1} h_k.$$

You can guess how we multiply: concatenation. Of course, if we multiply $g_1 h_1$ by $h_1' g_1' h_2'$, then we need to combine $h_1 h_1'$ into a single element:

$$(g_1 h_1) \cdot (h_1' g_1' h_2') = g_1 h_1'' g_1' h_2',$$

where $h_1'' = h_1 h_1'$ in H.

What is the presentation for a free product? If $G \cong \langle S_G \mid R_G \rangle$ and $H \cong \langle S_H \mid R_H \rangle$, then

$$G * H \cong \langle S_G \sqcup S_H \mid R_G \sqcup R_H \rangle.$$

In other words, the relations in the free product only include the relations from the original groups, and no other relations.

Example: The free product of two cyclic groups. A key example of a free product is the case where G and H are both cyclic groups. Say $G \cong \mathbb{Z}/m\mathbb{Z}$ and $H \cong \mathbb{Z}/n\mathbb{Z}$. Then, following the previous discussion, we have:

$$\mathbb{Z}/m\mathbb{Z} * \mathbb{Z}/n\mathbb{Z} \cong \langle a, b \mid a^m = b^n = 1 \rangle.$$

Let's think about the easiest case, where $m = n = 2$. The elements of this group are alternating words in a and b; there are no powers allowed since $a^2 = b^2 = 1$. Very simple. Later we will see that this group is the group of symmetries of a very familiar graph.

The next simplest case is $m = 2, n = 3$. Later in this office hour, we will show that $\mathbb{Z}/2\mathbb{Z} * \mathbb{Z}/3\mathbb{Z}$ is isomorphic to $\mathrm{PSL}(2, \mathbb{Z})$.

We can also consider the case where one or both of the cyclic groups are infinite. For example, it is a good exercise to check that $\mathbb{Z} * \mathbb{Z}$ is isomorphic to the free group F_2.

Exercise 7. Show that $\mathbb{Z} * \mathbb{Z}$ is isomorphic to F_2.

Groups acting on trees with nontrivial vertex stabilizers.   Earlier we proved that a group acting freely on a tree is free. We will now see that even if the action is not free, we can often still describe the group using our newfound notion of a free product.

THEOREM 3.5. *Suppose that a group G acts without inversions on a tree T in such a way that G acts freely and transitively on edges. Choose one edge e of T and say that the stabilizers of its vertices are H_1 and H_2. Then*

$$G \cong H_1 * H_2.$$

OK, let's see if we can prove this in the same way we proved that a group acting freely on a tree is a free group. What should the G–tilings be? In this case, since G acts without inversions we can get away without the barycentric subdivision, so the definition of the tiling is the same as before except that the tiles are subgraphs of T itself. Since G acts transitively on the edges, we can take each edge to be a tile. So far so good.

Next we need to show that G is generated by H_1 and H_2. Again, let's try the same tactic as before. Let v_i be the vertex of e with stabilizer H_i.

Let g be any element of G. Last time we considered the path from gv to v for some vertex v. This time our edge e plays a special role, so let's use that. Connect ge to e by a path; this is the unique path obtained by joining any vertex of ge to any vertex of e by a path.

Following along the path, we obtain a sequence of edges

$$ge = e_n, \ldots, e_1, e_0 = e.$$

By assumption, each e_i can be written as $g_i e$ for some $g_i \in G$. It follows from the fact that G acts freely on edges that each g_i is unique; in particular, $g_n = g$. We will show by induction that g_i can be written as a product of elements of H_1 and H_2. The base case is $i = 0$, in which case there is nothing to do.

As a warm-up for the inductive step, let's consider the case $i = 1$. Now, e_1 and $e_0 = e$ share one vertex, either v_1 or v_2; say it is v_1. By definition, $g_1^{-1} e_1 = e_0$. Therefore, the vertex v_1 (which is a vertex of e_1) must get mapped to a vertex of e, either v_1 or v_2. But we know v_1 and v_2 are in different orbits, so we must have $g_1^{-1}(v_1) = v_1$. This is the same as saying that $g_1^{-1} \in H_1$, or $g_1 \in H_1$, and we are done.

The general inductive step is basically the same. We assume that g_k can be written as a product of elements of H_1 and H_2. Then we consider $e' = g_k^{-1} e_{k+1}$. Since e_k and e_{k+1} share a vertex, the edges e' and e share a vertex, say v_1. Also,

$$\left(g_{k+1}^{-1} g_k \right) e' = \left(g_{k+1}^{-1} g_k \right) \left(g_k^{-1} e_{k+1} \right) = e.$$

As in the previous paragraph, it follows that $g_{k+1}^{-1} g_k$ lies in H_1, from which it follows that g_{k+1} can be written as a product of elements of H_1 and H_2, as desired.

(In the previous paragraph, the vertex shared by e' and e depends only the parity of k. Why?)

We just showed that we can write any $g \in G$ as an alternating product of elements of H_1 and H_2, as per the definition of a free product. To show that G really is a free product of H_1 with H_2, we need to show that this is the only such expression for g. Again, we will mimic what we did in the free group case. We will show that if we are given a finite product

$$h_1 k_1 \cdots = g,$$

then we can recover the path from e to ge we studied above. The first edge in the path, of course, is e. The second edge is $h_1 e$. Because $h_1 v_1 = v_1$ by definition, the edge $h_1 e$ shares the vertex v_1 with e. Next, we show that $(h_1 k_1)e$ shares a vertex with the previous edge, $h_1 e$. Applying h_1^{-1} to both, we obtain $k_1 e$ and e, which share the vertex v_2. Therefore, $(h_1 k_1)e$ and $h_1 e$ share the vertex $h_1 v_2$. Continuing inductively, the sequence of edges e, $h_1 e$, $(h_1 k_1)e$, etc. is a path in T starting with e and ending with ge.

To summarize, a path in T from e to ge gives a unique alternating word in the elements of H_1 and K_1, and an alternating word in the elements of H_1 and K_1 gives a unique path in T from e to gv. But there is only one path from e to ge, and so it follows that there is only one word!

Example: the infinite dihedral group. The most basic group that illustrates the previous theorem is the group of symmetries of the graph Γ whose vertices are in bijection with the integers and whose edges connect integers that differ by ± 1.

This group is called the infinite dihedral group, and is denoted by D_∞. In analogy with the finite dihedral groups, we can think of this group as the group of symmetries of the ∞–gon Γ.

Note that D_∞ does not act on Γ without inversions (rather, it acts with inversions), since reflection about the midpoint of any edge is a symmetry of Γ inverting that edge (we can alternately think of this symmetry as rotation by π in the plane). However, whenever a group acts on a tree, it automatically acts without inversions on the barycentric subdivision. So we instead consider the action of D_∞ on the barycentric subdivision Γ'.

Now D_∞ acts on Γ' without inversions, and it is easy to see that D_∞ acts freely and transitively on the edges of Γ' (check both of these!).

As per the theorem, we consider the edge connecting the vertices 0 and 1/2. The stabilizer of each is isomorphic to $Z/2\mathbb{Z}$: there is the identity, and reflection through that vertex. So our theorem immediately implies that

$$D_\infty \cong \mathbb{Z}/2\mathbb{Z} * \mathbb{Z}/2\mathbb{Z}.$$

If we write this as a group presentation, it is:

$$D_\infty \cong \langle a, b \mid a^2 = b^2 = 1 \rangle.$$

Recall that the dihedral group D_n, the group of symmetries of a regular n–gon, has the presentation

$$D_n \cong \langle a, b \mid a^2 = b^2 = 1, (ab)^n = 1 \rangle.$$

We obtained the infinite dihedral group by replacing the n–gon with the ∞–gon, and so we could be greedy and hope that we can obtain a presentation for D_∞ by replacing the n with an ∞ in the above presentation. This actually makes sense! We should think of the relation $(ab)^\infty = 1$ as saying that no power of ab is equal to the identity, and hence this relation can be deleted. After deleting this relator from the presentation for D_n, we indeed obtain the presentation for D_∞.

Exercise 8. Show that D_∞ contains an index-2 subgroup isomorphic to $\mathbb{Z}$. Which symmetries of the ∞–gon does this group describe? What are the corresponding freely reduced words?

Checking that PSL$(2, \mathbb{Z})$ is a free product.   Remember that we wanted to show that the group PSL$(2, \mathbb{Z})$ is isomorphic to $\mathbb{Z}/2\mathbb{Z} * \mathbb{Z}/3\mathbb{Z}$. We would like to use our action of PSL$(2, \mathbb{Z})$ on the Farey tree. But wait—we do not even have such an action! We only have an action of SL$(2, \mathbb{Z})$. Fortunately, though, the action we defined for SL$(2, \mathbb{Z})$ on the Farey tree *descends* to an action of the quotient PSL$(2, \mathbb{Z})$ on the Farey tree.

Here is what I mean. First, we see that the negative of the identity in SL$(2, \mathbb{Z})$ acts trivially on the Farey tree (meaning it fixes every vertex and edge), since (m, n) and $-(m, n)$ represent the same vertex of the Farey tree. Next, given an element A of PSL$(2, \mathbb{Z})$ and a vertex v of the Farey tree, we can define Av as $\tilde{A}v$, where $\tilde{A}$ is any coset representative for A in SL$(2, \mathbb{Z})$. There was some ambiguity: we could have swapped $\tilde{A}$ with $-\tilde{A}$. But since $-I$ acts trivially on the Farey tree, $\tilde{A}v$ and $-\tilde{A}v$ are the same thing, and we have a well-defined action.

(The previous paragraph is a special case of a general fact: if a group G acts on a set X, and each element of the normal subgroup $H \leqslant G$ acts trivially on X, i.e., fixes every point, then there is an induced action of G/H on X. The proof of the general fact is the same as the one we just gave.)

Since SL$(2, \mathbb{Z})$ acts on the Farey tree without inversions, it follows that PSL$(2, \mathbb{Z})$ acts on the Farey tree without inversions (the latter action was defined in terms of the former, after all).

Next, we need to check that SL$(2, \mathbb{Z})$, hence PSL$(2, \mathbb{Z})$, acts transitively on the edges of the Farey tree. Recall that an edge of the Farey tree connects one vertex corresponding to an edge of the Farey complex to one vertex corresponding to a triangle of the Farey complex. We have a favorite edge e_0, namely the one connecting the vertex v_0 corresponding to $\{\pm(1, 0), \pm(0, 1)\}$ to the vertex w_0 corresponding to $\{\pm(1, 0), \pm(0, 1), \pm(1, 1)\}$. We will show that we can take any other edge to this one using an element of SL$(2, \mathbb{Z})$.

Consider an edge e connecting vertices v and w of the Farey tree. And say that v corresponds to $\{\pm(a, b), \pm(c, d)\}$. The fact that $\pm(a, b)$ and $\pm(c, d)$ span an edge of the Farey complex means that one of the matrices

$$A = \begin{pmatrix} a & c \\ b & d \end{pmatrix} \quad \text{or} \quad A' = \begin{pmatrix} a & -c \\ b & -d \end{pmatrix}$$

has determinant 1; say A does. Then $A^{-1}v = v_0$. We are halfway there—we just need to find a matrix B that stabilizes v_0 and takes w_0 to $A^{-1}w$. Then $B^{-1}A^{-1}e$ will be equal to e_0, as we wanted.

So what is $A^{-1}w$? All we know is that it is some vertex of the Farey tree that is connected to v_0. Well, there are only two of those, since an edge of the Farey complex is contained in exactly two triangles. There is w_0, and the vertex corresponding to $\{\pm(1, 0), \pm(0, 1), \pm(1, -1)\}$; call it w_1. If $A^{-1}w = w_0$ there is nothing to do. But there is the possibility that $A^{-1}w = w_1$, and so we need to show that there is a matrix B in $\mathrm{SL}(2, \mathbb{Z})$ taking w_0 to w_1 (in other words, we are showing that the stabilizer of v_0 acts transitively on the vertices adjacent to v_0).

Remember that we already computed the stabilizer of v_0 earlier. It is the group of matrices consisting of

$$\begin{pmatrix} 1 & 0 \\ 0 & 1 \end{pmatrix}, \begin{pmatrix} 0 & 1 \\ -1 & 0 \end{pmatrix},$$

and their negatives. But the second matrix takes the vertex w_0 to w_1, and so we have succeeded in showing that $\mathrm{SL}(2, \mathbb{Z})$, hence $\mathrm{PSL}(2, \mathbb{Z})$, acts transitively on the edges of the Farey tree.

Before we apply our theorem, we need to check one last thing: that $\mathrm{PSL}(2, \mathbb{Z})$ acts freely on the edges of the Farey tree. If an element of $\mathrm{SL}(2, \mathbb{Z})$ fixes the edge e_0 we were just looking at, then it must lie in the intersection of the stabilizers of v_0 and w_0. But this intersection is just the identity matrix and its negative, which both represent the trivial element of $\mathrm{PSL}(2, \mathbb{Z})$.

Now comes the fun part. We know that $\mathrm{PSL}(2, \mathbb{Z})$ acts on the Farey tree without inversions and freely and transitively on the edges. So our theorem tells us that $\mathrm{PSL}(2, \mathbb{Z})$ is isomorphic to the free product of the stabilizers of v_0 and w_0. In $\mathrm{SL}(2, \mathbb{Z})$, the stabilizers are isomorphic to $\mathbb{Z}/4\mathbb{Z}$ and $\mathbb{Z}/6\mathbb{Z}$. The negative of the identity corresponds to 2 and 3 in these two groups, and so the images of these stabilizers in $\mathrm{PSL}(2, \mathbb{Z})$ are isomorphic to $\mathbb{Z}/2\mathbb{Z}$ and $\mathbb{Z}/3\mathbb{Z}$. And finally, we have proven that

$$\mathrm{PSL}(2, \mathbb{Z}) \cong \mathbb{Z}/2\mathbb{Z} * \mathbb{Z}/3\mathbb{Z}.$$

Very nice!

Free products with amalgamation. We need one more algebraic construct before we get to our description of $\mathrm{SL}(2, \mathbb{Z})$ that we advertised in the beginning, and that is the notion of a free product with amalgamation. This is like a free product, but we allow some amount of mixing between the two groups being combined.

Let G, H, and K be groups. Fix injective homomorphisms $i_G : K \to G$ and $i_H : K \to H$. Let N be the normal subgroup of $G * H$ generated by all elements of the form $i_G(k)i_H(k)^{-1}$ for $k \in K$. By this we mean the group generated by these elements and all of their conjugates in $G * H$. Finally, the free product of G and H amalgamated along K is

$$G *_K H = (G * H) / N.$$

The homomorphisms i_G and i_H are very important; if we change them, the group can change drastically. However, we usually leave these out of the notation so that things do not get too cluttered.

As with the other constructions, there is a simple description of $G *_K H$ via presentations. If $G \cong \langle S_G \mid R_G \rangle$ and $H \cong \langle S_H \mid R_H \rangle$, then

$$G *_K H \cong \langle S_G \sqcup S_H \mid R_G \sqcup R_H \cup \{i_G(k)i_H(k)^{-1} \mid k \in K\} \rangle.$$

If K is finite, then there are finitely many relations of the form $i_G(k)i_H(k)^{-1}$. But actually, we can replace $\{i_G(k)i_H(k)^{-1} \mid k \in K\}$ with $\{i_G(k)i_H(k)^{-1} \mid k \in S_K\}$, where S_K is any generating set for K. So as long as K is finitely generated, we are still only adding finitely many extra relations at the end.

Free products with amalgamation first appeared in a paper of Otto Schreier in 1927. His idea was generalized by Hanna Neumann in 1948. While this notion might be a little hard to digest at first, it comes up in many places in math; for instance, in algebraic topology it is featured in the Seifert–van Kampen theorem, which gives instructions for computing the fundamental group of a space obtained by gluing two spaces along a subspace in each. Even if you don't know this theorem, this should resonate with the idea that the free product with amalgamation can be thought of as the group obtained by gluing two groups along a subgroup of each.

Example: A free product of cyclic groups with amalgamation.   Let's consider one example of a free product with amalgamation. Choose positive integers m and n, and say that they have a common divisor d. There are natural inclusions $\mathbb{Z}/d\mathbb{Z} \to \mathbb{Z}/m\mathbb{Z}$ and $\mathbb{Z}/d\mathbb{Z} \to \mathbb{Z}/n\mathbb{Z}$ given by $1 \mapsto m/d$ and $1 \mapsto n/d$. With these maps in hand, we can form the free product with amalgamation:

$$\mathbb{Z}/m\mathbb{Z} *_{\mathbb{Z}/d\mathbb{Z}} \mathbb{Z}/n\mathbb{Z} \cong \langle a, b, c \mid a^m = b^n = 1, a^{m/d} = b^{n/d} = c \rangle.$$

Notice that the generator c is not really needed, since is can be written as a power of a (or b). Therefore, we can rewrite this presentation as

$$\mathbb{Z}/m\mathbb{Z} *_{\mathbb{Z}/d\mathbb{Z}} \mathbb{Z}/n\mathbb{Z} \cong \langle a, b \mid a^m = b^n = 1, a^{m/d} = b^{n/d} \rangle.$$

(Convince yourself these are isomorphic groups!) This simplification of our presentation is an example of a Tietze transformation of a presentation, after Heinrich Franz Friedrich Tietze, who introduced these moves in 1908.

Notice that if we set $m = 4$, $n = 6$, and $d = 2$, we obtain the presentation for $\mathrm{SL}(2, \mathbb{Z})$ from the beginning of the office hour. We are almost there!

Group actions on trees and free products with amalgamation.   The action of $\mathrm{SL}(2, \mathbb{Z})$ on the Farey tree does not satisfy the hypotheses of our last theorem, since the negative of the identity matrix acts trivially. In particular, $\mathrm{SL}(2, \mathbb{Z})$ does not act freely on the edges of the Farey tree. Fortunately, this is not a big problem. Since $\mathrm{SL}(2, \mathbb{Z})$ acts on the Farey tree without inversions, the stabilizer of any edge includes into the stabilizers of its two vertices. It is hopefully not too surprising then that we can write $SL(2, \mathbb{Z})$ as a free product with amalgamation. Here is the general theorem we will need.

THEOREM 3.6. *Suppose that a group G acts without inversions on a tree T in such a way that G acts transitively on edges. Choose one edge of T and say that the stabilizer of this edge is K and that the stabilizers of its vertices are H_1 and H_2. Then*

$$G \cong H_1 *_K H_2,$$

where the maps $K \to H_i$ are the inclusions of the edge stabilizer into the two vertex stabilizers.

This theorem can be proven along the exact same lines as the previous versions, so I'll leave it to you!

Exercise 9. Prove Theorem 3.6.

Presenting $SL(2, \mathbb{Z})$. We can now write down our presentation for $SL(2, \mathbb{Z})$. The hard part is already done: we have already said that $SL(2, \mathbb{Z})$ acts without inversions on the Farey tree and that it acts transitively on edges, and we have already computed the stabilizers of e_0, v_0, and w_0: they are isomorphic to $\mathbb{Z}/2\mathbb{Z}$, $\mathbb{Z}/4\mathbb{Z}$, and $\mathbb{Z}/6\mathbb{Z}$. Therefore, applying the above theorem, we have that

$$SL(2, \mathbb{Z}) \cong \mathbb{Z}/4\mathbb{Z} *_{\mathbb{Z}/2\mathbb{Z}} \mathbb{Z}/6\mathbb{Z}.$$

As above, the group on the right-hand side has the presentation we gave at the beginning of the office hour. So we have accomplished our main goal!

Exercise 10. Find a presentation for the level 2 group $SL(2, \mathbb{Z})[2]$.

Exercise 11. Find a presentation for $GL(2, \mathbb{Z})$, the group of 2×2 invertible matrices with determinant ± 1.

Property FA. We have spent a lot of time talking about groups that act on trees. What about groups that do not act on a tree in any kind of nice way? Following Jean-Pierre Serre, we say that a group G has *property FA* if for every action without inversions of G on a tree T, there is a vertex of T that is fixed by every element of G. These actions are not so interesting in the context of what we are doing, since the entire group G is contained in the stabilizer of a single vertex. Serre proved the following theorem in 1974.

THEOREM 3.7. *A countable group G has property FA if and only if the following three conditions hold:*

1. *G is finitely generated,*
2. *there are no surjective homomorphisms $G \to \mathbb{Z}$, and*
3. *G is not isomorphic to a free product with amalgamation.*

Why should this be true? Well, we can imagine that if a group does decompose as a free product with amalgamation, then we can reverse engineer a tree that it acts on: just reverse the process from before (see Project 1).

So, just as we obtained algebraic information about our groups when they did act on trees, we again gain algebraic insight from the *non*-existence of nice actions on trees.

We'll end with a famous theorem about property FA. In the statement, $SL(n, \mathbb{Z})$ is the group of $n \times n$ integral matrices with determinant 1. Serre proved the following in 1974 [236].

THEOREM 3.8. $SL(n, \mathbb{Z})$ *has property FA when* $n \geq 3$.

So we cannot hope to present $SL(3, \mathbb{Z})$ in quite the same way as we did $SL(2, \mathbb{Z})$.

Exercise 12. Show that every finite group has property FA.

Exercise 13. Show that the quotient of a group with property FA also has property FA.

FURTHER READING

If you want to read more about group actions on trees, there is no better place to look than Serre's elegant classic, *Trees* [237]. Meier's excellent text *Groups, Graphs, and Trees* has a chapter on group actions on trees and also a chapter on Baumslag–Solitar groups (see the projects). Finally, if you want to know more about the mathematics related to the Farey graph, you should check out Allen Hatcher's unfinished textbook *Topology of Numbers* [163], whose topics range from Pythagorean triples to continued fractions to Diophantine equations to quadratic forms.

PROJECTS

Project 1. Theorem 3.1 says that if a group acts freely on a tree, then it is free. But the converse is also true because we have the action on the Cayley graph. Prove converses of Theorems 3.5 and 3.6. Specifically, given a decomposition of a group G into a free product, or a free product of amalgamation, build a tree that it acts on so the quotient is an edge. Note that this tree is different from (but related to) the Cayley graph for G.

Project 2. In Theorem 3.6, we assumed that a group G acts transitively on the edges of T. In other words, T/G is a single edge. What if the quotient T/G is a more complicated graph? For instance, what if T/G is a loop, that is, the graph with one vertex and one edge? Can you give a presentation for G in this case that is analogous to the presentation we gave when T/G is an edge with two vertices? If you do this you will discover a kind of group called an HNN extension (named after Graham Higman, B. H. Neumann, and Hanna Neumann)!

Project 3. The *Baumslag–Solitar group* $BS(m, n)$ is the group with the presentation

$$BS(m, n) = \langle a, b \mid ba^m b^{-1} = a^n \rangle.$$

Let's also define a tree $T(m, n)$ to be the tree embedded in the $\mathbb{R}^2$ where each vertex has m edges going up and n edges going down. Draw a picture of $T(m, n)$ and show that $BS(m, n)$ acts on $T(m, n)$. Describe the stabilizers. Compare with the previous project.

Project 4. Prove the following classification of automorphisms of a tree: if g is an automorphism of a tree T then either

1. there is a vertex or edge of T preserved by g or
2. there is a bi-infinite (non-backtracking) path L in T fixed by g and g acts on L by nonzero translation.

In the first case g is called *elliptic* and in the second case g is called *hyperbolic* and L is called the *axis* for L. The amount that g translates along L is called the *translation length* of g.

Project 5. When is the product of two elliptic automorphisms of a tree another elliptic automorphism? When is the product hyperbolic? What about the product of two hyperbolic automorphisms? What about the product of an elliptic automorphism with a hyperbolic one? What can you say about the resulting translation length when the result is hyperbolic?

Project 6. There is a way to turn a tree T into a metric space where the distance is defined on points that are not necessarily vertices. This construction is called the *geometric realization* of the tree. Actually this works for any graph $\Gamma = \Gamma(V, E)$. Here is the basic idea: we start with one point for each element of V, and one copy of the unit interval $[0, 1]$ for each element of E; then we form the geometric realization X of Γ by identifying the endpoints of each interval with the vertices according to the endpoint function; in other words, as a set, X is the quotient of a disjoint union of intervals and points. Work out the details of this construction and define a metric X on Γ that agrees with the path metric on pairs of vertices.

Project 7. Prove that automorphisms of a tree correspond to isometries of the geometric realization. Prove that a group action on a tree is free if and only if no nontrivial element of the group fixes a point of the geometric realization. Then give an alternate statement of the classification of automorphisms of a tree in terms of isometries of the geometric realization.

Project 8. Show that if the trace of $A \in \mathrm{SL}(2, \mathbb{Z})$ is $-1, 0$, or 1, then the action of A on the Farey tree is elliptic. Here are the steps:

1. If the trace of an element A of $\mathrm{SL}(2, \mathbb{Z})$ is $-1, 0$, or 1, show that the eigenvalues of A are complex.
2. Then use the Cayley–Hamilton theorem to show that A has finite order.

3. Then consider the finite cyclic subgroup of $SL(2, \mathbb{Z})$ generated by A and consider the (finite) orbit of some particular point in the Farey graph under the action of this finite group. Show that this orbit has a well-defined barycenter in the geometric realization of the Farey graph and that the barycenter is fixed by A, and so A is elliptic.

Can you find a formula for the vertex/edge fixed by A in terms of the entries of A?

Project 9. Show that if the trace of $A \in SL(2, \mathbb{Z})$ has absolute value greater than 1, then the action of A on the Farey tree is hyperbolic. (One way to do this is to show it is not elliptic! And you can accomplish this by showing that A has infinite order.) Can you find a recipe for the corresponding axis in the Farey tree? Or the translation length?

Project 10. An $\mathbb{R}$–*tree* is a metric space where between any two points there is a unique (non-backtracking) path. (Note that the geometric realization of a tree is an $\mathbb{R}$–tree!) Learn about $\mathbb{R}$–trees. What can we say about a group if it acts freely on an $\mathbb{R}$–tree?

Project 11. Learn (from Serre's book) the proof that $SL(3, \mathbb{Z})$ has property FA.

Project 12. Show that if a finite index subgroup of a group has property FA, then the larger group has property FA.

Project 13. Carter and Keller [75] proved the amazing theorem that $SL(n, \mathbb{Z})$ is boundedly generated for $n \geq 3$. More specifically, each element of $SL(n, \mathbb{Z})$ is a product of (at most) 48 elementary matrices. Use our description of $SL(2, \mathbb{Z})$ to show that it is not boundedly generated.

Project 14. Read the paper by Carter and Keller.

Project 15. What can you say about group actions on higher-dimensional complexes? For example, we created the Farey complex from the Farey graph by gluing in triangles. Could we deduce our presentation for $SL(2, \mathbb{Z})$ using directly the action on the Farey complex, and an analogous theory? For starters, think about how the usual presentation for $\mathbb{Z}^2$ is related to the action of $\mathbb{Z}^2$ on the usual tiling of $\mathbb{R}^2$ by squares.

Office Hour Four

Free Groups and Folding

Matt Clay

In this office hour, we will use the combinatorics of graphs and a simple operation called folding to study subgroups of free groups. The free group of rank n, F_n, was introduced in Office Hour 1 as the set of reduced words in the symbols $\{a_1, a_1^{-1}, a_2, a_2^{-1}, \ldots, a_n, a_n^{-1}\}$. This is a simple and concise description, one that you could write a computer program to implement, but it is hard to see the finer algebraic structure of free groups with this model. In Office Hour 3 it was shown that every subgroup of a free group is free (we will see this again here). But what if you wanted more information? For example, if you list k elements of F_n, how do you know if they generate a free subgroup of rank k?

As an example to keep in mind, you can consider the subgroup $H \subseteq F_2$ generated by the elements in $\{abab^{-1}, ab^2, bab, ba^3b^{-1}\}$. In other words, H consists of all reduced words made by concatenating the elements $abab^{-1}$, ab^2, bab, ba^3b^{-1}, and their inverses. For instance, $aba^{-1}b^{-1}$ is an element of H since

$$(ab^2)(bab)^{-1} = (ab^2)(b^{-1}a^{-1}b^{-1}) = aba^{-1}b^{-1}.$$

Here are four questions you will be able to answer at the end of this office hour about this subgroup and most any other subgroup of F_n:

1. What is the rank of the free group H?
2. Is $ab^2a^{-2}ba^3b^{-1}$ an element of H? What about b?
3. Does H have finite index in F_2?
4. Is H a normal subgroup of F_2?

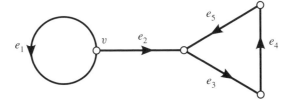

Figure 4.1 The graph Γ.

For those of you who are impatient, the answers are:

1. H is isomorphic to F_3.
2. Yes; no.
3. Yes, it has index 2.
4. Yes.

The yes answer to (4) follows immediately from the yes answer in (3) since subgroups of index 2 are always normal (for any group). I will give a direct way to see that H is normal. More interestingly, we will see that the yes answer in (4) here implies the yes answer in (3): every finitely generated nontrivial normal subgroup of a free group has finite index.

4.1 TOPOLOGICAL MODEL FOR THE FREE GROUP

How did I come by these answers? It is hard to see these conclusions just by thinking about reduced words. We are going to build a topological model for free groups and use this model to answer these questions.

Edge paths.   Let Γ be a directed graph. Since each edge has a preferred direction (usually indicated by an arrow), it makes sense to talk about its initial vertex (the one the arrow is pointing from) and the terminal vertex (the one the arrow is pointing to). For the edge with the opposite direction, the roles of initial and terminal swap.

An edge path in Γ is simply a string of edges $\alpha = e_0 \cdots e_k$ of Γ such that the terminal vertex of e_{i-1} is the initial vertex of e_i for $i = 1, \ldots, k$. You can think of an edge path as a path in the graph where we travel along one edge into a vertex and out along an adjacent edge, and repeat at the next vertex. For example, in the graph Γ in Figure 4.1, $e_1 e_2 e_3$ is an edge path, as is $e_1^{-1} e_2 e_5^{-1}$. Here e^{-1} denotes e with the opposite direction. Some mathematicatians use the notation $\bar{e}$, but let's use the exponent notation here as it makes the relation to free groups more transparent. The string of edges $e_2 e_4$ is not an edge path, nor is $e_2 e_3^{-1}$.

An edge path is closed if it starts and ends at the same vertex. In this case, the closed edge path is based at this single vertex. Referring to the graph Γ in Figure 4.1 again, $e_2 e_3 e_4 e_5 e_2^{-1}$ is a closed edge path based at v, but $e_1 e_2 e_3$ is not a closed edge path.

One last definition: an edge path $e_0 \cdots e_k$ is *tight* if $e_{i-1} \neq e_i^{-1}$ for all $i = 1, \ldots, k$. This is similar to reduced words in a free group. For example, again referring to Figure 4.1, $e_1 e_2 e_3$ is a tight edge path, but $e_1 e_2 e_2^{-1} e_1$ is not tight. Of course, any edge path can be tightened to a tight edge path by repeatedly removing ee^{-1} pairs. The edge path $e_1 e_2 e_2^{-1} e_1$ tightens to $e_1 e_1$. If you think about the original edge path in Γ, it has you traveling along e_2 just to turn around and travel back along e_2^{-1}. Tightening removes this extra trip.

One more visualization aid: imagine taking a router (as used in woodworking, not computing) and carving out the graph Γ on a table. You are left with a curved trough in the shape of Γ. Now take a piece of string and lay it along the edge path $e_1 e_2 e_2^{-1} e_1$. If two ants holding the initial part of the string and the terminal part of the string respectively start pulling, the piece of the string that is doubled back along e_2 will vanish and the string will trace out the edge path $e_1 e_1$. The string is now tight.

Hopefully, tightening edge paths reminds you a lot of reducing words, and you are starting to see that there might be a connection between edge paths in graphs and elements of free groups.

The fundamental group of a graph.   Here is our topological model for free groups. Suppose Γ is a directed graph and v is some vertex of Γ. By $\pi_1(\Gamma, v)$ we denote the set of tight closed edge paths in Γ based at v. As we will be referring to closed edge paths a lot, let's just call them *loops*. The trivial edge path allowed, this consists of the empty string of edges (like the empty word in free groups).

Given loops $\alpha, \beta \in \pi_1(\Gamma, v)$, both the initial and the terminal vertices of α are v, and the same is true for β, so the concatenation $\alpha\beta$ is a loop based at v. This is the loop that first traverses the edges in α and then the edges in β. By tightening, $\alpha\beta \rightsquigarrow \gamma$, an element of $\pi_1(\Gamma, v)$.

Exercise 1. This last claim deserves a little bit of thought; you should take a moment and write down that argument. What needs to be checked is that after tightening the loop, γ is (still) based at v.

We use these observations to define a multiplication on $\pi_1(\Gamma, v)$: concatenation followed by tightening. For example, referring to the graph in Figure 4.1, in $\pi_1(\Gamma, v)$ we have:

$$e_1 e_1 \cdot e_1^{-1} e_2 e_3 e_4 e_5 e_2^{-1} = e_1 e_1 e_1^{-1} e_2 e_3 e_4 e_5 e_2^{-1} \rightsquigarrow e_1 e_2 e_3 e_4 e_5 e_2^{-1}.$$

The trivial edge path is the identity and inverse of the loop $e_0 \cdots e_k \in \pi_1(\Gamma, v)$ is $e_k^{-1} \cdots e_0^{-1}$. Associativity is proven in the same way that one proves the associativity for multiplication in the free group. In summary, we have the following theorem.

THEOREM 4.1. $\pi_1(\Gamma, v)$ *is a group.*

Exercise 2. Provide the details for a proof of this theorem.

Exercise 3. Show that if two directed graphs come from the same undirected graph, then the corresponding fundamental groups are isomorphic.

Figure 4.2 "What's in a name? That which we call a rose by any other name would smell as sweet."—Juliet

The group $\pi_1(\Gamma, v)$ is very important and prevalent across topology and is aptly called the fundamental group. This group can be defined for any topological space, not just graphs. See any book on algebraic topology for details, Allen Hatcher's for instance [164]. More on this group in Office Hours 8 and 9.

THEOREM 4.2. *Suppose Γ is a connected graph with finitely many edges. Then for any vertex v of Γ:*

$$\pi_1(\Gamma, v) \cong F_n,$$

where n is 1 plus the number of edges of Γ minus the number of vertices of Γ.

The graph Γ in Figure 4.1 has five edges and four vertices; thus $\pi_1(\Gamma, v) \cong F_2$.

Proof of Theorem 4.2. Before we look at the general case, let's consider the special case when Γ is R_n, the *n–rose*. This is the graph with a single vertex v and n edges labeled $\{e_1, \ldots, e_n\}$. Necessarily, both the initial and the terminal vertices for each edge are the unique vertex v, so each edge looks like a small oval or petal; thus the name rose. See Figure 4.2, where R_2, R_3, and R_4 are shown.

In this case, $1 + n - 1 = n$, and so we want to demonstrate that $\pi_1(R_n, v) \cong F_n$. This is essentially just changing a's into e's and vice versa! Given a reduced word

$$g = a_{i_1}^{\epsilon_1} a_{i_2}^{\epsilon_2} \cdots a_{i_k}^{\epsilon_k} \in F_n,$$

where $i_j \in \{1, \ldots, n\}$ and $\epsilon_j \in \{-1, 1\}$, we assign the tight loop

$$\alpha = e_{i_1}^{\epsilon_1} e_{i_2}^{\epsilon_2} \cdots e_{i_k}^{\epsilon_k} \in \pi_1(R_n, v).$$

For example, given $g = a_1 a_2^{-1} a_2^{-1} a_4 a_3 a_1^{-1} \in F_4$, we assign $\alpha = e_1 e_2^{-1} e_2^{-1} e_4 e_3 e_1^{-1} \in \pi_1(R_4, v)$. Is this function injective? Surjective? Of course, just consider the inverse function, replacing the e's with a's. It is also fairly straightforward to check that this function respects the group structure: reduction in F_n is the same thing as tightening in $\pi_1(R_n, v)$.

To demonstrate the general case (where the graph is not a rose) we will use an operation called *collapse*. Suppose that e is an edge of a graph Γ that has initial and terminal vertices v_0 and v_1 where $v_0 \neq v_1$. Every edge of the graph in Figure 4.1

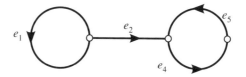

Figure 4.3 The collapse, $\Gamma_{\downarrow e_3}$, of Γ from Figure 4.1.

besides e_1 has this property. The collapse of Γ, denoted by $\Gamma_{\downarrow e}$, has the same collection of edges as Γ except e and the same collection of vertices as Γ except v_1. If e' is an edge of Γ whose initial (or terminal) vertex is v_1, then in $\Gamma_{\downarrow e}$ its initial (or terminal) vertex is v_0. All other edges are unaffected. In other words, collapse the edge e to a vertex. See Figure 4.3.

Notice that $\pi_1(\Gamma, v) \cong \pi_1(\Gamma_{\downarrow e}, v)$. (We are being a little sloppy here as v *could* be the vertex removed from Γ, but in that case just collapse e^{-1} and convince yourself that $\Gamma_{\downarrow e} = \Gamma_{\downarrow e^{-1}}$.) Indeed, just removing the occurrences of e and e^{-1} from any tight loop in Γ results in a tight loop in $\Gamma_{\downarrow e}$. (No additional tightening required! Do you see why?) Such a function is injective, is surjective, and respects the group structure.

Since we removed both an edge and a vertex from Γ, the quantity

number of edges minus the number of vertices

is the same for both Γ and $\Gamma_{\downarrow e}$. This quantity is the negative of the so-called *Euler characteristic* of Γ, written as $\chi(\Gamma)$. We can continue to collapse until we arrive at a rose where we can collapse no further. Thus:

$$\pi_1(\Gamma, v) \cong \pi_1(\Gamma_{\downarrow e_1}, v) \cong \pi_1(\Gamma_{\downarrow e_1 \downarrow e_2}, v) \cong \cdots \cong \pi_1(R_n, v) \cong F_n,$$

where n is 1 plus the number of edges of Γ minus the number of vertices of Γ.

4.2 SUBGROUPS VIA GRAPHS

The fundamental group of a graph was basically cooked up to be the same as a free group. So how does any of this help? Let's go back to the example $H \subseteq F_2$ mentioned earlier. We will now think of F_2 as $\pi_1(R_2, v)$, and so let's go ahead and call the edges of R_2 a and b respectively. So a tight loop in R_2 corresponds to the reduced word in F_2 obtained by reading off the edges the loop traverses. The subgroup H now corresponds to loops in R_2 that are formed by tightening concatenations of the four loops $abab^{-1}$, ab^2, bab, and ba^3b^{-1}.

Consider the graph Γ_H on the left in Figure 4.4. This graph has 15 edges and 12 vertices. Therefore we have $\pi_1(\Gamma_H, w) \cong F_4$. The edges of Γ_H are labeled to suggest a function $\Gamma_H \to R_2$. Just send the red edges with the black triangles of Γ_H to the a edge of R_2 preserving the direction, and likewise send the blue edges with white triangles of Γ_H to the b edge of R_2. This sends loops in Γ_H based at the black vertex w to loops in R_2 based at v. Once we tighten, this defines a homomorphism $\rho: \pi_1(\Gamma_H, w) \to \pi_1(R_2, v)$.

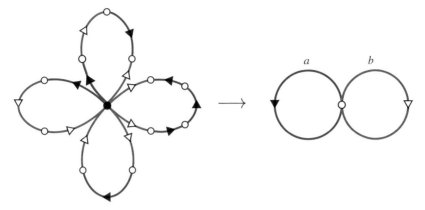

Figure 4.4 The graph Γ_H. The decoration specifies a map to R_2 by sending the red edges with black triangles to the a edge and the blue edges with white triangles to the b edge. The vertex w is colored black.

By the setup, $\rho(\pi_1(\Gamma_H, w)) = H$. If ρ is an isomorphism, then we can conclude $H \cong \pi_1(\Gamma_H, w) \cong F_4$. This leads to two questions:

1. How can we tell if this induced homomorphism $\rho\colon \pi_1(\Gamma_H, w) \to \pi_1(R_2, v)$ is injective (and hence an isomorphism onto its image)?
2. If ρ is not injective, can we modify Γ_H to get a new homomorphism that is injective?

The answer to both questions: *fold.*

Folding.   John R. Stallings described an extremely simple and elegant method for understanding maps between graphs $\Gamma \to \Delta$ and their induced homomorphisms on the fundamental groups $\pi_1(\Gamma, w) \to \pi_1(\Delta, v)$. By a graph map $\Gamma \to \Delta$ we mean a function that sends the vertices of Γ to vertices of Δ, that sends the edges of Γ to edges of Δ, and that preserves adjacency of vertices and edges. In other words a graph map is a function that takes edge paths to edge paths. The function described in Figure 4.4 is such an example.

To describe Stallings' method we only need a particularly simple type of map between graphs called a fold.

Let e_1 and e_2 be two edges of Γ that have the same initial vertex where $e_2 \neq e_1, e_1^{-1}$. We build a new graph $\Gamma_{e_1=e_2}$ by the following procedure. Remove the edges e_1 and e_2 from Γ and add a new edge e whose initial vertex is the initial vertex of e_1 and e_2. Edges of Γ whose initial (or terminal) vertex was one of the terminal vertices of e_1 or e_2 now have the terminal vertex of e as their initial (or terminal) vertex. There are two different pictures depending on whether or not the terminal vertices of e_1 and e_2 are the same. See Figure 4.5. The two types are called *type I* and *type II*, respectively.

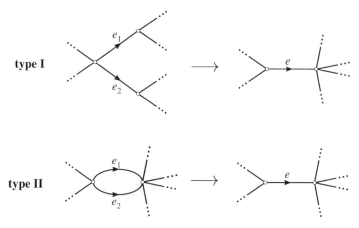

Figure 4.5 Folding edges e_1 and e_2.

There is a graph map $\Gamma \to \Gamma_{e_1=e_2}$ sending both e_1 and e_2 to e and defined on the rest of the edges in the obvious manner. Such a map is called a *fold*. We make two observations:

(type I) If $\Gamma \to \Delta$ is a type I fold, then $\pi_1(\Gamma, w) \to \pi_1(\Delta, v)$ is an isomorphism.

(type II) If $\Gamma \to \Delta$ is a type II fold, then $\pi_1(\Gamma, w) \to \pi_1(\Delta, v)$ is surjective but not injective.

Exercise 4. Prove these observations.

Suppose that $\Gamma \to \Delta$ is a graph map and that e_1 and e_2 are two edges in Γ with the same initial vertex and that have the same image in Δ. Then we can factor the map as follows:

$$\Gamma \to \Gamma_{e_1=e_2} \to \Delta.$$

See Figure 4.6. The edge decorations describe a graph map from the graph in the the upper left to the 3–rose in the upper right. As the two red edges share an initial vertex and have the same image in R_3, we can first fold them to get the graph in the middle and then map this resulting graph to R_3.

What if we cannot factor a graph map $\Gamma \to \Delta$ through a fold as above? This means that for any two edges e_1 and e_2 in Γ that share an initial vertex, their images in Δ are distinct. Such a map is called an *immersion*. An immersion sends tight edge paths in Γ to tight edge paths in Δ! In particular, nontrivial tight loops in Γ are sent to nontrivial tight loops in Δ. We just proved:

LEMMA 4.3. *If $\Gamma \to \Delta$ is an immersion, then the induced homomorphism on the fundamental groups $\pi_1(\Gamma, w) \to \pi_1(\Delta, v)$ is injective.*

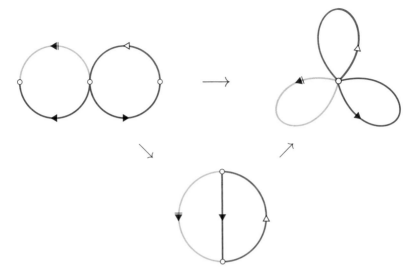

Figure 4.6 Factoring a graph map through a fold.

Exercise 5. Provide the details for the proof of Lemma 4.3.

Putting these concepts together, we have the following theorem of Stallings [244].

THEOREM 4.4. *If $\Gamma \to \Delta$ is a graph map between finite graphs, there is a factorization*

$$\Gamma = \Gamma_0 \to \Gamma_1 \to \cdots \to \Gamma_k \to \Delta$$

where each $\Gamma_{i-1} \to \Gamma_i$ is a fold and $\Gamma_k \to \Delta$ is an immersion.

Let's start applying this technique to answer the questions we posed in the beginning.

4.3 APPLICATIONS OF FOLDING

I will now give five(!) applications of folding. The first one is the Nielsen–Schreier subgroup theorem, which we already saw in Office Hour 3. The proof here does not mention trees.

Application 1: Subgroups of free groups are free.   Figure 4.4 shows a graph map $\Gamma_H \to R_2$. As I said previously, the induced map on the fundamental groups surjects $\pi_1(\Gamma_H, w)$ onto H. Is this map an immersion? No! Several edges adjacent to v have the same color and direction. This means they are sent to the same edge in R_2. Let's factor into folds to get an immersion. This process is shown in Figure 4.7.

In some steps several edges are folded together at once. There are some choices to be made along the way. If we make other choices of edges to fold, we will get a

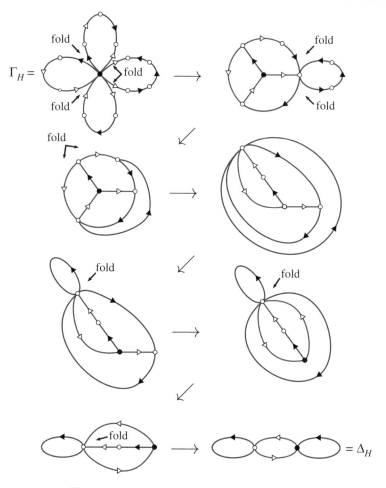

Figure 4.7 Factoring the map $\Gamma_H \to R_2$ into folds.

different sequence of graphs, but the final immersion will be the same. We will use the terms *a* edge and *b* edge to talk about the red edges with black triangles and the blue edges with white triangles, respectively.

Step 1. We start at the black vertex in Γ_H and identify all adjacent edges with similar decoration and orientation. There are three such groups, corresponding to a, b, and b^{-1}.

Step 2. We separately fold two pairs of a edges.

Step 3. We fold a single pair of b edges.

Step 4. We flip the a loop over the remainder of the graph. This doesn't change the graph at all, but it will make our next fold easier to see.

Step 5. We perform a fold that wraps an a edge over the a petal.

All of the folds we have performed so far are type I; thus the fundamental groups of these graphs are all isomorphic to F_4.

Step 6. We perform a type II fold, by folding together the two *a* petals based at the same vertex. This reduces the rank by 1 and so the fundamental group of the resulting graph is F_3.

Step 7. We perform a final fold, which is of type I. Folding the two *b* edges results in the final graph.

Step 8. The induced map is now an immersion.

We have factored the map $\Gamma_H \to R_2$ into:

$$\Gamma_H \to \Gamma_1 \to \cdots \to \Delta_H \to R_2,$$

where each of the maps in $\Gamma_H \to \cdots \to \Delta_H$ is a fold and $\Delta_H \to R_2$ is an immersion. The original map $\Gamma_H \to R_2$ induced a surjection from $\pi_1(\Gamma_H, w)$ onto $H \subseteq \pi_1(R_2, v)$; thus the homomorphism $\pi_1(\Delta_H, w) \to \pi_1(R_2, v)$ surjects onto H as well. As the map $\Delta_H \to R_2$ is an immersion, by Lemma 4.3, the homomorphism $\pi_1(\Delta_H, w) \to \pi_1(R_2, v)$ is injective too. Therefore $H \cong \pi_1(\Delta_H, w) \cong F_3$! This answers question (1) from the beginning of the office hour.

If you've taken linear algebra, this result might strike you as odd. Here we have a subgroup of a free group that itself is free of larger rank. This never happens in linear algebra, where any proper subspace of a finite-dimensional vector space has strictly smaller dimension. This is one illustration of the intricacies in the universe of nonabelian groups.

This example can be generalized to show the following.

THEOREM 4.5. *Every finitely generated subgroup of a free group is free.*

The stronger statement is true too, that every subgroup of a free group is free. This is called the Nielsen–Schreier Subgroup theorem; see Office Hour 3.

There is additional information to be gained from Δ_H. We have that $\{a, b^2, bab^{-1}\}$ is a generating set for $\pi_1(\Delta_H, w)$ and hence for H too. To see this, collapse the bottom *b* edge in Δ_H to get R_3 and observe that these three loops go to the three petals of R_3. This generating set looks much easier to work with than the original one. If you trace through the folding, you can work out how to write the original generators in terms of the new ones, and vice versa. For instance:

$$bab = (bab^{-1})(b^2) \qquad \text{old in terms of new}$$

$$a = (abab^{-1})(ba^3b^{-1})^{-1}(bab)(ab^2)^{-1}(abab^{-1}) \qquad \text{new in terms of old}$$

Writing the old generators in terms of the new generators is straightforward; just follow the corresponding loop in Γ_H for the old generator throughout the folding and write this loop in terms of the generators in Δ_H. The reverse is a bit trickier. You will need to take the new generator in Δ_H and unfold it back to the original graph. Referring to Figure 4.7, here is the unfolding sequence for *a*:

$$a \leftsquigarrow a(bb^{-1}) \leftsquigarrow ab(a^{-1}a)b^{-1} \leftsquigarrow aba^{-1}(b^{-1}b)ab^{-1}$$

$$\leftsquigarrow ab(aa^{-1})a^{-1}(a^{-1}a)b^{-1}bab^{-1}$$

$$\leftsquigarrow aba(b^{-1}b)a^{-3}(b^{-1}b)a(bb^{-1})b^{-1}(a^{-1}a)bab^{-1}$$

$$= (abab^{-1})(ba^3b^{-1})^{-1}(bab)(ab^2)^{-1}(abab^{-1}).$$

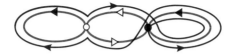

Figure 4.8 Tracing out the element $ab^{-1}a^{-1}b^{-1}a$ in Δ_H.

Application 2: Membership problem. Now we turn to our attention to question (2) from the beginning of the office hour. How can we determine if an element of F_2 belongs to H? This is called the membership problem and for finitely generated subgroups of free groups there is a solution:

> *Start at $w \in \Delta_H$ and attempt to trace out a given element g as a tight edge path. If we can trace out g and we end back at w, then $g \in H$. Otherwise $g \notin H$.*

Let me explain what is meant by "trace out." Recall that edges in Δ_H are called a or b depending on their image in R_2. Thus following an edge path in Δ_H reads off a reduced word in F_2. Traversing an a edge in the positive direction reads a, in the opposite direction reads a^{-1}. Similarly for b edges.

By "trace out an element g," I mean: follow an edge path that reads off the reduced word g. The image of this edge path in R_2 is the loop corresponding to g. Notice that since $\Delta_H \to R_2$ is an immersion, there are at most two a edges of opposite direction at a given vertex. Same for the b edges. So if we can trace out an element, there is only one way to do so. See Figure 4.8, where it is shown how to trace out $ab^{-1}a^{-1}b^{-1}a$ in Δ_H.

Let's use folding to show why this algorithm works. Consider the graph $\Delta_{H,g}$ that is obtained from Δ_H by attaching a loop at w. We get a map $\Delta_{H,g} \to R_2$ by sending this new loop to the tight loop representing g in R_2 and by using the immersion on Δ_H. In other words, the new loop reads off g. This induces a surjection onto $\langle H, g \rangle$, the subgroup generated by H and g. This map might not be an immersion. If not, fold.

If the result of folding terminates at Δ_H, then $H \cong \pi_1(\Delta_H, w) \cong \langle H, g \rangle$ and so $g \in H$. Also following the folds we see how to trace out g as a loop in Δ_H. If the resulting of folding terminates at something besides Δ_H, then we cannot trace out g as a loop in Δ_H, and H is contained as a proper subgroup of $\langle H, g \rangle$, so $g \notin H$.

We can now answer question (2) and see that $ab^2a^{-2}ba^3b^{-1} \in H$. Trace out this element in Δ_H as a tight loop to verify. In terms of our new generating set, we see:

$$ab^2a^{-2}ba^3b^{-1} = (a)(b^2)(a)^{-2}(bab^{-1})^3.$$

Considering how hard it was to write a in terms of the old generating set, it would be hard to write this word in the old generating set as well. However, you can always take the loop representing this element in Δ_H and unfold it back to Γ_H.

Additionally, we see that $b \notin H$. We can trace out b as a tight edge path in Δ_H starting at w, but it does not end at w; it ends at the other vertex of Δ_H. In the next part we will see that such paths are related to coset representatives.

Application 3: Index. We will now tackle the question (3) we posed at the beginning of this office hour and show that the index of $H \subseteq F_2$ is 2. I claim that this can be computed simply by counting the number of vertices of Δ_H. One vertex of Δ_H is already labeled as w; let w' denote the other vertex.

Let's start by explaining how cosets can be thought of topologically.

$$H = \{\text{tight edge paths in } \Delta_H \text{ that start at } w \text{ and end at } w\}$$

$$Hb = \{\text{tight edge paths in } \Delta_H \text{ that start at } w \text{ and end at } w'\}$$

The first equality is the isomorphism $H \cong \pi_1(\Delta_H, w)$.

The second takes a little bit more thought. Decompose an element of Hb as hb where $h \in H$. By the isomorphism $H \cong \pi_1(\Delta_H, w)$, there is a loop α based at w corresponding to h. The edge path that is the concatenation of α and the edge b in the positive direction based at w traces out hb. Tightening does not affect the initial or terminal vertices of an edge path, and so after tightening we have a tight edge path that starts at w and ends at w' whose image in R_2 is the loop representing hb. This shows "$\subseteq$".

To show "$\supseteq$" we also need to show that the image in R_2 of any tight edge path that starts at w and ends at w' is an element of Hb. If a tight edge path has the form of a loop based at w followed by the edge b in the positive direction, then this is clear. Otherwise, if a tight edge path α, which starts at w and ends at w', is not of this form, then αb^{-1} is a loop based at w. So the image in R_2 of the edge path $(\alpha b^{-1})b$ is an element of Hb. Since $(\alpha b^{-1})b \rightsquigarrow \alpha$, this is the same element as the image in R_2 of α.

We can conclude now that the index of H in F_2 is 2 by showing that these are all of the cosets of H, that is, $H \cup Hb = F_2$. In terms of our description of cosets, we want to show that every element of F_2 can be traced out as a tight edge path in Δ_H that starts at w. For this, we make use of a special property of Δ_H:

$$\text{Each vertex of } \Delta_H \text{ has four adjacent edges. } (\star)$$

As the map $\Delta_H \to R_2$ is an immersion, this implies by the pigeonhole principle that the four adjacent edges are: an incoming and outgoing a and an incoming and outgoing b. This condition tells us that every element of F_2 can be traced out as a tight edge path in Δ_H based at w. To build the tight edge path, just begin at w and start tracing. As we can go along an incoming or outgoing a edge and an incoming or outgoing b edge at every vertex, you'll never get stuck! Such an edge path is a *lift* of the loop.

What we just proved is that every loop in R_2 lifts to a tight edge path in Δ_H that starts at w. If the lift terminates at w, the loop corresponds to an element of H. If the lifts terminates at w', the loop corresponds to an element of Hb. As there are only two vertices, these are the only two possibilities; hence $H \cup Hb = F_2$ and so the index of H in F_2 is 2.

The first step of our computation above works for any immersion. Suppose $\Delta \to R_n$ is an immersion and $H \subseteq F_n$ is the corresponding subgroup $H \cong \pi_1(\Delta, w)$. Then the set of reduced edge paths in Δ from w to w' corresponds to the right coset of $Hg \subseteq F_n$, where g is the image of any tight edge path from w to w'.

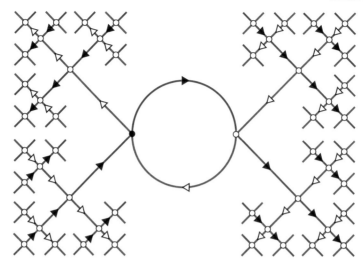

Figure 4.9 Promoting an immersion into a covering. A right coset of $\langle ab \rangle \subset F_2$ corresponds to the sets of reduced edge paths from the black vertex to some fixed vertex.

Exercise 6. Prove this assertion.

For the second step, that every coset arises in this fashion, we need to additionally assume that our immersion satisfies something like the special property ($\star$). An immersion $\Delta \to R_n$ is a *covering* if each vertex of Δ has $2n$ adjacent edges.

Exercise 7. Show that if $\Delta \to R_n$ is a covering, then every tight loop in R_n lifts to a reduced edge path.

Exercise 8. Put these two steps together now to prove the following theorem.

THEOREM 4.6. *Suppose Δ is a finite graph and $\Delta \to R_n$ is a covering. Then $\pi_1(\Delta, w)$ has finite index and this index is the number of vertices of Δ.*

What do we do if $\Delta \to R_n$ is an immersion, but not a covering? In other words, there is a vertex of Δ that has fewer than $2n$ adjacent edges. So, missing from this vertex is an a_i edge either incoming or outgoing. Add this edge to Δ at this vertex and let the opposite end of this edge dangle. As we added an edge and a vertex, the fundamental group does not change. Of course, the result will not be a covering either. The dangling vertex will only have one adjacent edge. But we can repeat this procedure. Doing so inductively, results in a covering $\widehat{\Delta} \to R_n$ where $H \cong \pi_1(\widehat{\Delta}, w)$. The graph $\widehat{\Delta}$ is Δ with some infinite trees attached. See Figure 4.9, where this process is carried out for the cyclic subgroup $\langle ab \rangle \subseteq F_2$. The two edge loops in the center depict an immersion $\Delta \to R_2$ representing the subgroup $\langle ab \rangle$. By attaching the four infinite trees, we get a covering $\widehat{\Delta} \to R_2$ representing the same subgroup.

Since $\widehat{\Delta} \to R_n$ is a covering, we can proceed as before to associate to *any* coset of H the set of tight edge paths in $\widehat{\Delta}$ from w to some vertex. As there are infinitely many vertices in $\widehat{\Delta}$, there are infinitely many cosets and so the index of H in

F_n is infinite. Referring again to Figure 4.9, this shows that the index of $\langle ab \rangle \subset F_2$ is infinite.

This observation plus Theorem 4.6 gives us:

THEOREM 4.7. *Suppose Δ is a finite graph and $\Delta \to R_n$ is an immersion. Then $\pi_1(\Delta, w)$ has finite index in $\pi_1(R_n, v)$ if and only if the map is a covering. In this case, the index is the number of vertices of Δ.*

Application 4: Normality.   The notion of a covering map can be used to answer question (4) from the beginning of the office hour.

Suppose $\Delta \to R_n$ is a covering and $H \subseteq F_n$ is the corresponding subgroup $H \cong \pi_1(\Delta, w)$. What we saw in the preceding section is that every right coset of H corresponds to the set of tight edge paths that start at w and end at some fixed vertex. In the same fashion you can show that every left coset of H corresponds to the set of tight edge paths that start at some fixed vertex and end at w.

Combining these two ideas, you can convince yourself that a *conjugate* of H, a subgroup of the form $g^{-1}Hg$, corresponds to the set of tight loops based at some fixed vertex w'. This vertex is the endpoint of the lift of the loop in R_n representing g that is based at w.

A subgroup is normal if all of its conjugates are equal to the subgroup. What does it mean in our setup? It means that for any vertex w' of Δ, any element that can be traced out as a tight loop based at w also can be traced out as a tight loop based at w'. This can only happen if Δ is vertex transitive, i.e., for any vertex w', there is a graph automorphism $\Delta \to \Delta$ that sends w to w', sending a_i edges to a_i edges, preserving direction. Loosely speaking, the graph looks the same whether you are standing at w or at w'.

Conversely, if Δ is vertex transitive, then loops based at w can be identified with loops based at any other vertex; hence H is normal. This gives us the following theorem.

THEOREM 4.8. *Suppose $\Delta \to R_n$ is a covering. Then $\pi_1(\Delta, w)$ is a normal subgroup of $\pi_1(R_n, v)$ if and only if Δ is vertex transitive.*

The graph Δ_H in Figure 4.7 is vertex transitive; hence H is normal as claimed. The graph in the lower right corner of Figure 4.10 depicts a covering that is not vertex transitive and hence the corresponding subgroup F_2 is not normal.

This criterion allows us to prove a strong property for finitely generated normal subgroups for free groups.

THEOREM 4.9. *If $H \subseteq F_n$ is a finitely generated nontrivial normal subgroup, then H has finite index in F_n.*

Let's prove Theorem 4.9. By folding we can represent the subgroup H by an immersion $\Delta \to R_n$ where Δ is a finite graph that has a loop (this is where we use "finitely generated" and "nontrivial"). If this is a covering, then H has finite index. Done. If not, then by attaching trees we get a covering $\widehat{\Delta} \to R_n$ that still represents H. As H is normal, the graph $\widehat{\Delta}$ is vertex transitive. This is a contradiction! In Δ, there is an embedded loop that contains some vertex; call it w. As none of the new vertices in the attached trees in $\widehat{\Delta}$ lie on an embedded loop, there is no graph automorphism $\widehat{\Delta} \to \widehat{\Delta}$ that sends w outside of Δ.

If we drop either of the adjectives for H, the conclusion is false.

Exercise 9. Show that the subgroup of F_2 generated by $\{b^n a b^{-n}\}_{n \in \mathbb{Z}}$ is normal but not finitely generated. *Hint: Try to build a covering that represents this subgroup. Notice that the latter condition, "not finitely generated," implies that the subgroup does not have finite index. Do you see why?*

Application 5: Residual finiteness. A group G is *residually finite* if for every nontrivial element g of G there is a normal subgroup N of finite index in G so that g is not in N. By the first isomorphism theorem, this is the same as saying that there is a finite group H and a homomorphism $G \to H$ so that the image of g is nontrivial. (To go between the two definitions, take H to be G/N.)

This seems like a clunky definition; why do we care about it? Well, remember from Office Hour 1 that we can sometimes understand a lot about groups by understanding their quotients, since often the quotients are simpler than the original groups. For instance, $\mathbb{Z}/2\mathbb{Z}$ is a quotient of $\mathbb{Z}$ and $\mathbb{Z}^2$ is a quotient of F_2. Finite groups are especially simple because, being finite, they can be completely understood, or even completely entered into a computer. For that reason it is nice when there are finite quotients of the group you are trying to understand.

That is all well and good. But if you want to say something nontrivial about the element g, you would be happier if its image in the finite quotient were nontrivial. Through this reasonsing, we have led ourselves exactly to the notion of residual finiteness.

For our last application of folding we will prove the following classical theorem.

THEOREM 4.10. *For any $n \geq 1$ the group F_n is residually finite.*

Let's start out, as many proofs in mathematics do, by making our work a bit easier. We only need to construct a finite index subgroup $H \subseteq F_n$ such that $g \notin H$, i.e., we can disregard the adjective "normal." For if we can find such an H, then the subgroup

$$N = \bigcap_{k \in F_n} k H k^{-1}$$

satisfies the conclusion of the theorem. This follows as it is normal by design, and since H has finite index, H has only finitely many conjugates and so the intersection is really just a finite intersection of finite index subgroups and hence of finite index too. Finally, if $g \notin H$, then g is not contained in any subgroup of H, so $g \notin N$.

Here is our strategy:

> *Build a finite graph Γ and a covering $\Gamma \to R_n$ such that the lift of g to Γ does not close up, i.e., we cannot trace out g in Γ.*

If you've made it this far through the office hour, it should be clear that this suffices to prove the theorem. The subgroup $H \cong \pi_1(\Gamma, w)$ has finite index by

Application 3, and $g \notin H$ by Application 2. Let's build the covering! Along the way, refer to Figure 4.10, where this construction is carried for the element $aba^{-1}b \in F_2$. The general construction is intermixed with this specific example.

General construction. Start with a graph that consists of a single cycle C divided into $d + 1$ edges where d is the number of edges in the tight loop representing g in R_n. Define an immersion $C \to R_n$ by sending the first d edges to the loop representing g and the last edge to any edge as long at the resulting map is an immersion. (Just don't send it to the opposite of the first or dth edge.)

Our example. In our example, C has five edges, and we sent the last edge to a. As our element does not start with a^{-1} nor end with a, this map is an immersion.

At this point we have an immersion $C \to R_n$ where the lift of g does not close up, but it is not going to be a covering. If we run the procedure from Application 3, we will get an infinite graph, which is not good for our purposes.

If we add edges in a clever way to C, we can build a covering. By not adding vertices, we can guarantee that the resulting group has finite index. Of course, the fundamental group will change, but that's fine. The lift of g still will not close up and so g will not be in the resulting group.

How do we add edges? One at a time.

General construction. Start at any vertex, w_0, where the number of adjacent edges is less than $2n$. In the first step, this is any of the vertices. One of the edges is missing, say an incoming a_1. Now if there is a vertex, w', without an outgoing a_1, then we can add the edge in the obvious fashion and still have an immersion.

Our example. This is how we started in Figure 4.10. There is no incoming edge b at this vertex. Nor is there an outgoing b edge, so we added the one edge loop. At this point there are four edges adjacent to this vertex and we move on and consider the next vertex going counterclockwise.

General construction. Hold on. In general, how do we know there is such a vertex w'? Well, for starters, if there is no outgoing a_1 edge at w_0, then take $w' = w_0$.

Now suppose there is an outgoing a_1 edge at w_0. Travel along that edge to a vertex w_1. If there is no outgoing a_1 edge at w_1, take $w' = w_1$. Otherwise continue producing a list of vertices $w_0, w_1, \ldots$. Since our graph has only finitely many vertices, at some point either we reach a vertex w_k where there is no outgoing a_1 edge and so we take $w' = w_k$, or we reach a vertex w_k that we already considered. But this means that there are two edges incoming a_1 edges at w_k, which contradicts the fact that the map to R_n is an immersion. Therefore we can always find such a vertex w'.

Our example. This is what happens next in our example. There is no outgoing a edge at this vertex. We need to find some vertex without an incoming a edge. Traveling back along the incoming a edge we reach the black vertex that too has an incoming a edge. We travel back along this one to reach a vertex without such an edge. We then add the a edge accordingly.

General construction. Continue to add edges in this fashion to get a covering.

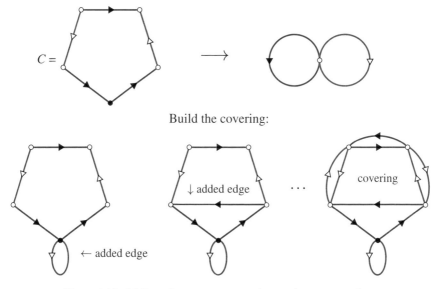

Build the covering:

Figure 4.10 Adding edges to promote an immersion to a covering.

Our example. The final covering is shown on the bottom right.

The result is a finite graph with a covering $\Gamma \to R_n$ such that the lift of the loop representing g does not close up. This proves the theorem. Very nice!

FURTHER READING

Stallings' method of folding is an example of transformative mathematics: simple and elegant, yet unifying several previous results and opening the door to many, many more. Folding will pop up again in Office Hour 6 on automorphisms of free groups. A search on Google Scholar results in almost 100 articles. I hope that this office hour has convinced you that the next time you hear someone say the phrase

"Let $H \subseteq F_n$ be a finitely generated subgroup,"

you should replace it with

"Let $\Delta \to R_n$ be an immersion of a finite graph."

The original article on folding by Stallings [244] is a great place to look first for more information and examples. Bogopolski's text [43] contains a section describing the folding algorithm. Section 1.3 of Hatcher's text [164] shows many examples of covers of graphs. Folding is applied in the setting of groups acting on trees (the subject of Office Hour 3) by Stallings [245]. The article by Wade [258] applies folding to study free group automorphisms (the subject of Office Hour 6).

PROJECTS

Project 1. Practice, practice, practice! Take a collection of elements in F_n and apply the folding algorithm to find a basis for the subgroup they generate. Here are a couple to get started:

$$\{a^2b^{-1}, a^3b, a^{-1}b^2, aba, a^3b^{-2}a^{-1}\}$$
$$\{a^3b, a^2b^2a^{-1}, ab^{-2}ab, a^{-2}ba^{-1}, b^{-2}a^{-1}ba^{-1}\}$$

Come up with your own.

Project 2. Write a computer program that implements the folding algorithm.

Project 3. Look at the examples of covers of R_2 given in Section 1.3 of Hatcher's text [164]. Which ones correspond to finite index subgroups? Which to normal subgroups?

Project 4. Given a subgroup $H \subseteq G$, the normalizer is the subgroup:

$$N(H) = \{g \in G \mid gHg^{-1} = H\}.$$

It is the largest subgroup of G that contains H as a normal subgroup. If H is a finitely generated subgroup of F_n, can you compute $N(H)$? What can you say about the index $[N(H) : H]$? Try $H = \langle a^2, b^2, aba^{-1}, ba^2b^{-1}, baba^{-1}b^{-1} \rangle \subseteq F_2$ to get started.

Project 5. Describe an algorithm that, given a finitely generated subgroup $H \subseteq F_n$ and an element $g \in F_n$, decides if there is an $h \in F_n$ such that $hgh^{-1} \in H$.

Project 6. Show that if H is a finite index subgroup of F_n then $\mathrm{rank}(H) - 1 = [F_n : H](n - 1)$.

Project 7. Let H be a finitely generated subgroup of F_n with the property that for all $g \in F_n$ there is an $n \in \mathbb{N}$ such that $g^n \in H$. Show that H must have finite index. *Hint: Show the associated immersion $\Delta \to R_n$ is a cover.* Now look up Burnside groups and observe that this statement does not hold for an arbitrary finitely generated group G.

Project 8. This project generalizes Theorem 4.10. A group is locally extended residually finite (LERF) if for any finite collection of nontrivial elements $g_1, \ldots, g_k \in G$, there is a finite index normal subgroup N so that none of the g_i's are in N. Prove that for any $n \geq 1$, F_n is locally extended residually finite.

Project 9. This generalizes the preceding project and is known as Marshal Hall's theorem [160]. Suppose H is a finitely generated subgroup of F_n and $g_1, \ldots, g_k \in F_n - H$. Then there is a finite index subgroup H' so that $H \subseteq H'$ and none of the g_i's are in H'. Moreover, show that we can choose H' such that $H' = H * A$. (Free products are discussed in Office Hour 3.)

Project 10. Intersections of finitely generated subgroups are represented by pullbacks. What this means is that given immersions $f_1 \colon \Delta_1 \to R_n$ and $f_2 \colon \Delta_2 \to R_n$ of finite graphs, if we can find a graph Γ and immersions $g_1 \colon \Gamma \to \Delta_1$ and $g_2 \colon \Gamma \to \Delta_2$ where $f_1 g_1 = f_2 g_2$, then the intersection of the subgroups $\pi_1(\Delta_1, w_1), \pi_1(\Delta_2, w_2) \subseteq \pi_1(R_n, v)$ is $\pi_1(\Gamma, w)$. This fits into what is called a pullback diagram:

$$
\begin{array}{ccc}
\Gamma & \xrightarrow{\ g_2\ } & \Delta_2 \\
\Big\downarrow{\scriptstyle g_1} & & \Big\downarrow{\scriptstyle f_2} \\
\Delta_1 & \xrightarrow[\ f_1\ ]{} & R_n
\end{array}
$$

First, show how to construct Γ and the immersions $g_1 \colon \Gamma \to \Delta_1$ and $g_2 \colon \Gamma \to \Delta_2$. Then show that $\pi_1(\Gamma, w)$ is the intersection $\pi_1(\Delta_1, w_1) \cap \pi_1(\Delta_2, w_2)$ in $\pi_1(R_n, v)$.

Project 11. Read Section 7.7 of Stallings' paper [244] to see how these ideas can prove the Hanna Neumann bound [210]:

$$\operatorname{rank}(A \cap B) - 1 \le 2(\operatorname{rank}(A) - 1)(\operatorname{rank}(B) - 1)$$

whenever A and B are finitely generated subgroups of F_n.

Project 12. If A and B are finite index subgroups of F_n, show that:

$$\operatorname{rank}(A \cap B) - 1 \le (\operatorname{rank}(A) - 1)(\operatorname{rank}(B) - 1).$$

Find examples where you get equality.

Project 13. Read Igor Mineyev's paper [205] or Warren Dicks' paper [111] for a proof of the Hannah Neumann theorem:

$$\operatorname{rank}(A \cap B) - 1 \le (\operatorname{rank}(A) - 1)(\operatorname{rank}(B) - 1)$$

whenever A and B are finitely generated subgroups of F_n.

Office Hour Five

The Ping-Pong Lemma

Johanna Mangahas

It may happen in mathematics that you come upon a group without immediately knowing which group it is. As when recognizing a favorite kind of bird in a forest, it is gratifying to discover that a group one encounters in the mathematical wilderness is actually like a familiar friend. Fortunately, you can tell a group by how it acts. That is to say, a good group action (for example, action by isometries on a metric space) can reveal a lot about the group itself. This theme, which we saw in Office Hour 3, is central to geometric group theory and is prevalent throughout this book. Here we focus on an identifying feature of free groups F_n: only these groups play ping-pong.

5.1 STATEMENT, PROOF, AND FIRST EXAMPLES USING PING-PONG

The ping-pong lemma, sometimes called the Schottky lemma or Klein's criterion, gives a set of circumstances under which you know a group is a free group. This elegant lemma has several variants. Let's start with a way to detect that an unknown group is F_2.

Statement and proof. First, a heads-up: some may find the proof below too maddeningly concise. It is professional, pitiless, expertly expedient, and contains only slightly more words than this sentence. If that troubles you, relax and find relief by reading just beyond the QED box.

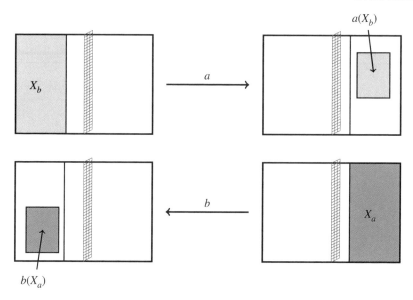

Figure 5.1 The ping-pong action on X.

Lemma 5.1 (Ping-pong for two players). *Suppose a and b generate a group G that acts on a set X. If*

1. *X has disjoint nonempty subsets X_a and X_b, and*
2. *$a^k(X_b) \subset X_a$ and $b^k(X_a) \subset (X_b)$ for all nonzero powers k,*

then G is isomorphic to a free group of rank 2.

Proof. No element g of G represented by a word of the form $a^*b^*a^* \cdots b^*a^*$ is the identity since $g(X_b) \subseteq X_a$ (the asterisks represent nonzero exponents). This suffices, because any element of G is conjugate to one of that form. $\qquad\square$

That was short! Let's draw some pictures and expand on the details. Figure 5.1 illustrates the requirements (1) and (2): the square object is the set X, with X_a and X_b taking up portions of the right and left, respectively. The net is added for flare.

The first sentence of the proof can be illustrated as follows. Figure 5.2 gives the idea. You start with any point p in X_b. Where does g take p? Let's apply the syllables of g one by one (by a syllable we mean one of the subwords, a^* or b^*). In the way group actions work, we proceed from right to left. Applying condition (2) of the lemma, the first a^* takes p to X_a. Then the next syllable b^* takes the image of p under the first syllable back to X_b. And then the image goes back to X_a. And back to X_b, etcetera, etcetera. Feel like you are watching a ping-pong match? At the very end, the last syllable a^* takes the image of p to X_a. Since X_a is disjoint from X_b (condition (1) of the lemma), we conclude that $g(p) \neq p$. This immediately implies that $g \neq 1$.

On to the second sentence. Recall, the conjugate of a group element x by another element y is the product yxy^{-1}. The observation here is that, given any nontrivial

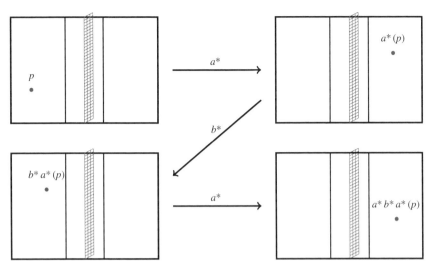

Figure 5.2 See why it's called ping-pong?

reduced word x in $\{a, b, a^{-1}, b^{-1}\}$, there's a y which makes yxy^{-1} take the form $a^*b^*a^* \cdots b^*a^*$. For example, if $x = b^8ab^2a$, then we can conjugate by $y = a^2$ to get $yxy^{-1} = a^2b^8ab^2a^{-1}$. Our example isn't a proof, but it illustrates the underlying mechanism. How does this help? Well, suppose we have a nontrivial reduced word in $\{a, b, a^{-1}, b^{-1}\}$. Let g be the corresponding element of G. By what we just discussed, g is conjugate to the type of element discussed in the previous paragraph. In other words, g is conjugate to a nontrivial element of G. But then it must be that g is nontrivial, since all conjugates of the identity are equal to the identity.

This completes the proof, as we have shown that G has no relations with respect to the generating set $\{a, b\}$: no nonempty reduced words equal the identity. This means G is the free group of rank 2 freely generated by a and b.

Exercise 1. Provide the proof that every element of G is conjugate to one of the form $a^*b^*a^* \cdots b^*a^*$.

Does this work for free groups of higher rank? Yes! The proof is essentially the same.

LEMMA 5.2 (Ping-pong for *n* players). *Suppose $\{g_1, g_2, \ldots, g_n\}$ generates a group G, which acts on a set X. If*

1. *X has pairwise disjoint nonempty subsets $\{X_1, \ldots, X_n\}$, and*
2. *$g_i^k(X_j) \subset X_i$ for all nonzero powers k and $i \neq j$,*

then G is a free group of rank n.

Exercise 2. Prove ping-pong for n–players.

Exercise 3. Prove a slight variant of the ping-pong lemmas stated: remove the requirement that the sets X_i be disjoint, and add the requirement that some point in X not be contained in $\bigcup_i X_i$.

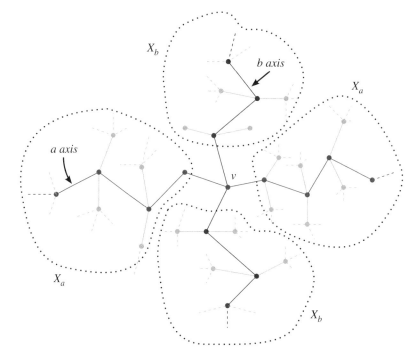

Figure 5.3 The tree T in the example.

Exercise 4. Prove a converse to the ping-pong lemma: find an action of F_n on some set X with subsets X_i fulfilling the ping-pong conditions.

An example using trees.   Suppose that T is a tree and that T has two automorphisms a and b that are hyperbolic. In Project 4 in Office Hour 3 we said what it means for an automorphism of T to be hyperbolic: it preserves some bi-infinite (non-backtracking) path in T (called an axis) and acts as a nontrivial translation along this path. Let us suppose further that the axes for a and b intersect in a single point, a vertex v. We would like to use ping-pong to show that a and b generate a free subgroup of rank 2 inside the group of automorphisms of T.

What are the sets X_a and X_b? Well, when we remove v from T and remove all edges incident to v, we are left with some number of pieces. There are at least four pieces, due to the fact that the axes for a and b cross at T. Let X_a be the union of the two pieces containing what remains of the axis of a and define X_b similarly (we think of X_a and X_b as subsets of T, not the graph we got by removing v). See Figure 5.3 for an illustration. It is straightforward to check that these sets satisfy the hypotheses of the ping-pong lemma. Indeed, v itself certainly gets sucked into one of the two sets X_a and X_b after applying a or b, respectively. And when v gets sucked into X_b, say, it carries all of X_a along for the ride. That does it!

Exercise 5. Verify the last claim.

What is more, it is easy to cook up explicit examples. For instance, we can take T to be a regular k-valent tree (meaning every vertex has k edges coming out) for any $k \geq 4$; we can choose a vertex v and two bi-infinite paths in T that cross at v, and then define a and b to be hyperbolic automorphisms that preserve those axes and perform nontrivial translations on them. There is more than one way to define a and b, but the important thing is that there is no obstruction to defining them because T is homogeneous.

This is pretty amazing. With very little effort, we can now recognize lots of different free groups in automorphism groups of trees!

Exercise 6. Generalize the tree example by allowing the overlap of the axes for a and b to be a finite union of edges. What happens if the axes are disjoint? *Hint: Take powers!*

Exercise 7. Generalize the tree example to find free groups of higher rank.

An example using matrices. One common time groups appear in the wild is when we study a known group and become interested in its subgroups. Subgroups can be vastly stranger than the group they sit in; for example, F_2 is finitely generated and has infinitely generated subgroups, such as the one generated by all words of the form $b^k a b^k$ with nonzero k. See Office Hour 4 for other examples.

Recall that $GL(2, \mathbb{Z})$ is the group of 2×2 integer matrices with determinant equal to ± 1. For one thing, it's the group of automorphisms of $\mathbb{Z}^2$, as explained in Office Hour 6. The following example identifies some free subgroups of $GL(2, \mathbb{Z})$ using its action on $\mathbb{R}^2$.

Here is the example: for an integer $m \geq 2$, the matrices

$$A = \begin{bmatrix} 1 & m \\ 0 & 1 \end{bmatrix} \quad \text{and} \quad B = \begin{bmatrix} 1 & 0 \\ m & 1 \end{bmatrix}$$

generate a free subgroup of $GL(2, \mathbb{Z})$ of rank 2. To prove this assertion using ping-pong, first you can verify by induction that

$$A^k = \begin{bmatrix} 1 & km \\ 0 & 1 \end{bmatrix} \quad \text{and} \quad B^k = \begin{bmatrix} 1 & 0 \\ km & 1 \end{bmatrix}.$$

After that, you only need to check that the subsets of $\mathbb{R}^2$

$$X_A = \left\{ \begin{bmatrix} x \\ y \end{bmatrix} \in \mathbb{R}^2 \mid |x| > |y| \right\} \quad \text{and} \quad X_B = \left\{ \begin{bmatrix} x \\ y \end{bmatrix} \in \mathbb{R}^2 \mid |x| < |y| \right\}$$

fulfill the conditions of the ping-pong lemma; see Figure 5.4. Note we may replace $GL(2, \mathbb{Z})$ with $GL(2, \mathbb{R})$ above, and then m need not be an integer.

Exercise 8. Complete this example by verifying that $A^k(X_B) \subseteq X_A$ and $B^k(X_A) \subseteq X_B$ for all nonzero powers k.

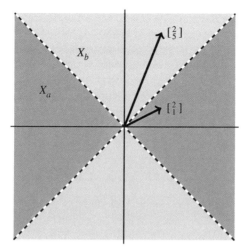

Figure 5.4 The (x, y)–plane with regions X_B and X_A. The dashed lines cover the remaining points where $x = \pm y$. If $m = 2$ in the matrix B, then $B\left[\begin{smallmatrix} 2 \\ 1 \end{smallmatrix}\right] = \left[\begin{smallmatrix} 2 \\ 5 \end{smallmatrix}\right]$ so $\left[\begin{smallmatrix} 2 \\ 1 \end{smallmatrix}\right]$ is an example of a vector that B moves from X_A into X_B.

A second example using matrices. Given that $\left\langle \left[\begin{smallmatrix} 1 & m \\ 0 & 1 \end{smallmatrix}\right], \left[\begin{smallmatrix} 1 & 0 \\ m & 1 \end{smallmatrix}\right] \right\rangle \cong F_2$, it takes a bit of inspiration to discover the proof. First, you have to decide that $\mathbb{R}^2$ is a good place to observe the action of A and B. Then, you have to find sets X_A and X_B that reveal the ping-pong nature of the action.

For a different example, consider the matrices

$$A = \begin{bmatrix} 3 & 0 \\ 0 & 1/3 \end{bmatrix} \quad \text{and} \quad B = \begin{bmatrix} 5/3 & 4/3 \\ 4/3 & 5/3 \end{bmatrix}.$$

As before, you can consider the linear action of these on $\mathbb{R}^2$. You can check that A has eigenvectors $\left[\begin{smallmatrix} 1 \\ 0 \end{smallmatrix}\right]$ and $\left[\begin{smallmatrix} 0 \\ 1 \end{smallmatrix}\right]$, while B has eigenvectors $\left[\begin{smallmatrix} 1 \\ 1 \end{smallmatrix}\right]$ and $\left[\begin{smallmatrix} -1 \\ 1 \end{smallmatrix}\right]$. Each triples the length of one eigenvector and contracts the other eigenvector by a third. Even given this information, you may find it more challenging to find a neat pair of ping-pong sets in $\mathbb{R}^2$ and prove that $\langle A, B \rangle$ is a free group.

Exercise 9. Find ping-pong subsets of $\mathbb{R}^2$ for the action of

$$A = \begin{bmatrix} 3 & 0 \\ 0 & 1/3 \end{bmatrix} \quad \text{and} \quad B = \begin{bmatrix} 5/3 & 4/3 \\ 4/3 & 5/3 \end{bmatrix}.$$

There's an alternative, rather marvelous setting where A and B also act and where the ping-pong sets are simpler to describe. This is the subject of the next section.

5.2 PING-PONG WITH MÖBIUS TRANSFORMATIONS

Historically, mathematical ping-pong arose in the study of subgroups of $\text{PSL}(2, \mathbb{C})$, a group with multiple claims to fame. This group is almost the same as $\text{SL}(2, \mathbb{C})$, the group of 2×2 complex matrices with determinant 1, except that we declare

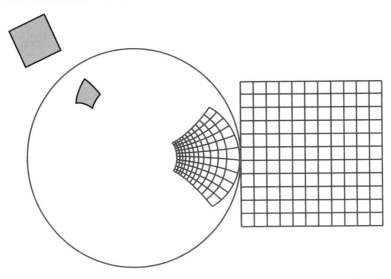

Figure 5.5 There is a Möbius transformation that preserves the circle and takes the checkerboard and square outside the circle to the warped checkerboard and warped
square inside the circle.

any matrix M to be equivalent to $-M$. In group theory parlance, PSL$(2, \mathbb{C})$ is
the quotient SL$(2, \mathbb{C})/N$, where N is the two-element subgroup consisting of the
identity matrix I and its negative $-I$.

Another, more useful, interpretation of PSL$(2, \mathbb{C})$ involves conformal geometry.
The quickest way to see this connection uses complex numbers and the complex
plane.

A crash course on Möbius transformations.   If you have taken a course in complex analysis, you have seen PSL$(2, \mathbb{C})$ in a different guise. *Möbius transformations* are automorphisms (i.e., holomorphic maps with holomorphic inverses) of the
extended complex plane $\widehat{\mathbb{C}} = \mathbb{C} \cup \{\infty\}$. Perhaps you have heard these maps called
linear fractional transformations, and maybe you like thinking of the extended
plane as its conformal equivalent, the Riemann sphere. In case you are less familiar
with these objects, I will spend a few paragraphs on a condensed explanation.

Concretely, a Möbius transformation is a map $\widehat{\mathbb{C}} \to \widehat{\mathbb{C}}$ of the form

$$z \mapsto \frac{az + b}{cz + d},$$

where a, b, c, and d are also complex numbers, and $ad - bc \neq 0$ (to exclude constant maps). The set of these maps forms a group under function composition. You
can think of complex numbers $z = x + iy$ as points on the (x, y)–plane. As maps
of the plane, Möbius transformations have the geometric distinction of being *conformal*, meaning they preserve oriented angles between line segments. Figure 5.5
shows what this can look like (notice in the picture that all of the angles in the

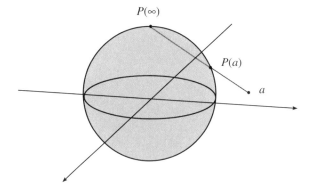

Figure 5.6 Stereographic projection.

warped checkerboard are right angles!). The figure also illustrates another useful property of these maps: lines and circles are always sent to lines and/or circles.

Exercise 10. Find a formula for the Möbius transformation illustrated in Figure 5.5

In fact, the group of Möbius transformations consists of exactly the one-to-one, onto conformal maps from the extended plane to itself. The domain must be the *extended* plane because the complex number $z = -d/c$ is sent to infinity and likewise the point $z = a/c$ is the image of infinity (or, if c is zero, infinity is a fixed point).

The Riemann sphere.   There's an elegant way to envision the extended complex plane $\widehat{\mathbb{C}}$ as the 2–sphere S^2. Specifically, we will find a conformal map $P : \widehat{\mathbb{C}} \to S^2$ using a technique called stereographic projection.

The definition of P is illustrated in Figure 5.6. First, define $P(\infty)$ to be $(0, 0, 1)$, the north pole. Then think of the complex plane $\mathbb{C}$ as the set of points $\{(x, y, 0)\}$ in $\mathbb{R}^3$. For $a = (x, y, 0)$, the way to find $P(a)$ is to draw the line segment from $(0, 0, 1)$ to a; $P(a)$ is where the line segment intersects the unit sphere S^2. This map P is bijective and conformal from $\widehat{\mathbb{C}}$ to the unit sphere S^2, so conformal automorphisms of the sphere correspond to conformal automorphisms of the extended plane.

Exercise 11. Derive a formula for P and show that it preserves angles.

When we think of $\widehat{\mathbb{C}}$ as a sphere, this object is often called the Riemman sphere. What did we gain from this point of view? On the sphere, it's easier to think of infinity as a point like any other. The real line in $\mathbb{C}$, represented by the x–axis in $\mathbb{R}^3$, corresponds to a great circle closing up at $P(\infty)$ in S^2. In fact, any line on the plane is projected by P to a circle through $P(\infty)$ on the sphere. Therefore, you may think of a line in $\mathbb{C}$ as actually a type of circle, specifically, a circle through infinity. Thus if we think of a line as a generalized circle, we can give a more elegant description of the Möbius transformation: they are the maps that preserve generalized circles.

Exercise 12. Draw the images of some circles in $\widehat{\mathbb{C}}$ for the Möbius transformations $a(z) = 9z$ and $b(z) = \frac{5z+4}{4z+5}$.

Möbius transformations as matrices.   So what do these Möbius transformations have to do with matrices? Perhaps you guessed it:

$$\begin{bmatrix} a & b \\ c & d \end{bmatrix} \quad \longleftrightarrow \quad z \mapsto \frac{az+b}{cz+d}.$$

Using this correspondence between PSL$(2, \mathbb{C})$ and the group of Möbius transformations, you can deduce that the two groups are isomorphic.

Exercise 13. Verify the last statement.

Exercise 14. What matrices in PSL$(2, \mathbb{C})$ correspond to the Möbius transformations $a(z) = 9z$ and $b(z) = \frac{5z+4}{4z+5}$? Does this look familiar?

Aside. Besides corresponding to the group of Möbius transformations, PSL$(2, \mathbb{C})$ is also exactly the group of orientation-preserving isometries of three-dimensional *hyperbolic space*, a point of view that was first studied more than a century ago by Felix Klein and Henri Poincaré. We'll get a better feeling for two-dimensional hyperbolic space, called the hypbolic plane, later in this office hour.

Why did we spend all this time on Möbius transformations? In 1883, Klein formulated the ping-pong lemma to identify free subgroups of PSL$(2, \mathbb{C})$, called Schottky groups [187]. Let's look at an example of one of these, before giving the general idea.

The second example again.   Consider the Möbius transformations $a(z) = 9z$ and $b(z) = \frac{5z+4}{4z+5}$. I have expressed these as Möbius transformations because I want to use their action on the extended complex plane $\widehat{\mathbb{C}}$.

Before even doing anything with a and b we are going to change our viewpoint on $\widehat{\mathbb{C}}$. The middle part of Figure 5.7 shows a view of part of $\widehat{\mathbb{C}}$ after the transformation $z \mapsto \frac{z-i}{1-iz}$, so that the images of the real line and the point at infinity lie on the unit circle, the inside of which contains the images of all complex numbers with positive imaginary part. We'll use this as our picture of $\widehat{\mathbb{C}}$. By our identification of $\widehat{\mathbb{C}}$ with S^2, this is totally legal; we have just rotated S^2 and are looking head-on at i instead of looking head on at 0 as usual.

I want to emphasize one more time that the linear transformation we just used will have no interaction with a and b. It just changes the way that $\widehat{\mathbb{C}}$ looks, so that ∞ is a point in the plane instead of, well, the point at infinity.

So what do a and b do? The element a fixes 0 and ∞ and moves all other points closer to ∞; its inverse $a^{-1}(z) = z/9$ fixes the same two points and moves the rest closer to 0. Powers of a iterate the forward or backward action, moving points even closer to ∞ or 0 if the power is positive or negative, respectively.

The element b acts similarly, except that it fixes 1 and -1, with the former attracting and the latter repelling, while b^{-1} just switches the roles of attractor and repeller.

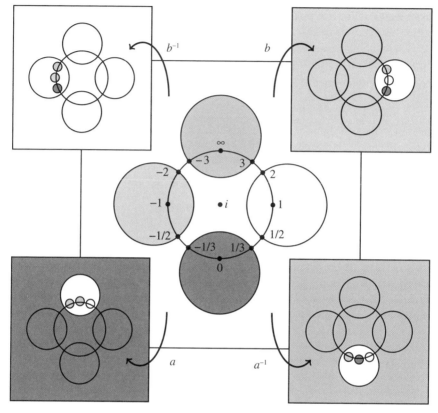

Figure 5.7 Generators of a Schottky group. Here we used the transformation $z \mapsto \frac{z-i}{1-iz}$, illustrated in Figure 6, to map the extended real line to the unit circle and put the upper half-plane inside the unit disk. This makes it easy to fit the colored neighborhoods into one picture.

In Figure 5.7, we placed colored disk-shaped neighborhoods X_a^+ and X_a^- about the attracting and repelling points of a, respectively, and similarly X_b^+ and X_b^- for b. In the pictures at the four corners, you can see the images of the neighborhoods after the actions of a, b, and their inverses. Choosing $X_a = X_a^+ \cup X_a^-$ and $X_b = X_b^+ \cup X_b^-$, we can apply ping-pong to show that a and b generate a free subgroup of $\mathrm{PSL}(2, \mathbb{C})$. If you believe the pictures, there is nothing left to check! By the way, did you notice that these two Möbius transformations correspond to the matrices A and B at the end of the previous section? We have proven that they generate a free subgroup of $\mathrm{PSL}(2, \mathbb{C})$, and hence they generate a free subgroup of $\mathrm{PSL}(2, \mathbb{R})$.

Exercise 15. Verify Figure 5.7.

Exercise 16. Can you determine if A and B generate a free subgroup of $\mathrm{SL}(2, \mathbb{R})$?

Generally, suppose you can find in the extended plane n pairs of disks X_i^+ and X_i^-, so that any two of $\{X_i^{\pm}\}_{i=1}^n$ are disjoint, and you also have corresponding Möbius transformations g_i so that g_i maps the outside of X_i^- to the inside of X_i^+.

Then the multiplayer ping-pong lemma (Lemma 5.2) tells you that $\{g_1, \ldots, g_n\}$ generates a group isomorphic to F_n. Groups coming from this construction are the classical Schottky subgroups of $\mathrm{PSL}(2, \mathbb{C})$.

To end this section, let's celebrate what just happened: we wanted find a space where two *real* matrices have simple-to-describe ping-pong sets to show that they generate a free group. In order to do this, we expanded our viewpoint to consider complex matrices and their action on the extended complex plane, which is a sphere. With this expanded viewpoint, we all of a sudden have a clear view of the answer. Very satisfying!

5.3 HYPERBOLIC GEOMETRY

We have seen ping-pong in action in the worlds of trees, matrices, and Möbius transformations. There is one more arena where ping-pong plays a huge role: hyperbolic geometry.

Group actions by isometries on hyperbolic space provide a fundamental model for many ideas in geometric group theory, so we give them more attention here. We will only glimpse the tip of the iceberg; however, Office Hour 9 further explores these ideas.

The hyperbolic plane.   Let's take an informal tour of two-dimensional hyperbolic space $\mathbb{H}^2$, also known as the hyperbolic plane. There are several geometrically equivalent ways to define $\mathbb{H}^2$, each with its own advantages, but we'll only use make use of two.

Here is our first definition of the hyperbolic plane, called the upper half-plane model. We start with the upper half-plane $\mathbb{U}$ in $\mathbb{R}^2$, namely, $\{(x, y) \in \mathbb{R}^2 \mid y > 0\}$. Then we put the following Riemannian metric on $\mathbb{U}$:

$$ds^2 = \frac{dx^2 + dy^2}{y^2}.$$

Wait. What is a Riemannian metric? You are already familiar with the usual Riemannian metric on $\mathbb{R}^2$, the Euclidean metric:

$$d\mathrm{Euc}^2 = dx^2 + dy^2.$$

You used this Riemannian metric in calculus when you were computing the lengths of parametrized curves (although you probably didn't call it a Riemannian metric). Just to remind you, if $f(t) = (x(t), y(t))$ is a parametrized curve, then the integral

$$\int_0^1 \sqrt{\left(\frac{dx}{dt}\right)^2 + \left(\frac{dy}{dt}\right)^2}\, dt$$

is the Euclidean length of the path from $f(0)$ to $f(1)$.

There's more. We can use the Euclidean Riemannian metric to compute the distance between any two points in the Euclidean plane. First, a shortest path between two points is called a *geodesic*. Then the distance between any two points is

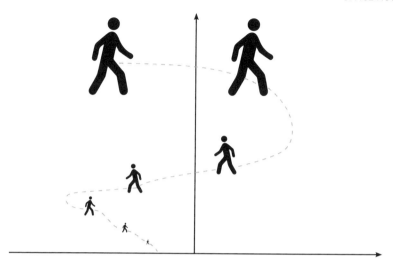

Figure 5.8 Walking around in hyperbolic space.

defined to be the length of any geodesic connecting them. We know that the answer in Euclidean space is always a straight line. (Can you see why this is true, just from the Riemannian metric?) So in this way, by finding geodesics, the Riemannian metric gives rise to a metric (as in the definition of a metric space).

Let's do the same thing with the metric ds^2 we defined on $\mathbb{U}$ above. But actually before we do that, let's just look at the formula for ds^2 and think about what it means qualitatively. We can rewrite ds^2 as

$$ds^2 = \frac{d\mathrm{Euc}^2}{y^2}.$$

In other words, the hyperbolic metric on $\mathbb{U}$ is, at each point, a scalar multiple of the Euclidean metric. And we can compute the length of a parametrized curve $f(t) = (x(t), y(t))$ as

$$\int_0^1 \frac{1}{y(t)} \sqrt{\left(\frac{dx}{dt}\right)^2 + \left(\frac{dy}{dt}\right)^2}\, dt.$$

Of course, $1/y$ goes to ∞ as y tends to 0. What does that mean? It means that if we draw a path in $\mathbb{U}$ that is very close to the x–axis, then its length in $\mathbb{U}$ (in the hyperbolic metric) is much larger than the Euclidean length. And similarly if we draw a path very far from the x–axis, its length in $\mathbb{U}$ is much smaller than the Euclidean length.

Here is another way to think about it. Say we are watching a two-dimensional person walking around in $\mathbb{U}$. As she starts walking towards the x–axis, she appears smaller and smaller to us, with our Euclidean eyes, and in fact she can keep walking towards the x–axis forever, and she will never get there! See Figure 5.8. As she gets smaller (to our eyes) the steps that she takes get smaller (to our eyes). But from the point of view of our walker (or anyone else living in the hyperbolic plane), she is staying the same size and walking at constant speed.

Exercise 17. Pause to think about the shortest path for our walker to get from $(0, 0.1)$ to $(10, 0.1)$. It is not the horizontal path! *Hint: To travel a large Euclidean length, the walker should make her legs look bigger from the Euclidean point of view.*

Examples of geodesics.   OK, let's get back to the quantitative. We want to see if we can explicitly find the geodesics between any two points; again the length of such a path is the distance between the two points.

There's one case where it is not too hard to do the calculus directly and find geodesics. Consider two points on the positive y–axis, say, $(0, a)$ and $(0, b)$. We may as well assume that $a < b$. Let $f(t) = (x(t), y(t))$ be a path from $(0, a)$ to $(0, b)$ with $0 \leq t \leq 1$. We gave the formula for the hyperbolic length of this path:

$$\int_0^1 \frac{1}{y(t)} \sqrt{\left(\frac{dx}{dt}\right)^2 + \left(\frac{dy}{dt}\right)^2} \, dt.$$

In this case it pays to be lazy and ask, taking a path from $(0, a)$ to $(0, b)$, why move at all in the x–direction? It's not necessary, since we have to start and end at the same x–value. Furthermore, it only adds to arc length, by putting that $(dx/dt)^2$ term through some nonzero values.

Therefore a shortest path between $(0, a)$ and $(0, b)$ better be directly vertical, with $dx/dt = 0$ throughout. Any backtracking in such a path will also make it needlessly longer, so we may decide our path goes straight upwards, so $dy/dt \geq 0$. This ensures that $\sqrt{(dy/dt)^2} = dy/dt$ rather than just its absolute value. Once we've committed to a vertical, non-backtracking path, there's only one path (up to reparametrization): $f(t) = (0, tb + (1 - t)a)$.

Now that we have our path, it's easy to take our integral:

$$\int_0^1 \frac{1}{y} \frac{dy}{dt} \, dt$$

(notice we can avoid absolute values here because both y and dy/dt are positive here). After changing variables this is simply

$$\int_a^b \frac{dy}{y}.$$

That evaluates to

$$\ln(b/a).$$

We have found the hyperbolic distance between $(0, a)$ and $(0, b)$ in $\mathbb{U}$!

We can now measure distances between points on the imaginary axis, and, by the same logic, between any two points with the same x–coordinate. Moreover, we know that lines in $\mathbb{U}$ perpendicular to the real axis are infinite geodesics, and our reasoning also establishes that these are the unique geodesics through points with the same x–value.

The rest of the geodesics. At this point we have found the geodesics only between very special pairs of points. The tricks we used don't seem to apply to other pairs of points. For the other cases, I am now going to give you a tremendous shortcut. First, think of $\mathbb{U}$ as lying in $\mathbb{C}$ instead of $\mathbb{R}^2$; again complex numbers are coming into play! We have the following very useful fact:

> *Every Möbius transformation that preserves $\mathbb{U}$ is an isometry of $\mathbb{U}$ with its hyperbolic metric.*

You can check this with a bit of computation, after you convince yourself that the Möbius transformations preserving $\mathbb{U}$ are exactly the ones with *real* parameters $a, b, c,$ and d where $ab - cd > 0$.

This group of Möbius transformations has a nice interpretation. Specifically, the group of Möbius transformations preserving the upper half-plane is isomorphic to $\text{PSL}(2, \mathbb{R})$, the group of 2×2 matrices with real coefficients, determinant equal to 1, and each matrix M declared equivalent to $-M$. The next exercise asks you to prove this.

Exercise 18. Prove that the group of Möbius transformations preserving the upper half-plane is isomorphic to $\text{PSL}(2, \mathbb{R})$. First observe that, to preserve $\mathbb{U}$, a Möbius map must send the extended real line $\mathbb{R} \cup \{\infty\}$ to itself, and this happens if and only if it can be written as $f(z) = \frac{az+b}{cz+d}$ using *real* parameters $a, b, c,$ and d. Once you know f preserves the extended real line, it either sends $\mathbb{U}$ to itself or swaps the upper and lower half-planes. Just by testing where $z = i$ is sent, you can check that a Möbius transformation with real parameters preserves $\mathbb{U}$ if and only if $ad - bc > 0$. The rest of your work is similar to showing that $\text{PSL}(2, \mathbb{R})$ is isomorphic to the entire collection of Möbius transformations.

Exercise 19. Prove that the Möbius transformations preserving $\mathbb{U}$ act as isometries of $\mathbb{U}$ with the hyperbolic metric. To do this, you'll have to make sense of how a Riemannian metric transforms under a differentiable function.

This is great news, because we've already heard that Möbius transformations preserve angles, and also send circles (which, remember, include lines) to circles. Let's use these facts to our advantage. If any two points p and q in $\mathbb{U}$ are not already on the same vertical line, then there is a unique circle C containing both these points and perpendicular to the real line; see Figure 5.9. Say this circle intersects the real axis at r and s, where $r > s$. Because it has real parameters with $r - s > 0$, the Möbius transformation $f(z) = \frac{z-r}{z-s}$ satisfies $f(\mathbb{U}) = \mathbb{U}$. Moreover, since Möbius transformations are conformal, $f(C)$ is a generalized circle—actually it is a line— that is perpendicular to the real axis. This line is exactly the imaginary axis, and $f(p)$ and $f(q)$ must be on the positive part.

Exercise 20. Can you compute what r and s must be in Figure 5.9?

We also know that the positive imaginary axis is the unique infinite geodesic through $f(p)$ and $f(q)$; thus its isometric image under f^{-1}, namely, $C \cap \mathbb{U}$, is the

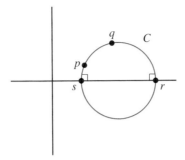

Figure 5.9 The circle C is centered on the unique point on $\mathbb{R}$ equidistant from p and q.

unique infinite geodesic through p and q. Since we can play this game with any pair of points, we can deduce the following fact:

> *The infinite geodesics in $\mathbb{U}$ are exactly the semicircles orthogonal to $\mathbb{R}$ (this includes the vertical semi-lines).*

Exercise 21. Compute the hyperbolic distance between $(0, 1)$ and $(1/\sqrt{2}, 1/\sqrt{2})$ in $\mathbb{U}$ (equivalently i and $1/\sqrt{2} + i/\sqrt{2}$ as complex numbers). *Hint: Both points are on the unit circle.*

We have come far. Starting from just the formula for the hyperbolic metric ds^2, we have found all of the geodesics in the hyperbolic plane!

The Poincaré disk model. Let's now look at the hyperbolic plane from a different point of view. Consider the Möbius transformation

$$z \mapsto \frac{i - z}{i + z}.$$

This transformation takes the upper half-plane $\mathbb{U}$ to the open unit disk $\mathbb{D}$ in $\mathbb{C}$, and it takes the real line to the unit circle; see Figure 5.10 for an illustration of this mapping. Since we have already identified $\mathbb{U}$ with the hyperbolic plane, we now have an identification of the hyperbolic plane with $\mathbb{D}$. We refer to $\mathbb{D}$ as the *Poincaré disk*.

What does the Riemannian metric look like from this point of view? If we denote the hyperbolic metric on $\mathbb{D}$ by du^2, it turns out that

$$du^2 = 4 \frac{d\mathrm{Euc}^2}{(1 - r^2)^2},$$

where r here denotes distance from the origin. This is similar to the hyperbolic metric ds in the upper half-plane: at each point, du is a scalar multiple of the Euclidean metric.

Here is another way to explain what is happening: we have the metric ds^2 on $\mathbb{U}$ and we have the metric du^2 on $\mathbb{D}$, and the Möbius transformation $z \mapsto \frac{i-z}{i+z}$ takes one to the other, that is, it is an isometry from one to the other. So $\mathbb{U}$ and $\mathbb{D}$

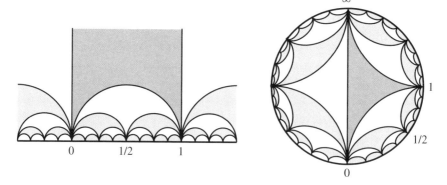

Figure 5.10 The Farey tessellation of a piece of $\mathbb{U}$, on the left, and of $\mathbb{D}$, on the right.

are isometric with their respective metrics; they are both isometric *models* for the hyperbolic plane.

Exercise 22. Prove that the given Möbius transformation is an isometry between the two metrics.

If we again imagine a person walking around in the disk $\mathbb{D}$, the metric goes to infinity as the person approaches the boundary of the disk, and so to our eyes she appears to get smaller and smaller, and she never reaches the boundary. Sound familiar?

Since the given Möbius transformation is an isometry, and since it is conformal, and since we already know all the geodesics in $\mathbb{U}$, we can immediately list all of the geodesics in $\mathbb{D}$: they are the circles/lines in the plane that are orthogonal to the unit circle at two points (well, the intersections of these circles with $\mathbb{D}$). That was easy!

Both models of $\mathbb{H}^2$ are useful. In $\mathbb{U}$ we have the advantage that the formulas are often simpler (for instance, the formula for the Riemannian metric). In $\mathbb{D}$ the main advantage is that there is more symmetry (for instance, in this model there are no special geodesics, such as the vertical lines in the $\mathbb{U}$ model).

Farey tesselations and hyperbolic geometry.   Let's take all the hard work we just did and have some fun with it (no ping-pong yet; that's coming). In Figure 5.10 we have drawn two pictures of the Farey tesselation of the hyperbolic plane. We already saw the version on the right in Figure 3.1 from Office Hour 3. What do we see? The edges of the Farey graph are actually geodesics in the hyperbolic plane! If that's the case, we should be able to transport the whole picture to the upper half-plane, and that's exactly what we see on the left. These are really two views of the same picture.

Exercise 23. Practice going back and forth between the two pictures of the Farey tesselation of the hyperbolic plane. For instance, take a triangle in one picture

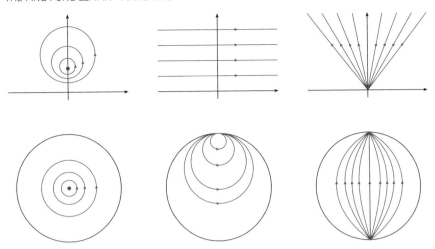

Figure 5.11 Elliptic, parabolic, and hyperbolic isometries in the two models for $\mathbb{H}^2$.

and find the corresponding triangle in the other. What are the coordinates of the vertices?

Exercise 24. Find an image of M. C. Escher's art work titled *Angels & Devils*. What do you notice?

Isometries of the hyperbolic plane.   Now that we're a little at home in hyperbolic space, let's try to understand its isometries. Why? Because we are going to play ping-pong with them.

You would be right to wonder which isometries of $\mathbb{U}$ are *not* Möbius transformations. Here's one: $z \mapsto -\bar{z}$. In either model, this map reflects around the imaginary axis, and you can convince yourself that this map sends geodesics to geodesics, and is therefore an isometry. But it's not Möbius, because it reverses orientation. It happens that every orientation-reversing isometry is the product of an orientation-preserving isometry and the one reflection we just gave. Also, squaring any orientation-reversing map yields one that preserves orientation (to say it another way, the group of orientation-preserving isometries has index 2 in the full group of isometries). So we'll mainly focus on the orientation-preserving isometries. It turns out that these orientation-preserving isometries are exactly the Möbius transformations that preserve $\mathbb{U}$, and we already said that this group can be identified with the group PSL$(2, \mathbb{R})$.

There are three kinds of isometries of $\mathbb{H}^2$, distinguished by their behavior. In order to understand these isometries, we must again expand our viewpoint. It is not enough to just look at the hyperbolic plane; we must also look at its *boundary*. We'll deal with boundaries more formally in Office Hour 9, but for now the boundary is what you think it is: in the $\mathbb{U}$–model it is $\mathbb{R} \cup \{\infty\}$ and in the $\mathbb{D}$–model it is the unit circle.

Here are the three types of isometries, with examples. Throughout, refer to Figure 5.11.

- An *elliptic isometry* fixes a point in $\mathbb{H}^2$ and acts as a rotation about that point. For example, the map $z \mapsto \frac{-1}{z-1}$ fixes $(1 + \sqrt{3}i)/2$ in $\mathbb{U}$.
- A *parabolic isometry* does not fix a point in the hyperbolic plane, but does fix a single point in its boundary. For example, in the model $\mathbb{U}$ the map $z \mapsto z + 1$ fixes no points of $\mathbb{U}$ but does fix the point ∞. In $\mathbb{U}$, this map preserves horizontal lines. Remember, these are not geodesics! They're called *horocycles*. In $\mathbb{D}$ the horocycles look like circles tangent to the fixed point.
- A *hyperbolic isometry* fixes exactly *two* points in the boundary of the hyperbolic plane. For example, the map $z \mapsto 9z$ fixes 0 and ∞. A hyperbolic isometry has an *axis*: a (unique) geodesic on which it acts like a nontrivial translation. For the hyperbolic isometry $z \mapsto 9z$, the axis is the positive imaginary axis in $\mathbb{U}$ and the isometry translated by $2 \ln 3$ along this axis. The other lines preserved by the isometry (as shown in the figure) are not geodesics; they are simply the points of a given fixed distance from the axis.

Exercise 25. Consider the Möbius transformation $f(z) = \frac{az+b}{cz+d}$, where $a, b, c,$ and d are all real numbers, and assume we've already made sure that $ad - bc = 1$. Show that f is elliptic if and only if $|a + d| < 2$, is parabolic if and only if $|a + d| = 2$, and is hyperbolic if and only if $|a + d| > 2$.

Exercise 26. In this exercise you will show that every orientation-preserving isometry of the hyperbolic plane is conjugate to one in a standard form.

1. In the Poincaré disk model, show that an elliptic isometry is conjugate to $z \mapsto e^{i\theta}z$ for some angle θ.
2. If $f(z) = \frac{az+b}{cz+d}$ is parabolic, find the unique fixed point and show that f is conjugate to the Möbius transformation $z \mapsto z + k$ for some real number k.
3. If $f(z) = \frac{az+b}{cz+d}$ is hyperbolic, show that it is conjugate to the Möbius transformation $z \mapsto \lambda z$ for some positive real number λ. What is the axis and by how much does f shift along its axis?

Back to ping-pong. Let's play ping-pong in the hyperbolic plane! First of all, we can reinterpret one of our old examples in terms of isometries of the hyperbolic plane. We already used Möbius transformations to show that the matrices

$$A = \begin{bmatrix} 3 & 0 \\ 0 & 1/3 \end{bmatrix} \quad \text{and} \quad B = \begin{bmatrix} 5/3 & 4/3 \\ 4/3 & 5/3 \end{bmatrix}$$

generate a free subgroup of $\mathrm{PSL}(2, \mathbb{C})$. Hence they also generate a free subgroup of $\mathrm{PSL}(2, \mathbb{R})$, the group of isometries of the hyperbolic plane (in the $\mathbb{U}$ model)! In our new context, these are both isometries of hyperbolic type, and the endpoints of the axes are 0 and ∞ (for A) and 1 and -1 (for B). What are the ping-pong sets? We may take the intersections with $\mathbb{U}$ of the four colored circles in Figure 5.7. The same proof applies; cut from the picture everything outside of the open unit disk.

You can find other explicit examples of isometries of $\mathbb{H}^2$ that generate a free group. Since $\mathrm{PSL}(2, \mathbb{R})$ is a subgroup of $\mathrm{PSL}(2, \mathbb{C})$, you might not be able to find examples that you couldn't just consider from the point of view of Möbius transformations. Still, it is worth understanding the hyperbolic point of view. The hyperbolic plane is ubiquitous in geometric group theory, and we'll definitely spend more time with it later in the book.

Exercise 27. Let g and h be two hyperbolic isometries of the hyperbolic plane whose axes cross in $\mathbb{H}^2$. Show that there is an N so that g^N and h^N generate a free group of rank 2 in the group of isometries of $\mathbb{H}^2$.

5.4 FINAL REMARKS

Throughout this book you should find many more examples of group actions in which one finds prime conditions for ping-pong. These include trees (Office Hour 3), automorphisms of free groups (Office Hour 6), word hyperbolic groups (Office Hour 9), and subgroups of mapping class groups (Office Hour 17).

Perhaps the most famous application of ping-pong was by Jacques Tits, in the context of *linear groups*: subgroups of $n \times n$ matrices with nonzero determinant and coefficients in $\mathbb{C}$ or some other field of characteristic zero. For example, $\mathrm{SL}(n, \mathbb{Z})$ is a linear group. Tits proved that if a linear group is not virtually solvable (meaning, it does not have a finite-index solvable subgroup), then one can find in it a nonabelian free subgroup via ping-pong [256]. Many other mathematicians have established a "Tits alternative" for other classes of groups, including all of those mentioned above, except, notably, automorphisms of trees, where a Tits alternative is false!

We end with a generalization of the ping-pong lemma that detects right-angled Artin groups. A *right-angled Artin group* is a group admitting a presentation with finitely many generators and only commutation relations, that is,

$$\langle g_1, g_2, \ldots, g_n \mid g_i g_j = g_j g_i \text{ for some } i, j \rangle.$$

These are examined in more depth in Office Hour 14, but perhaps you can see already how they form a bridge between F_n (no relations) and $\mathbb{Z}^n$ (commutation relations between all generators).

LEMMA 5.3 (Ping-pong in right-angled Artin groups). *Suppose* $\{g_1, \ldots, g_k\}$ *generates G, and suppose we know certain pairs of generators g_i, g_j commute. If G acts on a set X and each generator g_i corresponds to a subset X_i of X such that*

1. $g_i^k(X_j) \subset X_j$ *for all nonzero k if g_i and g_j commute,*
2. $g_i^k(X_j) \subset X_i$ *for all nonzero k if otherwise, and*
3. *there exists $x \in X - \bigcup_i X_i$ such that $g_i(x) \in X_i$ for all i,*

then G is a right-angled Artin group.

This lemma was first observed by Crisp and Farb [95]; you can find a published proof by Koberda [188].

Exercise 28. Prove the Lemma 5.3.

FURTHER READING

For history and examples, we've leaned heavily on the book [104], which makes excellent further reading for this chapter and geometric group theory in general. Other beautiful resources for further reading include the books by Thurston [255] and Katok [181], and well as the book by Mumford, Series, and Wright [207], which gives many beautifully illustrated examples of Möbius transformations and Schottky groups. Finally we mention the great article by Cannon, Floyd, and Parry that describes the various models of hyperbolic space, their isometries and geodesics [74].

PROJECTS

Project 1. Apply ping-pong to the action of $SL(2, \mathbb{Z})$ on the Farey tree. First find the axes of some elements and then use the ideas from this office hour to find the ping-pong sets.

Project 2. Suppose that A and B are matrices in $SL(2, \mathbb{Z})$ each with two distinct real eigenvectors. Further suppose that eigenvectors for A are not eigenvectors for B. Use ping-pong to show that $\langle A, B \rangle$ is isomorphic to F_2.

Project 3. Learn about three-dimensional hyperbolic space and learn why its group of (orientation-preserving) isometries is $PSL(2, \mathbb{C})$.

Project 4. This project is about distances and geodesics in the hyperbolic plane.

1. Compute the distance between arbitrary points (x, y) and (x', y') in the upper half-plane model of the hyperbolic plane.
2. Consider the "mesa path" from (x, y) to (x', y') defined as follows. Let h be the radius of the semicircle in $\mathbb{R}^2$ corresponding to the geodesic between (x, y). The mesa path consists of three segments, two vertical and one horizontal, where the height of the horizontal path is h (cf. Office Hour 12). Compute the length of the mesa path.
3. Show that mesa paths are always quasi-geodesics (see Office Hour 9 for the definition).
4. Show further that mesa paths are uniform quasi-geodesics (that is, the constants in the definition of a quasi-geodesic don't depend on the two points you started with).
5. For the special case $y = y'$ show that the length of the mesa path is at most 2 greater than distance between (x, y) and (x', y').

Project 5. Learn about the hyperboloid model and the Klein model for the hyperbolic plane.

Project 6. Learn about higher-dimensional hyperbolic spaces.

Project 7. Prove that the space of geodesics in the hyperbolic plane can be identified with a Möbius strip. Here are the steps.

1. The set of geodesics in $\mathbb{H}^2$ are in bijection with the set of unordered pairs of distinct points on the boundary.
2. Given an unordered pair of distinct, non-antipodal points on the boundary of the unit disk in $\mathbb{R}^2$, associate a point in the exterior of the unit disk by taking the intersection point of the tangent lines to the unit circle at those points; show that this association is a bijection.
3. Finally, consider in $\mathbb{R}^2$ the complement of the interior of the unit disk, and take the quotient of this set by identifying antipodal points on the unit circle. Show that this quotient space can be identified with an open Möbius strip (open means the boundary is absent). Use this to conclude the proof.

Office Hour Six

Automorphisms of Free Groups

Matt Clay

In this office hour we will look at the automorphisms of free groups. In some regards a free group behaves like a discrete non-commutative vector space. As such, tools and techniques from linear algebra are applicable, at least philosophically. We will focus on an example showing how the notions of eigenvalue and eigenvector relate to free group automorphisms.

6.1 AUTOMORPHISMS OF GROUPS: FIRST EXAMPLES

As we saw in Office Hour 2, every group is the collection of symmetries of some object, namely, its Cayley graph. In this office hour we are not concerned with symmetries of a geometric object, but instead we are concerned with the symmetries of the group itself! A symmetry of a group is called an *automorphism*; it is merely an isomorphism of the group to itself. The collection of all of the automorphisms is a group too, called the *automorphism group* and denoted by $\text{Aut}(G)$. Let's look at our basic examples of groups to get a feel for what an automorphism is.

The symmetric group. First let's consider automorphisms of S_3, the symmetric group on three elements. An automorphism of the group S_3 must preserve the order of each element. Remember, the *order* of $g \in G$ is the smallest integer $n > 0$ such that g^n is the identity. Thus under any automorphism the identity is fixed; 2–cycles go to 2–cycles and 3–cycles go to 3–cycles. In Office Hour 1, we saw that any permutation can be written as a product of 2–cycles; hence an automorphism of S_3 is completely determined by where it sends the 2–cycles. Since there are three 2–cycles in S_3, we can have at most six automorphisms. This follows as each automorphism is a permutation of the three 2–cycles.

Now for each element $\sigma \in S_3$, conjugation defines a function $\rho_\sigma \colon S_3 \to S_3$ by $\rho_\sigma(\tau) = \sigma\tau\sigma^{-1}$. This function is a homomorphism, as we can check:

$$\rho_\sigma(\tau_1)\rho_\sigma(\tau_2) = (\sigma\tau_1\sigma^{-1})(\sigma\tau_2\sigma^{-1}) = \sigma\tau_1\tau_2\sigma^{-1} = \rho_\sigma(\tau_1\tau_2).$$

Since $\rho_{\sigma^{-1}}$ is the inverse of ρ_σ, the homomorphism ρ_σ is an isomorphism. In other words, ρ_σ is an automorphism of S_3. For any group G and element $g \in G$, we can define $\rho_g \colon G \to G$ as above. This always defines an automorphism of the group G, called an *inner automorphism*.

If we think of ρ_σ as a permutation of the set $\{\mathbf{1} = (2\,3), \mathbf{2} = (1\,3), \mathbf{3} = (1\,2)\}$ we have that $\rho_\sigma = \sigma$ in the sense that $\rho_{(1\,2)}$ interchanges $(2\,3) \leftrightarrow (1\,3)$ and fixes $(1\,2)$. It follows that the different inner automorphisms give rise to all of the six possible permutations of the set of 3–cycles. Hence there are exactly six automorphisms of S_3 and, moreover, we can identify the group of all automorphisms of S_3 with the group S_3 itself. In other words, we have just shown: $\mathrm{Aut}(S_3) \cong S_3$.

The above proof doesn't apply to S_4. One of the differences between 3 and 4 is that there are elements of order 2 in S_4 that are not 2–cycles, e.g., $(1\,2)(3\,4)$. However, it does turn out that $\mathrm{Aut}(S_4) \cong S_4$.

Exercise 1. Show that $\mathrm{Aut}(S_4) \cong S_4$. To do this, let $C_2 \subset S_4$ denote the set of 2–cycles and let $C_{2,2} \subset S_4$ denote the set of permutations of the form $(a\,b)(c\,d)$. Show the following:

1. $\#|C_2| = 6$ and $\#|C_{2,2}| = 3$.
2. If $\rho \in \mathrm{Aut}(S_4)$, then $\rho(C_2) = C_2$.

Conclude that $\mathrm{Aut}(S_4) \cong S_4$. This last step is a bit tricky; you may want to seek a hint.

The argument sketched in the above exercise works for all $n \neq 2, 6$ to show that $\mathrm{Aut}(S_n) \cong S_n$, specifically that every automorphism of S_n is an inner automorphism. Try to fill in the details for $n \neq 2, 6$.

Exercise 2. In fact, $\mathrm{Aut}(S_6) \neq S_6$. Let's follow the method of Janusz and Rotman to construct an automorphism of S_6 that is not inner, i.e., is not conjugation by some element. Show the following:

1. There are six distinct cyclic subgroups of S_5 generated by a 5–cycle.
2. S_5 acts faithfully and transitively on these six subgroups by conjugation, inducing an injective homomorphism $\phi \colon S_5 \to S_6$.
3. S_6 acts faithfully and transitively on the six left cosets of $\phi(S_5)$ by left multiplication, inducing an automorphism $\psi \colon S_6 \to S_6$.
4. The automorphism ψ is not inner. (*Hint: Check what happens to a 2–cycle.*)

See the article of Janusz and Rotman [175] for more details on the exotic case of $n = 6$.

Exercise 3. Compute the automorphism groups $\mathrm{Aut}(\mathbb{Z}/n\mathbb{Z})$ and $\mathrm{Aut}(D_n)$.

The free abelian group.   Moving along to a more interesting group, let's now consider automorphisms of $\mathbb{Z}^2$, the free abelian group of rank 2. We can use vector notation and write:

$$\mathbb{Z}^2 = \left\{ x \begin{bmatrix} 1 \\ 0 \end{bmatrix} + y \begin{bmatrix} 0 \\ 1 \end{bmatrix} \mid x, y \in \mathbb{Z} \right\}.$$

Since the group structure on $\mathbb{Z}^2$ is addition, an automorphism ϕ of $\mathbb{Z}^2$ satisfies $\phi(\mathbf{u} + \mathbf{v}) = \phi(\mathbf{u}) + \phi(\mathbf{v})$ for all $\mathbf{u}, \mathbf{v} \in \mathbb{Z}^2$. This equation should look familiar to you; this is one-half of the definition of a linear transformation of a vector space. Since $\mathbb{Z}^2$ is generated by $\mathbf{a} = \begin{bmatrix} 1 \\ 0 \end{bmatrix}$ and $\mathbf{b} = \begin{bmatrix} 0 \\ 1 \end{bmatrix}$, an automorphism of $\mathbb{Z}^2$ is determined by where it sends these two elements. This is just like what we saw for S_3; every automorphism of S_3 is determined by where it sends the 2–cycles.

For an automorphism $\phi \colon \mathbb{Z}^2 \to \mathbb{Z}^2$ and an element $x\mathbf{a} + y\mathbf{b} \in \mathbb{Z}^2$ we have:

$$\phi(x\mathbf{a} + y\mathbf{b}) = \phi(x\mathbf{a}) + \phi(y\mathbf{b})$$

$$= \phi(\underbrace{\mathbf{a} + \cdots + \mathbf{a}}_{x \text{ copies}}) + \phi(\underbrace{\mathbf{b} + \cdots + \mathbf{b}}_{y \text{ copies}})$$

$$= \underbrace{\phi(\mathbf{a}) + \cdots + \phi(\mathbf{a})}_{x \text{ copies}} + \underbrace{\phi(\mathbf{b}) + \cdots + \phi(\mathbf{b})}_{y \text{ copies}}$$

$$= x\phi(\mathbf{a}) + y\phi(\mathbf{b})$$

$$= \begin{bmatrix} \phi(\mathbf{a}) & \phi(\mathbf{b}) \end{bmatrix}(x\mathbf{a} + y\mathbf{b}).$$

Here, $\begin{bmatrix} \phi(\mathbf{a}) & \phi(\mathbf{b}) \end{bmatrix}$ is the 2×2 matrix with columns given by the coordinates of the vectors $\phi(\mathbf{a})$ and $\phi(\mathbf{b})$ respectively, and $x\mathbf{a} + y\mathbf{b}$ is treated as a column vector (technically, $\mathbb{Z}^2$ is not a vector space, but it is a fruitful analogy here). Thus any automorphism $\phi \colon \mathbb{Z}^2 \to \mathbb{Z}^2$ is represented by a 2×2 matrix, just as for a linear isomorphism of $\mathbb{R}^2$. Since we are taking integer coordinates to integer coordinates, all of the entries in the matrix are integers and hence the determinant of the matrix is an integer too. The same is true for the matrix associated with the inverse of ϕ. As the determinant of the inverse is the reciprocal of the determinant of the original matrix, in order for both of these determinants to be integers they must be ± 1. In other words, an automorphism of $\mathbb{Z}^2$ is given by a 2×2 integer matrix with determinant equal to ± 1. We have now shown $\mathrm{Aut}(\mathbb{Z}^2) \cong \mathrm{GL}(2, \mathbb{Z})$, the *general linear group*.

Exercise 4.   Show that $\mathrm{Aut}(\mathbb{Z}^n) \cong \mathrm{GL}(n, \mathbb{Z})$.

6.2 AUTOMORPHISMS OF FREE GROUPS: A FIRST LOOK

In contrast with S_3, where every automorphism is an inner automorphism, an inner automorphism of $\mathbb{Z}^2$ is always the identity function! We can check this easily since vector addition is commutative: $\rho_{\mathbf{u}}(\mathbf{v}) = \mathbf{u} + \mathbf{v} + (-\mathbf{u}) = \mathbf{v}$. Automorphisms of free groups, the heart of this chapter, fit between these two extremes. Nontrivial elements give rise to nontrivial inner automorphisms, the opposite of $\mathbb{Z}^2$, but not every automorphism is an inner automorphism, as it was for S_3.

The free group.   Let's start by considering the free group of rank 2, F_2, with generating set $\{a, b\}$. So elements in F_2 are reduced words in $\{a, a^{-1}, b, b^{-1}\}$.

Keeping with the vector space analogy, we refer to the (minimal) generating set $\{a, b\}$ as a *basis* for F_2. If you have never thought about it before, you might be tempted to think that this is the only basis for F_2. Remember our analogy with vector spaces! Is the standard basis $\left\{\begin{bmatrix} 1 \\ 0 \end{bmatrix}, \begin{bmatrix} 0 \\ 1 \end{bmatrix}\right\}$ the only basis for $\mathbb{R}^2$? Of course not! How do we get another basis? One way is to multiply each vector in the standard basis by a nonsingular matrix, i.e., apply an automorphism. The same is true for free groups. Ordered bases of the free group are in one-to-one correspondence with automorphisms.

Exercise 5.   Before you read on, you should take a moment and convince yourself that $\{a, ab\}$ is a basis for F_2, i.e., show that every word in $\{a, a^{-1}, b, b^{-1}\}$ can be uniquely expressed as a reduced word in $\{a, a^{-1}, ab, (ab)^{-1}\}$, and vice versa. Is the same true for the set $\{a^2, b\}$? Can you give examples of other bases?

Just as for S_n and $\mathbb{Z}^n$, an automorphism of F_n is an isomorphism $\phi \colon F_n \to F_n$. As for S_n, since F_n is nonabelian, there are inner automorphisms $\rho_x \colon F_n \to F_n$ given by $\rho_x(g) = xgx^{-1}$ that are not the identity function, as in general $g \neq xgx^{-1}$ for $x, g \in F_n$.

Here's an example of an automorphism of F_3 that is not an inner automorphism. As with linear maps, I will first define the automorphism by saying what it does to the basis $\{a, b, c\}$ of F_3.

$$\phi \colon \quad \begin{aligned} a &\mapsto a^{-1}b \\ b &\mapsto c \\ c &\mapsto ca \end{aligned} \qquad (6.1)$$

This defines ϕ on the inverses of the basis elements in the usual way:

$$\phi(a^{-1}) = \phi(a)^{-1} = (a^{-1}b)^{-1} = b^{-1}a$$

$$\phi(b^{-1}) = \phi(b)^{-1} = (c)^{-1} = c^{-1}$$

$$\phi(c^{-1}) = \phi(c)^{-1} = (ca)^{-1} = a^{-1}c^{-1}$$

Then extend this to all of F_3 by concatenation and free reduction if necessary. For example:

$$\phi(ab^{-1}c) = (a^{-1}b)(c^{-1})(ca) = a^{-1}bc^{-1}ca \rightsquigarrow a^{-1}ba$$

$$\phi(ca) = (ca)(a^{-1}b) = caa^{-1}b \rightsquigarrow cb$$

Then, by construction, ϕ preserves the group multiplication, the concatenation operation. We can verify that ϕ is a bijection by producing an inverse.

$$\phi^{-1} \colon \quad \begin{aligned} a &\mapsto b^{-1}c \\ b &\mapsto b^{-1}ca \\ c &\mapsto b \end{aligned} \qquad (6.2)$$

Here we check that ϕ and ϕ^{-1} are inverse functions by looking at their composition on the basis $\{a, b, c\}$.

$$a \xrightarrow{\phi^{-1}} b^{-1}c \xrightarrow{\phi} (c^{-1})(ca) \rightsquigarrow a$$

$$b \xrightarrow{\phi^{-1}} b^{-1}ca \xrightarrow{\phi} (c^{-1})(ca)(a^{-1}b) \rightsquigarrow b$$

$$c \xrightarrow{\phi^{-1}} b \xrightarrow{\phi} c$$

I said that automorphisms and ordered bases were in one-to-one correspondence for free groups. Hence $\{a^{-1}b, c, ca\}$ is a basis for F_3. Using ϕ and ϕ^{-1}, do you see how to go back and forth between words in $\{a, a^{-1}, b, b^{-1}, c, c^{-1}\}$ and words in $\{a^{-1}b, b^{-1}a, c, c^{-1}, ca, a^{-1}c^{-1}\}$?

Exercise 6. Earlier, I claimed that $\{a, ab\}$ is a basis for F_2. What is the corresponding automorphism and its inverse?

Main question.   This brings us to the main question of this chapter:

How do we understand the dynamics of an automorphism of a free group?

In other words, can we understand how the automorphism

$$\phi^m = \underbrace{\phi \circ \phi \circ \cdots \circ \phi}_{m \text{ copies}}$$

behaves as we increase m? (In general, a dynamical system consists of some object X and a self-map $X \to X$ that we iterate many times.)

Keeping with the vector space analogy, let us try to use some tools from linear algebra. How do we understand the dynamics of a square matrix (equivalently, a linear transformation) acting on $\mathbb{R}^n$? By using eigenvalues and eigenvectors! Recall that $\lambda \in \mathbb{R}$ is an *eigenvalue* of a linear transformation $\psi : \mathbb{R}^n \to \mathbb{R}^n$ if there is a nonzero vector $\mathbf{u} \in \mathbb{R}^n$ (the *eigenvector*) such that $\psi(\mathbf{u}) = \lambda \mathbf{u}$. In other words, $\mathbf{u}$ is a vector that is stretched by the map ψ. If $\mathbf{u}$ is an eigenvector with eigenvalue λ, then $\psi^m(\mathbf{u}) = \lambda^m \mathbf{u}$ for all $m \geq 1$, i.e., the vector $\mathbf{u}$ is repeatedly stretched by powers of ψ. And if we are lucky enough to have a basis for $\mathbb{R}^n$ that consists of eigenvectors, then we know the dynamics of ψ entirely. All we have to do is write an arbitrary vector $\mathbf{v}$ in terms of the basis of eigenvectors and then to find the image $\psi^m(\mathbf{v})$ we need to multiply each coefficient by the mth power of the respective eigenvalue.

Can we find some notion of eigenvalues and eigenvectors for ϕ?

6.3 TRAIN TRACKS

To see the automorphism and the stretching more easily, we use a topological model for the free group F_n: the n–rose R_n. I introduced this model in Office Hour 4, so I'll only briefly recall it here.

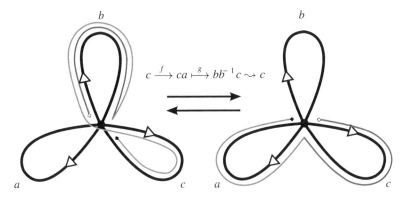

Figure 6.1 Tightening the composition gf: $f(c)$ is the blue-brown path on the right, which is mapped by g to the blue-brown path on the left. This path tightens to c.

A topological model for the free group.   Recall that the n–rose is the graph with a single vertex v and n edges; see Figure 4.2. Also recall that there is a correspondence between tight edge paths in R_n and elements in F_n (Theorem 4.2). A *marking* of the rose is a labeling of the petals of the n–rose with some basis $\{a_1, a_2, \dots, a_n\}$. This specifies the isomorphism:

$$\text{tight edge paths in } R_n \leftrightarrow \text{reduced words in } \{a_1^{\pm 1}, a_2^{\pm 1}, \dots, a_n^{\pm 1}\}$$

Given a word, we can create an edge path that traverses the edges specified by the word; given a path, we can read off the labels of the edges traversed by the path. Because of this, we often use the $a_i^{\pm 1}$ to denote both the edges in R_n and the elements of the basis. This is slightly confusing notation, but hopefully the usage of $a_i^{\pm 1}$ as edges or elements will be clear from the context in which it appears.

We can now interpret our automorphism $\phi\colon F_3 \to F_3$ from (6.1) and its inverse from (6.2) as a pair of maps f, g on the 3–rose marked with the basis $\{a, b, c\}$.

$$f\colon \begin{aligned} a &\mapsto a^{-1}b \\ b &\mapsto c \\ c &\mapsto ca \end{aligned} \qquad g\colon \begin{aligned} a &\mapsto b^{-1}c \\ b &\mapsto b^{-1}ca \\ c &\mapsto b \end{aligned} \qquad (6.3)$$

We read this as "f maps the edge a over the edge path $a^{-1}b$," and so on for the other images.

We might hope that these maps are inverses (as functions on the 3–rose) and give us a symmetry of the 3–rose, but this is really too much to expect. But we do still get a nice feature, namely, that the maps are inverses once we tighten. We can see how $g \circ f(c)$ is a loop that tightens to the loop c in Figure 6.1. The loop $f(c)$ is the blue-brown path on the right, which is mapped by g to the blue-brown path on the left. This path tightens to c. The same holds for the loops $g \circ f(a)$ and $g \circ f(b)$.

The same is true for any automorphism $\phi \colon F_n \to F_n$: there is a pair of maps $f, g \colon \mathbb{R}_n \to \mathbb{R}_n$ that are inverses once we tighten. In sophisticated math lingo, the maps are *homotopy inverses*.

Exercise 7. Find an example of an automorphism ϕ of F_3 that has order 6. It might help to use the topological model introduced above. Can you find one of order 5?

Stretching, round one: Perron–Frobenius.   Great, but how does any of this help in understanding the dynamics of ϕ acting on F_3? Well, for starters, we can place a metric on the 3-rose (i.e., assign lengths to the edges) such that map f in (6.3) stretches every edge by the same factor λ; this is a sort of eigenvalue for the function. Denote the length of the edge a by α, the length of the edge b by β, and the length of the edge c by γ. We find this special metric and the factor λ by solving the following linear system:

$$\begin{cases} \alpha + \beta & = \lambda\alpha \\ \gamma & = \lambda\beta \\ \alpha \quad + \gamma & = \lambda\gamma \end{cases} \tag{6.4}$$

Put differently, we are trying to find an eigenvalue and eigenvector for the matrix

$$\begin{bmatrix} 1 & 1 & 0 \\ 0 & 0 & 1 \\ 1 & 0 & 1 \end{bmatrix} \tag{6.5}$$

such that the coordinates of the eigenvector are positive numbers (remember, these coordinates α, β, and γ are all supposed to be lengths of edges). This matrix is called the *transition matrix* for the map f.

How did this linear system arise? Look back at the definition of f in (6.3). If we want f to stretch the edge a by some number λ, then we need the length of $f(a) = a^{-1}b$ to have length $\lambda\alpha$. The length of the edge path $a^{-1}b$ is $\alpha + \beta$; thus we get the equation $\alpha + \beta = \lambda\alpha$. Similarly for the other two equations.

The matrix in (6.5) is a particularly nice type of matrix: all of the entries are nonnegative integers and the matrix is *irreducible*. Irreducibility of an $n \times n$ nonnegative matrix $\mathbf{M}$ has a few equivalent definitions. One is that for each $1 \leq i$, $j \leq n$, there is an m such that the ijth entry of $\mathbf{M}^m$ is nonzero. The other definition uses a directed graph which has n vertices and a directed edge from the ith vertex to the jth vertex whenever the ijth entry of $\mathbf{M}$ is nonzero. Then $\mathbf{M}$ is irreducible if for every $1 \leq i, j \leq n$ there is a directed path from the ith vertex to the jth vertex. Most linear algebra books discuss the connection between these two views; for instance, see the text by Meyer [203].

For example, the directed graph for the matrix in (6.5) consists of edges $1 \to 2$, $2 \to 3$, and $3 \to 1$ and loops based at 1 and 3. It is clear that we can always follow the directed edges to get between any two of the vertices. See below. Hence the matrix in (6.5) is irreducible.

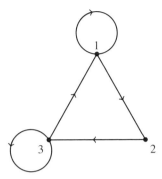

These types of matrices show up all over the place in mathematics and modern culture. For instance, they form the backbone of ranking systems such as the (now defunct) BCS in College Football and the Google search engine [191]. If you watch carefully, in the movie *Good Will Hunting*, it is exactly this theory that Matt Damon expounds leading to mathematical glory. We need the following important theorem about such matrices, which tells us there is always a nice stretching metric when the map has an irreducible transition matrix.

THEOREM 6.1. *Let* **M** *be a nonnegative integer irreducible matrix. Then* **M** *has a (unique up to scalar multiplication) eigenvector with positive coordinates. The associated eigenvalue is a real number greater than or equal to 1. Moreover, this eigenvalue is the largest real root of the characteristic polynomial of* **M**.

This theorem is often refered to as the Perron–Frobenius theorem. Recall the *characteristic polynomial* of a matrix **M** is $p(x) = \det(\mathbf{M} - x\mathbf{I})$. The roots of this polynomial are the eigenvalues of **M**. We call the eigenvalue guaranteed by the theorem the *Perron–Frobenius eigenvalue*. Consult [203] for a proof of the Perron–Frobenius theorem.

In our example, we can find the nice metric by just assigning a length to one of the edges. Starting with $\beta = 1$, the second equation tells us $\gamma = \lambda\beta = \lambda$ and the third equation tells us $\alpha = \lambda\gamma - \gamma = \lambda^2 - \lambda$. Then the first equation of (6.4) gives us:

$$(\lambda^2 - \lambda) + 1 = \lambda(\lambda^2 - \lambda).$$

Rewriting, we have that λ is the largest real root of $x^3 - 2x^2 + x - 1 = 0$; this gives $\lambda \approx 1.75488$. With this metric every edge is stretched by exactly λ. See Figure 6.2.

Cool, almost. Notice what happens when we apply f twice to the loop c. With the first iteration we get

$$c \xmapsto{\ f\ } ca$$
$$\lambda \rightsquigarrow \lambda + (\lambda^2 - \lambda) = \lambda^2,$$

as we expected. But with the second iteration we need to tighten the image of ca:

$$ca \xmapsto{\ f\ } caAb \qquad \text{tightens to} \qquad \rightarrow cb$$
$$\lambda^2 \xmapsto{\ \lambda\cdot\ } \lambda + (\lambda^2 - \lambda) + (\lambda^2 - \lambda) + 1 = \lambda^3 > 1 + \lambda$$

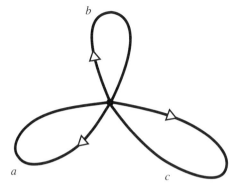

Figure 6.2 The marked 3–rose with edge lengths $\lambda^2 - \lambda$ (a), 1 (b) and λ (c). With these lengths, f streches each edge by λ.

Therefore f^2 does not stretch the loop c by λ^2 as we hoped.

Returning to the underlying analogy with vector spaces, quite often the standard basis vectors for $\mathbb{R}^n$ are not eigenvectors for a particular linear isomorphism. Normally we have to make a change of basis before the dynamics of the linear isomorphism are apparent. (Think diagonalization or more generally Jordan normal form.) The same is true here. We need to make a change of basis before the dynamics of ϕ are apparent.

Stretching, round two: train tracks.   There is an algorithm that performs the required change of basis. The algorithm is due to Mladen Bestvina and Michael Handel [31]. I will give the output here for our example and explain why it does what we want. In the last section, I will say a bit about the steps in the algorithm.

We will denote the new basis by $\{x, y, z\}$. The relation with our previous basis $\{a, b, c\}$ is:

$$
\begin{aligned}
x &= ca & a &= z^{-1}x \\
y &= b & b &= y \\
z &= c & c &= z
\end{aligned}
\tag{6.6}
$$

With this new basis $\{x, y, z\}$, the automorphism ϕ from (6.1) is represented by:

$$
\begin{aligned}
x = ca &\xmapsto{\phi} cb = zy \\
y = b &\xmapsto{\phi} c = z \\
z = c &\xmapsto{\phi} ca = x
\end{aligned}
$$

Now let f be the corresponding map in the 3–rose marked by $\{x, y, z\}$. Our linear system akin to (6.4) for finding the natural metric on the marked 3–rose is:

$$
\begin{cases}
\beta + \gamma = \lambda\alpha \\
\quad\quad \gamma = \lambda\beta \\
\alpha \quad\quad = \lambda\gamma
\end{cases}
$$

where, as before, α is the length of the edge labeled x, β is the length of the edge labeled y, and γ is the length of the edge labeled z. We find a solution as before (starting with $\beta = 1$):

$$\alpha = \lambda^2, \quad \beta = 1, \quad \gamma = \lambda,$$

where λ is the largest real root of $x^3 - x - 1 = 0$; this gives $\lambda \approx 1.32472$.

We claim that for any $m \geq 1$ and edge e with length ℓ, that after tightening the loop $f^m(e)$, its length is $\lambda^m \ell$. As in round one, the metric is precisely chosen so that the length of $f^m(e)$ before tightening is $\lambda^m \ell$. Hence our claim is actually that we never need to tighten $f^m(e)$. Let's check this for some small m and for the edge $z (= c)$ which gave us problems before. Again, we want to show that we never need to tighten the image of $f^m(z)$.

$$z \overset{f}{\longmapsto} x \overset{f}{\longmapsto} zy \overset{f}{\longmapsto} xz \overset{f}{\longmapsto} zyx \overset{f}{\longmapsto} xzzy \overset{f}{\longmapsto} zyxxz \overset{f}{\longmapsto} \cdots$$

OK, good evidence; why does $f^m(e)$ not require tightening in general? (There is a simple reason why for this example, which we explore in the projects; but for now we present a more general argument.) The reason has to do with the images zy, z, x of the basis elements x, y, z. The only way to make a word using zy, z, x that needs tightening is to combine the pair $(zy)^{-1}$ and z or the inverse of this pair, z^{-1} and zy. If we step back by f, we see that these bad combinations arise from a path that uses the turn $\{x, y\}$. A *turn* is the little "V"–shaped object formed by two edges at the vertex.

Let's be more precise. The map f induces a map on the set of turns, which we denote by Df, that takes a turn $\{e_1, e_2\}$ to the turn $\{f_1(e_1), f_1(e_2)\}$, where $f_1(e)$ denotes the first edge crossed by $f(e)$. For example

$$\{x, y\} \overset{Df}{\longmapsto} \{z, z\}$$

$$\{x, z\} \overset{Df}{\longmapsto} \{z, x\}$$

$$\{x^{-1}, y\} \overset{Df}{\longmapsto} \{y^{-1}, z\}$$

$$\{x^{-1}, z^{-1}\} \overset{Df}{\longmapsto} \{y^{-1}, x^{-1}\},$$

and so on. We see here why $\{x, y\}$ was a bad turn; its image under Df is a degenerate turn $\{z, z\}$, which does not look like a true V, but rather a |.

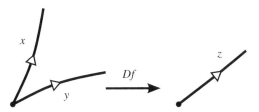

This means the image of a path that crosses the turn $\{x, y\}$ would contain $z^{-1}z$ and therefore would need to be tightened. As this turn $\{x, y\}$ creates problems, we call it *illegal*.

If we check the images of all of the nondegenerate turns, only $\{x, y\}$ is mapped to a degenerate turn. (There are $12 = \binom{6}{2} - 3$ nondegenerate turns; can you think of an effective way to check this without enumerating all possibilities?) Moreover, we see that the illegal turn $\{x, y\}$ never appears as the image of a turn. We call all of the other nondegenerate turns *legal*. Now as Df maps legal turns to legal turns, which are nondegenerate, their images do not need tightening. Iterating this logic we see that if a turn $\{e_1, e_2\}$ is legal, then $Df^m(\{e_1, e_2\})$ is nondegenerate for all $m \geq 1$. Notice, the turn $\{z^{-1}, y\}$ crossed by $f(x) = zy$ is legal.

Let's pause for a moment and see what we have:

(**TT1**) A collection $\mathcal{L}$ of nondegenerate turns (which we call legal), such that $Df(\mathcal{L}) \subseteq \mathcal{L}$, i.e., legal turns go to legal turns.

(**TT2**) Every turn crossed by the image of an edge under f is in $\mathcal{L}$. (There is only one turn in our example that is in the image of an edge, namely $\{z^{-1}, y\}$, and this turn is legal.)

Any map that satisfies the above is called a *train track map* and the collection of turns is called a *train track structure*. Explanation of the name will come in a second. First I will show that f^m stretches every edge by λ^m. In fact, I will show more. Call a path *legal* if it only crosses legal turns. Condition (**TT2**) can be restated as saying that the images of the edges are legal paths.

Claim. Suppose $(f, \mathcal{L})$ is a train track map whose transition matrix has Perron–Frobenius eigenvalue λ. If p is a legal path with length ℓ, then for all $m \geq 1$, the length of $f^m(p)$ is $\lambda^m \ell$.

Proof. Since λ is the Perron–Frobenius eigenvalue for the transition matrix, as we have seen above, this implies that for every edge e, the map f stretches the edge by λ. Thus, the map f stretches every path by exactly λ too. However, we might need to tighten the image in order to find the actual length.

Since the image of a legal turn is nondegenerate, the image of a legal path does not need tightening and hence the length of a legal path is multiplied by exactly λ. Also, since the image of a legal turn is a legal turn (**TT1**) and only legal turns appear in the image of an edge (**TT2**), the image of a legal path is a legal path! Hence, as p is a legal path, the path $f^m(p)$ is always legal and never needs tightening. Thus $f^m(p)$ has length $\lambda^m \ell$. $\square$

The 3–rose is drawn to reflect the structure of legal and illegal turns in Figure 6.3. It is called a *train track representative* of ϕ. The vertex is enlarged and the illegal turn $\{x, y\}$ is drawn sharply as this is a turn a train cannot make. Any path that a train can take consists only of legal turns and thus f^m stretches the path by λ^m.

The algorithm.   Now I'll say a bit about where the basis $\{x, y, z\}$ came from. We need to use folding, which I introduced in Office Hour 4. If we look back at the original map f in (6.3) defined on the 3–rose marked with $\{a, b, c\}$, the edges c^{-1} and a both map to an edge path that starts with a^{-1}, and it is exactly this word ca that causes problems in f^2. Using the language of turns, $f(c)$ contains a turn, $\{c^{-1}, a\}$, that is mapped by Df to a degenerate turn.

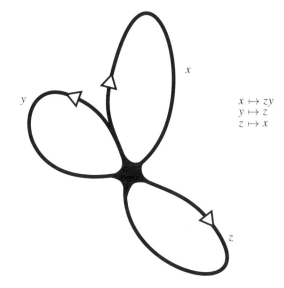

$$x \mapsto zy$$
$$y \mapsto z$$
$$z \mapsto x$$

Figure 6.3 A train track structure for ϕ.

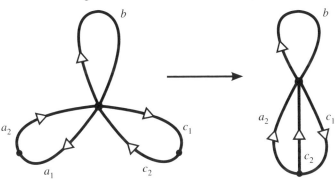

Figure 6.4 Folding a_1 and c_2^{-1} to c_2^{-1}.

Let's subdivide these edges in the problem-causing turn and then fold. Specifically, add vertices to these edges and assign labels: $a = a_1 a_2$ and $c = c_1 c_2$. See the graph on the left in Figure 6.4. With this, f becomes:

$$
\begin{aligned}
a_1 &\mapsto a_2^{-1} \\
a_2 &\mapsto a_1^{-1} b \\
f: \quad b &\mapsto c_1 c_2 \\
c_1 &\mapsto c_1 c_2 a_1 \\
c_2 &\mapsto a_2
\end{aligned}
$$

Since a_1 and c_2^{-1} have the same image by f, we will fold them together.

Call the resulting graph Γ the denote the result of folding a_1 and c_2^{-1} by c_2^{-1}, as depicted in Figure 6.4. In Office Hour 4 we would have considered the induced map $\Gamma \to R_3$. Here though, we want to fold the target graph as well. Doing so, we have that f induces a map on the folded graph $f' \colon \Gamma \to \Gamma$:

$$f' \colon \begin{aligned} a_2 &\mapsto c_2 b \\ b &\mapsto c_1 c_2 \\ c_1 &\mapsto c_1 c_2 c_2^{-1} \rightsquigarrow c_1 \\ c_2 &\mapsto a_2 \end{aligned}$$

To get f' all that we did was replace occurrences of a_1 by c_2^{-1}. Notice, we did the obvious tightening in the image of c_1 to get $f'(c_1) = c_1$.

Now, since $f'(c_1) = c_1$ we are free to collapse this edge (another operation introduced in Office Hour 4). The resulting graph $\Gamma_{\downarrow c_1}$ is isomorphic to R_3 and the three edges are labeled a_2, b, and c_2. The map f' now induces a map on the collapsed graph $f'' \colon \Gamma_{\downarrow c_1} \to \Gamma_{\downarrow c_1}$:

$$f'' \colon \begin{aligned} a_2 &\mapsto c_2 b \\ b &\mapsto c_2 \\ c_2 &\mapsto a_2 \end{aligned}$$

All that we did here was delete occurrences of c_1. This is the map we want; it satisfies the train track properties. You can see this is the same one we considered earlier by setting $x = a_2$, $y = b$, and $z = c_2$.

How do we find the new marking, i.e., the new basis? For this we need to unfold. Doing so we find that $c_1 c_2 a_1 a_2$ in the original R_3 is folded to $c_1 c_2 c_2^{-1} a_2 \rightsquigarrow c_1 a_2$ which is then collapsed to a_2. Thus the petal a_2 corresponds to ca. Likewise, b maps to b and $c_1 c_2$ is collapsed to c_2. This gives the basis we used before.

In general, this process will have to be iterated several times. At each step we find an illegal turn that is crossed by the image of some edge. Subdivide the edges in this illegal turn and then fold, collapsing edges if possible. The final resulting graph may not be a rose, so in this sense our use of the word basis was a little misleading. The output of the algorithm will be a graph with a self-map that satisfies the train track properties TT1 and TT2. There is another subtlety that we omitted from the conversation so far. This is the notion of reducibility for an automorphism. All of the statments we made only apply to automorphisms that are irreducible, that is, not reducible. These notions are explored in the projects.

FURTHER READING

Have we answered the main question we posed earlier? Do we now understand the dynamics of a given free group automorphism? Well, yes and no. Given an element $x \in F_3$, we still cannot easily determine which element $\phi^m(x)$ actually is, but we can estimate how long it is. Answer: for $x \in F_n$, $\phi^m(x)$ has length approximately λ^m. Do you see why, even for elements where the associated loop is not legal? You will need a lemma called Cooper's bounded cancellation lemma; see the article by Bestvina and Handel [31, Remark 1.8] for more details.

More is true, too. As in the case of a Perron–Frobenius matrix where there is one dominant eigenvector that attracts all of the vectors in $(\mathbb{R}_{>0})^n$, the elements

$\phi^m(x)$ are attracted to *legal lines*: certain bi-infinite paths that only contain legal turns. Just as for a typical Perron–Frobenius matrix, where the coordinates of the dominant eigenvector are not all integers (hence the vector is not in $\mathbb{Z}^n$), these legal lines are not actually elements of F_n but are stretched like eigenvectors.

To read more about train tracks, Bestvina and Handel's original article [31] is a good place to start, especially its Section 1. The book by Bogopolski [43] is another source. Bestvina [27] has a new proof of the existence of train tracks that parallels a proof about train tracks for surface homeomorphisms (introduced in Office Hour 17).

PROJECTS

Project 1. This office hour is concerned with a pretty advanced topic, the dynamics of automorphisms. Take some time to think about more basic questions. What do automorphisms of a free group look like anyway? Is the group $\mathrm{Aut}(F_n)$ finitely generated? The answer to the preceding question is yes and, moreover, a generating set can be found using the folding technique from Office Hour 4. Read Chapter I.4 in Lyndon and Schupp's book [194] or Wade's artice [258] to get a better idea of the basics of free group automorphisms.

Project 2. We usually describe a homomorphism of F_n by listing the images of some basis. The homomorphism is an automorphism if and only if the images form a basis for F_n. Therefore we are led to ask for a way to determine if n elements in F_n are a basis. One way to answer this question is to use folding as described in Office Hour 4. Provide the details. Another way is to use what is known as Whitehead's algorithm. Look at this algorithm and how it applies to this question.

Project 3. There is a homomorphism $F_n \to \mathbb{Z}^n$ akin to the homomorphism $F_2 \to \mathbb{Z}^2$ mentioned in Office Hour 1. Explicitly, the ith generator a_i of F_n is mapped to the vector $(0, \ldots, 0, 1, 0, \ldots, 0)$ of $\mathbb{Z}^n$ where the 1 is in the ith position. In other words, this homomorphism records the exponent sums of the generators. This map is called the *abelianization* and plays a role in later office hours (cf. Office Hour 16). Show that this map induces a surjective homomorphism $\mathrm{Aut}(F_n) \to \mathrm{Aut}(\mathbb{Z}^n) = \mathrm{GL}(n, \mathbb{Z})$. (In fact, once you read the definition of abelianization in Office Hour 16, you should prove that there is a homomorphism $\mathrm{Aut}(G) \to \mathrm{Aut}(G^{ab})$ for any group G.) Show that inner automorphisms are in the kernel of this homomorphism. For $n = 2$, this is the entire kernel. See the article by Cohen, Metzler, and Zimmermann for a proof [87]. This is no longer true when $n \geq 3$. Find an automorphism of F_3 in the kernel of the map $\mathrm{Aut}(F_3) \to \mathrm{GL}(3, \mathbb{Z})$ that is not an inner automorphism.

Project 4. Several people have programmed the Bestvina–Handel train track algorithm. Two implementations were written by Peter Brinkmann [59] and Thierry Coulbois [91], respectively. Find, familiarize yourself with, and use some of the existing software. For example, you might start with the example in this office hour and the examples in projects; or after you've looked at Project 1, you can start writing down your own automorphisms and testing these.

Project 5. One notion we glossed over in the above discussion was the idea of *irreducibility* for an automorphism. Train tracks as described above only exist for irreducible automorphisms. An automorphism is *reducible* if there is a basis for F_n such that the automorphism restricts to an automorphism on the group generated by some proper subset of the basis; else it is *irreducible*. Show that if an automorphism ϕ is reducible, then there is a marked rose and map f representing ϕ whose transition matrix is reducible, i.e., not irreducible as we defined it above.

Project 6. Consider the following automorphisms for $n \geq 1$:

$$\phi_n : \begin{array}{l} a \longmapsto ac^n \\ b \longmapsto c \\ c \longmapsto ab \end{array}$$

Compute ϕ_n^{-1}. Find the transition matrix and show that it is irreducible. Argue as we did above that the obvious representative on the rose (without changing bases) is a train track representative. In other words, identify all of the illegal turns and show that all of the turns in the images ac^n, c, and ab are legal. Find the length of a, b, c, and λ as we did above. Draw the train track graph as in the above figure with the appropriate edge lengths. What happens to the length of b?

Project 7. If a linear isomorphism $T \colon \mathbb{R}^n \to \mathbb{R}^n$ is represented by a diagonal matrix, then the elements of the standard basis are eigenvectors. There are certain automorphisms where the obvious map on the rose is a train track map. *Positive automorphisms* fit into this category. A positive automorphism is an automorphism $\phi \colon F_n \to F_n$ such that the images $\phi(a_i)$ only consist of a_j's with positive exponents. Or in other words, thinking of the basis with lowercase letters and using uppercase letters for inverses, the images $\phi(a_i)$ only consist of lower case letters. For example, the automorphism

$$\begin{array}{l} a \longmapsto b \\ b \longmapsto c \\ c \longmapsto ab \end{array}$$

is a positive automorphism, as are the automorphisms in Exercise 6. The automorphism ϕ we looked at in depth earlier is not positive (with respect to the basis $\{a, b, c\}$) as $\phi(a) = a^{-1}b$. However, with respect to the basis $\{ca, b, c\}$, ϕ is positive. Prove the above assertion about train track representatives for positive automorphisms. That is, prove that when iterating the obvious representative on the rose, we never need to tighten the images of an edge. As in Exercise 6, you will want to show that all the turns in the images $\phi(a_i)$ are legal.

Project 8. Find an automorphism $\phi \in \mathrm{Aut}(F_3)$ that is not positive but such that the obvious representative on the 3–rose is a train track representative.

Project 9. Consider the automorphism $\phi \colon F_3 \to F_3$:

$$\begin{array}{l} a \longmapsto b \\ b \longmapsto c \\ c \longmapsto c^{-1}b^{-1}a^{-1} \end{array}$$

Show that the obvious representative of this automorphism is not a train track map (*Hint: What happens to* $\phi^2(c)$*?*). Run the algorithm to find a train track representative for this example. What's the corresponding λ? Try the same for the automorphism $\phi : F_3 \to F_3$:

$$a \longmapsto b$$
$$b \longmapsto caba^{-1}b^{-1}$$
$$c \longmapsto a$$

Try more examples.

Project 10. This projects asks you to develop further the analogy between the stretching we saw in this office hour and the stretching you see with eigenvalues of matrices. Suppose you already have a train track representative for an automorphism of F_n. The graph for your representative has a covering that is a tree. It is called the universal cover. The train track structure on the graph gives you a train track structure on the tree (just make the tree look locally like the graph). Then the legal lines you found in the graph correspond to legal lines in the tree. But in the tree the legal lines really look like lines—they never come back to where they started. The train track map you had on the graph gives you a train track map on the tree. Can you see that most paths get pulled toward the legal lines, just as most vectors of a Perron–Frobenius matrix get pulled toward the Perron–Frobenius eigenvector? Try some examples. For instance, look at what happens to the edge z under iteration by f^2 for the example considered in this office hour.

PART 3

Large Scale Geometry

Office Hour Seven

Quasi-isometries

Dan Margalit and Anne Thomas

We have already seen that there is a natural way to think of a group as a metric space, namely, by replacing the group with its Cayley graph. By restricting attention to the vertices of the Cayley graph, we get a metric on the original group. While this construction is natural, it is not canonical—there are typically lots of generating sets for a group, and it is usually impossible to find a single best generating set (even for $\mathbb{Z}^2$). So which metric on a group is the right one?

The main point of this office hour is that if we come up with the right notion of equivalence on metrics, then all of the word metrics we can put on a group are essentially the same. But we want to be careful—we want our notion of equivalence to be fine enough to distinguish between groups that have essential differences.

We will accomplish our goal with the notion of quasi-isometry. A quasi-isometry is sometimes called a "coarse isometry," since it is like an isometry, but allows for bounded multiplicative and additive errors. Although at first the definition of quasi-isometry might seem much too loose to be useful, we'll see in this and later office hours that a whole suite of important algebraic and geometric properties is preserved by quasi-isometries.

We can apply the notion of quasi-isometry to the algebraic structure of groups. A sample result, which shows that quasi-isometries can have powerful algebraic consequences, is the following theorem of Gromov [156].

THEOREM 7.1. *If G is quasi-isometric to $\mathbb{Z}^n$ then G has a finite index subgroup isomorphic to $\mathbb{Z}^n$.*

In other words, from coarse geometric information about a group G we obtain fairly fine algebraic information. This is quite amazing! We'll conclude the office hour with a proof of this theorem in the (vastly easier) case $n = 1$, first proven (in a stronger form) by Hopf [169] in 1943.

Along the way to this theorem, we will prove the Milnor–Schwarz lemma, sometimes called the fundamental lemma of geometric group theory. This result tells us that if a group G acts nicely enough on a space X then G is quasi-isometric to X. So if the group G has some important property preserved by quasi-isometries, so does the space X, and vice versa. We can take X to be the Cayley graph of another group, and so the Milnor–Schwarz lemma neatly ties together the notions of quasi-isometry of metric spaces, quasi-isometries of groups, and actions of groups on metric spaces. This is really fundamental to geometric group theory: we use groups to understand spaces, and spaces to understand groups.

7.1 EXAMPLE: THE INTEGERS

We'll first discuss Cayley graphs and review path metrics and word metrics for the simplest infinite group, the integers $\mathbb{Z}$. Recall that for a group G with finite generating set S we denote by $\Gamma(G, S)$ the corresponding Cayley graph and by $d_S(\cdot, \cdot)$ the corresponding word metric on G.

Two different Cayley graphs for $\mathbb{Z}$. Recall from Office Hour 2 that the Cayley graph $\Gamma(\mathbb{Z}, \{1\})$ is a line (with directed edges). For each integer $n \in \mathbb{Z}$, there is an edge connecting n to $n + 1$:

Now $\mathbb{Z}$ can also be generated by the set $\{2, 3\}$. What does the Cayley graph for $\mathbb{Z}$ with respect to the generating set $\{2, 3\}$ look like? The vertices of $\Gamma(\mathbb{Z}, \{2, 3\})$ will be the same as those of $\Gamma(\mathbb{Z}, \{1\})$, since the vertices of *any* Cayley graph for $\mathbb{Z}$ are just the elements of $\mathbb{Z}$. But instead of an edge connecting n to $n + 1$, there will be an edge connecting n to $n + 2$ and an edge connecting n to $n + 3$, for every $n \in \mathbb{Z}$. The sketch below shows part of the Cayley graph $\Gamma(\mathbb{Z}, \{2, 3\})$. This Cayley graph is not a line, but it's not that different from a line either—especially if you squint!

Two different word metrics on $\mathbb{Z}$. In the path metric on a graph, each edge has length 1. So although it looks like some of the edges in the sketch of $\Gamma(\mathbb{Z}, \{2, 3\})$ above are longer than others, really they are all the same length, 1. (This would be hard to draw nicely.)

Let's consider the distance between two vertices, say -2 and 5, in these two different Cayley graphs equipped with the path metric.

- In $\Gamma(\mathbb{Z}, \{1\})$, the distance from -2 to 5 is 7. There is a unique shortest path, and this path has exactly seven edges, so this path has length 7.

- In $\Gamma(\mathbb{Z}, \{2, 3\})$, the distance from -2 to 5 is only 3. To see this, we first find a path containing three edges from -2 to 5, and then show that there is no shorter path. There are actually a few different choices of path of length 3 from the vertex -2 to the vertex 5, which you should find. Now do the first exercise below, from which it follows that there is no shorter path from -2 to 5. Hence -2 and 5 are at distance 3 from each other in $\Gamma(\mathbb{Z}, \{2, 3\})$.

To summarize, we have shown that

- $d_{\{1\}}(-2, 5) = 7$ and
- $d_{\{2,3\}}(-2, 5) = 3$.

In particular, the distances between -2 and 5 in the two word metrics are not the same!

Now let $f \colon \mathbb{Z} \to \mathbb{Z}$ be the identity map. Then we can think of f as being a (not very interesting) function from the vertex set of $\Gamma(\mathbb{Z}, \{1\})$ to the vertex set of $\Gamma(\mathbb{Z}, \{2, 3\})$. The function f does *not* preserve distances, since $d_{\{1\}}(-2, 5) \neq d_{\{2,3\}}(-2, 5)$, but we will prove in the next section that f can't distort distances by too much either.

Exercise 1. Show that the set of vertices at distances less than 3 from -2 in $\Gamma(\mathbb{Z}, \{2, 3\})$ does *not* include the vertex 5.

Exercise 2. Given any $g \in \mathbb{Z}$, how many vertices of $\Gamma(\mathbb{Z}, \{1\})$ and of $\Gamma(\mathbb{Z}, \{2, 3\})$ are at distance 1 from g? Distance 2, 3? Distance n, for all $n \geq 1$? How many vertices of these graphs are at distance at most n from g? Why is it enough to consider $g = 0$?

Exercise 3. Given $g, h \in \mathbb{Z}$ with $g \neq h$, how many shortest paths are there from g to h in $\Gamma(\mathbb{Z}, \{1\})$ and in $\Gamma(\mathbb{Z}, \{2, 3\})$? Why is it enough to consider $g = 0$?

For a broader discussion relating to the last two exercises, go to Office Hour 12.

7.2 BI-LIPSCHITZ EQUIVALENCE OF WORD METRICS

The next thing we would like to do is to develop a language to describe exactly how different (or similar) are our two word metrics on $\mathbb{Z}$. We already showed that the identity map on $\mathbb{Z}$ does not give an isometry between the two metrics. So what we will do is weaken the notion of isometry to find the right relationship between the metrics, namely, bi-Lipschitz equivalence. The first theorem here will be that two different word metrics on a finitely generated group are bi-Lipschitz equivalent. After that, we will define quasi-isometric equivalence, a further weakening of bi-Lipschitz equivalence, and show that quasi-isometric equivalence is the right notion for capturing many large-scale properties of groups.

Isometric embeddings and isometries.   First we need to expand on the definition of isometry given in Office Hour 2. Let X be a metric space with distance function (or metric) $d_X \colon X \times X \to \mathbb{R}$. For short, we will say: let (X, d_X) be a metric space.

Then if (X, d_X) and (Y, d_Y) are metric spaces, a function $f\colon X \to Y$ is called an *isometric embedding* if f preserves distances, that is, for all $x_1, x_2 \in X$,

$$d_Y(f(x_1), f(x_2)) = d_X(x_1, x_2).$$

An isometric embedding f is called an *isometry* if it is also surjective. When we want to specify the metric carefully, we may write f as a function $f\colon (X, d_X) \to (Y, d_Y)$ instead of just $f\colon X \to Y$. Here are two examples:

1. The inclusion map $\iota\colon \mathbb{Z}^2 \to \mathbb{R}^2$ is an isometric embedding when $\mathbb{Z}^2$ and $\mathbb{R}^2$ both have the taxicab metric.
2. The identity map $f\colon \mathbb{Z} \to \mathbb{Z}$ from the set of vertices of $\Gamma(\mathbb{Z}, \{1\})$ to the set of vertices of $\Gamma(\mathbb{Z}, \{2, 3\})$ is *not* an isometry—even though it is surjective—because it does not preserve distances.

Bi-Lipschitz embeddings and bi-Lipschitz equivalences. We're now going to weaken the notion of isometry by allowing distances to be stretched and compressed by bounded amounts.

Let (X, d_X) and (Y, d_Y) be metric spaces. A function $f\colon X \to Y$ is called a *bi-Lipschitz embedding* if there is some constant $K \geq 1$ such that for all $x_1, x_2 \in X$,

$$\frac{1}{K} d_X(x_1, x_2) \leq d_Y(f(x_1), f(x_2)) \leq K d_X(x_1, x_2).$$

Notice that if $K = 1$ this is the same thing as an isometric embedding. If $K > 1$, the function f is allowed to dilate distances, but only by a factor of K (this is the right-hand inequality), and to shrink distances, but only by a factor of $1/K$ (this is the left-hand inequality). A bi-Lipschitz embedding f is a *bi-Lipschitz equivalence* if it is also surjective. As the name suggests, bi-Lipschitz equivalence is an equivalence relation on the collection of all metric spaces.

Exercise 4. Show that bi-Lipschitz equivalence is an equivalence relation on the collection of all metric spaces.

Bi-Lipschitz equivalence for word metrics. We can now state and prove our first main result.

THEOREM 7.2. *Let G be a finitely generated group and let S and S' be two finite generating sets for G. Then the identity map $f\colon G \to G$ is a bi-Lipschitz equivalence from the metric space (G, d_S) to the metric space $(G, d_{S'})$.*

That is, changing the generating set can stretch or compress distances between elements of G, but only by a bounded amount.

Proof. We begin by reducing the theorem to a special case. Since G acts by isometries on itself with respect to any word metric d_S, we have that $d_S(g, h)$ is equal to $d_S(1, g^{-1}h)$, the word length of $g^{-1}h$ with respect to S. Hence to prove the theorem, it's enough to show that there is a constant $K \geq 1$ so that for all $g \in G$,

$$\frac{1}{K} d_S(1, g) \leq d_{S'}(1, g) \leq K d_S(1, g). \tag{7.1}$$

That is, we will be comparing the word lengths of each element $g \in G$ with respect to the two different generating sets S and S'.

Since S is finite, we can define a constant $M \geq 1$ by

$$M = \max\{d_{S'}(1, s) \mid s \in S \cup S^{-1}\}.$$

Now consider any $g \in G$. Suppose g has word length n with respect to S, so we can write $g = s_1 s_2 \cdots s_n$, where each s_i is in $S \cup S^{-1}$. Using the triangle inequality, we get:

$$d_{S'}(1, g) = d_{S'}(1, s_1 s_2 \cdots s_n)$$

$$\leq d_{S'}(1, s_1) + d_{S'}(s_1, s_1 s_2 \cdots s_n)$$

$$\leq d_{S'}(1, s_1) + d_{S'}(s_1, s_1 s_2) + d_{S'}(s_1 s_2, s_1 s_2 \cdots s_n)$$

$$\vdots$$

$$\leq d_{S'}(1, s_1) + d_{S'}(s_1, s_1 s_2) + \cdots + d_{S'}(s_1 s_2 \cdots s_{n-1}, s_1 s_2 \cdots s_n).$$

For each $1 \leq k < n$, the distance between $s_1 s_2 \cdots s_k$ and $s_1 s_2 \cdots s_k s_{k+1}$ is equal to the distance between 1 and $(s_1 s_2 \cdots s_k)^{-1} s_1 s_2 \cdots s_k s_{k+1} = s_{k+1}$. So we obtain

$$d_{S'}(1, g) \leq d_{S'}(1, s_1) + d_{S'}(1, s_2) + \cdots + d_{S'}(1, s_n)$$

$$\leq M + M + \cdots + M$$

$$= Mn.$$

But n is the word length of g with respect to S, that is, $d_S(1, g) = n$. Putting it all together, we have shown that for all $g \in G$,

$$d_{S'}(1, g) \leq M d_S(1, g).$$

Now swap the roles of S and S' to complete the proof.   □

Returning to the example $G = \mathbb{Z}$ with $S = \{1\}$ and $S' = \{2, 3\}$, Theorem 7.2 shows that the identity map $f \colon \mathbb{Z} \to \mathbb{Z}$ is a bi-Lipschitz equivalence from $\mathbb{Z}$ with the word metric with respect to $\{1\}$, to $\mathbb{Z}$ with the word metric with respect to $\{2, 3\}$. This is the same thing as saying that the identity map $f \colon \mathbb{Z} \to \mathbb{Z}$ is a bi-Lipschitz equivalence from the vertex set of the Cayley graph $\Gamma(\mathbb{Z}, \{1\})$ to the vertex set of the Cayley graph $\Gamma(\mathbb{Z}, \{2, 3\})$, where these Cayley graphs are equipped with the path metric. What about the edges? We can't hope to find a terribly nice map from the edge set of $\Gamma(\mathbb{Z}, \{1\})$ to the edge set of $\Gamma(\mathbb{Z}, \{2, 3\})$. So to construct a map on the set of edges, we will allow the map to be bad on small scales. It then turns out that the Cayley graphs $\Gamma(\mathbb{Z}, \{1\})$ and $\Gamma(\mathbb{Z}, \{2, 3\})$ are equivalent in a looser sense than bi-Lipschitz equivalence: they are *quasi-isometric*.

7.3 QUASI-ISOMETRIC EQUIVALENCE OF CAYLEY GRAPHS

We're now ready to define quasi-isometries, and to use Theorem 7.2 to prove the second main result of this office hour, Theorem 7.3, which says that any two Cayley graphs for a finitely generated group are quasi-isometric.

Quasi-isometric embeddings and quasi-isometric equivalences.   Let (X, d_X) and (Y, d_Y) be metric spaces. A function $f \colon X \to Y$ is called a *quasi-isometric embedding* if there are some constants $K \geq 1$ and $C \geq 0$ so that

$$\frac{1}{K} d_X(x_1, x_2) - C \leq d_Y(f(x_1), f(x_2)) \leq K d_X(x_1, x_2) + C \qquad (7.2)$$

for all $x_1, x_2 \in X$. Notice that if $C = 0$ this is the same thing as a bi-Lipschitz embedding, and if $C = 0$ and $K = 1$ this is the same thing as an isometric embedding. You can think of $C > 0$ as being an error term that allows f to be bad on small scales.

Although we use the word "embedding," a quasi-isometric embedding does not have to be injective. A quasi-isometric embedding also doesn't have to be continuous (so it certainly doesn't have to be differentiable). The examples in Exercise 5 below show that some pretty strange functions can be quasi-isometric embeddings.

A quasi-isometric embedding $f \colon X \to Y$ is called a *quasi-isometric equivalence*, or just a *quasi-isometry*, if there is a constant $D > 0$ so that for every point $y \in Y$, there is an $x \in X$ so that $d_Y(f(x), y) \leq D$. That is, every point in Y is within distance D of some point in the image of f. Notice that if $D = 0$, this is exactly surjectivity. For $D > 0$, you can think of f as being "coarsely surjective": f hits a set of points whose D–neighborhoods cover Y.

Finally, we say that metric spaces (X, d_X) and (Y, d_Y) are *quasi-isometric* if there is a quasi-isometry $f \colon X \to Y$. If X and Y are quasi-isometric, then there's a sense in which anything small-scale doesn't really matter, so that X and Y "look the same from far away." As the name suggests, quasi-isometric equivalence is an equivalence relation on the collection of all metric spaces.

For a basic example, let (X, d_X) be $\mathbb{Z}^2$ with the taxicab metric and (Y, d_Y) be $\mathbb{R}^2$ with the taxicab metric. Then X and Y are quasi-isometric. The notion of quasi-isometry captures the idea that when you look at the lattice of points $\mathbb{Z}^2$ from a long way away, these points blur together to form the plane $\mathbb{R}^2$. We leave the details as an exercise.

Exercise 5. Show that each of the following maps $f \colon \mathbb{R} \to \mathbb{R}$ is a quasi-isometric embedding and a quasi-isometry.

1. Let $f(x) = n$ where n is the unique integer such that $n \leq x < n + 1$. This is called the floor function.
2. Let $f(x) = (\sin x + 2)x + 3$ if x is an integer, and $f(x) = x + \cos x$ if x is not an integer.
3. For $x = \pm n$, where n is a positive integer, let $f(x) = x + d_n$ where d_n is the nth digit of the decimal expansion of π. For all other x, let $f(x) = x$.

What are the smallest possible constants K, C, and D for these maps?

Exercise 6. Equip $\mathbb{Z}^2$ and $\mathbb{R}^2$ with the taxicab metric. Let $\iota\colon \mathbb{Z}^2 \to \mathbb{R}^2$ be the inclusion. Define $\kappa\colon \mathbb{R}^2 \to \mathbb{Z}^2$ by sending each point in the plane to a closest point in the lattice $\mathbb{Z}^2$; if there's a choice, just pick one direction to go in. Define $F\colon \mathbb{R}^2 \to \mathbb{Z}^2$ by $F(x, y) = (f(x), f(y))$, where $f\colon \mathbb{R} \to \mathbb{Z}$ is the floor function. Show that the maps ι, κ, and F are all quasi-isometries. What are the smallest possible constants K, C, and D for these maps?

Exercise 7. Show that $\mathbb{R}^2$ with the taxicab metric is quasi-isometric to $\mathbb{R}^2$ with the Euclidean metric.

Exercise 8. Show that the composition of two quasi-isometries is a quasi-isometry.

Exercise 9. Let (X, d_X) and (Y, d_Y) be metric spaces. A *quasi-inverse* of a function $f\colon X \to Y$ is a function $g\colon Y \to X$ so that, for some constant $k \geq 0$, for all $x \in X$ we have $d_X(g(f(x)), x) \leq k$, and for all $y \in Y$ we have $d_Y(f(g(y)), y) \leq k$.

1. Prove that a quasi-isometric embedding $f\colon X \to Y$ is a quasi-isometry if and only if f has a quasi-inverse. Where did you use the axiom of choice?
2. Prove that if $f\colon X \to Y$ is a quasi-isometry, and $g\colon Y \to X$ is a quasi-inverse of f, then g is a quasi-isometry.

Exercise 10. Show that quasi-isometric equivalence is an equivalence relation on the collection of all metric spaces.

Exercise 11. Let (X, d) be a metric space.

1. Show that the collection of all isometries $f\colon X \to X$ forms a group under composition of maps. This group is called the *isometry group* of X, and is denoted by $\mathrm{Isom}(X)$.
2. Define a relation $\sim$ on the collection of all maps $f, g\colon X \to X$ by:
$$f \sim g \quad \text{if} \quad \sup_{x \in X} d(f(x), g(x)) \text{ is finite.}$$
   (a) Check that $\sim$ is an equivalence relation, and write $[f]$ for the equivalence class of f under $\sim$.
   (b) Show that the collection of all equivalence classes $[f]$ of quasi-isometries $f\colon X \to X$ forms a group under composition of maps. This group is called the *quasi-isometry group* of X, and is denoted by $\mathcal{QI}(X)$.

Exercise 12. Show that if $f\colon X \to Y$ is a quasi-isometry then $\mathcal{QI}(X) \cong \mathcal{QI}(Y)$. Hence give an example of spaces X and Y that are not isometric but have isomorphic quasi-isometry groups.

Exercise 13. Suppose X and Y are metric spaces so that $\mathcal{QI}(X) \cong \mathcal{QI}(Y)$. Are X and Y quasi-isometric?

Quasi-isometric equivalence of Cayley graphs.   The second main result for this office hour, Theorem 7.3 below, says that the geometric realizations of any two Cayley graphs for a group G are quasi-isometric to each other. This means that

when we want to see whether a Cayley graph for G—or its geometric realization—
is quasi-isometric to some metric space X, it won't matter which Cayley graph we
consider (or if we just consider G with its word metric), and so we can just talk
about the group G being quasi-isometric to the space X. (See Office Hour 2 for a
discussion of geometric realizations of graphs.)

THEOREM 7.3. *Let G be a finitely generated group and let S and S' be two finite
generating sets for G. Then the geometric realization of the Cayley graph $\Gamma(G, S)$
is quasi-isometric to the geometric realization of the Cayley graph $\Gamma(G, S')$.*

Proof. First, there is a quasi-isometry from the geometric realization of any graph
to its set of vertices (with the path metric) obtained by sending every point on an
edge to a nearest vertex (there are sometimes two choices). We already showed in
Theorem 7.2 that the identity map $G \to G$ induces a bi-Lipschitz map—hence a
quasi-isometry—between the vertex sets of $\Gamma(G, S)$ and $\Gamma(G, S')$. Therefore using
Exercises 8 and 9 we obtain a quasi-isometry $\Gamma(G, S) \to \Gamma(G, S')$ as a composition of three quasi-isometries.   □

Exercise 14. Show that the isometry groups of $\mathbb{Z}$ and $\mathbb{Z}^2$ are countable, but that
$\mathcal{QI}(\mathbb{Z})$ and $\mathcal{QI}(\mathbb{Z}^2)$ are uncountable. How could you show that $\mathcal{QI}(\mathbb{R})$ and
$\mathcal{QI}(\mathbb{R}^2)$ contain elements that are not represented by isometries of $\mathbb{R}$ and $\mathbb{R}^2$,
respectively?

Let's now apply Theorem 7.3 to some of the finitely generated groups seen
so far.

1. Take $G = \mathbb{Z}$. There is a quasi-isometry from the Cayley graph $\Gamma(\mathbb{Z}, \{1\})$
   to the Cayley graph $\Gamma(\mathbb{Z}, \{2, 3\})$, and in fact, *any* Cayley graph for $\mathbb{Z}$ (with
   respect to a finite generating set) will be quasi-isometric to $\Gamma(\mathbb{Z}, \{1\})$. That
   is, *any Cayley graph for $\mathbb{Z}$ is quasi-isometric to $\mathbb{R}$.*
2. Now consider $G = \mathbb{Z}^2$. The Cayley graph for $\mathbb{Z}^2$ with respect to the generating set $S = \{(1, 0), (0, 1)\}$ is a grid (with directed edges). As you can
   easily see from this grid, the word metric d_S is the same thing as the taxicab
   metric on $\mathbb{Z}^2$. We discussed in the example in Section 7.3 how $\mathbb{Z}^2$ is quasi-
   isometric to the plane $\mathbb{R}^2$ when both spaces have the taxicab metric (or the
   Euclidean metric). So Theorem 7.3 means that *any* Cayley graph for $\mathbb{Z}^2$ is
   quasi-isometric to $\mathbb{R}^2$. That is, *any Cayley graph for $\mathbb{Z}^2$ is quasi-isometric to
   the plane.*
3. Let G be F_2, the free group of rank 2. The Cayley graph for F_2 with generating set $\{a, b\}$ was constructed in Office Hour 2, and is a tree. So *any Cayley
   graph for F_2 is quasi-isometric to a tree.*

An important consequence of Theorem 7.3 is:

COROLLARY 7.4. *Let G be a finitely generated group. For any two finite generating sets S and S' for G, the group G with the word metric d_S is quasi-isometric to
the group G with the word metric $d_{S'}$.*

7.4 QUASI-ISOMETRIES BETWEEN GROUPS AND SPACES

We now come to a fundamental result in geometric group theory which is called the Milnor–Schwarz lemma (sometimes big results get referred to as lemmas). This result is what allows us to go back and forth between groups and spaces, carrying useful properties from one to the other.

Geodesics and properness. We first need to define some important properties of metric spaces. We'll also give key examples of metric spaces that have these properties.

Let (X, d) be a metric space. A *geodesic segment* is an isometric embedding $\gamma \colon [a, b] \to X$, where $a, b \in \mathbb{R}$ and $a \leq b$, and a *geodesic* is an isometric embedding $\gamma \colon \mathbb{R} \to X$. We often identify a geodesic segment or a geodesic line with its image in X. Geodesic lines are isometrically embedded copies of $\mathbb{R}$ in X.

We then define (X, d) to be a *geodesic metric space* if for any $x_1, x_2 \in X$, there is a geodesic segment $\gamma \colon [a, b] \to X$ such that $\gamma(a) = x_1$ and $\gamma(b) = x_2$. That is, any two points in X are connected by a geodesic segment that realizes their distance. Many important spaces are geodesic metric spaces. Here are some examples:

1. The plane $\mathbb{R}^2$ with the Euclidean metric is a geodesic metric space, in which geodesics are straight lines, and so there is a unique geodesic segment realizing the distance between any two points.
2. Similarly, any convex set in $\mathbb{R}^2$, such as the open unit disk, is a geodesic metric space.
3. If $\mathbb{R}^2$ has the taxicab metric, then vertical and horizontal lines are geodesics, and unless points (x_1, y_1) and (x_2, y_2) are on the same vertical or horizontal line, there are infinitely many geodesic segments that realize their distance.
4. The hyperbolic plane is a geodesic metric space, with a unique geodesic segment realizing the distance between any two points. In the Poincaré disk model, the geodesic lines are arcs of circles that are orthogonal to the boundary. (See Office Hour 5.)
5. An example of a metric space that is *not* a geodesic metric space is the plane minus the origin, that is, $X = \mathbb{R}^2 - \{(0, 0)\}$, with the metric given by restricting the Euclidean metric. This is not a geodesic metric space since, for example, there is no geodesic in X from $(1, 0)$ to $(-1, 0)$.
6. For G a finitely generated group, the Cayley graph $\Gamma(G, S)$, equipped with the path metric, is a geodesic metric space. The set of geodesic segments between vertices $g, h \in G$ is the set of shortest paths in $\Gamma(G, S)$ from g to h. In $\Gamma(\mathbb{Z}, \{2, 3\})$, for instance, there are three geodesic segments from -2 to 5. On the other hand, in the Cayley graph for F_2 with generating set $\{a, b\}$, there is a unique shortest path between any two points, since this graph is a tree.

A metric space (X, d) is *proper* if for all $x \in X$ and all $r > 0$, the closed ball centered at x of radius r, denoted by $B(x, r)$, is a compact subset of X. Most of the above examples of geodesic metric spaces are proper except for $\mathbb{R}^2 - \{(0, 0)\}$ and the open unit disk in the plane.

Another interesting type of geodesic metric space that is not proper is a Hilbert space (this is a type of infinite-dimensional vector space). In such a space even a small closed ball is not compact because it is infinite dimensional!

Exercise 15. Let (X, d) be a metric space. A *quasi-geodesic segment* in X is a quasi-isometric embedding $c \colon [a, b] \to X$, where $a, b \in \mathbb{R}$ and $a \leq b$, and a *quasi-geodesic* is a quasi-isometric embedding $c \colon \mathbb{R} \to X$. That is, a quasi-geodesic is a quasi-isometrically embedded copy of $\mathbb{R}$ in X.

1. Show that in $\mathbb{R}^2$ with the Euclidean metric, for every $b > 0$ the map $c \colon [0, b] \to \mathbb{R}^2$ given by $c(t) = (t \cos(\ln(1 + t)), t \sin(\ln(1 + t)))$ is a quasi-geodesic segment.
2. Sketch the image of c.
3. Deduce that for every $R > 0$, there is a $b > 0$ so that the image of c is not contained in the R–neighborhood of the geodesic segment connecting $p = c(0)$ and $q = c(b)$. That is, in Euclidean space, quasi-geodesics do not have to stay close to geodesics.

Group actions. Let (X, d) be a proper geodesic metric space. The Milnor–Schwarz lemma concerns groups G that act "nicely" on such spaces X. We're now going to explain what we mean by "nicely."

We first require that the action of G on X preserve distances, that is, preserve the metric structure of X. As in Office Hour 2, a distance-preserving action is called an *action by isometries*. Formally, this means that there is a homomorphism from G to the group of isometries of X (see Exercise 11).

Now there are some pretty trivial group actions by isometries, for example if every element of G fixes every point of X. This action tells us absolutely nothing about G or X! Less obviously, we also want to avoid cases where some point of X is fixed by every element of G, since that also won't tell us much about the structure of G. In order to pull apart the group G, we will consider actions by isometries that are *properly discontinuous*, meaning that for each compact set $K \subseteq X$, the set

$$\{g \in G \mid gK \cap K \neq \emptyset\}$$

is finite. For instance, for each $x \in X$ the set $\{x\}$ is compact, so the G–stabilizers of points in X are finite. Also, since X is proper, for each $x \in X$ and each $r > 0$, there are only finitely many $g \in G$ so that $B(gx, r)$ has some overlap with $B(x, r)$. Equivalently, no matter how big you take r, all but finitely many elements of G will move $B(x, r)$ off of itself.

We also want the action of G to fill out the space X, rather than have the G–orbits restricted to just a small slice (which would tell us nothing about what was going on far away). We say that the action of G on X is *cocompact* if for any base point $x_0 \in X$ there is an $R > 0$, so that for any $x \in X$, there is some $g \in G$ so that $B(gx_0, R)$ contains x. In other words, the closed R–neighborhood of the G–orbit of x_0 is all of X

We now define the action of G on X to be *geometric* if it is a properly discontinuous and cocompact action by isometries. These are the kind of nice actions to which the Milnor–Schwarz lemma will apply.

The following are all examples of geometric actions:

1. Let $G = \mathbb{Z}^2$ and $X = \mathbb{R}^2$ with the Euclidean metric, and consider the action of G on X by translations.
2. Let G be the fundamental group of a compact orientable surface Σ of genus $g \geq 2$ and let X be the hyperbolic plane, which is the universal cover of Σ. Then G acts on X by isometric deck transformations.
3. Let G be the group of isometries of the Euclidean plane X generated by reflections in the sides of an equilateral triangle. Note that some points (and lines) in X have nontrivial stabilizers in G.
4. Let G be $\mathrm{SL}(2, \mathbb{Z})$ and consider the action of G on the Farey tree (see Office Hour 3).
5. Let G be a finitely generated group with finite generating set S and let $X = \Gamma(G, S)$. Then G acts on X via left multiplication (see Office Hour 2).

Exercise 16. Verify the claim that each of the above examples is a geometric action.

We are now ready to state and prove one of our main results, the Milnor–Schwarz lemma, sometimes written as the Milnor–Švarc lemma or the Švarc–Milnor lemma, and also known as the fundamental lemma of geometric group theory. Naturally, this fact has its roots in the work of Milnor and Schwarz [204, 253].

THEOREM 7.5 (Milnor–Schwarz lemma). *Let G be a group and let (X, d) be a proper geodesic metric space. Suppose that G acts geometrically on X. Then G is finitely generated and G is quasi-isometric to X.*

Note that we start with no information about the group G except for its geometric action on X. We will first prove that G is finitely generated by a set S that we construct (this part of the proof will remind you of the proof of Theorem 3.1 in Office Hour 3), and then show that G with the word metric d_S is quasi-isometric to (X, d). By Theorem 7.3 and its corollary, saying that G and X are quasi-isometric is the same as saying that any Cayley graph for G is quasi-isometric to X.

Proof of Theorem 7.5. Fix a base point $x_0 \in X$. Choose $R > 0$ so that the G–translates of the closed ball $B(x_0, R)$ cover X. Let S be the set of nontrivial $g \in G$ so that $B(gx_0, R) \cap B(x_0, R)$ is nonempty. Then S is a finite set and $S = S^{-1}$. We claim that S generates G.

To simplify notation, write B for $B(x_0, R)$. Then B is compact, so if we define a constant

$$c = \inf\{d(B, gB) \mid g \in G, g \neq 1, g \notin S\}$$
$$= \inf\{d(x, gy) \mid x, y \in B, g \in G, g \neq 1, g \notin S\},$$

then $c > 0$. To see this, note that if we take any gB disjoint from B and $d(B, gB) = D$, then by proper discontinuity there are finitely many translates of B that have distance at most D from B. It follows that c is a minimum of finitely many positive numbers and hence is positive.

To show that S generates G, let $g \in G$ with $g \notin S \cup \{1\}$. Then $d(x_0, gx_0) \geq 2R + c$ and in particular $d(x_0, gx_0) \geq R + c$. So there is a $k \geq 2$ such that

$$R + (k - 1)c \leq d(x_0, gx_0) < R + kc.$$

In other words, k is the smallest positive integer so that $d(x_0, gx_0) < R + kc$ (or it is the largest positive integer so that $R + (k-1)c \le d(x_0, gx_0)$!). Hence we can choose points $x_1, x_2, \ldots, x_{k+1} = gx_0$ on a geodesic segment from x_0 to gx_0 such that $d(x_0, x_1) \le R$ and $d(x_i, x_{i+1}) < c$ for $1 \le i \le k$. Then there are elements $1 = g_0, g_1, \ldots, g_k = g$ in G such that $x_{i+1} \in g_i B$ for $0 \le i \le k$ (because we want $g_0 = 1$, we need the special assumption $d(x_0, x_1) \le R$ here).

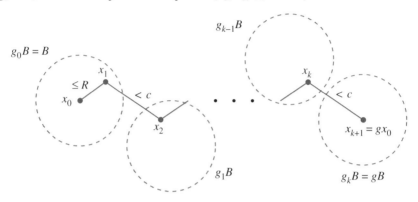

Let $s_i = g_{i-1}^{-1} g_i$ for $1 \le i \le k$. Then for $1 \le i \le k$ we have

$$d(B, s_i B) = d(B, (g_{i-1}^{-1} g_i)B) = d(g_{i-1}B, g_i B) \le d(x_i, x_{i+1}) < c,$$

so $s_i \in S$. But

$$s_1 s_2 \cdots s_k = (g_0^{-1} g_1)(g_1^{-1} g_2) \cdots (g_{k-1}^{-1} g_k) = g_k = g,$$

and thus the set S generates G (this is where we use the assumption $g_0 = 1$).

Now define a map $G \to X$ by $g \mapsto gx_0$. We will show that this map is a quasi-isometry when G is equipped with the word metric d_S. First of all, every point of X is within distance R of some point gx_0 and so our map $G \to X$ is coarsely surjective. It remains to show that there are constants $K \ge 1$ and $C \ge 0$ such that for all $g, h \in G$ we have

$$\frac{1}{K} d_S(g, h) - C \le d(gx_0, hx_0) \le K d_S(g, h) + C.$$

Actually, since

$$d(gx_0, hx_0) = d(x_0, (g^{-1}h)x_0)$$

and

$$d_S(g, h) = d_S(1, g^{-1}h),$$

it is enough to show that there are constants $K \ge 1$ and $C \ge 0$ such that for all $g \in G$ we have

$$\frac{1}{K} d_S(1, g) - C \le d(x_0, gx_0) \le K d_S(1, g) + C.$$

Let

$$L = \max\{d(x_0, sx_0) \mid s \in S\}$$

and define K and C by

$$K = \max\{1/c, L, 2R\} \text{ and } C = \max\{1/K, c\}.$$

We will now check that these choices of K and C satisfy the required inequalities for all $g \in G$. We pulled these choices of K and C out of thin air; the reasons for these choices will only become apparent as we complete the proof (think about the ϵ–δ proofs you did in calculus).

Let $g \in G$. If $g = 1$ then $d(x_0, gx_0) = d_S(1, g) = 0$ and so we indeed have

$$\frac{1}{K}d_S(1, g) - C \leq d(x_0, gx_0) \leq Kd_S(1, g) + C.$$

Now suppose that g is equal to some $s \in S$. By definition of S, we have

$$0 \leq d(x_0, sx_0) \leq 2R.$$

Also, $d_S(1, s) = 1$, and by definition of K and C we have $K \geq 2R$ and $C \geq 1/K$, and so the desired inequalities follow:

$$\frac{1}{K}d_S(1, g) - C = 1/K - C \leq 0 \leq d(x_0, sx_0) \leq 2R \leq K \leq Kd_S(1, g) + C.$$

Finally suppose that $g \notin S \cup \{1\}$. From the proof that S generates G we know that if k is the largest integer with $R + (k-1)c \leq d(x_0, gx_0)$ then $d_S(1, g) \leq k$. We can then combine the two inequalities in the last sentence to get

$$R + (d_S(1, g) - 1)c \leq d(x_0, gx_0).$$

Subtracting R from both sides and using the fact that $R \geq 0$ we then get

$$cd_S(1, g) - c \leq d(x_0, gx_0) - R \leq d(x_0, gx_0).$$

Also, remembering that L is the maximum of $d(x_0, sx_0)$ where s runs over S, the argument from the proof of Theorem 7.2 gives us

$$d(x_0, gx_0) \leq L\, d_S(1, g).$$

Combining the last two inequalities, we have

$$cd_S(1, g) - c \leq d(x_0, gx_0) \leq Ld_S(1, g).$$

As above, since $K \geq L$ and $K \geq 1/c$ (or $1/K \leq c$) and $C \geq c$, the desired inequalities hold. This completes the proof. □

Exercise 17. Identify where each of the assumptions on G and X gets used in the proof of the Milnor–Schwarz lemma (remember that quite a lot is packed into the definition of a geometric action).

Exercise 18. A topological space X is *connected* if it cannot be written as a disjoint union of two open subsets U and V. Use similar ideas to the proof of the Milnor–Schwarz lemma to establish the following theorem:

Let G be a group that acts properly discontinuously by isometries on a connected proper geodesic metric space X (note: the action is not assumed to be cocompact). For each $x_0 \in X$ and each $R > 0$, define

$$D(x_0, R) = \{x \in X \mid \exists g \in G \text{ such that } d(x, gx_0) \leq R\}.$$

Then G is finitely generated if and only if, for all $x_0 \in X$, there is some $R > 0$ so that $D(x_0, R)$ is connected.

Examples. We can apply the Milnor–Schwarz lemma to all of our above examples of geometric actions. So we immediately obtain that $\mathbb{Z}^2$ is quasi-isometric to $\mathbb{R}^2$, that the fundamental group of a closed orientable surface of genus two is quasi-isometric to the hyperbolic plane, that a free group with k generators is quasi-isometric to a $2k$–regular tree, and that $SL(2, \mathbb{Z})$ is quasi-isometric to a tree.

Application: Groups that differ by finite groups. We can also gain a lot of information from the Milnor–Schwarz lemma by allowing groups to act on Cayley graphs. For instance:

1. If G is a finitely generated group and H is a subgroup of finite index, then H is quasi-isometric to G.
2. Similarly, if N is a finite normal subgroup of a finitely generated group G, then G/N is quasi-isometric to G.

Exercise 19. Prove the last two facts.

We can think of the last two statements as saying that finite groups do not affect the quasi-isometry type of a group. Let's say that groups G and G' *differ by finite groups* if G is isomorphic to a finite-index subgroup of G' or G is isomorphic to a quotient of G' by a finite group (this relation is not an equivalence relation, but it generates one). In this language, the last two statements say that:

If two groups differ by finite groups then they are quasi-isometric.

As special cases, we have:

1. Every finite group is quasi-isometric to the trivial group.
2. The infinite dihedral group D_∞ is quasi-isometric to $\mathbb{Z}$.
3. For all $k \geq 2$, the free group F_k is quasi-isometric to F_2.

To see the second statement, recall that D_∞ has a subgroup of index 2 that is isomorphic to $\mathbb{Z}$, or apply the Milnor–Schwarz lemma directly to the action of D_∞ on $\mathbb{R}$.

The third statement is quite surprising, and requires some explanation. What we will show is that F_k is actually isomorphic to a finite-index subgroup of F_2. This is very different from what happens for free abelian groups $\mathbb{Z}^k$—there is no way that $\mathbb{Z}^3$ will fit inside $\mathbb{Z}^2$ (see the exercises below).

To get some idea of how these embeddings work for free groups, let $k = 3$ and consider the set $\{a, bab^{-1}, b^2\}$ in F_2. It requires some careful combinatorial arguments, but these elements generate a copy of F_3 that has index 2 in F_2. There is also a lovely short proof using algebraic topology that every F_k embeds with finite index in F_2.

Since F_k is quasi-isometric to F_2, it follows that all free groups of rank ≥ 2 are quasi-isometric to each other. Alternatively, the same conclusion can be reached by considering Cayley graphs. For suitable generating sets, the Cayley graph of F_k is the $2k$–regular tree, and the Cayley graph for F_2 is the 4–regular tree, so we can show that F_k is quasi-isometric to F_2 by constructing an explicit quasi-isometry between these trees (see the exercises).

Exercise 20. Show that if $m > n$ and $\phi \colon \mathbb{Z}^m \to \mathbb{Z}^n$ is a group homomorphism, then ϕ cannot be injective.

Exercise 21. For $m, n \geq 3$, let T_m be the m–regular tree and T_n the n–regular tree, equipped with the path metric. Construct an explicit quasi-isometry from T_m to T_n. For which groups other than free groups could you use this to establish quasi-isometric equivalence, without going via free groups?

Exercise 22. Let (X, d_X) and (Y, d_Y) be metric spaces and let $y_0 \in Y$. Show that the map $f \colon X \to Y$ given by $f(x) = y_0$ for all $x \in X$ is a quasi-isometry if and only if both X and Y have bounded diameter. Deduce that a group is quasi-isometric to the trivial group if and only if it is finite (we already showed one direction).

Exercise 23. Let $\triangle$ be an equilateral triangle in the Euclidean plane and let W be the group generated by the reflections r, s, and t in the sides of $\triangle$. Prove that W has a finite index subgroup isomorphic to $\mathbb{Z}^2$ (hence W is quasi-isometric to $\mathbb{Z}^2$, which you already knew from applying the Milnor–Schwarz lemma to the action of W on the plane). Give generators of this subgroup in terms of r, s, and t.

7.5 QUASI-ISOMETRIC RIGIDITY

We just saw that if two groups differ by finite groups then they are quasi-isometric. When does the converse hold? We say that:

> *A group G is quasi-isometrically rigid if every group quasi-isometric to G differs from G by finite groups.*

It follows from what we already said that the trivial group is quasi-isometrically rigid: a group is quasi-isometric to the trivial group if and only if it is finite and all finite groups differ by finite groups.

At the beginning of this office hour we stated Theorem 7.1, which states that free abelian groups are quasi-isometrically rigid:

> *If G is quasi-isometric to $\mathbb{Z}^n$, then G has a finite-index subgroup that is isomorphic to $\mathbb{Z}^n$.*

Given how loose the definition of quasi-isometry is, it is amazing that there are any quasi-isometrically rigid infinite groups (and more amazing if we can actually prove it!). Before we end, we will prove this theorem in the (vastly easier) case of $n = 1$. The proof of quasi-isometric rigidity for $\mathbb{Z}^n$ with $n > 1$ is much deeper. All known proofs use Gromov's polynomial growth theorem (see Office Hour 12).

There are other groups besides $\mathbb{Z}^n$ that are quasi-isometrically rigid, for instance, finitely generated free groups of rank at least 2 (see Project 2 below), the fundamental groups of closed surfaces (cf. Office Hour 9), and mapping class groups (cf. Office Hour 17) [20].

We can also have *classes* of groups that are quasi-isometrically rigid—meaning that any group quasi-isometric to a group in that class differs from some group in the class by finite groups—even when the particular groups in the class are not. Examples of this are the class of abelian groups [216], the class of nilpotent groups (this again follows from Gromov's polynomial growth theorem), the class of finitely presented groups [143], the class of finitely presented groups with solvable word problem, the class of hyperbolic groups (see Office Hour 9), and the class of fundamental groups of closed, hyperbolic 3–manifolds [251, 257].

There are also interesting classes of groups that are known to not be quasi-isometrically rigid, such as the class of simple groups, the class of solvable groups [120], and the class of residually finite groups. In other words, in each case, there is an example of a group that is quasi-isometric to a group in the class but does not differ by finite groups from any group in the class.

Finally, there are a number of classes of groups that are conjectured to be quasi-isometrically rigid, for instance, the class of polycyclic groups, the class of right-angled Artin groups, and the class of outer automorphism groups of free groups.

Quasi-isometric classification problem. More general than the question of quasi-isometric rigidity for a particular group is Gromov's very difficult problem of quasi-isometric classification:

Classify all finitely generated groups up to quasi-isometry.

For this, suppose that $\mathcal{P}$ is some property so that if metric spaces X and Y are quasi-isometric then X has property $\mathcal{P}$ if and only if Y has property $\mathcal{P}$. Such a property $\mathcal{P}$ is called a *quasi-isometry invariant*.

If we have two finitely generated groups, and there is a quasi-isometry invariant $\mathcal{P}$ that holds for one but not the other, then we can conclude that the two groups are *not* quasi-isometric. That is, quasi-isometry invariants allow us to distinguish between quasi-isometry classes of groups (but generally invariants do not allow us to show that two groups are quasi-isometric).

Examples of quasi-isometry invariants that you will see in later office hours include hyperbolicity, ends, and growth (see also Project 8).

Groups quasi-isometric to the integers. As promised, we will end by proving that $\mathbb{Z}$ is quasi-isometrically rigid:

THEOREM 7.6. *If a finitely generated group G is quasi-isometric to $\mathbb{Z}$, then G has a finite-index subgroup isomorphic to $\mathbb{Z}$.*

Hopf proved the more general theorem that any group with two ends (see Office Hour 10) is quasi-isometric to $\mathbb{Z}$. Since the number of ends is a quasi-isometry invariant, our theorem is implied by his.

Let's get right to the proof. Suppose there is a quasi-isometry $f\colon G \to \mathbb{Z}$ with constants K and C:

$$\frac{1}{K} d(x, y) - C \le |f(x) - f(y)| \le K\, d(x, y) + C$$

for all $x, y \in G$. Here $d(\cdot, \cdot)$ is the word metric on G and $|\cdot|$ is the usual metric on $\mathbb{Z}$. For simplicity, we may assume that $f(1) = 0$; indeed, this can be achieved by postcomposing the original f with a translation of $\mathbb{Z}$ (or, if we don't care about changing the constant C, we can just redefine $f(1)$ to be 0).

In order to prove that G has a finite index subgroup isomorphic to $\mathbb{Z}$, we proceed in two steps:

1. G has an infinite order element a, and
2. the group $\langle a \rangle \cong \mathbb{Z}$ has finite index in G.

The first step is much harder, and that is where we start.

Step 1. Our strategy is to find a set $A \subseteq G$ and an element $a \in G$ with $a(A)$ a proper subset of A. It immediately follows that A has infinite order, since $a^n(A) \ne A$ for all $n > 0$. We should think of this as a version of the ping-pong lemma for free groups with one generator (refer back to Office Hour 5).

Let $L = K + C$ and let $B = f^{-1}([-L/2, L/2])$. We would like to say that B divides G into more than one piece. To make this precise, we need some terminology. First, by a path in G we mean a path in the Cayley graph for G. We can use this to define an equivalence relation on $G \setminus B$ as follows: we say that two elements of $G \setminus B$ lie in the same component if there is a path from one to the other in $G \setminus B$. The components of $G \setminus B$ are then the equivalence classes under this relation.

The various components of $G \setminus B$ can have either finite or infinite diameter. If any components of $G \setminus B$ have finite diameter, then we add these components to B. After doing this, all components of $G \setminus B$ have infinite diameter (indeed, the set of components of the new $G \setminus B$ forms a subset of the set of components of the old $G \setminus B$).

Define

$$A_+ = f^{-1}(L/2, \infty) \setminus B \qquad \text{and}$$

$$A_- = f^{-1}(-\infty, -L/2) \setminus B.$$

(Both of these sets are nonempty!) In the end, we will take our set A to be equal to A_+. The goal of the next two claims is to prove that B, A_+, and A_- are configured as in the following picture:

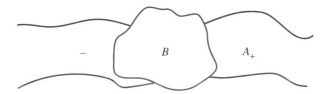

The idea here is that, if you squint, G looks like $\mathbb{R}$, B looks like 0, A_+ looks like $\mathbb{R}_{>0}$, and A_- looks like $\mathbb{R}_{<0}$. We will continue to develop this analogy as we go.

Claim 1. No component of $G \setminus B$ contains points of both A_+ and A_-.

Let $x \in A_+$ and $y \in A_-$. Choose a path $(x = g_1, \ldots, g_n = y)$ in G. Since f is a quasi-isometry with constants K and C, each $|f(g_i) - f(g_{i+1})|$ is at most $K \cdot 1 + C = L$. Since $f(g_0) > L/2$ and $f(g_n) < -L/2$, it follows that $f(g_i) \in [-L/2, L/2]$ for some i. But this means that $g_i \in B$, and the claim follows.

Claim 2. $G \setminus B$ has exactly two components, namely, A_+ and A_-.

By Claim 1 and the fact that $A_+ \cup A_- = G \setminus B$, it suffices to show that A_+ and A_- are each equal to a single component of $G \setminus B$. Suppose to the contrary that A_+ contains two distinct components $A_+^{(1)}$ and $A_+^{(2)}$ of $G \setminus B$. Since the diameter of $A_+^{(1)}$ is infinite (we assumed all the components of $G \setminus B$ had infinite diameter), we can find an infinite path $P_1 = (g_0, g_1, \ldots)$ whose distance from B gets arbitrarily large, that is, $\lim d(g_i, B) \to \infty$. Using the fact that f is a quasi-isometric embedding, it follows that $f(P_1)$ is a path in $\mathbb{Z}$ that tends to infinity, that is, $\lim f(g_i) = \infty$. Since each $|f(g_i) - f(g_{i+1})|$ is at most L, it follows that there is a ray $[n_1, \infty) \subseteq \mathbb{Z}$ contained in the $L/2$–neighborhood of $f(P_1)$.

Similarly, there is a path P_2 in $A_+^{(2)}$ and a ray $[n_2, \infty) \subseteq \mathbb{Z}$ contained in the $L/2$–neighborhood of $f(P_2)$. But then we can choose points x_1 and x_2 in P_1 and P_2 so that $d(x_i, B)$ is as large as we want and so that $|f(x_1) - f(x_2)| \leq L/2$. This is a contradiction, because by the definition of a component of $G \setminus B$, any path from x_1 to x_2 must pass through B (so the distance between x_1 and x_2 is as large as we want compared to $L/2$).

Before proceding to the final claim for Step 1, we choose elements $g, h \in G$ so that $g \in A_+$ with $d(1, g) > 2 \operatorname{diam}(B)$ and $h \in A_-$ with $d(1, h) > 2 \operatorname{diam}(B)$.

Claim 3. For some $a \in \{g, h, gh\}$, the image of A_+ under a is a proper subset of A_+.

By our choice of g it follows that $g(B)$ is contained in A_+ (and is disjoint from B). Notice that $G \setminus g(B)$ has exactly two components, namely $g(A_+)$ and $g(A_-)$, and so we have a picture like this:

If $g(A_+)$ is contained in A_+, then there is nothing to do since then $g(A_+) \subseteq (A_+ \setminus g(B)) \subsetneq A_+$. Similarly, if $h(A_-)$ is contained in A_- there is nothing to do. Therefore, we may assume that $g(A_+)$ is not contained in A_+ and $h(A_-)$ is not contained in A_-. It follows that $g(A_-) \subseteq A_+$ and $h(A_+) \subseteq A_-$.

We now want to show that $gh(A_+) \subsetneq A_+$. The picture you should have in mind is of the infinite dihedral group D_∞. If we look at the above picture and squint, then

G looks like a line, and B looks like the point 0. Then g is like a reflection D_∞ that takes 0 to $d(1, g)$ and h looks like a reflection that takes 0 to $d(1, h)$. Therefore, gh looks like a translation in the positive direction. In particular, $gh(A_+) \subsetneq A_+$. This element gh is precisely the infinite order element a we have been seeking.

Exercise 24. Make the last argument precise.

Step 2. We will now show that the group $\langle a \rangle \cong \mathbb{Z}$ has finite index in G.

Claim 1. $d(1, a^n) \to \infty$ as $n \to \infty$.

Since we chose the generating set to be finite, the degree of every vertex in the Cayley graph for G is finite. It follows that the ball of radius R about the identity is finite for every R. Since a has infinite order (Step 1), it follows that the a^i are all distinct, and so for every R there is some i with a^i outside the ball of radius R. The claim follows.

Claim 2. There is some D so that the D–neighborhood of $\langle a \rangle$ in G is all of G.

It follows from Claim 1 that $d(a^m, a^n)$ tends to infinity as $|n - m|$ tends to infinity (indeed, $d(a^m, a^n) = d(1, a^{n-m})$). From this and the fact that f is a quasi-isometry (more precisely, since f does not shrink distances too much) we obtain that one of the sequences $f(1), f(a), f(a^2), \ldots$ and $f(1), f(a^{-1}), f(a^{-2}), \ldots$ tends to ∞ and the other tends to $-\infty$. Again using the fact that f is a quasi-isometry (the fact that f does not stretch too much), the distance between $f(a^i)$ and $f(a^{i+1})$ is bounded by some number M that does not depend on i, namely, $M = Kd(1, a) + C$. In particular, each element of $\mathbb{Z}$ lies at a distance of at most $M/2$ from a point of $f(\langle a \rangle)$, that is, the $M/2$–neighborhood of $f(\langle a \rangle)$ is all of $\mathbb{Z}$.

From this and the fact that f does not shrink distances too much, we can conclude the claim. To this end, let $g \in G$. By the previous paragraph, there is an i so that $|f(g) - f(a^i)| \leq M/2$. But $Kd(g, a^i) - C \leq |f(g) - f(a^i)|$ and so

$$d(g, a^i) \leq (|f(g) - f(a^i)| + C)/K \leq (M/2 + C)/K.$$

Since this number does not depend on g, the claim is proven (take $D = (M/2 + C)/K$).

Claim 3. The quotient $G/\langle a \rangle$ is finite.

In the quotient of the Cayley graph Γ of G by $\langle a \rangle$, all of the elements of $\langle a \rangle$ are identified as a single vertex. An important feature of the way G acts on Γ is that whenever two vertices of Γ get identified, the corresponding edges emanating from the two vertices get identified as well. What this gives us is that the quotient $\Gamma/\langle a \rangle$ is locally finite—the degree of each vertex in the quotient is the same as the degree of any corresponding vertex in Γ.

Since $\langle a \rangle$ is cobounded in G (Claim 2) it follows that $\Gamma/\langle a \rangle$ has finite diameter (paths descend to paths of the same length or shorter). Since the quotient is locally finite (by the previous paragraph), the quotient is finite.

We are now done because the elements of $G/\langle a \rangle$ are exactly the cosets of $\langle a \rangle$ in G. In other words, by Claim 3 the subgroup $\langle a \rangle$—which is isomorphic to $\mathbb{Z}$ by Step 1—has finite index in G.

FURTHER READING

For more information on quasi-isometries, a good place to start is the article by Ghys and de la Harpe [143], which covers most of the material here and explains some connections to the other office hours as well.

Both Bridson and Haefliger's book [54] and de la Harpe's book [104] contain treatments on quasi-isometries that are quite accessible, rigorous, and full of examples. Cannon's survey article on geometric group theory [70] also spends some time discussing the notion of a quasi-isometry.

There is a large body of literature on quasi-isometric rigidity and classification for various classes of groups. One such article is by Cook, Freden, and McCann, who showed that all Baumslag–Solitar groups $BS(p, q)$ for $1 < p < q$ are quasi-isometric [90]. They give an elementary geometric proof of this result originally due to Whyte.

PROJECTS

Project 1. Prove that a regular 3–valent tree (that is, an infinite tree with three edges emanating from each vertex) is quasi-isometric to a regular 4–valent tree. Show that both are quasi-isometric to any k–regular tree with $k \geq 3$. What about a tree where some of the valences are 3 and some are 4? Or where all of the valences lie in $\{3, 4, \ldots, k\}$?

Project 2. Prove the quasi-isometric rigidity of non-abelian free groups using Theorem 3.1 from Office Hour 3.

Project 3. Draw a spiral in $\mathbb{R}^2$. Under what circumstances is it a quasi-isometric embedding $\mathbb{R} \to \mathbb{R}^2$? What about spirals in $\mathbb{H}^2$?

Project 4. Prove that $\mathbb{R}$ is not quasi-isometric to $[0, \infty)$. Prove that $\mathbb{R}^2$ is not quasi-isometric to a regular 3–valent tree. If you need a hint, peek ahead at Office Hour 10.

Project 5. Which of the following spaces are quasi-isometric to each other? $\mathbb{R}$, $\mathbb{R}^2$, a 3–valent tree, $\mathbb{H}^2$.

Project 6. How big can $\mathcal{QI}(X)$ be compared to $\mathrm{Isom}(X)$ for an unbounded metric space X? How small can it be in comparison?

Project 7. Consider the group $\mathcal{QI}(\mathbb{R})$.

1. Determine which (if any) of the maps $f: \mathbb{R} \to \mathbb{R}$ defined in the examples from Exercise 5 are in the same equivalence class in $\mathcal{QI}(\mathbb{R})$, and which (if any) of these maps is equivalent to the identity map on $\mathbb{R}$. Can you find a linear map of $\mathbb{R}$ equivalent to any of these maps?
2. Construct uncountably many distinct elements of $\mathcal{QI}(\mathbb{R})$ that can be represented by linear maps, and then uncountably many distinct elements that can be represented by piecewise linear maps but not by linear maps.

3. If $[f]$ is an element of $\mathcal{QI}(\mathbb{R})$, what are the possible orders of $[f]$?
4. Can you find a subgroup of $\mathcal{QI}(\mathbb{R})$ that is isomorphic to F_2?

Project 8. This project is about amenability of groups; see Office Hour 16 for a different take on this topic. Let G be a finitely generated group and let S be a finite generating set for G. For any subset A of vertices of $\Gamma(G, S)$, we define the *S–boundary* of A, denoted by $\partial_S A$, by

$$\partial_S A = \{g \in G \mid g \notin A \text{ and } d_S(g, a) = 1 \text{ for some } a \in A\}.$$

That is, $\partial_S A$ is the set of vertices of $\Gamma(G, S)$ that are adjacent to a vertex of A but not contained in A. The group G is *S–amenable* if for all $\varepsilon > 0$, there is a finite subset A of vertices so that

$$|\partial_S A| < \varepsilon |A|$$

(here the absolute value symbol means cardinality). That is, the S–boundary of A is "small" compared to A.

1. Let $G = \mathbb{Z}$, $S = \{1\}$, and $S' = \{2, 3\}$. Prove that G is S–amenable and S'–amenable.
2. Let G be any finite group. Show that G is S–amenable for any generating set S for G.
3. Let G_1 and G_2 be finitely generated groups with finite generating sets S_1 and S_2, respectively. Let e_1 be the identity element of G_1 and e_2 be the identity element of G_2. Prove that if G_1 is S_1–amenable and G_2 is S_2–amenable, then the direct product $G = G_1 \times G_2$ is S–amenable, where

$$S = \{(s_1, e_2) \mid s_1 \in S_1\} \cup \{(e_1, s_2) \mid s_2 \in S_2\}.$$

4. Let G be any finitely generated group and let S and S' be two different finite generating sets for G. Show that G is S–amenable if and only if G is S'–amenable. Thus we can define a finitely generated group G to be *amenable* if it is S–amenable for some (hence any) finite generating set S.
5. Use the first three parts of the project to show that all finitely generated abelian groups are amenable.
6. Prove that the free group F_2 is *not* amenable.
7. Show that amenability is a quasi-isometry invariant. That is, show that if G and H are quasi-isometric groups, then G is amenable if and only if H is amenable.

Office Hour Eight

Dehn Functions

Timothy Riley

Suppose you are given a group defined using finitely many generators and relations and you wish to compute with it. You express group elements as products of generators, which you can then multiply or invert, but you face the problem that there can be great redundancy: many different products can represent the same group element. We will see that this redundancy is governed by the Dehn function. The faster the Dehn function grows, the greater the number of times relations must be used to unravel the redundancy.

Suppose you are in a more geometric mood and you view your group as a space. For instance, you might look at the group elements as a cloud of points, the distances between them being given by a word metric, or you might view the group via some other metric space on which the group acts in a suitably nice way so that the shapes of the cloud and the metric space are in close agreement. This space may be exotic, for instance, high-dimensional or curving in complicated ways, and so may be hard to visualize. Well, imagine the space is awash with soap solution, and you probe it by lowering in it a wire loop. You lift the loop out and see a soap film spanning the loop, providing a two dimensional snapshot of the mysterious space. The Dehn function measures the wildness of these soap films by giving the optimal upper bound on their areas as a function of the lengths of the loops.

Suppose you have a collection of groups, each given using finitely many generators and relations, and you wish to tell them apart, or distinguish between them more crudely on a large scale. The Dehn function may be the invariant you need.

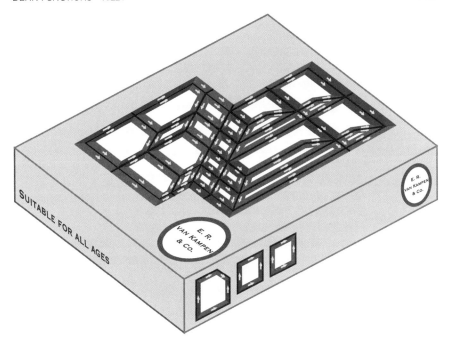

Figure 8.1 A puzzle kit.

We will explore all these perspectives and more in this office hour. But first, let us put aside groups and consider jigsaw puzzles.

8.1 JIGSAW PUZZLES REIMAGINED

I will describe jigsaw puzzles that are somewhat different from the familiar kind. A box, shown in Figure 8.1, contains an infinite supply of the three types of pieces pictured on its side: one five-sided and two four-sided, their edges colored green, blue, and red and directed by arrows. It also holds an infinite supply of red, green, and blue rods, again directed by arrows.

Solving puzzles. A good strategy for solving a standard jigsaw puzzle is to first assemble the pieces that make up its boundary, and then fill the interior. In our jigsaw puzzles the boundary—by which I mean a circle of colored rods end-to-end on a tabletop—is the starting point. A list, such as that in Figure 8.2, of boundaries that will make for good puzzles is supplied.

The aim of the puzzle is to fill the interior of the circle with the puzzle pieces in such a way that the edges of the pieces match up in color and in the direction of the arrows. The way the pieces can be used differs significantly from that of a standard jigsaw: our pieces can be flipped and can be stretched. Flipping a piece reverses the sequence of colored edges around its boundary; stretching it does not disturb their order or directions.

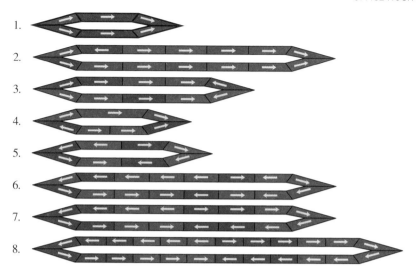

Figure 8.2 A list of puzzles accompanying the puzzle kit shown in Figure 8.1.

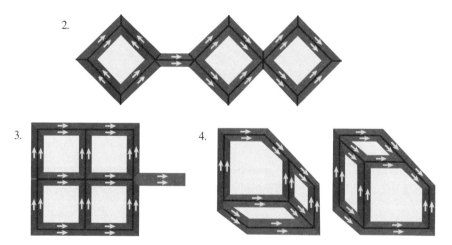

Figure 8.3 Solutions to three of the puzzles listed in Figure 8.2.

When solving a puzzle you are allowed to push together rods in the boundary circle as happens in our solutions to Puzzles 2 and 3 (Figure 8.3). All that is required for a valid solution is that the completed puzzle be flat on the tabletop (that is, be planar), the colors and arrows all match up, the boundary be the prescribed circuit of rods, and the interior be entirely filled with the supplied pieces.

Solutions to Puzzles 2, 3, and 4 from the above list are shown in Figure 8.3. The box in Figure 8.1 displays a solution to Puzzle 7.

Solutions are not unique in general. For example, Figure 8.3 shows two solutions to Puzzle 4. As will become apparent (in the light of Lemma 8.1, especially), there are circles of rods that give puzzles with no solution; but, assuming the manufacturer has been diligent, all on the supplied list should be solvable.

Exercise 1. Solve the remaining puzzles on the list in Figure 8.2.

The puzzle kit of Figure 8.1 is one of many possibilities. In general, a puzzle kit will have a finite number of types of pieces, each in infinite supply. Each piece will be a polygonal tile whose edges are colored and are directed by arrows. A polygonal tile is allowed to have any number of edges greater than or equal to 1. The possibility of a tile having only one or two edges is accommodated by allowing the edges to curve: a one- or two-sided tile could, for example, be circular with its perimeter divided into one or two edges, respectively. The kit will also include an infinite supply of rods that are also decorated by arrows and are given one of finitely many possible colors. The set of colors of the edges of the puzzle pieces will always be a subset of the set of colors of the rods.

(The flexibility of the pieces and their limitless number would surely prevent these puzzle kits from ever being manufactured, but a computer implementation would seem within the capacity of a skillful programmer.)

The sizes of puzzles.   So what, then, is a Dehn function? When buying a puzzle kit, you are likely to want to know how hard its puzzles can be. There are many ways of interpreting 'hard' here but, just as for standard jigsaw puzzles, a reasonable first consideration is how many pieces the puzzles require to complete. That is what a Dehn function measures. We look at all circles of at most n rods that give puzzles with solutions, and we ask for the minimum number N such that all those puzzles have solutions that use no more than N pieces. The *Dehn function* maps $n \mapsto N$.

8.2 A COMPLEXITY MEASURE FOR THE WORD PROBLEM

In Office Hour 1, group presentations were introduced: for a set S (which we'll call an *alphabet*) and a set of words R (termed *defining relators*) on $S \cup S^{-1}$, the expression $\langle S \mid R \rangle$ presents a group G when G is isomorphic to the quotient of the free group $F(S)$ by the smallest normal subgroup containing the elements of $F(S)$ represented by R. Another convenient way to express such a presentation is as $\langle S \mid u_1 = v_1, u_2 = v_2, \ldots \rangle$, which is taken to mean that $R = \{u_1 v_1^{-1}, u_2 v_2^{-1}, \ldots\}$. (The equalities $u_i = v_i$ are termed *defining relations*.) The presentation is finite when S and R are finite sets.

It would seem that a minimum standard for being able to work with a group G given by a finite presentation $\langle S \mid R \rangle$ is that we should be able to tell whether or not two words represent the same group element, or equivalently whether or not a word represents the identity. This is the known as the *word problem* for the presentation. The word problem is the first of three problems posed by Max Dehn providentially about one hundred years ago [105, 106].

A cyclic permutation of a word w is a word obtained from w by moving some of the letters at the end to the beginning—that is, if w is a concatenation of words

uv, then $w' = vu$ is a cyclic permutation. Note that w and w' represent conjugate elements in the free group.

A word on $S \cup S^{-1}$ represents the identity in G when it can be converted to the empty word via a finite sequence of free reductions $uss^{-1}v \rightsquigarrow uv$, free expansions $uv \rightsquigarrow uss^{-1}v$, and applications of defining relators $wuw' \rightsquigarrow wvw'$ where uv^{-1} or vu^{-1} is a cyclic permutation of an element of R. Such a sequence is called a null-sequence for w. Counting how many moves this takes gives a natural measure of how hard it is to work with the presentation. This is what the Dehn function does.

To be precise, the Dehn function $\mathbb{N} \to \mathbb{N}$ maps n to the minimal number N such that if w is a word of length at most n that represents the identity, then there is a null-sequence for w, involving at most N application-of-defining-relator moves. There are only finitely many words of length at most n since S is finite, and so N is well defined. This is essentially how Madlener and Otto introduced the Dehn function under the name "derivational complexity" [195] around the same time as Gromov [155] defined an equivalent geometric invariant in the manner we will discuss in Section 8.3.

(That we only count application-of-defining-relator moves here, rather than all the moves, is just a technicality that allows for the cleanest possible statement of Lemma 8.1 below; if we counted all the moves we would get an equivalent function in the sense defined in Section 8.3. This is because if there is a null-sequence for w that uses N application-of-defining-relator moves, then there is one that uses N application-of-defining-relator moves and at most $kN + \ell(w)$ free reduction moves, and no free expansion moves, where k is the length of the longest word in R.)

Exercise 2. As a warm-up for what follows, find null-sequences for the words $a^{-2}b^{-1}ab^2ab^{-1}$ and $a^{-4}b^{-2}a^2b^4a^2b^{-2}$ with respect to the presentation

$$\langle a, b, c \mid a^{-1}b^{-1}ab = c, \ ac = ca, \ bc = cb \rangle.$$

This is the three-dimensional integral Heisenberg group, H_3, which is the multiplicative group of 3×3 matrices of the form

$$\begin{pmatrix} 1 & x & z \\ 0 & 1 & y \\ 0 & 0 & 1 \end{pmatrix},$$

where $x, y, z \in \mathbb{Z}$. The matrices

$$\begin{pmatrix} 1 & 1 & 0 \\ 0 & 1 & 0 \\ 0 & 0 & 1 \end{pmatrix}, \quad \begin{pmatrix} 1 & 0 & 0 \\ 0 & 1 & 1 \\ 0 & 0 & 1 \end{pmatrix}, \quad \begin{pmatrix} 1 & 0 & 1 \\ 0 & 1 & 0 \\ 0 & 0 & 1 \end{pmatrix}$$

correspond to a, b, c, respectively.

Example: Dehn function of $\mathbb{Z}/m\mathbb{Z}$. The Dehn function $f(n)$ of the presentation $\langle a \mid a^m \rangle$ of the cyclic group of order m is the greatest integer less than or equal to n/m (sometimes denoted by $\lfloor n/m \rfloor$).

Here is a proof. A word on $\{a, a^{-1}\}$ represents the identity in $\langle a \mid a^m \rangle$ when it can be converted to a^{rm} for some $r \in \mathbb{Z}$ by free reductions. Doing so only shortens the word and costs nothing as far as Dehn function is concerned. Then, $|r|$ application-of-the-defining relation reduce the word to the empty word by deleting $a^{\pm m}$ substrings. So $f(n) \leq n/m$. Consideration of the effect of free reductions, free expansions, and application of one the defining relators on the sum of the exponents of the letters in a word leads to the lower bound. Free reductions and free expansions leave it unchanged and applying $a^m = 1$ changes it by at most m. So it is not possible to reduce a^{rm} to the empty word using fewer than $|r|$ applications of the defining relation.

Exercise 3. Show that the Dehn function of a finite presentation of a finite group is always bounded above by Cn for some constant C.

Exercise 4. Compute the Dehn function exactly for some finite presentations of finite groups.

Example: Dehn function of $\mathbb{Z} \times \mathbb{Z}$.   The Dehn function $f(n)$ of the presentation

$$\langle a, b \mid a^{-1}b^{-1}ab \rangle$$

of $\mathbb{Z} \times \mathbb{Z}$ grows quadratically. More precisely,

$$(n - 3)^2 \leq 16 f(n) \leq n^2$$

for all n.

For the upper bound, suppose w is a word on $\{a, b\}^{\pm 1}$ that has length n and represents the identity. Then the number of a's present in w equals the number of a^{-1}'s and the number of b's present equals the number of b^{-1}'s. The defining relator $a^{-1}b^{-1}ab$ can be expressed as $ab = ba$ or $ab^{-1} = b^{-1}a$ or $a^{-1}b = ba^{-1}$ or $a^{-1}b^{-1} = b^{-1}a^{-1}$, and so can be used to shuffle an $a^{\pm 1}$ past a $b^{\pm 1}$. So if we collect the $a^{\pm 1}$'s together by shuffling them past the past the $b^{\pm 1}$'s and then freely reduce, we will reach the empty word. There are at most n of each, and so $f(n) \leq n^2$.

Exercise 5. Sharpen these estimates to get $f(n) \leq n^2/16$ and show that this bound is realized by the words $a^{-k}b^{-k}a^k b^k$.

The lower bound comes from the fact that the word $w_k = a^{-k}b^{-k}a^k b^k$ has length n, where $n = 4k$, and represents the identity in $\mathbb{Z} \times \mathbb{Z}$, but any null-sequence carrying it to the empty word requires at least $k^2 = n^2/16$ applications of the defining relator $a^{-1}b^{-1}ab$. A direct proof in terms of null-sequences would be cumbersome and unenlightening. A more natural proof can be given using geometric techniques we will see in Section 8.5. Here is a somewhat surprising alternative approach (which you could skip as we will not need it later). It originates in an article by Baumslag, Miller, and Short [17].

Suppose u_i are words on $\{a, b\}^{\pm 1}$ and $\epsilon_i = \pm 1$ so that

$$W_k = \left(u_1^{-1}(a^{-1}b^{-1}ab)^{\epsilon_1} u_1 \right) \cdots \left(u_N^{-1}(a^{-1}b^{-1}ab)^{\epsilon_N} u_N \right)$$

equals w_k in $F(a,b)$ and N is the minimum number of times the defining relator has to be applied to reduce w_k to the empty word with respect to the presentation $\langle a,b \mid a^{-1}b^{-1}ab \rangle$. Then in the Heisenberg group H_3 of Exercise 2, the word W_k represents the same element as

$$\left(u_1^{-1}c^{\epsilon_1}u_1\right)\cdots\left(u_N^{-1}c^{\epsilon_N}u_N\right),$$

and so also as $c^{\epsilon_1}\cdots c^{\epsilon_N}$, since c commutes with a and b. But a calculation using the matrix representation of H_3 shows that w_k and c^{k^2} represent the same element in H_3, and that c has infinite order in H_3. So $N \geq k^2$, which proves that $n^2 \leq 16 f(n)$ when n is a multiple of 4. The stated lower bound of $(n-3)^2$ then follows because the largest multiple of 4 less than or equal to an integer n is at least $n-3$.

Here are some exercises concerning bounds for Dehn functions obtainable by the approach we used for $\mathbb{Z} \times \mathbb{Z}$.

Exercise 6. Show that for your favorite presentation of any finitely generated abelian group, the Dehn function is at most a constant times n^2.

Exercise 7. Use the fact that each element of the Heisenberg group H_3 can be expressed in a unique way as a word of the form $a^p b^q c^r$ for some integers p, q, and r to show that the Dehn function $f(n)$ of the presentation

$$\langle a,b,c \mid a^{-1}b^{-1}ab = c,\ ac = ca,\ bc = cb \rangle$$

of H_3 satisfies $f(n) \leq Cn^3$ for all n and a suitable constant C. In fact, $f(n)$ also admits a cubic lower bound; see [17].

Exercise 8. Find bounds on the Dehn function $f(n)$ for the presentation

$$\langle a,b,c \mid abc = 1,\ acb = 1 \rangle$$

for $\mathbb{Z} \times \mathbb{Z}$.

Puzzle kits are presentations.   The beginnings of how to reconcile the definition of the Dehn function in terms of null-sequences with that given in terms of puzzles may be evident. A finite presentation corresponds to a puzzle set. The colors of the rods and the edges of the puzzle pieces correspond to the generating set S. Each puzzle piece corresponds to a defining relator $r \in R$: following the boundary of the piece either clockwise or counterclockwise from some starting point, one reads r on translating the colors to generators and understanding that travel against the direction of an edge should mean an inverse letter. For example, the Heisenberg group, as presented in Exercise 2, corresponds to the puzzle kit of Figure 8.1.

I claim that for a word w the problem of finding a sequence of free reductions, free expansions, and applications of defining relators that carries it to the empty word is equivalent to solving the puzzle where (on translating colors to generators and taking into account the directions), starting from some vertex v, one reads w around the initial circle of rods. This is because of the way that null-sequences relate to solutions of puzzles, as I will explain.

Here is how to obtain a sequence of free reductions and applications of defining relators carrying w to the empty word from a solution to the corresponding puzzle.

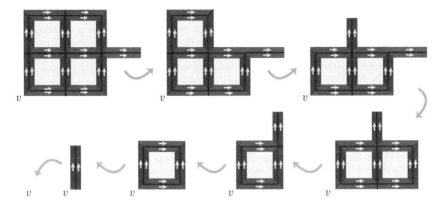

Figure 8.4 A disassembly of a completed puzzle.

Disassemble the completed puzzle, rod-by-rod and piece-by-piece in any way using the following moves, until all that remains is the vertex v:

1. remove any pair of rods that forms a spike coming out of the puzzle, that is, run side-by-side and only meet the rest of the puzzle at one end;
2. remove any piece that abuts the boundary circle of the puzzle, and then re-configure the boundary circle so as to close up the resulting hole.

An example of such a disassembly of a completed puzzle is shown in Figure 8.4.

The word corresponding to the circuit of rods around the perimeter changes by a free reduction in the first type of move and by an application of a relator in the second, and the number of application of a relator moves is equal to the area of the puzzle (area being measured by the number of puzzle pieces). In the example of the disassembly in Figure 8.4, the corresponding null-sequence (reading the diagrams anticlockwise from v) is

$$c^2 baa^{-1} bc^{-2} b^{-2} \ \leadsto\ c^2 baa^{-1} c^{-1} bc^{-1} b^{-2} \ \leadsto\ c^2 baa^{-1} c^{-1} bb^{-1} c^{-1} b^{-1}$$

$$\leadsto\ c^2 bc^{-1} bb^{-1} c^{-1} b^{-1} \ \leadsto\ cb^2 b^{-1} c^{-1} b^{-1} \ \leadsto\ cbc^{-1} b^{-1}$$

$$\leadsto\ bb^{-1} \ \leadsto\ empty\ word.$$

The natural way to translate from a null-sequence for w to a solution for the corresponding puzzle is to reverse the procedure: place successive pieces and rods on the tabletop as dictated by the null-sequence until a solution to the puzzle has been assembled. This is workable and the agreement between the number of puzzle pieces and the number of application-of-defining-relator moves becomes evident, but there is a technical problem. It may be that the puzzle cannot be kept on the tabletop when assembled in this manner, that is, planarity may break down. When this problem occurs, in some sense the null-sequence must have been inefficient, and so the difficulty can be avoided by improving the null-sequence.

Seen in this group-theoretic light, a completed puzzle is known as a *van Kampen diagram* for w. Modulo the difficulty mentioned above, we have established the following lemma which is closely related to a foundational lemma of van Kampen [180]. (A proof of this lemma that deals carefully with the planarity issue can be found in Bridson's text [52].)

LEMMA 8.1. *In a finite presentation for a group, the words that represent the identity are precisely those that correspond to puzzles which admit solutions. Moreover, the Dehn function defined in terms of puzzles agrees exactly with the Dehn function defined in terms of null-sequences.*

You may like to revisit Exercise 2 in the light of this lemma. The two words in that exercise correspond to Puzzles 5 and 7 of Figure 8.2.

Solving the word problem.   Our next result, found for example in Gersten's survey [142], gives a direct connection between the Dehn function of a finite presentation and its word problem. When we discuss algorithms in what follows, you can think of programs written in any reasonable programming language, running on any computer you like. But to be formal and precise, I mean an idealized computing device known as a *Turing machine*. A function $g \colon \mathbb{N} \to \mathbb{N}$ is *recursive* when there is an algorithm—implemented on the Turing machine—that on input n, outputs $g(n)$. There are functions $\mathbb{N} \to \mathbb{N}$ that are not recursive.

THEOREM 8.2. *For a finite presentation $\langle S \mid R \rangle$ of a group with Dehn function $f \colon \mathbb{N} \to \mathbb{N}$, the following are equivalent.*

1. *There is an algorithm that, given the input of a word on $S^{\pm 1}$, will declare whether or not that word represents the identity (i.e., the presentation has solvable word problem).*
2. *There is a recursive function $g \colon \mathbb{N} \to \mathbb{N}$ such that $f(n) \leq g(n)$ for all n.*
3. *f itself is a recursive function.*

Here is most of a proof. Given an upper bound $g(n)$ on the Dehn function, it is always possible to reduce a word of length w that represents the identity to the empty word using a null-sequence with at most $g(n)$ application-of-a-defining-relator moves and at most $kg(n) + n$ free-reduction moves, where k is the length of the longest defining relator. So if $g(n)$ is recursive, we can test whether a word of length n represents the identity by trying all null-sequences that use at most that number of moves. If, on the other hand, we have an algorithm that solves the word problem, then we can calculate $f(n)$ by the following arduous procedure. First list all words on $S^{\pm 1}$ of length at most n; discard from the list all that fail to represent the identity; and then, for each word w that remains, calculate the minimal number of application-of-defining-relation moves necessary to reduce w to the empty word or the minimal number of pieces in a solution to the puzzle corresponding to w.

Exercise 9. Complete this proof by explaining how to do the final step.

How hard is the word problem really?   To be honest, the Dehn function is not a good measure of the difficulty of the word problem. It is a worst-case measure of how long a direct attack on the word problem (by successively applying defining relators and free reductions and expansions) will take. But that attack is non-deterministic: in order to reduce a word to the empty word using the shortest possible null-sequence, the right choices need to be made about which moves to apply and where in the word to apply them. Making this a deterministic algorithm, that is, removing the need to make choices, appears to cost an exponential leap in running time. It could be done by exhaustively trying all possible sequences of moves of a given length specified by the Dehn function. But this rarely seems worth the trouble as there are usually far more efficient ways to tackle the word problem, as we will now see.

As a simple example, consider $\mathbb{Z} \times \mathbb{Z}$ presented by $\langle a, b \mid ab = ba \rangle$, which we already saw has a quadratically growing Dehn function. To tell whether or not a word on $\{a, b\}^{\pm 1}$ represents the identity, it is enough just to add up the exponents of the $a^{\pm 1}$ and $b^{\pm 1}$ present and check whether both are zero. Thus we can solve the word problem in linear time.

Another example is the Heisenberg group H_3. Its Dehn function grows like $n \mapsto n^3$. But viewing H_3 as a matrix group and calculating by multiplying matrices is an efficient means of checking whether a word represents the identity.

The same strategy works for $\langle a, s \mid s^{-1}as = a^2 \rangle$, whose Dehn function grows exponentially fast as we will see in Section 8.5. It can be represented by matrices via $a \mapsto \left(\begin{smallmatrix} 1 & 0 \\ 1 & 1 \end{smallmatrix} \right)$ and $s \mapsto \left(\begin{smallmatrix} 1/2 & 0 \\ 0 & 1 \end{smallmatrix} \right)$.

In Section 8.5 we will see examples of groups with even faster growing Dehn functions. The Dehn function of Baumslag's group

$$\langle a, t \mid (t^{-1}at)^{-1}a(t^{-1}at) = a^2 \rangle$$

grows like a tower of exponential functions, but a polynomial time solution to its word problem was recently found by Miasnikov, Ushakov, and Won [209]. The word problems of the hydra examples Γ_k, also introduced in Section 8.5, have similarly efficient solutions [113], despite having Dehn functions growing like Ackermann's extremely fast-growing functions $A_k(n)$.

There are examples where the gap between the Dehn function and the running time of the most efficient algorithm to solve the word problem is similarly large or even greater. Cohen, Madlener, and Otto [86] gave examples where the gap is like $n \mapsto A_n(n)$, which is non-primitive recursive, and recently Kharlampovich, Myasnikov, and Sapir [183] showed that for any given recursive function there is an example with Dehn function growing faster than that function, but with the word problem solvable in polynomial time. Their technique is to take algorithms that are known to halt always but, on some inputs, take an amount of time comparable to the values of these especially fast-growing functions, and embed them in the word problem for a suitable finite presentation of a group. Lots of work goes into making this precise. It is done in such a way that checking that certain words represent the identity by using direct applications of generators and relations is similar to running these algorithms, and so makes the Dehn function grow similarly quickly. But this is unnecessary work, as the issue of whether those words represent the identity really only hinges on whether those slow algorithms terminate ... and they always do. Cohen gives an entertaining description of the phenomenon with the help of a magical salmon (!) in [85].

All this is not to detract from the Dehn function. It represents a compelling link between algebraic computation (see Theorem 8.2) and, as we will see in Section 8.3, geometry.

The challenge of making demanding puzzles.   Here is an informal question. Recall that each puzzle kit comes with a list of suggested puzzles. Assuming they are to be challenging puzzles, we might wonder how the manufacturer can come up with the list. Ideally, generating the list should be much easier than solving the puzzle.

> Are there finite presentations in which it is easy to generate words that represent the identity, but hard to solve the puzzle?

The sense in which the puzzles are hard to solve need not be that they employ a particularly large number of pieces, but rather that it is hard to describe a solution, or even provide some extrinsic proof as to why a solution exists.

Infinite presentations.   It is possible to define groups with presentations where the set of defining relators R is infinite. The definition of the Dehn function given above remains well founded, but is less compelling. The Dehn functions of any two finite presentations of the same group grow in similar ways, as we will discuss in Section 8.4. But if we allow R to be infinite, they can be very different; after all, if we go to the extreme of including all the words that represent the identity in the set of defining relators, then the Dehn function will only take values 0 or 1.

Nevertheless, see [151] for an interesting study of Dehn functions of infinite presentations. And there are striking open problems; see Projects 2 and 3 at the end of this office hour.

8.3 ISOPERIMETRY

In Virgil's *Aeneid*, Dido is described as purchasing land on which to found the city of Carthage. For an agreed price, the sellers allow her all she can enclose with a single oxhide. She cunningly cuts the oxhide into thin strips and arranges it in a semicircular arc between two points on the (roughly straight) coastline, thereby claiming a far larger parcel of land than the sellers had envisaged. In addition to skillful oxhide slicing, her success is based on her ability to give the optimal solution to a form of *isoperimetric problem*, namely the problem of finding the arc of a given length that, when connecting two points on a straight line in the plane, will enclose the largest area with the line.

In general, an isoperimetric problem concerns determining the maximal area, volume, or the like that a shape can have when its boundary is constrained in some way. The Dehn function of a finitely presented group G relates to an isoperimetric problem concerning spanning loops with disks in a suitable space. This sort of isoperimetry has a familiar physical manifestation, discussed at the beginning of the office hour: a wire loop lifted out of soap solution emerges spanned by an area-minimizing surface in the form of a soap film.

What can serve as this suitable space? It should be *simply connected*—that is, every two points should be joined by a path and every loop should span a disk, by which I mean that every continuous map from the circle $S^1 = \{(x, y) \in \mathbb{R}^2 \mid x^2 + y^2 = 1\}$ to the space should extend to a continuous map from the disk $D^2 = \{(x, y) \in \mathbb{R}^2 \mid x^2 + y^2 \leq 1\}$; there should be reasonable notions of the lengths of paths and the areas of disks; and it should resemble G in some sense.

We know one metric space that resembles G: simply, G itself with the word metric. Unfortunately, with this metric, G appears as a sparse cloud of points, and falls short of our requirements. Gromov puts it colorfully in an inspiring introduction to his text [155]:

> *This space may appear boring and uneventful to a geometer's eye since it is discrete and the traditional local (e.g., topological and infinitesimal) machinery does not run in G.*

I will give two ways to flesh out this metric space when G has a finite presentation $\langle S \mid R \rangle$. First, I will describe a combinatorial space, the Cayley 2–complex, and explain how the Dehn function corresponds to an isoperimetric problem there, and then I will explain a continuous version of this, via group actions.

The Cayley 2–complex. The Cayley graph $\Gamma(G, S)$, introduced in Office Hour 2, adds some substance to G: the vertices are the group elements and there is a directed edge labeled s from g to gs for every $g \in G$ and $s \in S^{\pm 1}$. But this space also appears insufficient. It is a graph and so is not usually simply connected; and, anyway, to draw a connection with the Dehn function, we surely need to add structure reflecting the set R of defining relators.

So, for every $r \in R$, we add a family of $\ell(r)$–sided polygonal faces to the Cayley graph, where $\ell(r)$ denotes the length of the word r. For every loop that is made up of a succession of edges along which we read r, one such face is attached by gluing its boundary edge-by-edge to the loop. The resulting space is named $\widetilde{K}$ and is known as the *Cayley 2–complex* of $\langle S \mid R \rangle$. We will establish soon that it is simply connected.

Two examples of $\widetilde{K}$ are shown at the top of Figure 8.5. The Cayley graph of $\mathbb{Z} \times \mathbb{Z}$ with respect to a two-element generating set $\{a, b\}$ is a grid-like graph and we get the Cayley 2–complex for $\langle a, b \mid a^{-1}b^{-1}ab \rangle$ by filling in squares with faces, with the result that $\widetilde{K}$ is a plane. On the other hand, the Cayley 2–complex of the presentation $\langle a, b \mid \rangle$ of the free group $F(a, b)$ is simply the Cayley graph itself, an infinite tree in which every vertex has degree 4, as there are no defining relators and so no faces.

Aside: The Cayley 2–complex as a universal cover. Here is an useful additional perspective on $\widetilde{K}$ for those of you with some background in algebraic topology or for those who have already read Office Hour 9. The reason for the notation $\widetilde{K}$ is that it is the universal cover of a certain finite two-dimensional complex K, which is illustrated for our specific examples at the bottom of Figure 8.5 and in general

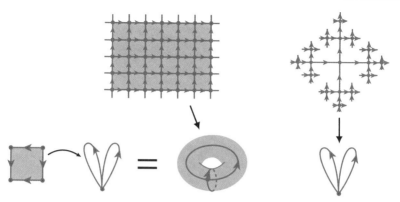

Figure 8.5 Portions of the Cayley 2–complexes for the presentations $\langle a, b \mid a^{-1}b^{-1}ab \rangle$ and $\langle a, b \mid \rangle$ of $\mathbb{Z} \times \mathbb{Z}$ and F_2.

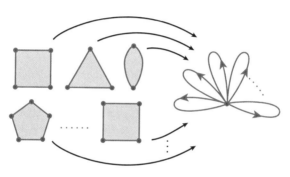

Figure 8.6 The 2–complex K associated to a presentation.

in Figure 8.6. This K has G as its fundamental group and is assembled as dictated by the presentation as follows. Start with a lone vertex. To that vertex, attach both ends of one directed edge for each element $s \in S$, labelling that edge by s. The result is called a *rose* (cf. Office Hour 4). Then for each $r \in R$, attach to the rose one $\ell(r)$–sided face, where $\ell(r)$ denotes the length of the word r, along the edge loop around which we read r. That K has fundamental group G is a consequence of the Seifert–van Kampen theorem. In one of the examples of Figure 8.5, K is a torus, and in the other it is a rose with two petals. Even if you don't already know algebraic topology, it hopefully seems intuitive that this K is closely related to the presentation we started with.

Exercise 10. Convince yourself that there is a natural action of G on $\widetilde{K}$ and that the quotient is nothing other than K.

For most presentations, $\widetilde{K}$ is hard to visualize. Try the example of the presentation for H_3 in Exercise 2, for instance.

We give $\widetilde{K}$ a metric by taking each of its edges to have length 1 and each of the faces to be a regular Euclidean polygon whose sides are all of length 1.

When we need a one-sided polygon, use a Euclidean disk of perimeter 1, and when we need a two-sided polygon use, a Euclidean disk whose perimeter has length 2 and is divided into two semicircular edges. We declare the distance between two points there to be the length of the shortest path connecting them.

Isoperimetry in the Cayley 2–complex. Here I will explain the connection between the Dehn function of a finite presentation $\langle S \mid R \rangle$ for G and isoperimetry in the Cayley 2–complex $\widetilde{K}$.

The outgoing edges from any given vertex in $\widetilde{K}$ are labeled in one-to-one correspondence with the elements of the generating set S. The same is true of the incoming edges. So words w on $S^{\pm 1}$ correspond bijectively with paths that traverse a succession of edges from any given starting vertex $g \in G$. Such a path ends at gw. Moreover, since gw and g are equal in G if and only if w represents the identity, we have the following fact.

LEMMA 8.3. *Words in $S^{\pm 1}$ that represent the identity in G correspond bijectively with edge loops in $\widetilde{K}$ based at some fixed vertex.*

Suppose ρ is an edge loop in $\widetilde{K}$ and w is the corresponding word. As w represents the identity, the associated puzzle has a solution; that is, there is a van Kampen diagram Δ for w. We will view Δ as a two dimensional planar complex

In general, a two-dimensional complex, or *2–complex*, is the space obtained from a graph (also known as a 1–complex) by gluing polygons (or faces) to the graph; each edge of the polygon is either glued onto an edge of the graph or it is collapsed to a point and glued to a vertex. (The topology on the 2–complex is the quotient topology.) The Cayley 2–complex is of course one example.

Back to our example Δ, the puzzle pieces are the faces, and the rods and the edges of the puzzle pieces form the edges. Figures 8.7, 8.10, and 8.11 are examples of van Kampen diagrams viewed as such complexes. The edges in this complex inherit orientations and labelings by generators from the rods and from the sides of the puzzle pieces.

It is possible to regard Δ as a disk spanning ρ in $\widetilde{K}$, as I will now explain. This disk may be singular in that it may have one-dimensional portions, as in Examples 2 and 3 in Figure 8.3.

Suppose we choose a vertex v in Δ and choose any vertex g in $\widetilde{K}$. I will explain that there is a unique map from all the edges and vertices of Δ to $\widetilde{K}$ that sends v to g and sends edges to edges in such a way as to match up edge orientations and labels. The point is that the image of any edge path in Δ emanating from v is determined by the matching of the edge orientations and labels. We might worry that this leads to inconsistencies (there might be many edge paths from v to any given edge, so why do we get a well-defined map?), but this concern would turn out to be unfounded. If two edge paths p_1 and p_2 in Δ emanate from v and have a common final edge (traversed in the same direction or in opposite directions), then p_1 and p_2 enclose a subcomplex of Δ that is itself a van Kampen diagram, so the word w' read around its boundary represents the identity, and therefore w' determines an edge loop in $\widetilde{K}$ starting from g; therefore the images of the final edges of p_1 and p_2 must agree. Finally, we can extend this map to the whole of Δ by sending faces in Δ to faces in $\widetilde{K}$. We can do so because the words read around the edge paths bounding the faces in Δ are defining relators, and so the images of those edge paths encircle faces $\widetilde{K}$. Note that the boundary circuit of Δ is carried to an edge loop in $\widetilde{K}$ around which we read w and that this edge loop would be ρ if we chose v and g suitably.

As a corollary, we learn that $\widetilde{K}$ is simply connected, just as we wanted.

In this combinatorial setting, the appropriate notion of area for Δ is the easiest one imaginable: the number of faces it has. Our discussion has established the following theorem.

THEOREM 8.4. *The Dehn function of a finite presentation* $P = \langle S \mid R \rangle$, *as defined in terms of puzzles, is a* minimal isoperimetric function *for* $\widetilde{K}$ *in that it maps* $n \mapsto N$, *where N is the minimal number such that any edge loop in $\widetilde{K}$ of length at most n can be spanned by a combinatorial disk (that is, a van Kampen diagram) with at most N faces.*

In light of this theorem, we denote the Dehn function of P by $\text{Area}_P \colon \mathbb{N} \to \mathbb{N}$.

We accomplished our goal of making an analogy between Dehn functions and isoperimetric problems, at least in the combinatorial setting. I also promised that I would give a continuous version, using group actions, and that is next. This material is a bit more advanced. If you are not familiar with Riemannian manifolds and covering spaces, you can safely skip the remainder of this section.

Riemannian manifolds for groups. The second type of space we associate to a finite presentation of a group G is a *Riemannian manifold* $\widetilde{M}$. A *manifold* is a space that on the small scale resembles Euclidean space $\mathbb{R}^n$ for some n; it is *Riemannian* when it is endowed with a certain structure which gives rise to notions such as lengths of paths, angles between paths, areas or volumes of subsets, and so on. We already saw an example of a Riemannian metric when we studied the hyperbolic plane in Office Hour 5. And we saw how that metric allowed us to compute lengths. A general Riemannian metric is the same kind of thing.

It turns out that there is a lot of flexibility over which Riemannian manifolds will work for our purposes. They should be simply connected and should coarsely resemble G. For example, for $\mathbb{Z} \times \mathbb{Z}$ we can take $\widetilde{M}$ to be the plane. The spaces $\mathbb{Z} \times \mathbb{Z}$ and $\widetilde{M}$ bear a coarse resemblance in that $\mathbb{Z} \times \mathbb{Z}$ is the set of points in the plane with integer coordinates; if you squint at $\mathbb{Z} \times \mathbb{Z}$ it looks like a plane (if you read Office Hour 7 you know that they are quasi-isometric).

To be more precise, the universal cover of any compact connected Riemannian manifold M that has G as its fundamental group can serve as $\widetilde{M}$. For the example of $\mathbb{Z} \times \mathbb{Z}$ presented by $\langle a, b \mid a^{-1}b^{-1}ab \rangle$, a torus can serve as M. Such an M always exists when G is finitely presented: in fact, we can take M to be the boundary of a small neighborhood of a copy of the complex K embedded in $\mathbb{R}^5$; this is a four-dimensional manifold (see [247]).

The manner in which G resembles this $\widetilde{M}$ stems from the action of G on $\widetilde{M}$ by deck transformations. Fix a base point $p \in \widetilde{M}$. The orbit map $G \to \widetilde{M}$ defined by $g \mapsto g \cdot p$ is a quasi-isometry. This is an application of the Milnor–Schwarz lemma from Office Hour 7, since the action of G on $\widetilde{M}$ is geometric.

Isoperimetry in Riemannian manifolds. It has long been known that any loop of length ℓ in the plane can be filled with a disk of area at most $\ell^2/(4\pi)$. As Queen Dido was probably aware, this bound is realized by a circle of perimeter ℓ. It is no coincidence that the Dehn function of $\mathbb{Z} \times \mathbb{Z}$ also grows quadratically as we saw earlier in an example. After all, the plane bears a coarse resemblance to $\mathbb{Z} \times \mathbb{Z}$. A similar connection can be drawn between the Dehn function of every finitely presented group and the isoperimetry of the associated Riemannian manifold $\widetilde{M}$.

By a *disk* in $\widetilde{M}$ I mean the image of a continuous map $D^2 \to \widetilde{M}$. As $\widetilde{M}$ is Riemannian, there is a notion of area for disks in $\widetilde{M}$. The *minimal isoperimetric function* $\text{Area}_{\widetilde{M}} \colon [0, \infty) \to [0, \infty)$ for $\widetilde{M}$ is defined so that $\text{Area}_{\widetilde{M}}(\ell)$ is the infimal real number such that every loop of length at most ℓ in $\widetilde{M}$ can be spanned by a disk of area at most $\text{Area}_{\widetilde{M}}(\ell)$. In fact, this infimum is a minimum; there is a long story here known as *Plateau's problem*.

The following theorem is generally attributed to Gromov, who is responsible for richly animating the study of finitely generated groups by drawing on analogies and connections with Riemannian geometry. Detailed proofs can be found in [52] and [69]. The fact that the fundamental group of a compact Riemannian manifold is always finitely presentable will be implicit in this theorem; we will not prove this here but the ideas involved are similar to those that establish Theorem 8.7; see Chapter I.8 of [54] for details.

The word *equivalent* in the theorem refers to the relation $\simeq$ which is commonly used in geometric group theory to capture the notion of functions growing at the same rate. For $f, g \colon [0, \infty) \to [0, \infty)$ we write $f \preceq g$ when there exists $C > 0$ such that $f(\ell) \le Cg(C\ell + C) + C\ell + C$ for all $\ell \ge 0$. And $f \simeq g$ when $f \preceq g$ and $g \preceq f$. This relation can be expanded to encompass functions with domain $\mathbb{N}$ by extending their domain to $[0, \infty)$ so that these functions become constant on the intervals $[n, n + 1)$ for all $n \in \mathbb{N}$.

Exercise 11. Show that all functions that grow at most linearly fast are equivalent. Show that $n \mapsto n^\alpha$ and $n \mapsto n^\beta$, where $\alpha, \beta \ge 1$, are equivalent if and only if $\alpha = \beta$. Show that polynomially growing functions are not equivalent to exponentially growing functions $n \mapsto c^n$ (where $c > 1$). Show that $n \mapsto c^n$ and $n \mapsto d^n$ are equivalent for all $c, d > 1$.

THEOREM 8.5 (The filling theorem). *Suppose M is a compact Riemannian manifold without boundary and that P is a finite presentation for its fundamental group G. The minimal isoperimetric function $\text{Area}_{\widetilde{M}}(\ell)$ for $\widetilde{M}$ is equivalent to the Dehn function $\text{Area}_P(n)$ of G.*

For an idea on how to prove the filling theorem let us look again at the example of $\mathbb{Z} \times \mathbb{Z}$, presented by $\langle a, b \mid a^{-1}b^{-1}ab \rangle$, with $\widetilde{M}$ being the plane. The infinite chessboard complex $\widetilde{K}$ is itself a plane and so maps to $\widetilde{M}$ in the natural way. So, in this instance, the theorem is about comparing filling general loops in the plane with disks to filling edge loops in the chessboard pattern with chessboard squares. The key points are firstly that an arbitrary loop in the plane can be pushed into the 1–skeleton of $\widetilde{K}$ without increasing its length too much, secondly that the number of squares it then encloses is comparable to the area it originally enclosed, and thirdly that an edge loop enclosing squares is the same thing as a van Kampen diagram.

This approach works in full generality: a disk spanning a loop in $\widetilde{M}$ is similar to a van Kampen diagram filling an edge loop in $\widetilde{K}$. To make sense of this we have to relate $\widetilde{K}$ and $\widetilde{M}$. We map $\widetilde{K}$ to $\widetilde{M}$ beginning with its vertices, for which we use the orbit map $G \to \widetilde{M}$, and then we extend to the edges of $\widetilde{K}$ by mapping the edge between a pair of vertices in $\widetilde{K}$ to a minimal length path between their images, and then we extend to the whole of $\widetilde{K}$ by mapping the interiors of faces in $\widetilde{K}$ to minimal area disks spanning the loops that are the images of their boundaries. The resulting map $\widetilde{K} \to \widetilde{M}$ can be used to carry edge loops and van Kampen diagrams into $\widetilde{M}$, while retaining control on their lengths and areas. Moreover, arbitrary loops or disks in $\widetilde{M}$ can be pushed to edge loops or combinatorial disks in the image of the $\widetilde{K}$ while maintaining similar control.

8.4 A LARGE-SCALE GEOMETRIC INVARIANT

Recall from Office Hour 7 that quasi-isometries are maps that carry one metric space almost onto another with a bounded amount of stretching and tearing. Also recall from Exercises 9 and 10 in Office Hour 7 that quasi-isometric equivalence is an equivalence relation. Also it is shown that, while a finitely generated group can have many different finite generating sets and so many different word metrics, all are equivalent in that the identity map is a quasi-isometry from the group with one word metric to the same group with another word metric (Theorem 7.3).

The Dehn function is a quasi-isometry invariant. We defined a Dehn function in terms of a finite presentation for a group, rather than simply in terms of a group. But if a group has a finite presentation, then it has many finite presentations. So the nagging question is how the Dehn function depends on the presentation. The relation $\simeq$ defined before Theorem 8.5 allows us a satisfying answer.

THEOREM 8.6. *The Dehn functions of any two finite presentations for the same group are equivalent in the sense of* $\simeq$.

This theorem highlights the need for the "$+Cn$" term in the definition of $\simeq$. Consider the presentations $\langle a \mid \; \rangle$ and $\langle a, b \mid b \rangle$ of $\mathbb{Z}$. The Dehn function of $\langle a \mid \; \rangle$ is constantly zero since there are no defining relators. But the Dehn function of $\langle a, b \mid b \rangle$ is $n \mapsto n$ since it takes n applications of the defining relation to reduce b^n to the empty word.

In fact, the Dehn function is an invariant in a broader sense. The next theorem says that the Dehn function is a quasi-isometry invariant. As in Office Hour 7, this means that if two groups have (sufficiently) different Dehn functions, then the groups are not quasi-isometric.

THEOREM 8.7. *If finitely generated groups G and H are quasi-isometric with respect to some word metrics, and G is finitely presented, then H is also finitely presentable and their Dehn functions, defined with respect to any finite presentations, are equivalent in the sense of* $\simeq$.

This theorem encompasses Theorem 8.6. It is proved by a follow-your-nose type of argument, remembering that, as quasi-isometries need not be continuous, it is better to examine their effect on configurations of points rather than on paths or disks. Here is a sketch. Consider an edge loop ρ in the Cayley 2–complex of H. Use a quasi-isometry $\Phi \colon H \to G$ to carry the vertices on this loop into the Cayley 2–complex for G. Join up the points there by minimal length edge paths in the order they appeared on ρ, to make a loop. Fill that loop with a minimal area van Kampen diagram Δ, and use a quasi-inverse $G \to H$ of Φ to carry the vertices of Δ back to H, which gives a configuration of points that coarsely fills ρ. Analyzing the distances between the vertices in this configuration by comparing them to the distances between the corresponding points in Δ leads to the result that H is finitely presentable. Moreover, the configuration can be fleshed out to give a genuine van Kampen diagram modeled on Δ filling ρ with respect to any finite presentation for H, and that shows the Dehn functions are equivalent. See the article by Alonso [7] or Bridson and Haefliger's book [54] for details.

Theorem 8.7 tells us that the Dehn function is an invariant of large-scale or coarse geometry and so can serve as a tool in the influential program of Gromov to understand discrete groups up to quasi-isometry (cf. Office Hour 7). In fact, the Dehn function is just the first of a number of filling functions discussed by Gromov in [155] alongside a wide assortment of other types of quasi-isometry invariants. Filling functions record geometric features of disks or other surfaces spanning loops in a space such as area, diameter, radius, the lengths that loops grow to in the course of nullhomotopies, and so on. They also have higher-dimensional analogs, concerning $(n + 1)$-dimensional balls spanning n-dimensional spheres rather than just disks (two dimensional balls) spanning loops (one-dimensional spheres).

8.5 THE DEHN FUNCTION LANDSCAPE

So, what functions arise as Dehn functions? What can we say about the class of groups with some given Dehn function? How wild can Dehn functions be?

Hyperbolic groups, the subject of Office Hour 9, are characterized as the finitely presented groups whose Dehn functions are equivalent to n. The linear upper bound on their Dehn functions is proved in Section 9.5. Hyperbolic groups stand isolated in that if a finitely presentable group has Dehn function not bounded below by a Cn^2 for some $C > 0$, then that group is hyperbolic [45, 54, 154, 213, 217]. This jump from linear to quadratic is known as "Gromov's gap."

Finitely generated abelian groups with $\mathbb{Z} \times \mathbb{Z}$ subgroups have Dehn functions equivalent to n^2. More generally, the classes of semi-hyperbolic groups, CAT(0) groups, and automatic groups have Dehn functions $\preceq n^2$. These groups all display features of non-positive curvature, and we might suspect that having Dehn function $\preceq n^2$ might be a reasonable characterization of non-positive curvature amongst finitely presentable groups. However, the class of groups with Dehn function equivalent to n^2 is broad. It contains $SL(n, \mathbb{Z})$ for $n \geq 5$ (and conjecturally $SL(4, \mathbb{Z})$) [265]; Thompson's group F, the subject of Office Hour 16 ([158]); Stallings' group [114]; and examples of nilpotent groups of all nilpotency classes [266]; and none of these could reasonably be called non-positively curved.

In notes for a summer school in 1996 [141], Gersten wrote the following, which has since been proved prescient as most of the examples just listed were established later.

> *I call this a zoo, because I am unable to see any pattern in this bestiary of groups. It would be striking if there existed a reasonable characterization of groups with quadratic Dehn functions, which was more enlightening than saying that they have quadratic Dehn functions.*

Looking beyond quadratic, we come next to finitely presented groups with Dehn function equivalent to n^α. The two-generator groups that are free nilpotent of class c have Dehn functions equivalent to n^{c+1}; see [17, 155, 220]. (These are the groups that have only the relations necessary to make them nilpotent of class c.)

More generally, there is much known about for which $\alpha \geq 1$ there are finitely presented groups with Dehn function equivalent to n^α. The set of such α is countable as there are only countably many finite presentations, and, as I have indicated, it has no values in the open interval $(1, 2)$. But it is dense in the interval $[2, \infty)$ [46, 51]; and for the interval $[4, \infty)$, remarkably detailed information is provided in [233]: the α occurring are characterized in terms of whether there is a Turing machine capable of writing out the first n digits of the decimal expansion of α within a certain amount of time. Additionally, a wide variety of other functions $f : \mathbb{N} \to \mathbb{N}$ that are $\succeq n^4$ are shown to be equivalent to Dehn functions. Indeed, this is true for just about all common such functions that have the superadditivity property, $f(n + m) \geq f(n) + f(m)$ for all $n, m \in \mathbb{N}$.

We now turn to the extremes of the Dehn function landscape.

A group with exponential Dehn function. Establishing how a Dehn function grows presents two difficulties. The upper bound requires consideration of all words that represent the identity. And for the lower bound, while it suffices to consider only a suitable family of words w_n whose lengths grow like n, we must argue that all van Kampen diagrams for those w_n have at least some given area. The situation is analogous to the struggle to establish the (worst-case) time complexity of some computational problem: for an upper bound we need to show there is an algorithm that solves the problem within some given time on all inputs, and for the lower bound we have to show that on some worst family of inputs, every algorithm that solves the problem takes at least some given amount of time.

In this section I will prove the following theorem.

THEOREM 8.8. *The Dehn function $f(n)$ of the presentation $\langle a, s \mid s^{-1} a s = a^2 \rangle$ satisfies $f(n) \simeq 2^n$.*

This group $\langle a, s \mid s^{-1}as = a^2 \rangle$ is often known as the Baumslag–Solitar group $BS(1, 2)$ and is one of a family of groups discussed in detail in Office Hour 12. Subgroup distortion is one of its pertinent geometric features. In general, *subgroup distortion* concerns how a finitely generated subgroup H sits inside an ambient finitely generated group G. There are two natural word metrics on such an H: the intrinsic metric d_H coming from its own generating set and the extrinsic metric d_G coming from the generating set of G. Given $n \in \mathbb{N}$, the *distortion function* supplies the maximal distance $d_H(1, h)$ from the identity of all elements $h \in H$ such that $d_G(1, h) \leq n$. Roughly speaking, this function grows quickly when H sits severely scrunched up on itself inside G. It is, in a sense, a lower-dimensional filling function in that it concerns filling 0–spheres (that is, pairs of points) with 1–disks (that is, paths).

Exercise 12. Up to the equivalence $\simeq$, the growth of the distortion function of a finitely generated subgroup of a finitely generated group does not depend on the finite generating sets.

The distortion function for the subgroup $\mathbb{Z} = \langle a \rangle$ of $\langle a, s \mid s^{-1}as = a^2 \rangle$ grows exponentially. That this distortion function grows $\succeq 2^n$ is a consequence of the doubling effect that s has when it conjugates a, which leads to the relations $s^{-n}as^n = a^{2^n}$. I will give some explanation for the upper bound at the end of this section.

The large distortion translates into a large Dehn function because a copy of a van Kampen diagram for $s^{-n}as^n a^{-2^n}$, displaying the repeated doubling, can be joined to its mirror image along the side labeled a^{2^n}, offset by 1, to give a van Kampen diagram for

$$w_n = as^{-n}as^n a^{-1} s^{-n} a^{-1} s^n$$

as illustrated below in the case $n = 5$. This van Kampen diagram and those to come are drawn more economically as 2–complexes rather than as puzzles. Note that we had to join two mirror-image copies because one on its own would have both area and boundary length $\simeq 2^n$ on account of the exponentially large power of a.

This family of van Kampen diagrams has area equivalent to 2^n. Here are two strategies for proving the lower bound. The same strategies can be employed to give the lower bound for the Dehn function of $\mathbb{Z} \times \mathbb{Z}$. The first uses some concepts from algebraic topology; the second is more elementary.

The first strategy is to use *Gersten's lemma*: argue that the Cayley 2–complex is contractible[1] and therefore that amongst all van Kampen diagrams for a given word, a diagram Δ that is embedded is of minimal area. The point is that any other van Kampen diagram Δ' for the same word could combine with Δ to make a sphere in the Cayley 2–complex, which is contractible, and so all the faces of Δ must cancel with faces in Δ'.

Exercise 13. Show that the Cayley 2–complex of $\langle a, s \mid s^{-1}as = a^2 \rangle$ is homeomorphic to the direct product of an infinite 3-valent (that is, three edges meet at

---

[1] A space X is contractible if it can be deformed to a point; more formally, if there is a continuous map $X \times I \to X$ so that the restriction $X = X \times \{0\} \to X$ is the identity map and the restriction $X = X \times \{1\} \to X$ is a constant map. As basic examples: $\mathbb{R}^n$ is contractible for any n and S^n is not contractible for any n.

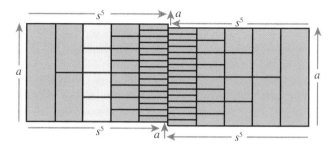

Figure 8.7 A van Kampen diagram for $as^{-5}as^5a^{-1}s^{-5}a^{-1}s^5$ with respect to $\langle a, s \mid s^{-1}as = a^2\rangle$. All vertical edges are labeled a and directed upwards. All horizontal edges are labeled s; those on the left half of the diagram are oriented to the right and those on the right half are oriented to the left. An example of a *corridor* is shown in green.

each vertex) tree with a line, and so is contractible. For a picture of the Cayley 2–complex for $BS(1, 3)$ see Figure 12.1, which is similar except that the tree there is 4-valent instead of 3-valent. Show that the maps from the diagrams of Figure 8.7 to the Cayley 2–complex are embeddings.

The second strategy uses *corridors* to understand van Kampen diagrams. As there is a sole defining relation $s^{-1}as = a^2$, adjoining each s in the boundary of a van Kampen diagram, there must be a face that has an edge labeled s on its far side; and that edge must adjoin another face with same property; and so on. So there is a chain of faces, joined one to the next by edges labeled s, proceeding through the diagram forming a corridor (or s–corridor) with the edges along its sides all labeled $a^{\pm 1}$. This corridor must terminate at some other edge labeled s elsewhere on the boundary of the diagram. Corridors therefore pair off edges labeled s in the boundary. The four green faces in Figure 8.7 comprise an example of a corridor.

Corridors corresponding to a letter s cannot cross themselves or each other. So corridors must stack up in any van Kampen diagram for w_n and it can be deduced that the exponent sums of the words on $a^{\pm 1}$ along their sides grow exponentially through this stack. In the example of Figure 8.7, they run vertically through the diagram, and grow exponentially in length towards the center of the diagram. So one corridor must include $\geq 2^n$ faces. Therefore any van Kampen diagram for w_n has area at least 2^n. A complication standing in the way of making this argument precise is that corridors that form rings instead of emerging on the boundary are possible. But such annuli are, in a sense, redundant and do not appear in diagrams of minimal area.

To complete a proof of Theorem 8.8, we need the exponential upper bound. Corridors are useful here also. Suppose we have a word w representing the identity, and Δ is a van Kampen diagram for w. If w contains any letters $s^{\pm 1}$, then since all such letters are paired off by corridors in Δ and no two corridors cross, there must be a subword $s^{\pm 1}us^{\mp 1}$ of w such that u is a word on $\{a, a^{-1}\}$ and the $s^{\pm 1}$ in this subword is joined to the $s^{\mp 1}$ by a corridor. It follows that there is a null-sequence for w that begins by replacing $s^{\pm 1}us^{\mp 1}$ with a word a^k where $|k|$ is either half or

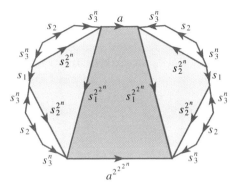

Figure 8.8 Threefold iterated exponential distortion in $\langle a, \ s_1, s_2, s_3 \mid s_1^{-1}as_1 = a^2,$ $s_2^{-1}s_1s_2 = s_1^2, \ s_3^{-1}s_2s_3 = s_2^2 \rangle$.

twice the sum of the exponent of the letters in u, and uses at most $|k|$ application-of-defining-relator moves in doing so. Repeat until all $s^{\pm 1}$ have been paired off and removed, and then freely reduce the resulting word on $\left\{ a, a^{-1} \right\}$ to the empty word. Summing the application-of-defining-relator moves we use along the way gives our exponential upper bound. The same approach leads to a proof of the exponential upper bound on the distortion function for $\langle a \rangle$ in $\langle a, s \mid s^{-1}as = a^2 \rangle$.

Iterated distortion and Baumslag's one-relator group.   Rename the previous example as $\langle a, s_1 \mid s_1^{-1}as_1 = a^2 \rangle$ and consider embellishing it by distorting $\langle s_1 \rangle$ by introducing a new letter s_2 acting on $\langle s_1 \rangle$ via $s_2^{-1}s_1s_2 = s_1^2$. The distortion function for $\mathbb{Z} = \langle a \rangle$ inside the resulting group

$$\langle a, s_1, s_2 \mid s_1^{-1}as_1 = a^2, \ s_2^{-1}s_1s_2 = s_1^2 \rangle$$

grows $\succeq \exp^{(2)}(n)$, since

$$(s_2^{-n}s_1s_2^n)^{-1}a(s_2^{-n}s_1s_2^n) \ = \ s_1^{-2^n}as_1^{2^n} \ = \ a^{2^{2^n}}.$$

(We write $\exp^{(l)}(n)$ for the l–fold iterate of the exponential function.)

Iterating, we find that the distortion function for $\mathbb{Z} = \langle a \rangle$ inside

$$\langle a, s_1, \cdots, s_l \mid s_1^{-1}as_1 = a^2, \ s_{i+1}^{-1}s_is_{i+1} = s_i^2 \ (i > 1) \rangle$$

grows $\succeq \exp^{(l)}(n)$. In fact, it grows equivalently to $\exp^{(l)}(n)$ for reasons similar to the case $l = 1$.

A schematic of a van Kampen diagram illustrating the calculation that leads to the 3–fold iterated exponential distortion is shown in Figure 8.8.

Exercise 14. Draw van Kampen diagrams for

$$(s_2^{-1}s_1s_2)^{-1}a^{-1}(s_2^{-1}s_1s_2)a^{-1}(s_2^{-1}s_1s_2)^{-1}a(s_2^{-1}s_1s_2)a$$

and

$$(s_2^{-2}s_1s_2^2)^{-1}a^{-1}(s_2^{-2}s_1s_2^2)a^{-1}(s_2^{-2}s_1s_2^2)^{-1}a(s_2^{-2}s_1s_2^2)a$$

in $\langle a, s_1, s_2 \mid s_1^{-1}as_1 = a^2, \ s_2^{-1}s_1s_2 = s_1^2 \rangle$.

As we did for a van Kampen diagram for $s^{-n}as^na^{-2^n}$ in $BS(1,2)$, we can glue a copy of such a diagram to its mirror image along the side labeled by the huge power of a and offset by 1, and the result is a family of diagrams with perimeter equivalent to n and area equivalent to $\exp^{(l)}(n)$. This is the beginning of a proof of the following theorem.

THEOREM 8.9. *The Dehn function $f(n)$ of*

$$\langle a, s_1, \cdots, s_l \mid s_1^{-1}as_1 = a^2, \; s_{i+1}^{-1}s_is_{i+1} = s_i^2 \; (i > 1)\rangle$$

satisfies $f(n) \simeq \exp^{(l)}(n)$.

This family of groups has a limit (loosely speaking), namely, the one-relator group

$$\langle a, t \mid (t^{-1}at)^{-1}a(t^{-1}at) = a^2\rangle$$

due to Baumslag [19]. Introducing s as shorthand for $t^{-1}at$, we can express this presentation as:

$$\langle a, s, t \mid s^{-1}as = a^2, \; s = t^{-1}at\rangle,$$

and we see that conjugation by s again has a doubling effect on a, and t conjugates a to s. This leads to a feedback effect whereby we get huge distortion of $\mathbb{Z} = \langle a\rangle$ on account of diagrams of the form shown schematically in Figure 8.9. If the portion of the perimeter of this diagram that excludes the huge power of a is to have length n, then the tree dual to the picture must have depth equivalent to $\lfloor \log_2 n\rfloor$, and this suggests the distortion grows as $n \mapsto \exp^{(\lfloor \log_2 n\rfloor)}(1)$, as is indeed the case. This is proved by Platonov [221], building on work of Gersten [140] and Bernasconi [25], en route to the following theorem.

THEOREM 8.10. *The Dehn function of*

$$\langle a, t \mid (t^{-1}at)^{-1}a(t^{-1}at) = a^2\rangle$$

is equivalent to the function $n \mapsto \exp^{(\lfloor \log_2 n\rfloor)}(1)$.

Again, diagrams with huge area can be obtained by joining two copies of the distortion diagrams along the large power of a, offset by 1.

Exercise 15. Draw detailed van Kampen diagrams in the manner of Figure 8.9 with the power of a along the horizontal path at the bottom being $2, 2^2, 2^{2^2}$, etc.

Exercise 16. What is the length of a shortest word equaling a^{100} in $\langle a, s, t \mid s^{-1}a s = a^2, \; s = t^{-1}at\rangle$? What about a^{1000}?

Hydra groups. Hydra groups were devised by Will Dison and me. Before we define these groups, let us revisit the legend of Hercules' fight with the Lernaean hydra. Define a *hydra* to be a positive word (that means no inverse letters are allowed) on the infinite alphabet $a_1, a_2, \ldots$ Hercules fights this hydra by striking off its first letter. It then regenerates according to the rule that each remaining a_i,

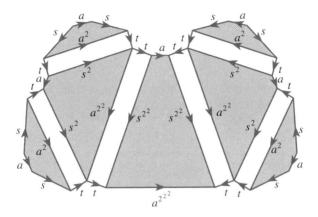

Figure 8.9 Distortion in Baumslag's group $\langle a, s, t \mid s^{-1}as = a^2, \; s = t^{-1}at \rangle$.

where $i > 1$, becomes $a_i a_{i-1}$ and each remaining a_1 remains as it is. This process— removal of the first letter and then growth—repeats, with Hercules victorious when (not *if!*) the hydra is reduced to the empty word.

Here is an example in which Hercules defeats $a_2 a_3 a_1$ in five strikes:

$$a_2 a_3 a_1 \; \rightarrow \; a_3 a_2 a_1 \rightarrow a_2 a_1 a_1 \rightarrow a_1 a_1 \rightarrow a_1 \rightarrow empty \; word.$$

Exercise 17. Prove that Hercules always wins.

Strikingly, battles can be of enormous duration, even against simple, short hydra. Define $\mathcal{H}(w)$ to be the number of strikes it takes Hercules to defeat the hydra w, and for integers $k \geq 1$, $n \geq 0$, define $\mathcal{H}_k(n) = \mathcal{H}(a_k^n)$.

Exercise 18. Show that $\mathcal{H}_1(n) = n$ and $\mathcal{H}_2(n) = 2^n - 1$.

Exercise 19. Give a formula for $\mathcal{H}_3(n + 1)$ in terms of $\mathcal{H}_3(n)$.

Exercise 20. For what values of n can you calculate $\mathcal{H}_4(n)$?

These functions $\mathcal{H}_k$ are a variation on Ackermann's famous fast-growing functions $\mathcal{A}_k : \mathbb{N} \to \mathbb{N}$ which are defined for integers $k, n \geq 0$ by:

$$\mathcal{A}_1(n) = 2n$$

$$\mathcal{A}_{k+1}(n) = \mathcal{A}_k^{(n)}(1).$$

So, in particular, $\mathcal{A}_1(n) = 2n$, $\mathcal{A}_2(n) = 2^n$, and $\mathcal{A}_3(n) = \exp_2^{(n)}(1)$, where $\exp_2^{(n)}$ denotes the n–fold iterate of $n \mapsto 2^n$. It turns out that $\mathcal{H}_k \simeq \mathcal{A}_k$; see [115].

The source of the extreme fast growth of Ackermann's functions is the recursion inherent in their definition. Such recursion is apparent in the battle with the hydra in that if $u a_k$ is a hydra that happens to end in the letter a_k, then in the time $\mathcal{H}(u)$ it takes to kill u, there appear $\mathcal{H}(u)$ letters a_{k-1} (and many other letters besides) after the final a_k which then have to be disposed of. So the time it takes to complete that initial task determines the size of the remaining challenge.

The *hydra group* G_k is the group presented by

$$\langle\, a_1, \ldots, a_k, t \ \mid\ t^{-1}a_1 t = a_1, \ t^{-1}a_i t = a_i a_{i-1}\ (\forall\, i > 1)\,\rangle$$

and H_k is the subgroup $\langle a_1 t, \ldots, a_k t\rangle$. The regeneration rules for the hydra are apparent in the defining relations for this presentation. These G_k are well behaved and straightforward in a number of respects, but nonetheless the distortion function of H_k in G_k grows equivalently to $\mathcal{A}_k$.

For an idea of why the distortion function grows so fast, consider this question: for what r (if any) is $a_k^n t^r$ in H_k? This is where the battle with the hydra comes in. To see how, look at the case where $k = 2$ and $n = 4$, for example. You can try to convert a_2^4 to a word on $a_1 t$ and $a_2 t$ times a power of t by introducing a $t t^{-1}$ to pair the first letter with t, and then carrying the accompanying t^{-1} to the end of the word by conjugating through the intervening letters; then pair off the next a_i likewise, and repeat:

$$
\begin{aligned}
a_2^4 &= a_2 t\ t^{-1} a_2^3 t\ t^{-1}\\
&= a_2 t\ a_2 a_1 a_2 a_1 a_2 a_1\ t^{-1}\\
&= a_2 t\ a_2 t\ t^{-1} a_1 a_2 a_1 a_2 a_1 t\ t^{-2}\\
&= a_2 t\ a_2 t\ a_1 a_2 a_1 a_1 a_2 a_1 a_1\ t^{-2}\\
&\qquad\vdots
\end{aligned}
$$

The hydra battle

$$a_2 a_2 a_2 a_2 \ \rightarrow\ a_2 a_1 a_2 a_1 a_2 a_1 \ \rightarrow\ a_1 a_2 a_1 a_1 a_2 a_1 a_1 \ \rightarrow\ \cdots$$

plays out within this calculation. Pairing off the first letter with a t corresponds to removing the first letter of a hydra and conjugating a t^{-1} through to the right-hand end corresponds to regenerating. So, as Hercules wins after $\mathcal{H}(a_2^4) = 15$ steps, the process eventually arrives at a word on $a_1 t$ and $a_2 t$ times t^{-15}. This calculation, run to its conclusion, gives rise to the van Kampen diagram in Figure 8.10. There is nothing special about the example $k = 2$ and $n = 4$ here. So, as we know Hercules triumphs in $\mathcal{H}(a_k^n) = \mathcal{H}_k(n)$ steps, we have the answer $r = \mathcal{H}_k(n)$ to our question.

The diagram in Figure 8.10 can be paired with its mirror image, with three corridors of 2–cells arranged between them to give a van Kampen diagram that demonstrates the equality in G_2 of $a_2^5 t a_1 a_2^{-5}$ and a reduced word on $a_1 t$ and $a_2 t$ of length $2\mathcal{H}_2(4) + 3$. (Two copies of this diagram are shown in blue within the van Kampen diagram in Figure 8.11.) Similar diagrams can be constructed for all battles between Hercules and the hydra a_k^n, thereby showing that for all n and k, there are words on $\{a_1, \ldots, a_k, t\}^{\pm 1}$ of length $2n + 3$ that represent the same elements in G_k as certain reduced words on $\{a_1 t, \ldots, a_k t\}^{\pm 1}$ of length $2\mathcal{H}_k(n) + 3$. Given that $\mathcal{H}_k \simeq \mathcal{A}_k$, this establishes the lower bound on the distortion. See the article by Dison and me [115] for more details and a proof of the upper bound.

Now, G_k does not have a large Dehn function. Distortion does not always lead to a large Dehn function. But there are standard methods for translating

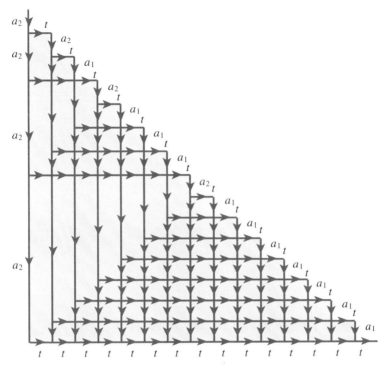

Figure 8.10 A van Kampen diagram showing that $a_2^4 t^{15}$ equals a word on $a_1 t$ and $a_2 t$
in $\langle a_1, a_2, t \mid t^{-1} a_2 t = a_2 a_1, t^{-1} a_1 t = a_1 \rangle$. Labels on the interior edges are
not shown; those on the horizontal edges are all t, and those on the vertical
edges are a_1 and a_2—which are which should be apparent from the defining
relations.

groups with heavily distorted subgroups to groups with a large Dehn function—see
Chapter III.Γ in [54] for a general discussion. We define Γ_k to be the group presented by $\langle G_k, p \mid [H_k, p] \rangle$, shorthand for the presentation obtained from our presentation of G_k by adding a new generator p and new defining relations $p\, a_i t =
a_i t\, p$ for each i, so that p commutes with all elements of H_k.

THEOREM 8.11. *The Dehn function of* Γ_k *presented by* $\langle G_k, p \mid [H_k, p] \rangle$ *is equivalent to the function* $\mathcal{A}_k(n)$.

The diagram in Figure 8.11 indicates how to get the lower bound. Any van
Kampen diagram with the same boundary must have a corridor of cells connecting
one of the edges labeled p to the another, since p only occurs in defining relations
of the form $p\, a_i t = a_i t\, p$. This corridor is shown in yellow in Figure 8.11. Along
the sides of this corridor we read words on $\{a_1 t, \ldots, a_k t\}^{\pm 1}$ that are necessarily
equal to words on $\{a_1, \ldots, a_k, t\}^{\pm 1}$ that we read around part of the boundary of the
diagram, since killing p maps $\Gamma_k \twoheadrightarrow G_k$. But as the distortion of H_k in G_k is $\succeq \mathcal{A}_k$,
the length of the corridor as a function of the length of the boundary circuit of the
diagram is $\succeq \mathcal{A}_k$.

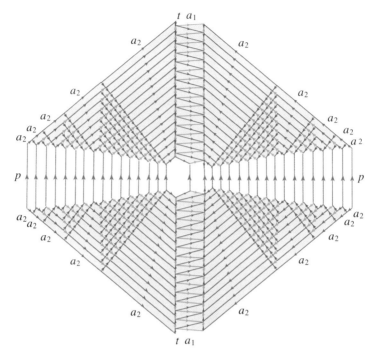

Figure 8.11 A van Kampen diagram for a word in $\langle G_2, p \mid [H_2, p] \rangle$, illustrating how heavy distortion of H_2 in G_2 leads to a large Dehn function. It is assembled from a pair of mirror image copies of diagrams that arise due to this distortion; they are separated by a corridor of faces (shown in yellow) which connects the two boundary edges labeled p.

Groups with undecidable word problem. One of the great achievements in twentieth century mathematics was the construction by Boone [44], Britton [60, 61], and Novikov [211] of finitely presented groups for which there can be no algorithm to solve the word problem. I recommend the account by Rotman [229].

The foundational non-computability result is Turing's proof that there is no algorithm for the *halting problem*: that is, no algorithm that will input a program p together with an input i for that program, and declare whether or not p eventually halts on input i. This can be proved by a Cantor-type diagonal argument; the existence of functions $\mathbb{N} \to \mathbb{N}$ that are not recursive immediately follows. Markoff [201] and Post [223] independently showed that there is no algorithm that solves the word problem for finitely presented semigroups. Roughly speaking, the point is that there is a strong analogy between the instructions of a Turing machine and a finite presentation for a semigroup. Moreover, running a calculation on the tape of a Turing machine resembles manipulating words in the generators of a semigroup using its defining relations. The idea for groups is the same, but there the result is much harder since the analogy is far more tenuous.

By Theorem 8.2, the Dehn function of a finitely presented group with unsolvable word problem is not bounded above by any recursive function. These are therefore examples for which, in this sense, the Dehn function must grow exceptionally quickly. (However, there are surprising subtleties here. Olshanskii [214] constructed an example of a finitely presented group for which there is no algorithm to solve the word problem, but on an infinite subset of $\mathbb{N}$ its Dehn function is bounded above by a quadratic function!)

FURTHER READING

A natural next step is Bridson's survey [52], which provides careful proofs of a number of the results discussed here, including the filling theorem and van Kampen's lemma, explains other techniques for establishing Dehn functions, and draws a variety of connections with other topics.

My article [47] on filling functions explores the interconnections between and applications of a variety of quasi-isometry invariants, including Dehn functions, that concern the geometry of van Kampen diagrams. The notes by N. Brady and by Short in the same volume address Dehn functions for groups displaying different manifestations of non-positive or negative curvature.

The book by Bridson and Haefliger [54] is a key resource for many topics in geometric group theory including Dehn functions, especially in the context of non-positive curvature.

Gromov instigated and inspired much of the explosion of work over the last thirty years on the geometry of discrete groups. Indeed, a substantial proportion of that research can be viewed as an exegesis of his book [155].

Gersten's survey [142] on isoperimetric functions and their analogs, isodiametric functions, which concern the diameters of disks rather than their areas, published in a companion volume of the same conference proceedings as Gromov's book, has also been influential and is where the term "Dehn function" was coined. It remains well worth reading. Also, Gersten's 1996 summer school notes [141] are readily accessible and include discussion of Dehn functions of hyperbolic and automatic groups.

Sapir's book [232] is a recent and wide-ranging survey that covers many areas of current research on Dehn functions and related topics.

PROJECTS

Project 1. Look ahead to the solution of the word problem for right-angled Artin groups given in Section 14.4 and use it to show that for each of these groups, there is a constant $C > 0$ such that the Dehn function $f(n)$ satisfies $f(n) \leq Cn^2$ for all n.

Project 2. Are there finite presentations in which it is easy to generate words that represent the identity, but hard to solve the puzzle?

Project 3. (I. Kapovich) Which functions are Dehn functions of infinite presentations of $\mathbb{Z} \times \mathbb{Z}$? In particular:

1. Can you get $n \mapsto n^3$? What general upper bound can you find for Dehn functions of infinite presentations of $\mathbb{Z} \times \mathbb{Z}$?
2. Can you get $n \mapsto n^{1/2}$?
3. How do your investigations change if you replace $\mathbb{Z} \times \mathbb{Z}$ with, for example, $\mathbb{Z} \times \mathbb{Z} \times \mathbb{Z}$?

Project 4. It is possible to vary the definition of the Dehn function by weighting the defining relators: assign a positive real number to each, which it contributes whenever it is applied. When all the weights are 1, this is the standard Dehn function.

1. Show that if $f(n)$ is the standard Dehn function and $f_w(n)$ is a weighted Dehn function of a finite presentation, then there exists $C > 0$ such that $f(n)/C \le f_w(n) \le Cf(n)$ for all n. (So weights do not significantly change Dehn functions of finite presentations.)
2. Revisit Project 2 admitting weighted Dehn functions.

Project 5. This project concerns $\langle x_1, x_2, \ldots, x_k \mid x_1, x_2, \ldots, x_k \rangle$ which is, of course, a presentation of the trivial group. Its Dehn function is easily seen to be $f(n) = n$ for all $n \in \mathbb{N}$, but determining the exact areas of individual words is not so straightforward.

For a word w on the generators and their inverses, let $f(w)$ denote the minimum number of application-of-defining-relator moves required to convert w to the empty word using a null-sequence. For example, for all $n > 0$, to convert x_1^n to the empty word the best one can do is remove one x_1 at a time, and so $f(x_1^n) = n$. However, $f(x_1 x_2 x_1^{-1} x_2^{-1}) = 2$ because after deleting the x_2 (say), the x_1 and x_1^{-1} can be canceled, and then the x_2^{-1} deleted. Deleting x_2 and x_2^{-1} costs 1 each, but canceling x_1 and x_1^{-1} is free.

1. Show that $f(x_1 x_2^2 x_1^{-1} x_2^{-2}) = 2$.
2. Show that $f(x_1^n x_2^n (x_1^{-1} x_2^{-1})^n) = 2n$ for all $n \in \mathbb{N}$.
3. Give an algorithm which on input w, will compute $f(w)$, halting in cubic time (as a function of the length of w).

(The method of dynamic programming can be used to answer to the third part. The same problem with some extra constraints that are driven by biological considerations (with no analog in group theory) goes by RNA secondary structure prediction. A dynamic programming algorithm solving it has been known to biologists for many years. The $x_1, \ldots, x_k$ correspond to nucleotides; an x_i and its inverse correspond to nucleotides that can form a matched base pair. The problem is to find a way for an RNA strand to fold against itself so that it maximizes the number of matched base pairs.)

Project 6. Define the *filling length function* $g \colon \mathbb{N} \to \mathbb{N}$ of a finitely presented group Γ to be the function mapping n to the minimal N such that for every word

w of length at most n that represents the identity, there is a null-sequence for w consisting of words all of which are of length at most N. (Compare this to how we defined the Dehn function $f(n)$ in Section 8.2.)

1. Show that for a finitely generated abelian group, there always exists some $C > 0$ such that $g(n) \leq Cn$ for all n.
2. Show that the same is true of $BS(1, 2)$ and H_3.
3. Show that for any finitely presented group, there exists $C > 0$ such that $g(n) \leq Cf(n) + n$ and $f(n) \leq C^{g(n)}$ for all n.
4. Deduce (i) that Γ has decidable word problem if and only if g is a recursive function, and (ii) that there are finitely presented groups for which g is not bounded above by a linear function but is recursive.
5. Give a geometric interpretation of g in the spirit of Section 8.3.

A further introduction to filling length can be found in [47].

Project 7. Explore the Dehn functions of one-relator groups (groups that can be presented with a single defining relation). Magnus [196] solved the word problem for any one-relator group. So their Dehn functions are recursive. But how fast can they grow? A longstanding problem of Gersten [140, 142] is whether the example of Theorem 8.10 is the fastest possible.

Project 8. Another challenging open problem due to Gersten is to determine how fast Dehn functions of finitely presentable linear groups can grow. The group of Theorem 8.8 is linear and has an exponential Dehn function. Is there a faster growing example?

Project 9. Here are two more problems that are open and likely very challenging. We saw (Theorem 8.11) how the hydra-phenomenon can be used to construct a group Γ_k whose Dehn function grows as $n \mapsto \mathcal{A}_k(n)$.

1. Elaborate on this construction so as to get an 'elementary' example of a finitely presented group whose Dehn function grows as $n \mapsto \mathcal{A}_n(n)$. (A first step might be to find an embedding of $\langle t, a_1, a_2, \ldots \mid t^{-1}a_1t = a_1, t^{-1}a_it = a_ia_{i-1} (\forall i > 0) \rangle$ in a finitely presented group.)
2. It follows from [37] and [113] that for all k, Γ_k embeds in a finitely presented group whose Dehn function grows no faster than polynomially. Find such an embedding explicitly. We can ask the same for the one-relator group of Theorem 8.11.

Office Hour Nine

Hyperbolic Groups

Moon Duchin

In most subjects of mathematics, the historical development starts with simple structures, which are later axiomatized, and then generations of mathematicians come along and layer on increasing structure and complexity. For instance, the theory of groups has early beginnings in the study of permutations (with important work on this by Cauchy in the 1810s), and the permutation group S_n is still one of the first examples you encounter in a twenty-first-century group theory class. The general axiomatic definition of a group as *a set where you can compose elements* came somewhat later (say with Cayley in the 1850s), and only still later does the record show people imposing extra structure on groups, such as topological or differentiable structure (notably by Lie in the 1870s).

The story of geometry goes the opposite way, somehow! The major foundational work in geometry in the modern tradition starts with surfaces and manifolds endowed with a topological and differential structure (Gauss and Riemann, particularly, in the 1820s–1850s). The idea of a geometric object as *a space where you can measure distance* doesn't enter the record until significantly later (early twentieth century), after the differential geometry had been very highly developed, and the extremely simple geometric models you've encountered in this book from the introduction on—such as the hub metric and taxicab metric, and graphs—could not have really been part of geometry until then. This book is in great part about developing this modern idea: that finitely generated groups can be regarded as geometric objects through the study of their Cayley graphs.[1] This office hour will explore those groups whose graphs are in an appropriate sense negatively curved.

Felix Klein made the prescient observation that the study of spaces and the groups that act on them represents really two sides of the same coin—but he was talking about the continuous examples that were already an accepted part of geometry in the 1880s. Current geometric group theory can perhaps be viewed as making good on the marriage of groups and spaces even when both are at their most wild, coarse, and unruly.

Curvature notions. *Curvature* is a fundamental way of understanding the intrinsic geometry of manifolds. There are three curvature regimes—positive, zero, and negative—and each has its poster child in the world of surfaces: the sphere, the plane, and the saddle.

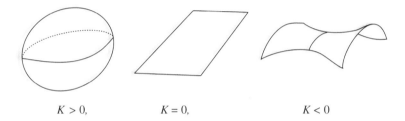

$$K > 0, \qquad K = 0, \qquad K < 0$$

Classically, curvature was defined through calculus on manifolds, then made very general through tensor calculus. This is extremely beautiful mathematics but it requires a lot of structure. The topic of this office hour, δ–hyperbolicity, is about exploring one way to make sense of curvature without derivatives or continuity, but directly from the distance function of a space under study. Before long we will be able to talk about the curvature of a fractal.

Consider the saddle surface above, in comparison to the plane or sphere. What are some of its qualitative geometric features? A crucial one is that, in a sense, the surface spreads out faster than a flat sheet. This can be measured in several ways; for instance, if you start at the saddle point and set out along straight (geodesic) paths in two different directions from it, those paths diverge quickly. Indeed, if one direction goes uphill on the saddle surface and another direction goes downhill, then there's no terribly efficient hypotenuse in the surface that connects those directions; the best connecting path travels back near the saddle point. (In a plane, geodesic paths diverge steadily or linearly; on a sphere, they diverge sublinearly, and in fact collide after a certain amount of time.)

There is a nice space that exhibits saddle point behavior at each of its points: the hyperbolic plane, $\mathbb{H}^2$, which you learned about in Office Hour 5. And it is a classical fact that all surfaces of negative total curvature admit a metric locally

---

[1] Working with groups as metric spaces got its start in *combinatorial group theory*, but I want to emphasize the extreme usefulness and flexibility of the metric space point of view beyond that initial context.

modeled on $\mathbb{H}^2$. But to study groups we want to capture this behavior in a discrete model, which earns them the name of "hyperbolic groups."

Below, we will see that we can mimic an extreme version of the qualitative behavior of a saddle in very simple metric spaces: trees. We will define δ–hyperbolic metric spaces to be those whose metric measurements are boundedly close to trees, with a discrepancy controlled by the number δ. One of the major features of this definition is that it achieves a strict generalization of classical notions of negative curvature: *a (complete, Riemannian) manifold is δ–hyperbolic for some $\delta > 0$ if there is a global negative upper bound κ on curvature (i.e., sectional curvature satisfies $K \leq \kappa < 0$).*

9.1 DEFINITION OF HYPERBOLICITY

Again, our definition of δ–hyperbolicity will be inspired by some of the extreme behavior exhibited by trees. So we'll start by exploring trees.

Triangles in trees. As in Office Hour 3, a *tree* is a graph with no cycles—for a path to close up, it is forced to backtrack. Trees can be regarded as metric spaces by just endowing each edge with length 1 and using the path metric. They have a very distinctive property as metric spaces: all (geodesic) triangles are tripods. What I mean by this is that if you take any three points in a tree and you take the three shortest paths connecting them pairwise, this forms either a line (if one of the points is between the other two) or, at its most complicated, a three-legged figure as seen below. (You can quickly satisfy yourself that all triangles must have this form, because if the edge $\overline{AC}$ were not totally contained in $\overline{AB} \cup \overline{BC}$, a cycle would be present.)

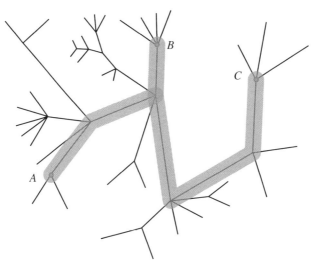

This means that every triangle has a vertex common to all three sides. That sets trees apart from most spaces you know, and it gives them some remarkable properties. Let's start with one: if you consider a pair of geodesic (meaning, non-backtracking) rays with a common base point, then they agree exactly up to a certain point; to connect later points on the geodesics you must pass back through that point of divergence. (You can see this in the figure considering the geodesics $\overline{BA}$ and $\overline{BC}$, say.) To summarize: *geodesics agree totally until they diverge completely, and then they start traveling in opposite directions along a new geodesic.* This is a more extreme version of the behavior observed on the saddle surface.

Rips and Gromov and others noticed that this property captures something fundamental about a big class of interesting spaces: let us call a space δ–*hyperbolic* (for some particular number $\delta \geq 0$) to mean that in some sense (to be detailed below) it has a metric that is δ–close to being a tree. If we talk about hyperbolicity without specifying a value of δ, then we just mean that it is δ–hyperbolic *for some* δ. (For the most part there is no very special value of δ, except for 0—in particular, you can simply rescale the metric on a space to change the δ while preserving the qualitative geometry.)

There are now a billion equivalent definitions of δ–hyperbolicity; let me give a common one, a cute one, and a general one. Later, I'll lay out some geometric consequences of hyperbolicity, but it is worth mentioning that almost every one of those can be formulated as an alternate definition! To me that indicates that this is a very natural class of spaces to study: the class has striking properties and is characterized by each.

Thin triangles. The most commonly seen definition of δ–hyperbolicity is that *triangles are thin.*[2] That is, if you have a geodesic triangle—three points, pairwise connected by geodesic segments α, β, and γ—it is said to be δ–thin if $\alpha \subset \mathcal{N}_\delta(\beta \cup \gamma)$ and the other two similar inclusions hold. (Here, $\mathcal{N}_r(A)$ denotes the r–neighborhood of the set A.)

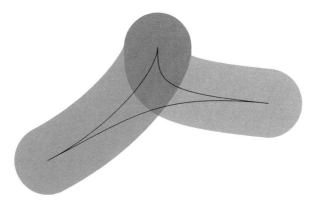

So this is a very natural relaxation of the observation we made above about trees: instead of the third side having to be totally contained within the union of the other

---

[2] Some authors use thin triangles, slim triangles, and fine triangles for different conditions and the terminology varies, but I think this is the most standard.

two, it's merely in the δ–neighborhood of that union. Thus, in particular, we see that trees are 0–hyperbolic.

Insize.   Back in Euclidean geometry, one defines the incircle of a triangle to be the largest inscribed circle. The points of tangency are called inpoints, and they cut the three sides of the triangle into six pieces that come in length-matched pairs.

Now in a more general space, we can't necessarily find an inscribed circle in any nice way, but we can generalize the other property: let the inpoints of a geodesic triangle be the uniquely determined three points that divide the sides into pairs of equal lengths, as shown in the figure below. (Why are they uniquely determined? Because you're just solving the system $a = r + s$, $b = s + t$, $c = r + t$, and the triangle inequality guarantees a solution.)

Recalling that the diameter of a set A in a metric space is defined as $\mathrm{diam}(A) = \sup\{d(x, y) : x, y \in A\}$, we can let the insize of a geodesic triangle be the diameter of its set of three inpoints. (That is, it's the largest of the three pairwise distances.)

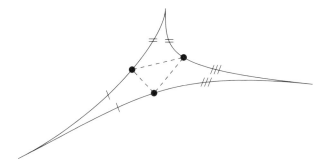

Now you can probably guess the insize definition of hyperbolicity. (Stop and try!) Tripods, of course, have insize 0 because all three inpoints are the same: the guaranteed point that is simultaneously on all three sides. So we'll call a space δ–hyperbolic if all geodesic triangles have insize $\leq \delta$.

Four-point condition.   One drawback of the definitions above, and all of the many tweaks on them that produce equivalent definitions, is that they only work in geodesic spaces (i.e., spaces where any two points are connected by a geodesic). Gromov had the insight that there are corresponding purely metric notions, definable directly from the distance function. Despite the fact that the prior definitions are all about triangles, hyperbolicity is actually a *four-point condition*, meaning that it can be detected in a metric space just from data about the pairwise distances between four-tuples of points. (Three points is not enough! You need the frame of a triangle plus at least some data about its middle to get anywhere.)

Define the Gromov product to be

$$(x \cdot y)_w = \frac{1}{2}\big(d(x, w) + d(y, w) - d(x, y)\big).$$

In a tree, this is the distance from w to the geodesic segment $\overline{xy}$, and it equals 0 if and only if w is between x and y. (Draw the picture!) More generally, it is proportional to the failure of the triangle inequality to be an equality: if the side lengths of a triangle are called a, b, and c, then the triangle inequality ensures that $a + b \geq c$, and its *defect* is the value $\Delta = a + b - c \geq 0$. The Gromov product is just half of this defect.

Once again, what's true in a tree? Let's think of x, y, and z as vertices of a triangle and w as some observation point. How can you get to the vertices from w? Well, w can only be connected by a path outside the triangle to one single point (say p) on the triangle—otherwise, we've formed a cycle, contradicting treeness. But that point p is on at least two of the tree sides of the triangle. And that means if you consider the distances from w to each of the three sides, there's guaranteed to be a tie for the smallest distance. So in trees, for all x, y, z, and w,

$$(x \cdot y)_w \geq \min\big((y \cdot z)_w, \ (x \cdot z)_w\big),$$

with equality if p is between x and y. (Draw the picture!) As usual, we pass to hyperbolic spaces by relaxing this condition. This might produce a circumstance where one of the Gromov products is now smallest of all, but it can't be smallest by too much! That is, we say a space is δ–hyperbolic if all four-tuples satisfy

$$(x \cdot y)_w \geq \min\big((y \cdot z)_w, \ (x \cdot z)_w\big) - \delta.$$

How big of an improvement is this over the geodesic-based definitions? Well, it lets us establish that all kinds of really metrically hideous spaces are in this class, so we get all the spoils of hyperbolicity (and there are many). For instance, a space is called *ultrametric* if all triples satisfy

$$d(x, y) \leq \max\big(d(y, z), \ d(x, z)\big),$$

which precisely says that in every triangle there must be a tie for the longest side. (In particular, all triangles are isosceles.) Ultrametric spaces come up all over number theory, and the key example is the p–adic numbers, a fractalesque space defined by completing the rationals with respect to a distance function keyed to a prime p. This metric is totally disconnected (every point has an arbitrarily small neighborhood that is both closed and open), so there are certainly no geodesics. But the ultrametric inequality ensures 0–hyperbolicity, so all ultrametric spaces are fundamentally tree-like!

Exercise 1. First show that every triangle in every ultrametric space is isosceles (in fact, something slightly stronger: of the pairwise distances among any three points, the top two are equal). Then, working with the four-point condition, show that ultrametric spaces are 0–hyperbolic.

Invariance. When one says that these definitions of δ–hyperbolcity are equivalent, it is important to realize that the value of δ can change when you maintain the same metric space but switch between the definitions above (except $\delta = 0$, which is a distinguished value). See Exercise 2 below.

A crucial property, moreover, is that hyperbolicity is stable under quasi-isometry.[3] Recall from Office Hour 7 that quasi-isometries are maps of metric spaces that are only boundedly far (additively and multiplicatively) from being isometries. That is, two spaces are (K, C)–quasi-isometric to each other if there are maps both ways satisfying the fundamental inequality

$$\frac{1}{K}d_2 - C \le d_1 \le Kd_2 + C,$$

whose composition moves points no more than a bounded amount. (Here, d_1, d_2 are the distance functions for the two spaces; an isometry would satisfy $d_1 = d_2$.) And accordingly a *quasi-geodesic* is the image of the real line under a quasi-isometric embedding.

If you fix a particular δ, then of course the property of being δ–hyperbolic with constant δ can be destroyed by quasi-isometry. For instance, 0–hyperbolicity can clearly be ruined by replacing the vertices in a tree by some finite graph (as in the figure on page 186, where that graph is a triangle)—but the spaces are clearly quasi-isometric. (Actually this is a good test of the basic concept: check that you can spell out what the maps are both ways.) However, the good news is that if X is δ–hyperbolic, then after applying a (K, C)–quasi-isometry, it is still hyperbolic with some new constant δ' depending only on δ, K, and C.

Thus we have finally arrived at a precise definition of a *hyperbolic group*: a finitely generated group is called hyperbolic if any of its Cayley graphs (for a finite generating set) is δ–hyperbolic. And invariance under quasi-isometry means we don't have to worry about whether we've chosen good generators; any finite set will do.

9.2 EXAMPLES AND NONEXAMPLES

Let's look at some examples of groups and spaces that are δ–hyperbolic and also look at some obstructions to δ–hyperbolicity.

Example: Trees and free groups.   Since trees are hyperbolic, we obtain our first examples of hyperbolic groups: free groups F_n of any rank, since their Cayley graphs with standard generators are just $(2n)$–regular trees.

Example: Finite groups.   All bounded spaces (and thus all finite groups) are trivially seen to be hyperbolic for an appropriately large δ; just choose it larger than the diameter of the space or Cayley graph. Thus there is a precise sense in which the theory of finite groups is trivial to a geometric group theorist. (Be careful about saying this out loud in mixed company.)

---

[3] Actually, there's an important technical caveat here! If the spaces are not length spaces (i.e., where the distance between two points is the infimal length of a path between them), this can fail. Bridson–Haefliger give the example of an embedded spiral in the Euclidean plane, with the restriction of the Euclidean metric. This is quasi-isometric to the real line, but you can place vertices on the spiral that form an arbitrarily fat triangle in the plane.

Example: The hyperbolic plane. This fundamental object was introduced in Office Hour 5. The space $\mathbb{H}^2$ puts the hyperbolic in hyperbolicity! Here are the facts about $\mathbb{H}^2$ that I would like you to have in mind for the rest of this office hour (see Office Hour 5 for more details):

- $\mathbb{H}^2$ has an upper half-plane model and a disk model, and in both cases the geodesics are *orthocircles*, which are lines/circles perpendicular to the boundary.
- $\mathrm{SL}(2, \mathbb{R})$ acts on the upper half-plane by fractional linear transformations

$$\begin{pmatrix} a & b \\ c & d \end{pmatrix} \cdot z = \frac{az + b}{cz + d},$$

  which are isometries of the hyperbolic metric, and composition of fractional linear transformations corresponds to multiplication of the associated matrices.
- In particular, this means that

$$\begin{pmatrix} k & 0 \\ 0 & 1/k \end{pmatrix} \cdot z = k^2 z,$$

  while it looks like a dilation to Euclidean eyes, is an isometry of this metric— called a hyperbolic isometry—for $k \neq 0$.
- $\mathrm{PSL}(2, \mathbb{R})$ acts triply transitively on the boundary: that is, for any two triples of distinct points on the boundary, there is an isometry of $\mathbb{H}^2$ carrying the first triple to the second. In particular, any geodesic can be mapped to any other geodesic (such as the imaginary axis of the half-plane model) by an isometry.
- Distance on the y–axis in the upper half-plane model is given by the simple formula $d(ai, bi) = |\ln b/a|$.
- In particular, the family of maps $z \mapsto k^2 z$ pushes points along the y–axis at a constant speed with respect to k, with one attracting and one repelling endpoint on the boundary (∞ and 0, respectively).
- Conjugates of this basic hyperbolic isometry have the same behavior with any two boundary points as fixed points and the geodesic between them as an axis.

To verify the thin triangles property for the hyperbolic plane $\mathbb{H}^2$, it suffices to consider ideal triangles: triangles with all three vertices on the boundary at infinity. (Make sure you see why! In particular, if one triangle is totally contained in another, the thinness constant for the larger triangle works for the smaller one too.) Hyperbolicity is then obvious because any ideal triangle is isometric to any other—just pick your favorite ideal triangle to find the constant.

And in fact the same argument works for insize, once you check that in $\mathbb{H}^2$ the inpoints are once again (as in $\mathbb{R}^2$) precisely the points of tangency of the largest inscribed circle. Thus it is the diameter of that circle that controls the insize. Now we can once again pass to any ideal triangle; it has some largest inscribed circle, so hyperbolicity is verified.

Exercise 2. First, prove "synthetically" (without too much calculation) that the following fact from $\mathbb{R}^2$ is still true in $\mathbb{H}^2$: *every geodesic triangle has a unique circle tangent to all three sides, and this circle cuts the sides into three pairs of equal-length segments.* This will help with insize arguments.

For the upper half-plane model of $\mathbb{H}^2$ with the metric given above, show that the optimal hyperbolicity constant is $\delta = \ln(1 + \sqrt{2})$ in the thin triangles definition, and $\delta = \ln\left(\frac{3+\sqrt{5}}{2}\right)$ in the insize definition. Find the optimal constant in the four-point definition.

Non-examples: Flats and dilation.   It's also easy to see that $\mathbb{R}^2$ itself can't possibly be δ–hyperbolic for any value of δ, no matter how large—that was quite possibly the first thing you noticed when you saw the definitions. If you just take any triangle and grow it by dilation, both the thinness constant and the insize grow without bound, as long as they weren't zero to begin with. So in fact this shows us one property that δ–hyperbolic spaces share: they have no dilations/similarities (that is, there is no family of maps $f_t : X \to X$ for $t \in (0, \infty)$ such that $d(f_t(x), f_t(y)) = t \cdot d(x, y)$).[4]

This argument, that dilations let you "blow up" nondegenerate triangles, shows that no δ–hyperbolic space can have an isometrically embedded copy of $\mathbb{R}^2$ (this is called a *flat*) or even a quasi-isometrically embedded $\mathbb{R}^2$ (a *quasi-flat*).

In the world of groups, $\mathbb{Z}^2$ is also non-hyperbolic, which you can see by its quasi-isometric equivalence to $\mathbb{R}^2$ or by drawing fat triangles in the Cayley graph. Let us consider what this says about free abelian subgroups of hyperbolic groups.

For a subgroup $H \leq G$, one can compare the word metric on H with the word metric on G. If you choose finite generating sets so that $S_H \subset S_G$, then in general lengths in the respective word metrics must satisfy $|h|_H \geq |h|_G$. If the ratio $|h|_H/|h|_G$ is bounded above, we will call the subgroup *undistorted*. The interpretation is that distant elements of H in the Cayley graph of G can be efficiently connected within the subgraph (cf. Office Hour 8). Note that changing generating sets might change the bound, but not the fact of boundedness!

You can see a quick example of distortion in the Heisenberg group

$$H(\mathbb{Z}) = \left\{ \begin{pmatrix} 1 & x & z \\ 0 & 1 & y \\ 0 & 0 & 1 \end{pmatrix} \right\} \cong \langle a, b, c \rangle,$$

$$a = \begin{pmatrix} 1 & 1 & 0 \\ 0 & 1 & 0 \\ 0 & 0 & 1 \end{pmatrix}, \quad b = \begin{pmatrix} 1 & 0 & 0 \\ 0 & 1 & 1 \\ 0 & 0 & 1 \end{pmatrix}, \quad c = \begin{pmatrix} 1 & 0 & 1 \\ 0 & 1 & 0 \\ 0 & 0 & 1 \end{pmatrix},$$

in which you can check that c commutes with a and b, but the commutator $[a, b] = aba^{-1}b^{-1}$ satisfies $[a, b] = c$. (Note that this relation tells us that c was a redundant generator and that $H(\mathbb{Z}) = \langle a, b \rangle$.) Just as $\mathbb{Z}^2$ sits in $\mathbb{R}^2$ as a lattice of points, this discrete Heisenberg group sits in an ambient matrix group $H(\mathbb{R})$, which is defined the same way except with real entries.

The cyclic subgroups $\langle a \rangle$ and $\langle b \rangle$ are each undistorted copies of $\mathbb{Z}$. On the other hand, the central cyclic subgroup $\langle c \rangle$ is quadratically distorted, because

---

[4] Of course, the real line $\mathbb{R}$ is an exception, but the only exceptions should be a small subset of 0–hyperbolic cases. Can you describe them completely?

$c^{n^2} = [a^n, b^n]$ (check this!), so what takes quadratic word length (n^2) in the letter c can be rewritten with linear length ($4n$) in the big group.

We will use the terms flat or quasi-flat in groups to refer to undistorted copies of $\mathbb{Z}^2$; this is clearly an obstruction to hyperbolicity because of its fat triangles, whose distances in the big group are still about as big.

Note that so far we found no such obstruction to hyperbolicity in the Heisenberg group $H(\mathbb{Z})$, and indeed its $\mathbb{Z}^2$ subgroups are all distorted. It's still not hyperbolic, however! One nice way to see this is by the presence of an almost-similarity:

$$\begin{pmatrix} 1 & x & z \\ 0 & 1 & y \\ 0 & 0 & 1 \end{pmatrix} \xrightarrow{\alpha_n} \begin{pmatrix} 1 & nx & n^2z \\ 0 & 1 & ny \\ 0 & 0 & 1 \end{pmatrix}.$$

What I mean by calling that an almost-similarity is that it comes boundedly close to satisfying the dilation condition given above; in fact, it turns out that any word metric on $H(\mathbb{Z})$ is quasi-isometric to a metric on $H(\mathbb{R})$ where the corresponding maps $\{\alpha_t\}_{t>0}$ are true similarities. And that is plenty good enough to rule out the possibility of hyperbolicity.[5]

Exercise 3. Consider the four points $1, a, b,$ and c and check that applying α_n gives the four points $1, a^n, b^n,$ and c^{n^2}. Compute (or closely estimate) the six pairwise distances between these points with respect to the generating set $H(\mathbb{Z}) = \langle a, b \rangle$ and use that information to conclude that $H(\mathbb{Z})$ is not hyperbolic.

A poison subgroup. Above we observed that undistorted copies of $\mathbb{Z}^2$ are obstructions to hyperbolicity. In fact, something much stronger is true: *any subgroup isomorphic to $\mathbb{Z}^2$, no matter how distorted, rules out the possibility of hyperbolicity.* This is somewhat more sophisticated, but here's an outline of the steps.

(1) Show that any infinite-order element g has $\langle g \rangle = \{g^n\}$ as a *quasi-axis*: a quasi-geodesic which is setwise preserved by g, so that g has the effect of pushing points along this axis. (Only the quasi-geodesity needs to be checked, and that step requires some care.)

(2) There may be a choice of different quasi-axes for g, but they are all mutually close together. (Showing this is just a careful application of thin triangles.)

(3) Now suppose h is another infinite-order element that commutes with g, so that $\langle g, h \rangle$ generates a $\mathbb{Z}^2$ subgroup of the hyperbolic group. Then h must permute the quasi-axes of g.

(4) That means that h takes every point on a quasi-axis uniformly close to any other quasi-axis, in particular to the one formed by powers of g, so h is close to $\langle g \rangle$. If we wrote things down precisely, we would state that the

---

[5] In fact, the Heisenberg group is a simple example of a class of groups called *nilpotent groups*; these turn out to be characterized by their having almost-similarities. Nilpotent and hyperbolic groups barely overlap: the only groups that are both are virtually cyclic!

centralizer $C(g)$—the set of all elements commuting with g—is virtually $\mathbb{Z}$, meaning in this case that $C(g)$ has $\langle g \rangle$ as a subgroup of finite index. But $\langle g \rangle$ has infinite index in $\langle g, h \rangle \cong \mathbb{Z}^2$, which is a contradiction, so we are done.

So $\mathbb{Z}^2$ is a "poison subgroup" for hyperbolicity. In some classes of groups, it is the only obstruction! For example, Moussong showed that Coxeter groups are hyperbolic if and only if they contain no $\mathbb{Z}^2$. This is now known to be true for non-positively curved 3–manifold groups as well (i.e., fundamental groups of three-dimensional manifolds, which I'll explain below). For higher-dimensional manifold groups, this is an open problem.

Example: SL(2, $\mathbb{Z}$). So far our only examples of hyperbolic groups are free groups (including $\mathbb{Z}$), which have Cayley graphs that are not just tree-like; they *are* trees.

Here is a group example with $\delta > 0$. As you've seen in Office Hour 3, $\mathrm{PSL}(2, \mathbb{Z})$ is generated by (the images of) $\sigma = \left(\begin{smallmatrix} 0 & -1 \\ 1 & 0 \end{smallmatrix} \right)$ and $\tau = \left(\begin{smallmatrix} 1 & 1 \\ 0 & 1 \end{smallmatrix} \right)$, via the presentation

$$\mathrm{PSL}(2, \mathbb{Z}) = \langle \sigma, \tau \mid \sigma^2, \ (\sigma\tau)^3 \rangle.$$

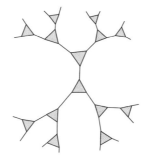

But this means we could modify the presentation by defining $\nu = \sigma\tau$ and taking $\langle \sigma, \nu \mid \sigma^2, \ \nu^3 \rangle$. (This connects us with Office Hour 3 again.) With this presentation, the Cayley graph looks just like a tree of triangles, a graph that clearly has the thin triangles property with $\delta = 1$.

Since $\mathrm{SL}(2, \mathbb{Z})$ has $\mathrm{PSL}(2, \mathbb{Z})$ as a quotient of degree 2, this means that the bigger group is hyperbolic as well. (As we saw in Office Hour 7, the quotient map induces a quasi-isometry of groups.)

9.3 SURFACE GROUPS

Next up is a major class of examples of hyperbolic groups: fundamental groups of closed hyperbolic surfaces, also known as surface groups. Each manifold S has an associated group $\pi_1(S)$, called the fundamental group, which can be thought of as group-theoretically encoding information about the topology of S. (And this works equally well for manifolds of dimension greater than 2.) Fundamental groups of graphs are defined in Office Hour 4, and it was mentioned then that any space has a fundamental group. We begin now with a brisk introduction to the fundamental group of an arbitrary space.

Classifying manifolds. The idea for the fundamental group comes from Henri Poincaré circa 1900. He was interested in studying manifolds, which are generalizations of surfaces to arbitrary dimension—an n–manifold is locally modeled on $\mathbb{R}^n$ just as a surface locally looks like $\mathbb{R}^2$. For instance, the sphere S^n in any dimension is obtained by taking all points at distance 1 from the origin in $\mathbb{R}^{n+1}$; the n–torus $T^n = S^1 \times \cdots \times S^1$ is a Cartesian product of n copies of the circle. In dimension 1, the 1–sphere and the 1–torus are both equal to the circle, but for $n > 1$ they give different manifolds.

Since it's very hard to visualize these high-dimensional manifolds, how can you even tell when two that are given to you with different descriptions are the same or different? How can you be sure that S^{16} isn't actually the same topological space as T^{16}?

Poincaré's fundamental idea here is to associate a *group* to each manifold, the fundamental group of the manifold, that depends only on the topology of the manifold, so that two manifolds with different groups are genuinely different. Such a contraption would then give us a way to distinguish between manifolds. The dream would be to have a result as clean and complete as the *classification of surfaces*, which says that *any (connected, closed, orientable) surface is homeomorphic to exactly one from this list*:

The number of holes is called the *genus* and is usually denoted by g; we will denote the surface of genus g by Σ_g. For reasons that will become clear below, the ones with genus $g \geq 2$ are called the hyperbolic surfaces. Each of the Σ_g has a different fundamental group $\pi_1(\Sigma_g)$, and we will compute these groups below. (Also see Office Hour 17 for more details about surfaces and the classification.)

In dimension 3 the situation is near-hopelessly more complicated. The Poincaré conjecture—*if a closed 3–manifold has the same fundamental group as S^3, then it is homeomorphic to S^3*—was open for one hundred years until Grisha Perelman settled it (and a great deal more about the structure of 3–manifolds) in his breakthrough work at about the turn of the millennium. (It is easy to build non-homeomorphic manifolds with the same fundamental group, which is why the Poincaré conjecture has content.)

The fundamental group. So now for the definition: the *fundamental group* of a topological space X with some chosen base point x_0, denoted by $\pi_1(X, x_0)$, is the group of loops, which are closed paths on X starting and ending at x_0. Of course, it's important to say when two of these are equivalent: we identify loops that are *homotopic*, which means that one loop can be continuously deformed to the other so that every intermediate loop along the way is still a loop based at x_0. The trivial element of the group is the constant loop that sits at x_0.

What makes π_1 into a group is the fact that loops can be concatenated by taking first one, then the other, obtaining another loop based at x_0. And in fact, if your space is path connected, then the base point doesn't matter; you can conjugate from one to the other. That is, suppose y_0 is some other base point, γ is some fixed (and directed) path from x_0 to y_0, α is any loop based at x_0, and $a * b$ denotes the concatenation path that follows a, then b. Then $\gamma^{-1} * \alpha * \gamma$ is a loop based at y_0 corresponding to α. So from now on we will drop the base point and just refer to $\pi_1(X)$. Once you've formalized what homotopy and homeomorphism mean, you can easily show that homeomorphic spaces have isomorphic fundamental groups.

Exercise 4. Make precise our definition of fundamental group (consult any text on algebraic topology if you need help).

Exercise 5. Show that the definition of the fundamental group of a graph from Office Hour 4 agrees with the general definition given above.

Example: Trees.   For example, for a tree T, the group $\pi_1(T)$ is the trivial group. That's because trees have no cycles, so any loop is degenerate (it just backtracks completely), so it can be continuously deformed down to the trivial loop. The same is true of the plane or the sphere S^2, where it's pretty easy to convince yourself that any loop is nullhomotopic (can be shrunk down to the trivial loop). Spaces with trivial π_1 are called *simply connected.*

Example: The circle.   The simplest example with nontrivial fundamental group is the circle S^1. There, if you take any loop, if a subarc backtracks along the circle, that detour can be deformed away. This gives us the idea that the homotopy class of a loop is characterized just by how many full turns around the circle it makes, keeping track of the positive and negative directions. (This can be made precise, for instance, in the same way that *winding number* is shown to be well defined in complex analysis.) So $\pi_1(S^1) \cong \mathbb{Z}$.

Exercise 6. Show that $\pi_1(S^1) \cong \mathbb{Z}$.

Example: Spheres.   Higher-dimensional spheres are simply connected again. Indeed, take any loop in S^n, and deform it so that it misses at least one point on the sphere, without loss of generality the north pole. Then puncture the sphere there; the rest of the sphere is homeomorphic to $\mathbb{R}^n$ by stereographic projection, and we can shrink a loop in $\mathbb{R}^n$ to the origin just by running the homotopy through dilations. (Note that this argument fails on S^1 because typical loops hit every point in the space—those that don't are indeed nullhomotopic.)

Example: The figure 8.   Another key example is the figure 8, or wedge of circles, $S^1 \vee S^1$. Suppose you call the simple loops around the circles a and b, respectively. Then it's pretty easy to see that you can realize any word in $\{a, b, a^{-1}, b^{-1}\}$ as a closed loop on the figure 8. Furthermore, distinct words give homotopically distinct loops, with the exception of the obvious cancellations that occur when a letter is followed by its inverse. Thus the fundamental group is the free group on two generators: $\pi_1(S^1 \vee S^1) \cong F_2$. (This agrees with Office Hour 4!)

Example: The torus. How about the torus $T^2 = S^1 \times S^1 = \mathbb{R}^2/\mathbb{Z}^2$? Let's think about it as the square with opposite sides identified, say the square S in $\mathbb{R}^2$ with vertices at $(0, 0)$, $(0, 1)$, $(1, 0)$, and $(1, 1)$; the images of these four points in the torus are all the same and give a natural choice of base point.

One nice way to understand all closed loops on the torus is to consider the close relationship between T^2 and the full plane $\mathbb{R}^2$. The plane is tiled with squares that are translated copies of S (secretly, if you know algebraic topology, $\mathbb{R}^2$ is the universal cover of $\mathbb{T}^2$).

Any path in $\mathbb{R}^2$ that starts at a corner of a copy of S and ends at another corner gives a loop in T^2. Actually, this process can be reversed: any based loop in T^2 can be lifted to such a path.

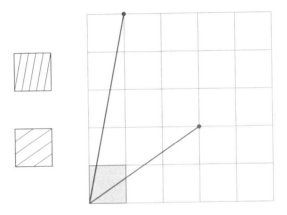

In particular, there's a loop on T^2 for every straight line segment in $\mathbb{R}^2$ from $(0, 0)$ to any $(p, q) \in \mathbb{Z}^2$. (And of course if you have an arbitrary curvy path from $(0, 0)$ to (p, q), you can straighten it out with a homotopy.)

So let's just check that there can't be a homotopy on the torus that identifies a path descended from $\overline{(0, 0)(p, q)}$ with one descended from $\overline{(0, 0)(p', q')}$ for $(p, q) \neq (p', q')$ (the orientation of the loop is important here!). The figure depicts such paths for $(1, 4)$ and $(3, 2)$. This argument isn't too bad. It basically reduces to the circle case in the following way: the path $\overline{(0, 0)(p, q)}$ can be projected to either axis, where it becomes $\overline{0p}$ on the x–axis and $\overline{0q}$ on the y–axis; on the torus, these axes are given by the horizontal and vertical sides of S, which are circles because of the gluings. But then if you had any homotopy between the two curves here, you could just look at its shadows on the axes to obtain a way to continuously deform a path that wraps p times around the circle to one that wraps around p' times, and likewise with q and q'. That can't happen! So we've seen that $\pi_1(T^2) \cong \mathbb{Z}^2$ (really, we found a bijection of the two sets; you can show that it is a homomorphism). Compare this with our construction of the complex K for $\mathbb{Z}^2$ in Office Hour 8.

Exercise 7. Check directly that the standard generators for $\pi_1(T^2)$ (namely, the elements corresponding to the standard generators for $\mathbb{Z}^2$) commute, by drawing their commutator and showing that it is homotopic to the identity. *Hint: Draw the torus as a square with sides identified.*

You can really see the $\mathbb{Z}^2$ fundamental group in the planar model: by using some combination of $(1, 0)$ and $(0, 1)$, you can build an arbitrary closed curve, up to homotopy. In fact, this argument generalizes in an obvious way to get that $\pi_1(T^n) \cong \mathbb{Z}^n$.

Incidentally we've shown that $T^{16} \not\cong S^{16}$ along the way, since they have different fundamental groups; $\pi_1(T^{16}) \cong \mathbb{Z}^{16}$ while $\pi_1(S^{16}) = 1$.

Tool: The Seifert–van Kampen theorem. We have discussed the simplest examples of fundamental groups, and now we introduce a powerful combination theorem for building up more complicated examples from these. Suppose we want to determine $\pi_1(X, x_0)$ for a topological space X that is covered by a collection of open sets U_i, all containing x_0, such that all threefold intersections $U_i \cap U_j \cap U_k$ are path-connected (i, j, k not necessarily distinct). Then

$$\pi_1(X) = \left\langle \cup S_i \mid \cup R_i \bigcup \cup R_{ij} \right\rangle,$$

where the S_i are the generators for the $\pi_1(U_i)$, the R_i are the relators for the $\pi_1(U_i)$, and the sets R_{ij} consist of relators as follows: for each loop in an overlap $U_i \cap U_j$, identify its representative in $\pi_1(U_i)$ with its representative in $\pi_1(U_j)$. This is the Seifert–van Kampen theorem. As a baby application, this gives another way to see that $\pi_1(S^2)$ is trivial: cover the sphere with two sets, namely fattened-up northern and southern hemispheres. The overlap is an annulus which without loss of generality contains the base point, and the other condition is (vacuously) met. But the disk has trivial fundamental group; so the sphere does too.

The Seifert–van Kampen theorem has a particularly useful corollary for the setting where a space is built up as a *cell complex*. Let's recall the definition of a 2–complex (refer to Office Hour 8). A graph is built up with vertices (zero-dimensional cells, or 0–cells) and edges (1–cells). We can rename it a 1–complex, and then proceed to define a 2–complex as a 1–complex (= graph) together with some polygonal faces (2–cells) glued in so that the edges of the boundary of the 2–cells map to an edge path in the graph. For instance, the tiling of $\mathbb{R}^2$ by unit squares realizes the plane as a 2–complex.

A corollary of the Seifert–van Kampen theorem says that if $X^{(1)}$ is a connected graph and X is a 2–complex obtained by gluing in 2–cells, then $\pi_1(X)$ is obtained from $\pi_1(X^{(1)})$ by adding relators for the boundary loops of the faces. (Think about how to derive this from the Seifert–van Kampen theorem. It begins with an appropriate fattening of the pieces.)

A simple application of this corollary: the torus, revisited. The square with gluings gives a 2–complex for which the underlying graph is the figure 8, with fundamental group $F_2 = \langle a, b \rangle$. The boundary loop of the square is $aba^{-1}b^{-1}$. So $\pi_1(T^2) = \langle a, b : aba^{-1}b^{-1} \rangle \cong \mathbb{Z}^2$.

Surface groups. Now let's dive in with something a bit harder: the surface of genus 2. It's not too hard to believe that the four loops in the picture are nonredundant generators for $\pi_1(\Sigma_2)$. What takes more work is to see that these four generators suffice, and that they have only one relation among them, which is a product of commutators.

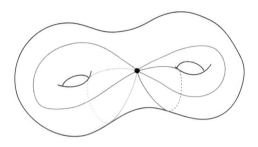

THEOREM 9.1. *For $g \geq 0$, the fundamental group of Σ_g is presented as:*

$$\pi_1(\Sigma_g) = \langle a_1, b_1, \ldots, a_g, b_g \mid [a_1, b_1][a_2, b_2] \cdots [a_g, b_g]\rangle.$$

These fundamental groups of Σ_g are called *surface groups*. In the special case $g = 1$, we recover the earlier fact that $\pi_1(T^2) \cong \mathbb{Z}^2$, and when $g = 2$, we obtain the presentation

$$\pi_1(\Sigma_2) = \langle a, b, c, d \mid [a, b][c, d]\rangle.$$

We'll give a Seifert–van Kampen proof in a moment, but first let's draw a picture: we want a model to draw the curves where we understand the geometry very well, and then we can classify the closed curves up to homotopy as we did above for the torus. In the case of a surface Σ_g of genus $g \geq 2$, that model is $\mathbb{H}^2$. (In this figure, $\mathbb{H}^2$ is drawn with its disk model rather than its upper half-plane model for pleasing visual symmetry.)

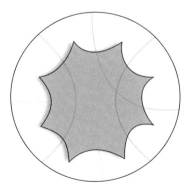 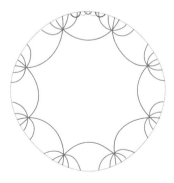

Just as the torus is given by a square with opposite sides identified, this model builds the surface Σ_2 from an octagon with side pairings, as in the picture on the left—the colored axes connect the sides to glue. This is not the only way to glue the sides of an octagon to obtain a surface of genus 2, but it works, as you should check.

All eight vertices are identified to a single point on the surface by these gluings, and the four colored arcs in the picture become loops on Σ_2 that are homotopic to the four basic loops pictured above. The four generators α, β, γ, and δ act on $\mathbb{H}^2$ by hyperbolic isometries pushing points along the colored axes on the left-hand side of the figure, and so each of the eight maps given by α, β, γ, δ, and their inverses pushes the octagon to a neighboring octagon sharing an edge with our original tile. This produces a highly symmetric octagonal tiling of the hyperbolic plane with eight octagons around each vertex. To make this work with eight octagons around a vertex, we should choose our original octagon to be regular with internal angles of $\pi/4$, and indeed up to congruence there is exactly one. This picture could be pursued to verify the presentation given above, but there are many things to check (see Project 1). Instead, let's revisit the Seifert–van Kampen theorem.

Now instead of focusing on those hyperbolic isometries, we'll focus on the edges of the fundamental octagon, which on the glued-up surface correspond to four loops sharing a single base point. Thus the graph for the 2–complex structure on Σ_2 is these four circles all glued at a single point, for which the fundamental group would be $F_4 = \langle a, b, c, d \rangle$. There is only one 2–cell, and the gluing pattern gives the label $[a, b][c, d]$ to its boundary word, because as you read around clockwise from an appropriate starting edge, it reads $aba^{-1}b^{-1}cdc^{-1}d^{-1}$. This proves the theorem.

Surface groups are hyperbolic. Remember, the whole reason we are doing all this work is that surface groups are important examples of hyperbolic groups, and now we'd like to see that. To this end, we will appeal to a fact from Office Hour 7, the Milnor–Schwarz lemma.

Let us first apply the Milnor–Schwarz lemma to the case of the torus. The group $\mathbb{Z}^2$ acts on $\mathbb{R}^2$ by translation, with generators $(1, 0)$ and $(0, 1)$, and the quotient is T^2. Under this action, the $\mathbb{Z}^2$–translates of the square S tile the plane and so we say that S serves as a *fundamental domain* for the action. The original square S has only eight neighboring squares, so only eight distinct translations fail to move the closed square off of itself, and that number is likewise finite for any bounded set, verifying proper discontinuity. Thus, by the Milnor–Schwarz lemma, $\mathbb{Z}^2$ is quasi-isometric to $\mathbb{R}^2$—they have the *same large-scale geometry*.

What is encoded in Poincaré's theorem (Theorem 9.1) is that $\pi_1(\Sigma_2)$ acts on $\mathbb{H}^2$ in just the same way. For each of the four basic loops, consider the element of $\mathrm{PSL}(2, \mathbb{R})$ corresponding to the hyperbolic isometry that translates along the appropriate axis in the figure on the left. The hyperbolic octagon is a fundamental domain for this action, because its images tile $\mathbb{H}^2$ nicely under the action of the group, as suggested in the figure on the right. (This figure quickly gets out of hand, so you have to use your imagination to see all the octagons, eight around each vertex!) This action is therefore cocompact and is also easily seen to be properly discontinuous. It follows from the Milnor–Schwarz lemma that $\pi_1(\Sigma_2)$, a subgroup of $\mathrm{PSL}(2, \mathbb{R})$, is quasi-isometric to the hyperbolic plane itself, and is therefore δ–hyperbolic in its own right. The same is true for all of the fundamental groups of compact surfaces, $\pi_1(\Sigma_g)$ with $g \geq 2$.

Exercise 8. Convince yourself of the last statement.

9.4 GEOMETRIC PROPERTIES

Hyperbolic spaces enjoy many qualitative geometric properties that other spaces, such as Euclidean spaces, do not. We'll discuss some of these here, emphasizing the analogy with trees.

Visual boundary. Here we meet an extraordinarily useful idea: fix a base point in a δ–hyperbolic space and think of all the geodesic rays based there as sight lines describing what you can see as you look around. This visual data can be thought of as describing the asymptotic or large-scale geometry of your space, and if the space is nice enough you actually get a topological boundary at infinity that compactifies it. "Nice enough" can mean several things, and, for instance, being a manifold with nonpositive curvature suffices to get a good boundary.[6] The problem is that the boundary is not in general quasi-isometry invariant, so one group can have several different visual boundaries.

However, for a δ–hyperbolic space or group, quasi-isometries induce *homeomorphisms* of the boundary, so this boundary (as a topological space) becomes a quasi-isometry invariant. Let's see why. In this discussion I will sometimes use a base-pointed notation and sometimes not—it's pretty easy to check that change of base point doesn't matter in nonpositively curved or δ–hyperbolic spaces.

As before, for a subset A of a metric space, let $\mathcal{N}_r(A)$ denote the neighborhood of A of radius r, that is, the set of points within distance r of A. Define

$$\partial_\infty(X) = \{\text{geodesic rays } \gamma \text{ based at } x_0\}/\sim,$$

where $\gamma \sim \gamma'$ if $\gamma' \subset \mathcal{N}_m(\gamma)$ for some $m < \infty$ (we identify two sight lines if they stay within some bounded distance forever, and we say that they are asymptotic). One often abuses notation by writing γ instead of $[\gamma]$ to represent an equivalence class.

This boundary has a nice topology: two rays are close if for every $m > 0$ they stay within distance m for a long time, so a sequence of rays γ_n converges to γ if the length of time for which γ_n fellow travels γ to within m goes to infinity as $n \to \infty$.

With this definition, we can see that in $\mathbb{R}^2$, as in $\mathbb{H}^2$, no distinct rays from a common base point are asymptotic, so the visual boundary is just the set of directions, parametrized by angle $\theta \in S^1$. The topology is just the standard topology on the circle, so $\partial_\infty(\mathbb{R}^2) = \partial_\infty(\mathbb{H}^2) = S^1$. The hyperbolic plane example makes a lot of sense, because the disk model makes it apparent that the horizon of infinitely far points is a topological circle, corresponding to "a circle's worth" of non-asymptotic directions. In the Euclidean case, we could stereographically project the plane onto an open hemisphere to see the circle as the boundary. In general, the sense in which $\partial_\infty(X)$ fits with X is more abstract.

In a tree, the situation is quite different, and any distinct non-backtracking paths give different boundary points.

---

[6] There's a powerful metric generalization of nonpositive curvature called CAT(0), but we don't have space to talk about that here.

Exercise 9. The topology on the boundary of the 4–regular tree is easily seen to be a modified Cantor set. The usual Cantor set construction begins with an interval and iteratively deletes middle thirds. Instead, here we begin with a circle, delete four intervals, and then begin to iteratively delete two subintervals from each remaining interval from then on. Show that this is homeomorphic to a standard Cantor set.

Boundary theory is amazingly useful in geometric group theory, with applications especially in dynamics and the classification of isometries.

Here is one quick application. Since the boundary of $\mathbb{H}^2$ is a circle and the boundary of $\mathbb{H}^3$ is a sphere, this gives a proof (and actually, probably the simplest one) that $\mathbb{H}^2$ is not quasi-isometric to $\mathbb{H}^3$. Further, just as fundamental groups of surfaces of genus $g \geq 2$ are quasi-isometric to $\mathbb{H}^2$ (by the Milnor–Schwarz lemma), there are many three-dimensional manifolds—hyperbolic 3–manifolds—whose fundamental groups are quasi-isometric to $\mathbb{H}^3$. So these surface groups cannot be quasi-isometric to hyperbolic 3–manifold groups.

Exponential divergence. A qualitative description of negative curvature that we saw in the saddle example at the beginning is that *geodesics diverge exponentially fast*; if you set up your terms right, the same thing holds in δ–hyperbolic spaces.

The naive thing to do would be to take two geodesics γ_1 and γ_2 with a common base point x_0 and study the distance between the time t points $\gamma_1(t)$ and $\gamma_2(t)$. This can't be quite right, however, because the triangle inequality ensures that that distance is at most $2t$, so it's surely not growing exponentially in t. To capture the sense in which geodesics are diverging exponentially fast, we should make a slightly different measurement: find the shortest path between those time t points that is outside $B(x_0, t)$, the ball of radius t about x_0. In a sense, this is like measuring the circumference of balls. Now that *does* distinguish between the different curvature regimes: on a sphere, flat plane, or hyperbolic plane, the circumference of a ball grows sublinearly, linearly, and exponentially, respectively, as a function of its radius.

Thus let us define

$$\mathrm{div}(\gamma_1, \gamma_2, t) = d_{X \setminus B(x_0, t)}\big(\gamma_1(t), \gamma_2(t)\big),$$

thought of as a function of t. It's basically measuring how hard it is to connect points on the rays γ_1 and γ_2 if you disallow backtracking.

Then a hyperbolic space is one in which *for any two geodesic rays that eventually separate past some threshold distance, their divergence grows at least exponentially.*

This is the appropriate relaxation of the much stronger fact in trees observed earlier, that pairs of rays fellow travel until they reach a threshold separation, then actually travel in opposite directions along a geodesic. That is, the distance outside $B(x_0, t)$ is infinite! We have already seen that trees have a totally disconnected boundary at infinity, which is somehow capturing the same idea. In general, δ–hyperbolic spaces can have better "connectivity at infinity," but one still has to pay an exponential penalty to close up the triangles without backtracking.

Geodesic stability: The Morse lemma. We begin with the following fact, which says that any two geodesics between a given pair of points have to be δ–close:

> *For any two geodesics α and β between the same endpoints, we have*
> $\alpha \subset \mathcal{N}_\delta(\beta)$.

The proof is super easy, taking δ from the thin triangles definition: by subdividing β at an arbitrary point on its interior and rewriting it as $\beta_1 \cup \beta_2$, we build a geodesic triangle, and we know that $\alpha \subset \mathcal{N}_\delta(\beta_1 \cup \beta_2) = \mathcal{N}_\delta(\beta)$.

This says that the most efficient path between two points may not be unique, but nearly—any two such paths are uniformly close together.

But something even better is true and not that much worse to prove: instead suppose α is merely a quasi-geodesic (that is, a quasi-isometric embedding of $\mathbb{R}$ with the usual metric) with constants (K, C) between the endpoints of β. Then there is some δ' depending only on δ, K, and C such that $\alpha \subset \mathcal{N}_{\delta'}(\beta)$.

This fundamental fact is sometimes called the *Morse lemma*. And actually, it follows fairly painlessly from the divergence property discussed above, which can be applied to show that there is a high "detour cost" associated to deviating from a geodesic.

And this, finally, is how we show the stability of hyperbolicity under quasi-isometry! If triangles are thin in one geodesic space X, and there is a quasi-isometry from X to another geodesic space Y, then start with a geodesic triangle in Y and pull it back to a quasi-geodesic triangle in X. Those sides are uniformly close to geodesic segments by the Morse lemma, and the thinness of this triangle ensures that our original triangle in Y is thin as well, with a possibly somewhat worse constant depending on K, C, and δ.

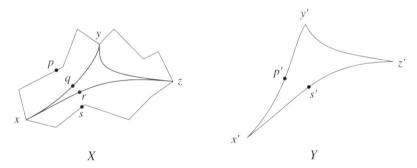

Strong contraction. As above, let $B(x, r)$ denote the ball of radius r centered at x (that is, all points of X within distance r from x). The strong contraction property sounds so strong you might have to check that you read it right:

> *For any geodesic γ, and any ball $B(x, r)$ of arbitrary radius that is disjoint from γ, the projection of the ball to the geodesic γ has uniformly bounded diameter.*

That's right—there's a constant M that does not depend on r that bounds the size of any such projection!

The projection here is the closest-point projection, defined as follows: for $b \in X$ and $A, B \subset X$, let

$$\mathrm{proj}_A(b) = \{a \in A : d(a, b) \le d(a', b) \quad \forall a' \in A\}, \quad \text{and}$$

$$\mathrm{proj}_A(B) = \bigcup_{b \in B} \mathrm{proj}_A(b).$$

So it's all points in A that are (at least tied for) smallest possible distance to B.

To test the definition, convince yourself that the projections to geodesics in $\mathbb{R}^2$ of metric balls in $\mathbb{R}^2$ that are disjoint from the geodesic can be arbitrarily large: for instance, if γ is the x–axis, what is the closest-point projection to γ of $B\big((0, 2r), r\big)$?

Matters are different in a tree, of course, because any ball disjoint from a geodesic line projects down to that line as a single point!

Let's prove this contraction property for the hyperbolic plane $\mathbb{H}^2$ using the upper half-plane model, where balls in the hyperbolic metric are round Euclidean disks. Without loss of generality, γ is the imaginary axis. It suffices to consider the projection to γ of balls that are tangent to both the imaginary axis and the real axis, since any ball disjoint from γ is contained in one of these. The proof is pleasingly simple.

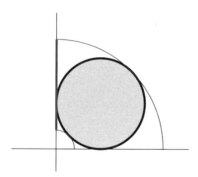

Exercise 10. Stare at this figure for a while and convince yourself that $M = \ln\left(\frac{\sqrt{2}+1}{\sqrt{2}-1}\right)$ is the optimal contraction constant for the hyperbolic plane.

This strong contraction phenomenon has several different phrasings that all sound remarkable. Here is another one: *As long as a geodesic β is disjoint from $\mathcal{N}_{2\delta}(\gamma)$, the projection $\mathrm{proj}_\gamma(\beta)$ has length $\le M$.*

Exercise 11. First prove this in the hyperbolic plane in a manner similar to the previous argument. Now prove this for a general δ–hyperbolic space X. (Choose δ large enough that X is hyperbolic for this constant in all three definitions given above.) Here are some steps. We prove the statement for a geodesic segment $\overline{xy}$ that avoids the (2δ)–neighborhood of γ. Let $\bar{x}$ and $\bar{y}$ be the projections to γ of x and y, respectively. We study the quadrilateral on x, $\bar{x}$, y, and $\bar{y}$; draw this with γ at the bottom. Consider the geodesic $\overline{x\bar{y}}$, which forms a triangle with the bottom and one of the sides of the quadrilateral. Consider the inpoints of this triangle: p (on the diagonal), p_1 (on the side), and $\bar{p}$ (on the bottom). Using the definition of $\bar{x}$, show that $d(p_1, \bar{x}) \leq \delta$. Now let p' be the projection of p to $\overline{y\bar{y}}$ and show that $d(p, p') \leq \delta$ using thin triangles. Now show $d(p', \bar{y}) \leq 2\delta$. Conclude that $M = 5\delta$ is a successful contraction constant.

9.5 HYPERBOLIC GROUPS HAVE SOLVABLE WORD PROBLEM

We will close this office hour with a discussion of a property of hyperbolic groups that seems combinatorial: every hyperbolic group possesses an algorithm to solve the word problem, i.e., to decide in finite time whether a particular word in the generators represents the trivial element of the group. In some groups, such as free abelian groups, this is very easy: just sort out the different generators and see if each one is canceled out. In other groups, this is hard or even in some cases provably *undecidable* (see Office Hour 8). By contrast, in hyperbolic groups, if a word has length n, then its triviality can be decided in linear time with respect to n. The geometry of hyperbolicity ensures the existence of a certain kind of presentation for the group that lends itself to a fast algorithm.

Dehn presentations.   Let us say that $G = \langle a_1, \ldots, a_n \mid r_1, \ldots, r_m \rangle$ is a *Dehn presentation* of G if the following very special set of circumstances is in place:

- There is a set of strings $u_1, v_1, \ldots, u_m, v_m$ and each relator r_i is of the form $r_i = u_i v_i^{-1}$. (Relator r_i encodes the equivalence in the group of u_i and v_i.)
- For each i, the word length of v_i is shorter than the word length of u_i.
- For any nonempty string w in $S = \{a_i\}$ that represents the identity element, if w has been reduced by canceling all occurrences $a_i a_i^{-1}$, then at least one of the u_i or u_i^{-1} must appear as a substring.

This last bullet point is a lot to ask! In general, given a presentation, the word problem is hard: you can look at a long word in the generators, and there's no good way to see whether it simplifies, because you might have to first use one relation in the group that makes the string longer before you can use another relation that makes it shorter. A Dehn presentation is special because any string can be simplified by using one of finitely many moves, each of which makes the string strictly shorter. Thus the number of replacements required is less than the initial length of the string, and the algorithm runs very fast.

As a trivial example, $F_2 = \langle a, b \mid \rangle$ is a Dehn presentation with $m = 0$. This works because there are no nonempty words representing the identity left after words are reduced.

As a less trivial example, consider the presentation $\mathrm{PSL}(2, \mathbb{Z}) = \langle \sigma, \nu \mid \sigma^2, \nu^3 \rangle$ from before. Here we can form a Dehn presentation with $u_1 = \sigma^2$, $v_1 = 1$, $u_2 = \nu^2$, and $v_2 = \nu^{-1}$, clearly satisfying the first two bullet points. For the third, one can easily use the Cayley graph to confirm that any nontrivial loop with no σ^2 or $\sigma\sigma^{-1}$ must contain a path around one of the ν triangles. Thus it has a ν^2 or ν^{-2} substring.

THEOREM 9.2. *Hyperbolic groups admit Dehn presentations.*

First, an exercise that we'll need in the proof.

Exercise 12. Let us call a path γ an m–*local geodesic* if every subpath of length m is geodesic. First, for practice with the concept: *(a)* Find a (nonhyperbolic) metric space such that for every $m > 0$, there are arbitrarily long m–local geodesic loops.

Now the main point: *(b)* Show that there do not exist (8δ)–local geodesic loops of length $\geq 8\delta$ in any δ–hyperbolic space. (One way to do this is to show that an (8δ)–local geodesic segment must stay within distance 2δ of any true geodesic between its endpoints.)

So these are nearly geodesic, but not always geodesic on the nose: *(c)* Come up with an example of an (8δ)–local geodesic in a δ–hyperbolic space that is not geodesic.

Proof of Theorem 9.2. Fix any $K > 8\delta$ and consider a Cayley graph for G with respect to a (finite) generating set $S = \{a_i\}$ for G. Now consider the list of *all* reduced words t_i with word length at most K. There are a lot of these, but only finitely many. You can check which of the t_i represent the same word by just following them in the graph. Let the u_i be the non-geodesic words from that list, and for each u_i, let v_i be some geodesic word from that list reaching the same point in the graph, so it is guaranteed to be strictly shorter. Then put $R = \{r_i = u_i v_i^{-1}\}$. I claim $G = \langle S \mid R \rangle$ is a Dehn presentation. The first two bullet points are satisfied by construction. For the third, use the previous exercise to rule out 8δ–local geodesic loops of length at least 8δ. So any long loop has a non-geodesic subsegment of length at most K, which is one of our u_i above. This verifies the last condition. $\square$

The word problem for surface groups. We already said that surface groups are hyperbolic when $g \geq 2$, and so it follows that these groups have Dehn presentations, and hence they have algorithms to solve the word problem. But actually, we can understand this case more directly and really see what is going on.

Let's stick with the case $g = 2$, although the general case is not any harder. The idea is that any loop includes some combination of backtracking and at least six letters of the length–8 string that represents a relator. So the shortening method works much the same as when dealing with the triangles in the $\text{PSL}(2, \mathbb{Z})$ case: begin with the length–8 string $aba^{-1}b^{-1}cdc^{-1}d^{-1}$ and consider all of its cyclic permutations ($ba^{-1}b^{-1}cdc^{-1}d^{-1}a$, etc.). Put all of the six-letter subwords of these as the u_i and their two-letter counterparts as the v_i. Starting with a word that may or may not represent the empty word, we seek these u_i subwords. Each time we can replace one with its corresponding v_i, we have shortened our word. If we find no u_i and no trivial cancellations $a_i a_i^{-1}$, we halt, and if there are any letters left, our original word was nontrivial in the group.

Now we need to convince ourselves that this is indeed a Dehn presentation. We saw from the Poincaré theorem above that the fundamental domain for the action of this group on $\mathbb{H}^2$ is an octagon in the hyperbolic plane. Each group element corresponds to pushing the octagonal tile (by a hyperbolic isometry) across one of its edges. Because there are eight octagons around each vertex, it takes eight moves to make a full "tour" around the vertex and come back where we started. The Cayley graph will therefore have an octagon corresponding to the basic relator, enclosing a vertex around which there are eight tiles. By symmetry, the whole Cayley graph can be drawn in the plane as a graph of octagons. (Another way to say this is that the Cayley graph is *dual* to the octagonal tiling by fundamental domains, and that it produces a combinatorially equivalent tiling by octagons.)

A nice way to visualize this combinatorics precisely is to draw a schematic that organizes these relator octagons by their distance from a central octagon in a certain precise sense. I am adapting this visualization directly from Stillwell [247, Figure 183]. It is an excellent way to understand the sense in which the structure of the genus 2 surface group is tree-like but not a tree.

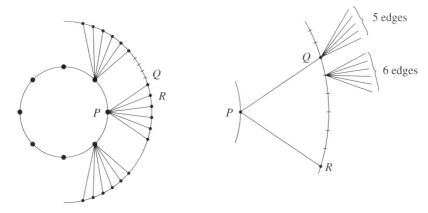

Begin with an arbitrary octagon in the Cayley graph and construct a diagram by arranging its vertices and edges around the unit circle C_1 in the plane. Then arrange the neighboring octagons radially around it, with the next ring (between C_1 and C_2) being formed by those that share either an edge or a vertex with the center tile; continue this way. Every octagon appears exactly once in this arrangement, and—aside from the central one—lies between the circle C_k and the circle C_{k+1} for some k. Any such octagon either has 0 or 1 edges on C_k, and accordingly it has either six or five edges on C_{k+1}. (The octagon containing the vertices marked P, Q, and R in the figure is the first case, and the one on the other side of edge PQ is the second.) Now consider an arbitrary loop along edges in the diagram. It touches some outermost circle C. If the loop ever traverses an edge and then immediately backtracks, then that represents a trivial cancellation $a_i a_i^{-1}$. Otherwise it reaches C along a radial edge, travels along C, and then returns along a different edge. But then that word contains six of the edges around a single octagon (the radial edge plus at least five edges on the outer circle).

Thus we can see that the tiling is rapidly branching, and though it is not a tree (there is more than one way to reach a vertex), alternate paths are costly. Actually, we could check all of the geometric properties of hyperbolicity in this picture!

Linear Dehn function and Gromov's gap.   Now we are in a position to prove, using Dehn presentations, a fact mentioned in Office Hour 8: *hyperbolic groups* *have cheap filling functions*. Recall that Riley defined the Dehn function to be the least area required to fill a loop of length n, where area is defined as the number of 2–cells you must slide the loop across in a null-homotopy.

We will explain why hyperbolic groups have linear Dehn function. The two other pieces of this story are somewhat harder: any group with a linear Dehn function is hyperbolic, and anything that is not hyperbolic has at least a quadratic Dehn function. This means there is a gap between linear and quadratic rates of filling—for instance, you can't engineer a group with $n \log n$ Dehn function, or $n^{3/2}$. And, taken together, these facts mean that linear (or indeed subquadratic) filling functions provide yet another alternative definition of hyperbolic groups.

To finish, let us observe that Dehn's algorithm for the word problem doubles as an algorithm for linear filling of loops in hyperbolic spaces, which proves the statement promised above. Begin with a loop—a word representing the trivial element of the group—of length n. Search for a shortenable substring and shorten it. Each time you replace a u_i substring with the corresponding v_i, geometrically you are sliding the word across the 2–cell corresponding to the relator $u_i v_i^{-1}$. Because each such move makes the word strictly shorter, you can do at most n such moves before the string disappears—the end!

FURTHER READING

There are many excellent sources introducing hyperbolicity, and I've enjoyed the books of Bridson–Haefliger [54], Burago–Burago–Ivanov [65], Druțu–Kapovich [119], and Stillwell [247], as well as the lecture notes by Gersten [141] and the article by Peifer [219].

PROJECTS

Here are a few projects, intended to be more open-ended than the exercises scattered above.

Project 1. Fill in all the details in the construction of the octagonal tiling used to explain the presentation of $\pi_1(\Sigma_2)$. Precisely how do the four axes of the generating hyperbolic isometries need to meet the sides of the octagon to make the construction work? Prove that the octagon is a fundamental domain for the action of the group on the hyperbolic plane: its images under the group elements cover the plane and intersect only along boundary segments.

Project 2. As we've seen, $\mathrm{PSL}(2, \mathbb{Z})$ is a hyperbolic group with exponential growth, generated by the matrices σ and τ described above, whose geometry is closely related to that of a 3–regular tree. It also acts on the hyperbolic plane—you can take $i \in \mathbb{H}^2$ as a base point and consider the orbit of the group on that point. First give upper and lower bounds for the number of words of length exactly n in the generators (that is, give bounds on the number of elements in the sphere of radius n in the Cayley graph). Now, where do you expect the orbit points of S_n to be in the hyperbolic plane? Write a computer program to draw the points of S_n in the disk model for as many n as you can. Repeat this exercise for a group with a compact fundamental domain, such as the surface group $\pi_1(\Sigma_2)$. What differences do you see? (Even if you don't carry this out, think about what to expect from the point of view of the Milnor–Schwarz lemma.)

Project 3. You might be wondering just how common hyperbolic groups are "in nature"! There is a sense in which they are very plentiful, and even a precise sense in which they are ubiquitous. Gromov defined a model (the so-called density model) of producing *random groups* as follows: begin with a free group on some fixed number of generators, say F_2. Now consider the set S_ℓ of all possible reduced words of length ℓ, of which there are roughly 4^ℓ. If you choose a density parameter $0 \le d < 1$, then consider all possible subsets R of S_ℓ of size $4^{d\ell}$. There are a lot of these, but only finitely many. Make them all equally likely and choose one at random. This picks out a group presentation $G = \langle a, b \mid R \rangle$. Say "a random group has property P" if the probability of P goes to 1 as $\ell \to \infty$. Gromov proved the remarkable theorem that $d = 1/2$ (that is, when you take the square root of the total possible number of generators) is a sharp phase transition: at $d > 1/2$, a random group has one or two elements, while for $d < 1/2$, *a random group is infinite and hyperbolic*. There are many cool theorems about random groups, but also many open questions. To whet your appetite, read some of the large literature on random groups, starting with Yann Ollivier's classic (but slightly dated) 2005 survey.

Project 4. Write some computer code to produce random groups, and explore their geometry. (Warning: Most computer algebra systems choke on long relators, so keep that in mind.)

Project 5. Here is a nice (but surely hard) question, suggested to me by Itai Benjamini. Take some density $d > 1/2$. Choose your random R. Now order its elements randomly. Consider the sequence of groups you get by starting with F_2 and adding the relations one at a time. Once you have added all of them, the group is almost surely trivial. What are some properties of the last infinite group?

Project 6. In a similar vein, one can study what happens to free groups when a small number of relators are added. General theory tells us that we should expect to create new free groups of smaller rank; i.e., if k random relators of length ℓ are added to F_m, this should almost surely produce a group isomorphic to $F_{m'}$, where m' depends on m and k. Study the relationships between k and $m \mapsto m'$.

To approach this, first explore the question: suppose a group is given to you via its Cayley graph or a presentation, but not with a free basis of generators. (For instance, F_2 could be presented with the generators a, b, and $c = abbb$, in which case it would have the relation $a^{-1}c = b^3$.) How could you detect that its rank is 2?

Project 7. Indira Chatterji and Graham Niblo have a beautiful paper called "A characterization of hyperbolic spaces" which loosely says that a space is hyperbolic if the intersection of two balls is itself almost a ball. (And it is 0–hyperbolic if the intersection of two balls is exactly a ball.) They do this with a notion of geodesic divergence that is different from the one in this office hour. This project is to prove the equivalence of their balls-intersect-in-balls property to the other definitions of hyperbolicity from this office hour, providing a new proof of the main result in their paper.

Project 8. Here is a problem to explore the metric geometry of $\mathbb{H}^2$. Consider the disk model, and fix a value $D > 0$. Take two rays out from the center, and suppose the angle between the rays is such that $d(\gamma_1(nD), \gamma_2(nD)) = 2D$. What is the distance between the rays if you backtrack by D? That is, this forms a quadrilateral with top side of length $2D$ and left and right sides of length D, and you're trying to find the last side length. The answer depends continuously on both n and D, but surprisingly enough there's a global upper bound for its length! Compute it and explain why it is that value.

What if the top side were kD with some $k \neq 2$?

Project 9. A classic problem at the intersection of combinatorics and geometry asks about the chromatic number of a metric space: if you fix a forbidden distance D, then for a metric space X let $\chi_D(X)$ be the smallest number of colors needed so that every point of X receives a color, but no two points at distance D have the same color. For the real line $\mathbb{R}$ the answer is easily seen to be 2 for any D; for the Euclidean plane $\mathbb{R}^2$ the answer does not depend on D, but the only known bounds are that $4 \leq \chi \leq 7$. In 2012 Matthew Kahle asked on MathOverflow what the answer would be for the hyperbolic plane $\mathbb{H}^2$. In this case the answer might depend on the chosen value of D, because, as discussed above, $\mathbb{H}^2$ does not admit any similarities that rescale the metric.

Check out the MathOverflow page

http://mathoverflow.net/questions/86234/
chromatic-number-of-the-hyperbolic-plane

to see what is currently known about the problem as you read this. As I write, there are many tantalizing (and accessible) open questions!

Project 10. Read the paper "Chromatic numbers of hyperbolic surfaces" by Hugo Parlier and Camille Petit. See whether you can use their ideas to say anything about chromatic numbers of the associated Cayley graphs. (The answer might depend on whether you choose the standard generators.)

Office Hour Ten

Ends of Groups

Nic Koban and John Meier

In a typical precalculus course where real-valued functions are studied, a natural question to ask about these functions is, "What is happening with the function at infinity?" This leads to our studying limits of the function as the independent variable x "approaches infinity." A similar question can be asked about a finitely generated group G. If S is a finite generating set, then the Cayley graph $\Gamma(G, S)$ gives us a picture of G, and we can wonder, "What is happening with this group at infinity?" As we approach infinity, are we approaching a space with a certain structure? The group G acts in a natural way on $\Gamma(G, S)$, so does this lead to an action on this space at infinity? In this office hour, we examine these ideas.

10.1 AN EXAMPLE

Given that the subject of this book is group theory, it may seem odd that we will begin this office hour with a discussion of the Cantor set, an object most commonly encountered in a real analysis course. We do this in order to construct a group action on the Cantor set that will serve as motivation for the ideas that follow.

The Cantor Set. Informally, the Cantor set is formed by repeatedly removing middle thirds from from the unit interval $[0, 1]$. More formally, let $\mathcal{C}_0 = [0, 1]$, then define

$$\mathcal{C}_1 = \mathcal{C}_0 - (1/3, 2/3) = [0, 1/3] \cup [2/3, 1].$$

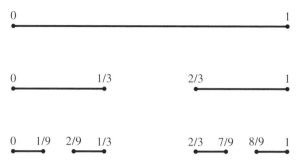

Figure 10.1 The first steps in constructing the Cantor set by removing middle thirds.

Remove $(1/9, 2/9)$ and $(7/9, 8/9)$ from $\mathcal{C}_1$ to form $\mathcal{C}_2$:

$$\mathcal{C}_2 = \mathcal{C}_1 - \{(1/9, 2/9) \cup (7/9, 8/9)\} = [0, 1/9] \cup [2/9, 3/9] \cup [6/9, 7/9] \cup [8/9, 1].$$

As this is already getting a bit awkward to present, notice that the step above can be expressed more compactly by:

$$\mathcal{C}_2 = \left\{\frac{1}{3}\mathcal{C}_1\right\} \cup \left\{\frac{2}{3} + \frac{1}{3}\mathcal{C}_1\right\} .$$

The sets $\mathcal{C}_0$, $\mathcal{C}_1$ and $\mathcal{C}_2$ are illustrated in Figure 10.1.

Continue this process by defining the set $\mathcal{C}_n$ to be

$$\mathcal{C}_n = \left\{\frac{1}{3}\mathcal{C}_{n-1}\right\} \cup \left\{\frac{2}{3} + \frac{1}{3}\mathcal{C}_{n-1}\right\} .$$

The set $\mathcal{C}_n$ consists of 2^n closed intervals, each of length $(1/3)^n$. The *Cantor set* is the result of applying this shrink-and-separate process forever:

$$\mathcal{C} = \bigcap_{n=1}^{\infty} \mathcal{C}_n.$$

The Cantor set is a totally disconnected subset of the unit interval that contains uncountably many points. In case you are familiar with the terms, we note that $\mathcal{C}$ is also a compact and perfect subset of $[0, 1]$.

The Cantor set $\mathcal{C}$ also has remarkable symmetry. For example, if f_L is the linear function $f_L(x) = \frac{1}{3}x$, then restricting the domain of f_L to the Cantor set produces a bijection between $\mathcal{C}$ and its left half:

$$f_L \colon \mathcal{C} \to \{\mathcal{C} \cap [0, 1/3]\} .$$

Similarly, if f_R is the linear function $f_R(x) = \frac{1}{3}x + \frac{2}{3}$, then restricting the domain of f_R to the Cantor set produces a bijection between $\mathcal{C}$ and its right half:

$$f_R \colon \mathcal{C} \to \{\mathcal{C} \cap [2/3, 1]\} .$$

For any string or word consisting of the letters L and R we can then define a bijection from $\mathcal{C}$ to a proper subset of $\mathcal{C}$. For example,

$$f_{LLR} = f_L \circ f_L \circ f_R = \frac{1}{3}\left(\frac{1}{3}\left(\frac{1}{3}x + \frac{2}{3}\right)\right) = \frac{1}{27}x + \frac{2}{27}.$$

Figure 10.2 A bijection from $\mathcal{C}$ to $\mathcal{C}$ formed by sliding copies of $\mathcal{C}$.

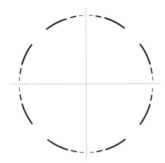

Figure 10.3 The Cantor set arranged on a circle.

This is a bijection between $\mathcal{C}$ and $\mathcal{C} \cap [2/27, 3/27]$.

For any finite sequence of lefts and rights, $LRL \cdots L$, define $\mathcal{C}_{LR\cdots L}$ by:

$$\mathcal{C}_{LR\cdots L} = f_{LR\cdots L}(\mathcal{C}).$$

Note that given any two such sub-Cantor sets, there is a bijection between them. For example,

$$f_{LLR} \circ f_{LRRL}^{-1} \colon \mathcal{C}_{LRRL} \to \mathcal{C}_{LLR}$$

is a bijection that can be expressed as the linear function $3x - \frac{22}{27}$ (restricted to $\mathcal{C}_{LRRL}$).

Exercise 1. Verify the above statements.

All of that is intended to help you understand the following bijection from $\mathcal{C}$ to itself. Think of the Cantor set as being decomposed into infinitely many copies of itself:

$$\mathcal{C} = \cdots \cup \mathcal{C}_{LLLR} \cup \mathcal{C}_{LLR} \cup \mathcal{C}_{LR} \cup \mathcal{C}_{RL} \cup \mathcal{C}_{RRL} \cup \mathcal{C}_{RRRL} \cup \cdots .$$

In this decomposition, any two adjacent copies of $\mathcal{C}$ have a bijection between them, and so we can "slide" these copies from left to right

$$\cdots \to \mathcal{C}_{LLLR} \to \mathcal{C}_{LLR} \to \mathcal{C}_{LR} \to \mathcal{C}_{RL} \to \mathcal{C}_{RRL} \to \mathcal{C}_{RRRL} \to \cdots ,$$

which is also indicated in Figure 10.2.

We are now finally able to introduce a group acting on the Cantor set. Arrange $\mathcal{C}$ around a circle, with one copy of $\mathcal{C}$ in each quadrant as shown in Figure 10.3. If the preceding discussion made sense to you, then you should be able to verify that there is no real difference between one copy of $\mathcal{C}$ and four copies of $\mathcal{C}$.

Define h as something of a horizontal slide, occurring in parallel to both the top and bottom copies of $\mathcal{C}$, as indicated in the left of Figure 10.4. Define v to be a similar, but vertical, slide occurring in parallel to the left and right copies of $\mathcal{C}$.

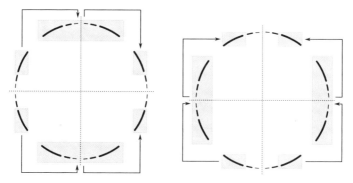

Figure 10.4 The slides h (on the left) and v (on the right).

The group of bijections of the Cantor set $\mathcal{C}$ generated by h and v is a group that you have already encountered.

THEOREM 10.1. *The group generated by h and v is a free group.*

The proof of this result is left as an exercise. As a hint, we will mention that you can use the ping-pong lemma (Office Hour 5) to establish this result, and the blocks shown in Figure 10.4 provide some guidance as to how to do this.

Exercise 2. Prove Theorem 10.1.

Having just seen a free group created from bijections of the Cantor set, the natural question is: where did this example come from? It came from the idea of trying to understand what is happening at infinity for an infinite group. This general idea is often captured by the phrase: *the ends of a group.*

10.2 THE NUMBER OF ENDS OF A GROUP

In order to discuss the notion of ends in the context of infinite groups, we would first like to consider the idea in the context of infinite graphs. Let Γ be a connected, locally finite graph. That is, in Γ there are only finitely many edges attached to any given vertex. Let C be any finite subgraph of Γ and define

$$||\Gamma \setminus C||$$

to be the number of unbounded, connected components of the graph formed by removing C from Γ. For example, let $\mathcal{T}$ be the standard Cayley graph of F_2, the infinite 4–valent tree. Let v be any vertex. Then $||\mathcal{T} \setminus v|| = 4$. Similarly, if e is an edge in $\mathcal{T}$, including both endpoints, then $||\mathcal{T} \setminus e|| = 6$.

Exercise 3. Prove the claim that if C consists of two non-adjacent vertices in $\mathcal{T}$ then $||\mathcal{T} \setminus C|| = 7$.

For every finite subgraph $C \subset \Gamma$, $||\Gamma \setminus C||$ is some nonnegative integer. Thus

$$\{||\Gamma \setminus C|| \mid C \text{ is a finite subgraph of } \Gamma\}$$

is a potentially infinite collection of integers and we can ask if there is a largest integer that appears in this collection.

For a locally finite graph Γ, the number of *ends of* Γ is

$$e(\Gamma) = \sup\{||\Gamma \setminus C|| \mid C \text{ is a finite subgraph of } \Gamma\}.$$

So $e(\Gamma)$ is either the maximum of the above set of numbers or it is ∞ if there is no maximum.

For example, let Γ be the grid connecting the integer points in the plane, that is, the standard Cayley graph of $\mathbb{Z} \times \mathbb{Z}$ (without the edge labels). If we remove a finite subgraph there may be more than one component. For example, if we remove the finite square with corners $(\pm n, \pm n)$, there are two components, the inside part and the outside part, but only one of them (the outside part) is unbounded. It is not hard to see that, as in this example, if we remove any finite graph from Γ, there is exactly one unbounded component, so $e(\Gamma) = 1$. Such a graph is called *one-ended*.

Exercise 4. Fill in the details of the proof that $e(\Gamma) = 1$.

Now we can give our main definition. Let G be a finitely generated group with finite generating set S and let $\Gamma(G, S)$ be its Cayley graph. Then the *number of ends of* G is defined to be the number of ends of $\Gamma(G, S)$.

In order for this definition to be a useful and well-defined notion for groups, we need to know that the number of ends does not depend on S. This is indeed true.

THEOREM 10.2. *The number of ends of a finitely generated group is independent of the choice of finite generating set.*

See Office Hour 7 for a hint on how to prove such a statement. Then prove it! You might instead trying proving the following stronger statement.

THEOREM 10.3. *If two finitely generated groups are quasi-isometric, then they have the same number of ends.*

Examples. The following four examples provide some insight into the number of ends of groups. Refer back to Office Hour 2 for pictures of the Cayley graphs if necessary.

- If G is a finite group, then $e(G) = 0$ because regardless of the choice of sub-graph C, $\Gamma \setminus C$ has no unbounded components.
- We previously described why $e(\mathbb{Z} \times \mathbb{Z}) = 1$.
- The Cayley graph Γ of $\mathbb{Z}$ (viewed as a cyclic group) is a combinatorial copy of the real line. Removing any subgraph from Γ leaves you with two unbounded components (plus any number of finite pieces). Thus $e(\mathbb{Z}) = 2$.
- The number of ends of a finitely generated group is not always finite. For example, the standard Cayley graph of the free group on two generators is a 4–valent tree $\mathcal{T}$. Removing larger and larger, finite subtrees from $\mathcal{T}$ results in more and more unbounded components. Thus F_2 has infinitely many ends.

Exercise 5. Verify this final claim that $e(F_2) = \infty$. Compare with Exercise 9 in Office Hour 9

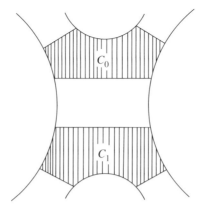

Figure 10.5 Three pieces gives you four.

How many ends can a group have?   These examples, it turns out, illustrate the full range of possibilities. The following result is due to Freudenthal and Hopf from the early part of the twentieth century.

THEOREM 10.4 (Freudenthal and Hopf). *If G is a finitely generated group, then G has zero, one, two, or infinitely many ends.*

The argument for Theorem 10.4 is rather intuitive. All that needs to be shown is that if you have a compact set C such that $\Gamma \setminus C$ has three or more unbounded, connected components, then there is a compact set D such that $\Gamma \setminus D$ has even more unbounded, connected components. To see why this happens, consider the following proof by picture. Let Γ be a Cayley graph for a group G and let C_0 be a finite subset of Γ such that $||\Gamma \setminus C_0|| = 3$, as in Figure 10.5 (the argument is essentially the same if $||\Gamma \setminus C_0|| \geq 3$). There is then an element g in the group so that $C_1 = g \cdot C_0$ is disjoint from C_0. Then if $D = C_0 \cup C_1$, it follows that $||\Gamma \setminus D|| > 3$.

Exercise 6. In the above argument, try to figure out why we didn't say $||\Gamma \setminus D|| = 4$, even though that is what is indicated in Figure 10.5.

You can find detailed proofs of Theorems 10.2 and 10.4 in Meier's text [202].

10.3 SEMIDIRECT PRODUCTS

As we previously mentioned, $\mathbb{Z} \times \mathbb{Z}$ has one end even though it is the direct product of two-ended groups. Using intuition from the case of $\mathbb{Z} \times \mathbb{Z}$ you might be able to establish the following general result about products of groups: If G and H are two finitely generated, infinite groups, then their direct product, $G \times H$, is one-ended.

Exercise 7. Prove the last statement. *Caution:* Moving from an intuitive argument for this claim to a rigorous one does require some subtle reasoning! A more realistic exercise would have you assume that $H \cong \mathbb{Z}$.

This result about direct products of finitely generated, infinite groups can be extended to a more general product that you might not have seen in your first course on algebra. We first illustrate this construction with the example of the Klein bottle group.

The Klein bottle group.   The group $\mathbb{Z} \times \mathbb{Z}$ has presentation $\langle a, b \mid ab = ba \rangle$. Presentations are introduced in Office Hour 1, but remember that this is shorthand for saying that two elements generate $\mathbb{Z} \times \mathbb{Z}$ and that the only fact you really need to know about these two elements is that they commute. You can imagine a more exotic interaction, and that is one way to view the Klein bottle group,

$$K = \langle x, y \mid xy = y^{-1}x \rangle.$$

So, for example:

$$
\begin{aligned}
x^2 yx y^2 &= xx yx yy \\
&= xx yy^{-1} xy \\
&= xxxy \\
&= xx y^{-1} x \\
&= xyxx \\
&= y^{-1} x^3.
\end{aligned}
$$

If you know a bit about fundamental groups, then it is worth noting that K is the fundamental group of the Klein bottle. Establishing this fact, via the Seifert–van Kampen theorem, is a very good exercise in algebraic topology. Looking at Office Hour 9 is a good way to start.

Exercise 8. Draw the Cayley graph of the Klein bottle group, K. Use this to show that K has one end. Be sure to describe a path between arbitrary elements $g, h \in \Gamma \setminus C$ for some finite subgraph $C \subset \Gamma$. *Hint: Exploit the fact that Γ contains multiple copies of the Cayley graph of $\mathbb{Z}$.*

Semidirect products.   The Klein bottle group is an example of a semidirect product. The basic idea of a semidirect product is that you start with two groups G and H and you combine them to get a bigger group. You already know one way to do this, namely, the direct product $G \times H$; the elements of this group are ordered pairs (g, h) where $g \in G$ and $h \in H$ and the product is the obvious one, namely, $(g, h) \cdot (g', h') = (g \cdot g', h \cdot h')$. The semidirect product $G \ltimes H$ is a generalization.

In order to define the semidirect product of G and H, you need one extra piece of information besides G and H: an action of G on H, in other words a homomorphism $\alpha \colon G \to \mathrm{Aut}(H)$. We will write $\alpha(g)$ as α_g. Note that α is more than just an action of G on the set H; the action respects the multiplication in H.

Here is one way to define the semidirect product $G \ltimes_\alpha H$, also written $G \ltimes H$ when there is no confusion (we can also write $H \rtimes_\alpha G$ or $H \rtimes G$). As a set, $G \ltimes_\alpha H$ is the set of pairs (g, h) where $g \in G$ and $h \in H$. The multiplication is

$$(g, h) \cdot (g', h') = (g \cdot g', h \cdot \alpha_g(h')).$$

Notice that if α is the trivial action (so each α_g is the identity), then the semidirect product $G \ltimes_\alpha H$ is isomorphic to the direct product $G \times H$. But if we can make α nontrivial, then we will definitely get some new and interesting groups.

Exercise 9. Verify that the semidirect product $G \ltimes H$ is a group.

We can also understand the semidirect product via group presentations. Suppose $G \cong \langle X \mid R \rangle$ and $H \cong \langle Y \mid S \rangle$. Then $G \ltimes_\alpha H$ has the presentation

$$G \ltimes_\alpha H \cong \langle X \cup Y \mid R \cup S \cup \{xyx^{-1} = \alpha_x(y) \mid x \in X, y \in Y\}\rangle.$$

In other words, we take all of the generators for G and H and all of the relations in G and H, and also relations $xyx^{-1} = \alpha_x(y)$.

Exercise 10. Verify that the two descriptions of the semidirect product are the same.

Semidirect products appear also in Office Hour 15.

The dihedral group as a semidirect product. The definition of the semidirect product probably seems awkward if you have never seen it before. However, it turns out that this construction comes up all over the place in mathematics. Here is a first example, the dihedral group. The group D_n has two nice subgroups, $\mathbb{Z}/2\mathbb{Z}$ and $\mathbb{Z}/n\mathbb{Z}$, where the former is generated by your favorite reflection s and the latter is generated by a rotation r by one click (in your favorite direction). Since r and s generate D_n, you might naively hope that D_n is isomorphic to the direct product $\mathbb{Z}/2\mathbb{Z} \times \mathbb{Z}/n\mathbb{Z}$, but it isn't. Specifically, this is because reflections do not commute with rotations (as they would in the direct product); more specifically, the conjugate of a clockwise rotation by k clicks by a reflection is a *counterclockwise* rotation by k clicks.

Our next best hope is that D_n is isomorphic to the semidirect product of $\mathbb{Z}/2\mathbb{Z}$ and $\mathbb{Z}/n\mathbb{Z}$. First we need a homomorphism $\alpha : \mathbb{Z}/2\mathbb{Z} \to \mathrm{Aut}(\mathbb{Z}/n\mathbb{Z})$. Since flips switch the direction of all rotations, we might guess that α should be the action by negation, that is, $\alpha_s(k)$ is the element of $\mathbb{Z}/n\mathbb{Z}$ corresponding to $-k$ (verify that this defines an action!). We claim that $\mathbb{Z}/2\mathbb{Z} \ltimes_\alpha \mathbb{Z}/n\mathbb{Z}$ is isomorphic to D_n. The isomorphism from our semidirect product to D_n takes the generator of $\mathbb{Z}/2\mathbb{Z}$ to s and the generator of $\mathbb{Z}/n\mathbb{Z}$ to r. By the way we set things up, our action α of $\mathbb{Z}/2\mathbb{Z}$ on $\mathbb{Z}/n\mathbb{Z}$ exactly encodes the action of a reflection on the group of rotations!

Exercise 11. Check the details of the isomorphism $\mathbb{Z}/2\mathbb{Z} \ltimes_\alpha \mathbb{Z}/n\mathbb{Z} \to D_n$.

Exercise 12. Express the Klein bottle group K as a semidirect product $\mathbb{Z} \rtimes_\alpha \mathbb{Z}$.

An infinite example. Another example might help. Let $F_2 \cong \langle a, b \mid \rangle$ be the free group on two generators, and let $\mathbb{Z} \cong \langle t \mid \rangle$ be the infinite cyclic group of integers. Define an action of $\mathbb{Z}$ on F_2 by $t \cdot a = b$ and $t \cdot b = a$. In other words, we have a homomorphism $\alpha : \mathbb{Z} \to \mathrm{Aut}(F_2)$ given by the rule that $\alpha(t^n)$ is the trivial automorphism if n is even and is the automorphism that swaps a and b if n is odd. Thus we have:

$$F_2 \rtimes \mathbb{Z} = \langle a, b, t \mid tat^{-1} = b, \ tbt^{-1} = a \rangle.$$

You should be able to verify that in this group $tab = bat$. The following exercises will help to strengthen your intuition on semidirect products.

Exercise 13. Sketch the Cayley graph Γ of $F_2 \rtimes \mathbb{Z}$.

Exercise 14. Show that every element in $F_2 \rtimes \mathbb{Z}$ can be uniquely written as hg for some $h \in F_2$ and $g \in \mathbb{Z}$.

Exercise 15. Fix an element $h \in F_2$. Let Γ_h be the subgraph of Γ generated by the vertices $\{hg \mid g \in \mathbb{Z}\}$ (i.e., keep these vertices and any edges that joined two of these vertices). Show that Γ_h is isomorphic to the Cayley graph of $\mathbb{Z}$. Likewise, fix $g \in \mathbb{Z}$, and let Γ_g be the subgraph generated by the vertices $\{hg \mid h \in F_2\}$. Show that Γ_g is isomorphic to the Cayley graph of F_2.

Exercise 16. Let C be a finite subgraph of Γ, and let $h_1 g_1$ and $h_2 g_2$ be vertices of $\Gamma \setminus C$. Describe a path in $\Gamma \setminus C$ from $h_1 g_1$ to $h_2 g_2$. Conclude that $F_2 \rtimes \mathbb{Z}$ has one end. *Hint: Show that $\Gamma_g \cap C = \emptyset$ and $\Gamma_h \cap C = \emptyset$ for infinitely many $g \in \mathbb{Z}$ and $h \in F_2$.*

Ends of a semidirect product.   These examples illustrate the following fact about ends of semidirect products, a proof of which is beyond the scope of our casual conversation!

THEOREM 10.5. *Suppose G and H are two finitely generated infinite groups, and suppose G acts on H. Then*

$$e(G \ltimes H) = 1.$$

Now let's examine another semidirect product. Let $\mathbb{Z}/2\mathbb{Z} \cong \langle t \mid t^2 \rangle$ be the group of order 2, and again, let $F_2 \cong \langle a, b \mid \rangle$ be a free group. Let $\mathbb{Z}/2\mathbb{Z}$ act on F_2 by $t \cdot a = a^{-1}$ and $t \cdot b = b^{-1}$. The resulting semidirect product has presentation

$$F_2 \rtimes \mathbb{Z}/2\mathbb{Z} \cong \langle a, b, t \mid t^2 = 1,\ tat^{-1} = a^{-1},\ tbt^{-1} = b^{-1} \rangle.$$

Exercise 17. Sketch the Cayley graph of $F_2 \rtimes \mathbb{Z}/2\mathbb{Z}$ and show that $e(F_2 \rtimes \mathbb{Z}/2\mathbb{Z}) = \infty$.

Now let's change the action of $\mathbb{Z}/2\mathbb{Z}$ on F_2 to $t \cdot a = b$ and $t \cdot b = a$, so the presentation for $F_2 \rtimes \mathbb{Z}/2\mathbb{Z}$ with this action is

$$F_2 \rtimes \mathbb{Z}/2\mathbb{Z} \cong \langle a, b, t \mid t^2 = 1,\ tat^{-1} = b,\ tbt^{-1} = a \rangle.$$

Exercise 18. Sketch the Cayley graph of this $F_2 \rtimes \mathbb{Z}/2\mathbb{Z}$ and show that $e(F_2 \rtimes \mathbb{Z}/2\mathbb{Z}) = \infty$.

Finite index subgroups and ends.   The result of the previous two exercises is that $F_2 \rtimes \mathbb{Z}/2\mathbb{Z}$ has infinitely many ends, even when the action that determines the semidirect product is different. Thus, it seems that $e(F_2 \rtimes \mathbb{Z}/2\mathbb{Z}) = e(F_2)$ regardless of the particular action of $\mathbb{Z}/2\mathbb{Z}$ one uses to define the semidirect product.

Instead of posing this as an exercise, let's look at another example. Consider an action of $\mathbb{Z}/2\mathbb{Z} = \langle t \mid t^2 = 1 \rangle$ on $\mathbb{Z} \times \mathbb{Z} = \langle a, b \mid ab = ba \rangle$ defined by $t \cdot a = a$ and $t \cdot b = b^{-1}$. This semidirect product is then

$$(\mathbb{Z} \times \mathbb{Z}) \rtimes \mathbb{Z}/2\mathbb{Z} \cong \langle a, b, t \mid ab = ba,\ t^2 = 1,\ tat^{-1} = a, tbt^{-1} = b^{-1} \rangle.$$

Exercise 19. Sketch the Cayley graph of $(\mathbb{Z} \times \mathbb{Z}) \rtimes \mathbb{Z}/2\mathbb{Z}$ and show that it has one end.

Once again, we see that the number of ends of the semidirect product $H \rtimes \mathbb{Z}/2\mathbb{Z}$ is the same as the number of ends of H. It would be natural to make the conjecture: *If G is a finite group, and H is a finitely generated group, then*

$$e(H \rtimes G) = e(H).$$

While this conjecture is true, it is actually not the best conjecture you could make. Notice that the group $(\mathbb{Z} \times \mathbb{Z}) \rtimes \mathbb{Z}/2\mathbb{Z}$ has some resemblance to the Klein bottle group, K. For one thing, the semidirect product has the same 'twisting' introduced by moving some generators to their inverses. However, K cannot be of the form $H \rtimes \mathbb{Z}/2\mathbb{Z}$, because such a semidirect product always contains $\mathbb{Z}/2\mathbb{Z}$ as a subgroup, but K has no $\mathbb{Z}/2\mathbb{Z}$ subgroup.

There is a way in which $(\mathbb{Z} \times \mathbb{Z}) \rtimes \mathbb{Z}/2\mathbb{Z}$ is very similar to the Klein bottle group. The semidirect product has an index 2 subgroup isomorphic to $\mathbb{Z} \times \mathbb{Z}$. Examining the Cayley graph of the Klein bottle group from Exercise 8, we see that $x^2 y = y x^2$, so the subgroup $\langle x^2, y \rangle$ is isomorphic to $\mathbb{Z} \times \mathbb{Z}$. This subgroup is the kernel of the homomorphism from the Klein bottle group onto $\mathbb{Z}/2\mathbb{Z} \cong \langle t \mid t^2 = 1 \rangle$ defined by $x \mapsto t$ and $y \mapsto 1$. Thus, $\langle x^2, y \rangle$ is a normal subgroup with finite index. To avoid confusion, let $\mathbb{Z} \times \mathbb{Z} \cong \langle a, b \mid ab = ba \rangle$, so we are identifying a with x^2 and b with y (also note that since $x \mapsto t$, we identify x and t). These identifications give the Klein bottle group the presentation $K \cong \langle a, b, t \mid ab = ba, tbt^{-1} = b^{-1},\ t^2 = a \rangle$. Each element of K can be uniquely written as (g, h), where $g \in \mathbb{Z} \times \mathbb{Z}$ and $h \in \mathbb{Z}/2\mathbb{Z}$.

We claim that since $\mathbb{Z} \times \mathbb{Z}$ is a finite index subgroup of K, $e(K) = e(\mathbb{Z} \times \mathbb{Z}) = 1$. Notice that since K is infinite, $e(K) \geq 1$. We now show that $e(K) \leq 1$.

The Cayley graph, Γ, of K with respect to the generating set $\{a, b, t\}$ (see Figure 10.6) contains two copies of the Cayley graph of $\mathbb{Z} \times \mathbb{Z}$, one for each element of $\mathbb{Z}/2\mathbb{Z}$. Denote these by Γ_1 and Γ_t. These subgraphs are joined by t–edges. Let C be a finite subgraph of Γ. We know that both $\Gamma_1 \cap C$ and $\Gamma_t \cap C$ each have one unbounded connected component. Since C is finite, only a finite number of t–edges are contained in C, so there must be a t–edge connecting these two unbounded components. Therefore, $\|\Gamma \setminus C\| = 1$, so $e(K) = 1 = e(\mathbb{Z} \times \mathbb{Z})$. (Notice that we have switched our generating set from the one we considered at the beginning of this section. Proposition 10.2 allows us to do this.)

These previous examples illustrate the following fact.

THEOREM 10.6. *Suppose that G is a finitely generated group with a finite index subgroup N. Then $e(G) = e(N)$.*

Exercise 20. Prove that our earlier conjecture follows as a corollary.

Exercise 21. Prove that this theorem follows from Theorem 10.3.

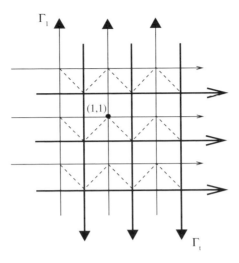

Figure 10.6 The Cayley graph of the Klein bottle group with generating set $\{a, b, t\}$. There
are two copies of the Cayley graph of $\mathbb{Z} \times \mathbb{Z}$ (one with darker lines, Γ_t, and
one with lighter lines, Γ_1). The horizontal edges are from multiplication by
a (directed toward the right), and the vertical edges are from multiplication
by b. The dashed edges are from multiplication by t. In Γ_1, the b–edges are
directed upward while in Γ_t, the b–edges are directed downward. The vertex
representing the identity is marked with a black dot and labeled $(1, 1)$.

10.4 CALCULATING THE NUMBER OF ENDS OF THE BRAID GROUPS

We will use the previous section to calculate the number of ends of the braid groups.
This is the most mathematically challenging part of our discussion of ends, so we
ask that you try to understand the general arc of the narrative, and bring sufficient
open-mindedness to believe that the details work the way we claim. Braid groups
are the topic of Office Hour 18, so we refer you there for more details about these
groups.

We can construct a braid by attaching n strings to a ceiling, braiding these strings,
and then attaching the braided strings to a floor. The collection of all braids forms
a group whose multiplication is 'stacking'. If we number the n points where the
strings are attached to the ceiling and the n points where the strings are attached
to the floor, then there is a surjective homomorphism from the braid group on n
strings, B_n, to the symmetric group, S_n. Indeed, the strands of any element of B_n
give an obvious bijection between the n points in the ceiling and the n points in the
floor, and since the points are numbered, this exactly gives an element of S_n. An
example is given in Figure 10.7.

The kernel of this homomorphism $B_n \to S_n$ consists of all braids that are sent
to the identity permutation, and thus, each string would have to be attached to the
same labeled point on the ceiling and floor. This normal subgroup of the braid
group is the pure braid group, PB_n. Thus, we have that the index of PB_n in B_n is
$n!$, so by Theorem 10.6, $e(B_n) = e(PB_n)$.

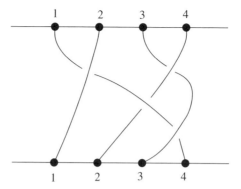

Figure 10.7 For this braid, the string attached at 1 on the ceiling is attached to 4 on the floor, so $1 \mapsto 4$ in the permutation. The string attached to 4 on the ceiling is attached to 2 on the floor, so $4 \mapsto 2$. Following the other strings, we see that $2 \mapsto 1$ and $3 \mapsto 3$, so this braid is mapped to the permutation (142).

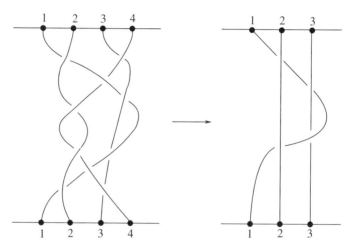

Figure 10.8 The pure braid on the left is mapped to the pure braid on the right via the "forgetful" homomorphism that removes the fourth string.

Still, how in the world are we going to compute $e(PB_n)$? It seems like a pretty complicated group, and we have no idea what the Cayley graph looks like. Well, it turns out that in Exercise 28 in Office Hour 18 you will use a technique called "combing" to prove the following isomorphism:

$$PB_n \cong F_{n-1} \rtimes PB_{n-1}.$$

Here is the quick idea. There is a surjective homomorphism from PB_n onto PB_{n-1} that maps a pure braid on n strings onto the pure braid on $n-1$ strings obtained by removing the nth string. Figure 10.8 gives an example of this. The kernel of this map consists of all pure braids for which only the nth string is braiding through the

Figure 10.9 A ray.

other strings, all of which go straight up and down, and (through combing) this kernel can be seen as the free group F_{n-1}. Since there is an injective homomorphism
from PB_{n-1} to PB_n that just adds an nth string that does not braid with the other
strings, we can view PB_{n-1} as a subgroup of PB_n, and discover that PB_{n-1} acts
on F_{n-1} by conjugation. It follows that $PB_n \cong F_{n-1} \rtimes PB_{n-1}$.

Exercise 22. Verify that if a group K has a surjective homomorphism $\phi : K \to G$
with kernel H and if ϕ has a right inverse $\psi : G \to K$, then indeed $K \cong H \rtimes G$,
where the action of G on h is given by $g \cdot h = \psi(g)h\psi(g)^{-1}$.

If $n \geq 3$, then PB_{n-1} is infinite, and thus, PB_n is the semidirect product of two
finitely generated infinite groups. By Theorem 10.5, we have:

THEOREM 10.7. *For* $n \geq 3$, $e(PB_n) = 1$, *and so* $e(B_n) = 1$ *as well.*

The remaining cases are straightforward. The group B_2 is isomorphic to $\mathbb{Z}$ and
B_1 is trivial, so $e(B_2) = 2$ and $e(B_1) = 0$.

10.5 MOVING BEYOND COUNTING

The discussion of the number of ends does not really explain where the introductory
example of F_2 acting on the Cantor set came from. Directly related to this issue,
the idea of the number of ends makes it sound like there is something called "ends"
that we have counted. Perhaps whatever it is that is being counted might come
with additional mathematical structure? To an extent it is true that we are counting
something, and that counting is the most elementary of mathematical techniques
that can be applied to that thing. In order to be able to skip over a number of
interesting and nontrivial issues, we will restrict this discussion to the standard
Cayley graph of F_2, that is, the 4–valent tree $\mathcal{T}$. Our goal is to hint at three things:

1. You can define the ends of a space, in addition to defining the number of ends;
2. If G is a finitely generated group, then G acts on the ends of its Cayley graph;
   and
3. In many circumstances you can give the set of ends additional structure, such
   as a metric or topology that describes which ends are close to each other and
   which are far apart.

The ends of $\mathcal{T}$. Call any subgraph of $\mathcal{T}$ that is an embedded copy of the graph
shown in Figure 10.9 a *ray*. The location of the initial vertex is called the *base
vertex*. Said another way, you can construct a ray in $\mathcal{T}$ by choosing any base vertex, and then continually traveling to adjacent vertices, never stepping on the same
vertex twice.

The instinct is that rays point toward the ends of $\mathcal{T}$. We can make this precise using an equivalence relation. If ρ_1 and ρ_2 are two rays in $\mathcal{T}$ where $\rho_1 \cap \rho_2$ is itself a ray, then we say that these two rays *point to the same end of* $\mathcal{T}$.

Exercise 23. Convince yourself that this is an equivalence relation.

We can denote equivalence classes of rays by:

$$[\rho] = \{\rho' \mid \rho' \text{ is a ray in } \mathcal{T} \text{ where } \rho \text{ and } \rho' \text{ point to the same end}\}.$$

The *ends of* $\mathcal{T}$ are these equivalence classes, and we denote their set by $\partial\mathcal{T}$. If you are uncomfortable working with equivalence classes, you can instead identify the ends of $\mathcal{T}$ with the collection of all rays that share the same, fixed base vertex.

Exercise 24. Prove that given an end $e \in \partial\mathcal{T}$ and a vertex $v \in \mathcal{T}$, there is a unique ray in $\mathcal{T}$ that represents e and is based at v.

We prefer not to define $\partial\mathcal{T}$ using rays based at a fixed vertex, as the exercise indicates that we could, because such a definition makes it difficult to see the action of F_2 on $\partial\mathcal{T}$. If $g \in F_2$ is any element of the free group, then g acts on $\mathcal{T}$. Thus if ρ is any ray in $\mathcal{T}$ then $g \cdot \rho$ is also a ray in $\mathcal{T}$. By definition, if $\rho_1 \sim \rho_2$, then $\rho_1 \cap \rho_2 = \rho_3$, where ρ_3 is a ray. Thus $g \cdot \rho_1 \cap g \cdot \rho_2 = g \cdot \rho_3$, and it follows that $g \cdot \rho_1 \sim g \cdot \rho_2$. Thus the action of F_2 on rays in $\mathcal{T}$ can be used to describe an action of F_2 on the equivalence classes of rays, yielding the following fact.

LEMMA 10.8. *The group F_2 acts on $\partial\mathcal{T}$.*

Exercise 25. Compare our definition of $\partial\mathcal{T}$ with the definition of the boundary of a hyperbolic space given in Office Hour 9.

Back to the beginning ...   Let a and b be generators of F_2 and let $\mathcal{T}$ be the corresponding Cayley graph. In terms of the standard picture, think of a as corresponding to the horizontal edges and b as corresponding to the vertical edges. There is an end of $\mathcal{T}$ that is represented by the ray whose vertices correspond to nonnegative powers of b:

$$\text{id} \longrightarrow b \longrightarrow b^2 \longrightarrow b^3 \longrightarrow \cdots$$

The element a moves this ray to the ray with vertices

$$a \longrightarrow ab \longrightarrow ab^2 \longrightarrow ab^3 \longrightarrow \cdots.$$

The end that we could denote by $[b^\infty]$ moves to $[ab^\infty]$ and, to be more general (and much more vague!), all the ends of $\mathcal{T}$ that are above the horizontal axis of $\mathcal{T}$ are shifted to the right. Similarly, all the ends of $\mathcal{T}$ that are below the horizontal axis of $\mathcal{T}$ are shifted to the right, as well. This action is the inspiration for the example from the beginning of this office hour.

There are two ends of $\mathcal{T}$ that we have not yet discussed: the one corresponding to nonnegative powers of a and the one corresponding to nonpositive powers of a. The ray

$$\text{id} \longrightarrow a \longrightarrow a^2 \longrightarrow a^3 \longrightarrow \cdots$$

is taken by a to the ray

$$a \longrightarrow a^2 \longrightarrow a^3 \longrightarrow a^4 \longrightarrow \cdots.$$

These are equivalent rays, and so a takes $[a^\infty]$ back to itself. In other words, this end is a fixed point for the action of a on $\partial\mathcal{T}$. The end $[a^{-\infty}]$ is also a fixed point for the action of a.

$\partial\mathcal{T}$ is a Cantor set.   We can define a distance between any two ends in $\partial\mathcal{T}$ using the fact that given a base vertex $v \in \partial\mathcal{T}$, every $e \in \partial\mathcal{T}$ can be represented by a unique ray ρ_e that begins at the vertex v. On the intuitive level, we think that two ends e and ϵ are close if their rays ρ_e and ρ_ϵ are the same for a large initial portion of the rays. We can make this precise by letting s denote the final vertex where ρ_e and ρ_ϵ overlap. Then set

$$d_\infty(e, \epsilon) = 2^{-d(v,s)},$$

where $d(v, s)$ is the combinatorial distance from v to s in $\mathcal{T}$.

Exercise 26. Draw $\mathcal{T}$ and select at least five ends of $\mathcal{T}$. Compute the distance between every pair of ends you selected.

Exercise 27. Prove that $\partial\mathcal{T}$ is homeomorphic with the standard Cantor set $\mathcal{C}$ described at the beginning of this office hour. *Hint: In most point-set topology textbooks the author will prove that the Cantor set is the only totally disconnected, perfect, compact metric space. Thus if you can prove that $\partial\mathcal{T}$ with the indicated metric has these properties, then $\partial\mathcal{T}$ is homeomorphic to $\mathcal{C}$.*

FURTHER READING

In addition to looking for metrics, there are a number of more strictly topological approaches for adding structure to the space of ends of an infinite group. For example, the free abelian groups $\mathbb{Z}^n = \mathbb{Z} \times \mathbb{Z} \times \cdots \times \mathbb{Z}$ are all one-ended for $n \geq 2$. At a purely intuitive level you would expect that "at infinity" $\mathbb{Z} \times \mathbb{Z}$ looks a lot like a circle while $\mathbb{Z} \times \mathbb{Z} \times \mathbb{Z}$ looks like a 2–sphere. And so the properties that can be used to distinguish circles from spheres might be applicable to defining topological invariants of ends. One such approach comes from taking the idea of simple connectivity and constructing the notion of being *simply connected at infinity*. That story, however, expects quite a bit of background, so maybe we should leave that for another day. If you would rather not wait, then we recommend reading Geoghegan's book [139], which does an excellent job of presenting topological invariants for the ends of infinite groups. As we indicated at the end of Section 10.2, Meier provides proofs of many of the theorems pertaining to the ends of a group [202], and Project 1 below refers you to Davis for more information about the ends of Coxeter groups [102]. Also, we would suggest reading Stallings' text for more information about the ends of a group [242].

PROJECTS

Project 1. Determine the number of ends of the Coxeter group

$$W = \langle a, b, c \mid a^2, \ b^2, \ c^2, \ (ab)^2, (bc)^3 \rangle.$$

The Cayley graph should look something like a "tree of hexagons and squares"; see Office Hour 13, in particular Figure 13.11. If you are interested in the ends of Coxeter groups, we recommend looking at the book by Davis [102].

Project 2. Determine the number of ends of the Baumslag–Solitar group $BS(1, 2) \cong \langle a, t \mid tat^{-1} = a^2 \rangle$; see Office Hour 12, in particular Figure 12.1.

Project 3. Determine the number of ends of the Lamplighter group L_2; see Office Hour 15.

Project 4. Find a group G that contains a subgroup H such that $e(H) = 1$ and $e(G) = \infty$.

Project 5. Given a group G and a subgroup H there are multiple definitions for the ends of the pair of groups (G, H). Find and explore the different definitions, and come up with your own illuminating examples that highlight the differences between the definitions.

Project 6. The definition of ends of a group applies even when using an infinite generating set.

1. Write down the definition of the number of ends of G with respect to any generating set S.
2. A clean but infinite presentation of Thompson's group F is given in Section 16.3. How many ends does Thompson's group F have with respect to this infinite generating set?
3. As is mentioned in Project 13 in Office Hour 3, for $n > 2$, $SL(n, \mathbb{Z})$ is boundedly generated by the elementary matrices. How many ends does $SL(n, \mathbb{Z})$ ($n > 2$) have with respect to the set of elementary matrices?
4. Show that the concept of ends with respect to an infinite set of generators is problematic by exploring what happens if, for any group G, you take the set of all elements of G as the generating set.

Project 7. We know that a group is zero-ended if and only if the group is finite. There is a similarly strong description of the two-ended groups. Find this classification. And while you are at it, look up John Stallings' theorem on groups with more than one end [243].

Project 8. The following theorem is true: If G is a finitely generated group and N is a finitely generated normal nontrivial subgroup of G with infinite index, then $e(G) = 1$. Investigate whether or not the converse is true.

Project 9. Right-angled Artin groups are discussed in Office Hour 14. Determine which right-angled Artin groups are one-ended.

Office Hour Eleven

Asymptotic Dimension

Greg Bell

The asymptotic topology of finitely generated groups is loosely described as a place where group theory, geometry, and topology meet. It is common for such interactions to have a synergistic effect so that the result of the interaction is greater than the sum of its parts. Asymptotic dimension is one such example. This is an interpretation of the basic topological notion of dimension for finitely generated groups that is both intuitive and interesting.

11.1 DIMENSION

Dimension is a fundamental concept in mathematics. Our first exposure is by example: you are accustomed to thinking that a point is zero dimensional, a line is one dimensional, a plane is two dimensional, etc. One way of thinking about dimension is in terms of the number of coordinates in the space that are necessary to describe a point in that space. Indeed, this is basically the definition you see in a linear algebra course. Thus, when the entire space is a single point, no coordinates are needed to specify that point. When the space is a line, only one coordinate is necessary, say, the signed distance to a fixed origin. In a plane, you could think in terms of rectangular coordinates (x, y) or polar coordinates (r, θ).

This definition is sufficient for many needs. For example, a circle in the plane is one dimensional, the surface of the earth is two dimensional, the ocean is three dimensional.

Figure 11.1 Define the map f as the limit of the process described by the pictures here
          working from left to right. The first step describes the image of [0, 1] as a kind
          of tent. This tent is copied four times, is reduced in size, and then the four small
          tents are rotated and translated to form the second picture; this is then copied
          four times, reduced in size, and rotated and translated to form the third picture,
          and so on. The remarkable fact is that in the limit, this (continuous!) curve
          touches every point in the square.

However, you can begin to see problems with this definition. We like to think that
the graph of a function of one variable or a parametrized curve is one dimensional
since it takes only one coordinate, the parameter value, to specify a point on the
curve. But, what if the image of the parametrized curve fills up every point in the
unit square? In fact, Peano's 1890 construction of a continuous map from [0, 1]
onto the unit square and the resulting possibility of parameterizing the unit square
with a single coordinate led mathematicians to begin to think seriously about how
to define dimension. In Figure 11.1 I give an indication of how such a map can be
constructed following the method given by Munkres [208, Section 44].

We would like to apply the notion of dimension to groups in a meaningful way,
but the idea of specifying a point with coordinates is insufficient for this purpose,
as often our groups are discrete objects. Instead we need a way of defining dimension that is independent of a choice of coordinates. One way to do this is to use
topological dimension (see Section 11.4). We will translate this topological notion
to the world of large-scale geometry and apply it to Cayley graphs of groups, since
Cayley graphs are geometric objects that can be associated to groups with specified
generating sets.

Although much of what I will say about asymptotic dimension applies equally
well to any metric space, most often we are interested in finitely generated groups
considered as metric spaces. Recall from Office Hour 2 that the Cayley graph of
a group G with fixed generating set S is a graph whose vertices are in one-to-one
correspondence with the elements of the group with a directed edge connecting g
to $g \cdot s$ for every $g \in G$ and $s \in S$.

We being our investigation by looking at our standard geometric group theory
examples: free abelian groups, $\mathbb{Z}^n$, and free nonabelian groups, F_n.

11.2 MOTIVATING EXAMPLES

In this section, we let our intuition take over and establish what the dimension of
our favorite groups should be.

The integers $\mathbb{Z}$.   Let's begin with the integers. When we consider the dimension of $\mathbb{R}$, we use the geometry of the space more than the algebraic structure. We draw the number line to represent $\mathbb{R}$ and say that things that look like lines should have dimension 1. So, we could approach the problem of defining the dimension of $\mathbb{Z}$ in the same way; thus, we need a geometric object that captures the essence of the integers. The problem with doing this is that $\mathbb{Z}$ is a discrete set and the usual geometric approach to the integers sees gaps in between them. This approach to the geometry of $\mathbb{Z}$ is therefore uninteresting: each point is isolated and discrete. On the other hand, when we think of how $\mathbb{Z}$ sits inside $\mathbb{R}$, a definite geometry appears.

In fact, on the macro scale, our picture of the integers is not dissimilar from our picture of the reals; see Figure 11.2. As integers increase, they wander off to infinity at unit gaps along the number line. Thus, it seems that $\mathbb{R}$ does a reasonable job of capturing the large-scale structure of $\mathbb{Z}$. There are just two directions to consider, positive and negative, and so it stands to reason that the dimension of this group should be 1.

The free abelian group of rank 2, $\mathbb{Z}^2$.   Next, let's consider the group $\mathbb{Z}^2$. Our geometric intuition tells us that the dimension of a product should be the sum of the dimensions of the factors. Therefore if we accept the fact that the dimension of $\mathbb{Z}$ is 1, then it is natural to conjecture that the dimension of $\mathbb{Z}^2$ is 2; see Figure 11.3. We can capture the geometry of this group by seeing that it sits inside the Euclidean

Figure 11.2 The integers' sitting inside the number line captures the large-scale one dimensionality of $\mathbb{Z}$, indicated by moving left or right along the number line.

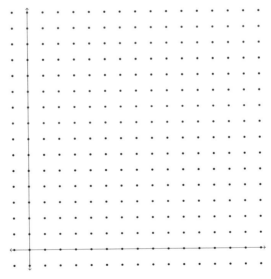

Figure 11.3 Ordered pairs of integers sitting in the plane indicate two dimensions on the large scale: up/down and left/right.

plane, $\mathbb{R}^2$, in much the same way that $\mathbb{Z}$ sits inside $\mathbb{R}$. Although we might be tempted to say this group's dimension is 2 since we see that it takes two coordinates to specify any point in it, this is not the best way to think of the situation. Indeed, like any countable set, there is a one-to-one correspondence between $\mathbb{Z}^2$ and the natural numbers $\mathbb{N}$, so the idea of dimension being a measure of the number of coordinates needed to specify a point isn't useful for this group, or any other finitely generated group.

By similarly considering the group $\mathbb{Z}^n$ sitting inside $\mathbb{R}^n$, we conjecture that the dimension of the free abelian group $\mathbb{Z}^n$ is n.

Free nonabelian groups, F_n.   So, what happens when we leave the world of free abelian groups and ask about the dimension of (nonabelian) free groups? Let us begin with the free group on two generators, F_2. Using these two generators, the Cayley graph of F_2 is a tree where every vertex is incident to four edges. If we try to place the vertices of such a tree in the Euclidean plane in such a way that the Euclidean distance is equal to the distance in the tree distance, we quickly run out of room. Therefore it is natural to think that the tree might have dimension more than 2. To see that this is not correct, we have to return to the pictures of $\mathbb{Z}$ and $\mathbb{Z}^2$ and think about the structure of these groups in a better way.

One way to regard the one-dimensionality of $\mathbb{R}$ is to consider how open sets, i.e., open intervals, overlap. You might be able to convince yourself, or maybe you saw in a real analysis class, that it is impossible to cover $\mathbb{R}$ by finite-length open intervals that are disjoint. Thus, no matter what you do, there are always points of $\mathbb{R}$ that meet at least two intervals in a cover. We describe this by saying that the cover has *order* 2. Then, the dimension is seen to be 1 less than the minimum order of these open interval covers.

Let's try to apply this reasoning to $\mathbb{Z}$. Clearly, this does not work as stated because each point is itself an open interval: for example, $\{n\} = (n-1, n+1) \cap \mathbb{Z}$. In the reals, the overlapping pieces of adjacent intervals can be made arbitrarily small, but the point is that they always have to overlap each other. In the integers, we cannot guarantee this because of the gaps in between adjacent intervals. In $\mathbb{R}$, no matter how small we make our perspective, the space always looks the same. Imagine a physical manifestation of the number line. It's an infinitely thin connected line, so regardless of whether you look at it with the naked eye, a magnifying glass, or an electron microscope, it will always look the same. Thus, arbitrarily small scales don't affect the space $\mathbb{R}$. In other words, with the reals, there is no scale so small that you only see one point.

If we try to achieve the same perspective independence with the integers, it is clear that smaller scales make quite a bit of difference. But, let us imagine that the only scales available to us are scales that are so large that you never see just one point, as was the case with $\mathbb{R}$. When we consider the multiplicities of points in $\mathbb{R}$ with respect to a cover by open sets, it is clear that the open sets do not just meet at a single point; they have to overlap a little bit. So, instead of the multiplicity of a point, we want to know how many open intervals meet a bounded set, say with radius much larger than 1. To complete the analogy with $\mathbb{R}$ and claim that the dimension of $\mathbb{Z}$ is 1, we want to be sure that the number of intervals meeting any bounded set can be arranged to be no more than 2.

Here is an alternative interpretation. Imagine that a fixed scale r, much larger than 1, is given. Now, your job is to color the integers with the fewest number of colors possible so that the following conditions are satisfied:

1. the number of consecutive integers that are the same color is always smaller than some fixed number and
2. if $m, n \in \mathbb{Z}$ have the same color and $|m - n| \leq r$, then all integers between m and n are the same color.

So, the task is to cover $\mathbb{Z}$ with bricks of different colors in such a way that each brick has diameter bounded by some fixed number such that no two bricks of the same color are within r units of each other.

Exercise 1. Show that this can be done with two colors, but it cannot be done with only one color.

So, we say that the dimension of $\mathbb{Z}$ is 1 less than the minimum order of the cover: $2 - 1 = 1$.

Exercise 2. Show that three colors can be used to cover $\mathbb{Z}^2$ in $\mathbb{R}^2$. That is, show how to cover $\mathbb{Z}^2$ by bricks of uniform size in such a way that any two bricks of the same color are sufficiently far away from each other. For a real challenge, try covering a bit of $\mathbb{Z}^3$ with four different colored blocks.

These exercises show that the dimension of $\mathbb{Z}^2$ is at most 2 and the dimension of $\mathbb{Z}^3$ is at most 3. We have not shown that these are the dimensions, since to do this would require us to show that three, or four, colors are the fewest number of colors of sets that cover with these properties.

Finally, let us return to the Cayley graph of F_2.

Exercise 3. Show that two colors can be used to cover the Cayley graph of F_2.

Thus, we conjecture that the dimension of F_2 is 1.

11.3 LARGE-SCALE GEOMETRY

In the examples we considered in the last section, you may have noticed that I cheated a bit. I said that the integers sit inside the real numbers in a certain way and that therefore their dimension should be 1. In the language of geometric group theory, we were viewing $\mathbb{R}$ as $\Gamma(\mathbb{Z}, \{1\})$, the Cayley graph of $\mathbb{Z}$ with the generating set $\{1\}$. In order for our notion of dimension to make sense for a finitely generated group, it should be an algebraic invariant. In other words, our computation should not depend on some choice of Cayley graph.

Different Cayley graphs for $\mathbb{Z}$. For example, $\langle a, b \mid a^2 = b^3, ab = ba \rangle$ is a presentation of a group isomorphic to $\mathbb{Z}$ (this presentation implicitly appears in

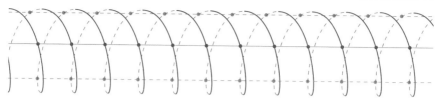

Figure 11.4 The Cayley graph of $\langle a, b \mid a^2 = b^3, ab = ba \rangle \cong \mathbb{Z}$. Multiplication by a is represented by blue edges and multiplication by b by red edges.

Office Hour 7). Indeed, every element can be written as a power of $c = ab^{-1}$, so the group is cyclic. For example, we see that $c^3 = ab^{-1}ab^{-1}ab^{-1} = a^3b^{-3} = a^3a^{-2} = a$. You can similarly check that $c^2 = b$. The Cayley graph for $\mathbb{Z}$ with this generating set is shown in Figure 11.4.

One can imagine this Cayley graph as a wire mesh wrapped around a cylinder of infinite length. It isn't immediately clear how to compute the dimension of $\mathbb{Z}$ using this space. Additionally, it's easy to see that a slight modification of the group presentation, changing the relation $a^2 = b^3$ to $a^2 = b^{2k+1}$ for larger and larger k, has the effect of increasing the radius of the cylinder in the Cayley graph. This, in turn, makes the Cayley graph look more and more $\mathbb{Z}^2$–like. In some sense it has the same look as the integer lattice $\mathbb{Z}^2$ in the Euclidean plane, so you might conjecture that it has dimension 2.

Quasi-isometries. Let's return once more to the integers sitting inside the real line. If an observer moves away from the real line, the perceived distance between distinct integers decreases. The farther the observer moves away from the real line, the closer the integers appear to become. Therefore, we can imagine that the observer has moved all of the way to infinity, whereby the observer is no longer able to distinguish the integers from the real line itself. Gromov [155] describes the result of this change in perspective this way:

> ... *the points in $\mathbb{Z}$ coalesce into a connected continuous solid unity which occupies the visual horizon without any gaps or holes and fills our geometer's heart with joy.*

Although it seems that this change in perspective will result in a loss of information, quite the opposite is true.

Instead of effecting this perspective change each time we want to study the large-scale geometry of a space, we could simply study properties of these metric spaces that are invariant under bounded perturbations, essentially ignoring small-scale features. There are several ways to make precise the notion of large-scale equivalence, but the most natural in the setting of groups is by using *quasi-isometries*.

This fundamental concept of geometric group is introduced and thoroughly developed in Office Hour 7. The most important aspect for the current discussion is Theorem 7.3: for a finitely generated group G and generating sets $S, S' \subset G$, the Cayley graphs $\Gamma(G, S)$ and $\Gamma(G, S')$ are quasi-isometric.

The upshot is that when we consider finitely generated groups and finite generating sets, the quasi-isometric equivalence class of a Cayley graph is a group property: it depends only on the group and not on the choice of the generating set. In other words, if we are interested only in the large-scale structure of a Cayley graph of a finitely generated group, then we don't need to specify a generating set. Any choice of a finite generating set will give rise to a Cayley graph with the same large-scale properties. Therefore, if we are going to use a Cayley graph to define the dimension of a group, then our definition needs to be well defined up to quasi-isometry.

11.4 TOPOLOGY AND DIMENSION

Before we get to the large-scale notion of dimension, we will review the usual topological notion of dimension. For complete details see Munkres' text [208, Section 50], the book by Aarts and Nishiura [1], or the book by Hurewicz and Wallman [173].

As mentioned previously, we often think of dimension as the number of coordinates needed to specify a point in the space. If we wish to describe dimension in a way that is independent of this process of choosing coordinates, we appeal to topology. There are several notions of dimension in topology, but we will only discuss covering dimension. For simplicity, we will always assume our spaces are metric spaces.

Covering dimension.   Let's recall a few definitions from topology; you may have also seen these definitions in a real analysis class. A collection $\mathcal{U}$ of subsets of a space X is said to have *order $n + 1$* if there is some point in X that lies in exactly $n + 1$ sets from $\mathcal{U}$ and there is no point that lies in $n + 2$ sets from $\mathcal{U}$. A collection $\mathcal{V}$ of subsets is said to *refine* the collection $\mathcal{U}$ if every set in $\mathcal{V}$ is a subset of some set from $\mathcal{U}$. Finally, call a collection $\mathcal{U}$ of subsets of X an *open cover* of a space X if every set $U \in \mathcal{U}$ is open and if every $x \in X$ is contained in some $U \in \mathcal{U}$.

With this setup, we can give the definition of (covering) dimension as alluded to previously in the context of the real line. We say that the *dimension of a metric space X does not exceed n* and write dim $X \leq n$ if for every open cover $\mathcal{U}$ of X there is a refinement $\mathcal{V}$ with order at most $n + 1$.

We see that this definition is a small-scale property of the space: computing dimension involves passing to a collection of smaller sets. Although it is possible to adapt this notion directly and arrive at a reasonable definition of large-scale dimension, we use an alternate characterization.

Computing dimension by coloring.   We will first characterize what it means to be zero dimensional and then use zero dimensional sets to compute higher dimensions. In what follows let's assume that all metric spaces are separable. This simply means that they contain a countable dense set. For example, the real numbers contain the countable dense set $\mathbb{Q}$ of rational numbers.

LEMMA 11.1. *Suppose X is a separable metric space that is nonempty. Then* $\dim X = 0$ *if and only if for all $p \in X$ and every open set U containing p, there is an open set $V \subset U$ so that $X - V$ is also open.*

Here is a sketch of a proof (also refer to the book by Hurewicz and Wallman [173]). Suppose $\dim X = 0$. Given $p \in X$ and an open set $U \subset X$ containing p, there is an open cover $\mathcal{U}$ containing U. As $\dim X = 0$, there is a refinement $\mathcal{V}$ with multiplicity at most 1. Then there is a (unique) $V \in \mathcal{V}$ such that $p \in V \subset U$ and $X - V = \bigcup_{V' \in \mathcal{V}, V' \neq V} V'$ is open. Conversely suppose X satisfies the latter property from the lemma. Given an open cover $\mathcal{U}$, we can describe a procedure for producing the refinement $\mathcal{V}$ with multiplicity at most 1. Given $p \in X$, let $U \in \mathcal{U}$ be such that $p \in U$. By the property of the lemma, there is an open set $V \subset U$ containing p for which $X - V$ is open. Add V to the refinement $\mathcal{V}$ and continue using $X - V$ with open cover $\{U \cap (X - V) \mid U \in \mathcal{U}\}$.

Example: $\dim \mathbb{Q} = 0$. Using Lemma 11.1, we can see that the set $\mathbb{Q}$ of rational numbers has dimension 0. Indeed, let $p \in \mathbb{Q}$ be given with $U \subset \mathbb{Q}$ an open set containing p. Since U is open, there is an $\epsilon > 0$ such that the (open) ball of radius ϵ about p, $B(p, \epsilon) = \{x \in Q \mid |x - p| < \epsilon\}$, is contained in U. Let $0 < \epsilon' \leq \epsilon$ be an irrational number and put $V = B(p, \epsilon') \subset B(p, \epsilon) \subset U$. Observe that if $x \in \mathbb{Q} - V$, then $B(x, \delta) \subset \mathbb{Q} - V$, where $\delta = d(x, p) - \epsilon'$. Notice that $\delta > 0$ since $d(x, p)$ is rational and at least ϵ'. Thus V and $\mathbb{Q} - V$ are both open.

Exercise 4. Show that the set of all pairs $(x, y) \in \mathbb{R}^2$ with exactly one rational coordinate is zero dimensional.

Now that we know what it means to be zero dimensional, we can use the following decomposition theorem of Aarts and Nishiura [1, Theorem 3.8] to compute dimension.

THEOREM 11.2. *For a separable metric space X, we have* $\dim X \leq n$ *if and only if there exist subsets $X_0, X_1, \ldots, X_n$ with $X = X_0 \cup X_1 \cup \cdots \cup X_n$ and $\dim X_i = 0$ for all i.*

In other words, $\dim X \leq n$ if and only if we can color X with $n + 1$ colors, such that monochromatic subsets are zero dimensional.

Exercise 5. Show that $\dim \mathbb{R} = 1$. *Hint: Argue that $\dim(\mathbb{R} - \mathbb{Q}) = 0$ and that* $\dim \mathbb{R} \neq 0$.

Exercise 6. Show that $\dim \mathbb{R}^2 \leq 2$ by finding three zero-dimensional sets that cover $\mathbb{R}^2$.

Motivated by this approach to computing dimension, to define a large-scale analog let's first define what it means to have large-scale dimension 0. Then, morally, we define a space to have dimension no more than n if it is a union of $n + 1$ subsets that are zero dimensional. This becomes precise in the next section.

11.5 LARGE-SCALE DIMENSION

We will now finally define asymptotic dimension—our large-scale analog of dimension—and give some of its basic properties. This definition of asymptotic dimension is due to Gromov [155].[1] Motivated by the previous section, to define a large-scale notion of dimension of a metric space, let's first define what it means to have dimension 0 at large scales. Before that though, let's talk about dimension at some fixed scale and then consider what happens as that scale becomes large.

Dimension at a fixed scale.   Let $r > 0$. A metric space X is said to have *dimension* 0 *at scale* r if it can be expressed as a union $X = \bigcup X_i$ where:

1. $\sup\{\mathrm{diam}(X_i)\} < \infty$ and
2. $\inf\{d(x, x') \mid x \in X_i, x' \in X_j\} > r$ whenever $i \neq j$.

A collection of sets X_i that satisfy the first property is called *uniformly bounded*. Sets that satisfy the second property are called *r–disjoint*.

We describe this notion succinctly by saying that X has dimension 0 at scale r if it admits a uniformly bounded cover by r–disjoint sets.

Example: Changing the scale.   Considering $\mathbb{Z}$ as a metric space with metric $d(m, n) = |m - n|$, we see that $\mathbb{Z}$ has dimension 0 at every scale $r < 1$. Indeed, for subsets we can use the singleton sets $X_i = \{i\}$, which all have diameter 0. Also, if $i \neq j$, we have $\inf\{|m - n| \mid m \in X_i, n \in X_j\} = |i - j| \geq 1 > r$. Therefore $\bigcup\{i\}$ is a uniformly bounded cover of $\mathbb{Z}$ by r–disjoint sets.

However, if $r \geq 1$, then $\mathbb{Z}$ does not have dimension 0 at scale r. This follows as whenever we express $\mathbb{Z} = \bigcup X_i$ as a union of bounded subsets, for any X_i there is some $k \in \mathbb{Z}$ such that $k \in X_i$ but $k + 1 \notin X_i$. We have $k + 1 \in X_j$ for some $j \neq i$ and so $\inf\{|m - n| \mid m \in X_i, n \in X_j\} \leq |k - (k + 1)| = 1 \leq r$.

This agrees with our earlier intuition: on small scales $\mathbb{Z}$ looks like a discrete collection of points and hence is zero dimensional, but on large scales $\mathbb{Z}$ no longer looks zero dimensional.

Asymptotic dimension of a metric space.   We can now define asymptotic dimension. We say that the *asymptotic dimension of a metric space* X *does not exceed* n, and write asdim $X \leq n$ if for each $r > 0$, there exist subsets $X_0, X_1, \ldots, X_n$ with $X = X_0 \cup X_1 \cup \cdots \cup X_n$, and for each i, X_i has dimension 0 at scale r.

We can rephrase this definition in several ways, but perhaps the most intuitive way is to say that asdim $X \leq n$ if for every r the entire space X can be painted with splotches of $n + 1$ colors of paint in such a way that each splotch has uniformly bounded diameter and any two splotches of the same color are at least r units away from each other (see Figure 11.5).

---

[1] The notion of asymptotic dimension that I describe here is commonly accepted to be the "right" one. In Gromov's paper, this notion of asymptotic dimension is denoted by asdim$_+$. What Gromov calls asdim is not equivalent to our definition and does not appear elsewhere in the literature.

Figure 11.5 asdim $\mathbb{Z}^2 \leq 2$.

We say that asdim $X = n$ if it is true that asdim $X \leq n$ and it is not true that asdim $X \leq n - 1$. We say that asdim $X = \infty$ if it is never true that asdim $X \leq n$ for any n.

Exercise 7. Considering $\mathbb{Z}$ as a metric space with metric $d(m, n) = |m - n|$, show that asdim $\mathbb{Z} = 1$. *Hint: See Exercise 1.*

Exercise 8. Suppose that two spaces have finite asymptotic dimension. What can be said about the asymptotic dimension of their direct product? Show that the dimension $X \times Y$ does not exceed $(\text{asdim } X + 1)(\text{asdim } Y + 1) - 1$ by constructing a cover.

Equivalent characterizations of asymptotic dimension. Using pictures as motivation, we walk through some equivalent ways to characterize asymptotic dimension. This material is not used in future sections, in case you want to skip it for now. A few definitions are in order.

Let $r > 0$. We say that a collection of subsets of a metric space X has r–*order* n if there is some $x \in X$ for which $B(x, r)$ has nonempty intersection with n subsets in the collection and there is no $y \in X$ for which $B(y, r)$ meets $n + 1$ subsets in the collection. With this definition, what we called *order* earlier is 0–order.

Given a cover $\mathcal{U}$ of a metric space X, we call $\lambda > 0$ a *Lebesgue number for* $\mathcal{U}$ if whenever $A \subset X$ and $\text{diam}(A) < \lambda$, there is some $U \in \mathcal{U}$ so that $A \subset U$. The Lebesgue number for a cover is a measure of how much the sets in the cover overlap, with higher Lebesgue numbers corresponding to greater overlap.

A finite-dimensional simplicial complex can be realized as a subset of $\ell^2(\mathbb{R})$, the set of square-summable sequences, by sending each vertex of the complex to a basis element $(0, \ldots, 0, 1, 0, \ldots)$ and interpolating on the higher-dimensional simplices. As a subset of $\ell^2(\mathbb{R})$, this complex inherits a metric, called the *uniform metric*. The metric on $\ell^2(\mathbb{R})$ is defined by $d((a_i), (b_i))^2 = \sum_{i=1}^{\infty}(a_i - b_i)^2$. A map from a metric space X to a simplicial complex is said to be *uniformly cobounded* if the diameter of the preimage of a simplex is uniformly bounded in X.

We can now state equivalent formulations of asymptotic dimension and provide some hints to walk through the equivalences.

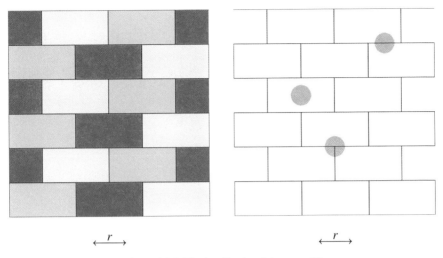

Figure 11.6 The implication $(a) \implies (b)$.

Exercise 9. For a metric space X, prove that following are equivalent:

(a) For every $r > 0$ there exist $n + 1$ uniformly bounded, r–disjoint families of subsets that cover X, i.e., asdim $X \leq n$.

(b) For every $m > 0$ there is a cover of X by uniformly bounded subsets with m–order $\leq n + 1$.

(c) For every $\lambda > 0$ there is a cover of X by uniformly bounded subsets with order $\leq n + 1$ and Lebesgue number $\geq \lambda$.

(d) For every $\epsilon > 0$ there is an ϵ–Lipschitz, uniformly cobounded map from X to an n–dimensional uniform simplicial complex.

Here are some hints to see the equivalences:

$(a \implies b)$ Figure 11.6 shows how three uniformly bounded, r–disjoint families covering $\mathbb{R}^2$ can give rise to a single cover of $\mathbb{R}^2$ with $r/2$–order. In the figure, several $r/2$–balls are shown indicating the range of possibilities.

$(b \implies c)$ Figure 11.7 indicates how the elements of the cover in part (b) can be expanded so that the resulting cover has large Lebesgue number. The colors are used to keep track of the expanded sets on the right-hand side. The disks indicate sets with diameter λ.

$(c \implies d)$ Figure 11.8 shows how a cover with large Lebesgue number can be mapped to a simplex of dimension 1 less than the order of the cover. This situation is also described as the mapping from the cover to its nerve. The idea of the map is that points in the space that belong to only one element of the cover are mapped to a vertex, points in two elements of the cover are mapped to an edge, and so on. The uniform coboundedness of the map comes from the uniform boundedness of the cover and the ϵ in the ϵ–Lipschitz map depends on the dimension and the Lebesgue number.

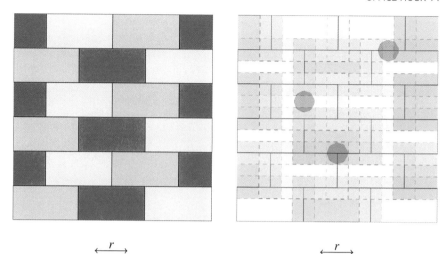

$$\xleftarrow{\quad r \quad}\xrightarrow{\quad}\qquad\qquad\xleftarrow{\quad r \quad}\xrightarrow{\quad}$$

Figure 11.7 The implication $(b) \implies (c)$.

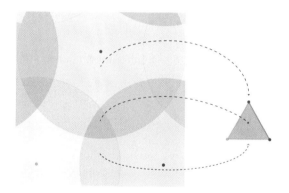

Figure 11.8 The implication $(c) \implies (d)$.

$(d \implies a)$ Finally we pass from a map to a simplicial complex back to a cover. If K denotes the uniform complex, let $\beta^1 K$ and $\beta^2 K$ denote the first and second barycentric subdivisions of K, respectively. To construct our cover, we take stars in $\beta^2 K$ of each barycenter $b_\sigma \in \beta^1 K$, where $\sigma \subset K$. In Figure 11.9 the complex K is shown at the left with barycenters indicated with vertices. The stars of these barycenters in $\beta^2 K$ are shown at the right. It remains to check that by pulling back these stars to the original space X you obtain a cover of X by $\dim(K) + 1$ families of uniformly bounded, r–disjoint families.

Asymptotic dimension of a group.   As is common in geometric group theory, we like to consider metric properties as group properties using a Cayley graph for the group. In order for this to result in a well-defined group property, we need the metric property to be invariant under quasi-isometries. That is precisely the context of the following theorem of Gromov [155] and Bell–Dranishnikov [23].

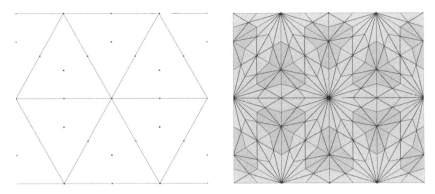

Figure 11.9 Passing from a map to a cover.

THEOREM 11.3. *Quasi-isometric metric spaces have the same asymptotic dimension.*

Exercise 10. Prove the theorem.

Combined with Theorem 7.3 (different word metrics on the same group are quasi-isometric), Theorem 11.3 says that the quantity asdim G is unambiguously defined for finitely generated groups. Indeed, we can define asdim G to be the asymptotic dimension of the metric space $\Gamma(G, S)$ where S is *any* finite generating set for G. Then any other choice of finite generating set will yield a metric space with the same asymptotic dimension. In the terminology of Office Hour 7, we see that asdim is a quasi-isometry invariant.

Exercise 11. Considering $\mathbb{Z}$ as a finitely generated group, show that asdim $\mathbb{Z} = 1$. *Hint: See Exercise 7.*

11.6 MOTIVATING EXAMPLES REVISITED

We have already seen that asdim $\mathbb{Z} = 1$, confirming our intuition from earlier. Let's visit our other motivating examples again, after first considering the circumstances under which asdim $= 0$.

Asymptotic dimension zero. Suppose that G is a finite group; so it is necessarily finitely generated. Then the diameter of the group's Cayley graph does not exceed its cardinality. Thus, for any $r > 0$ the Cayley graph itself is a uniformly bounded cover. The r–disjointness property is satisfied trivially, since there is only one set. Thus, the asymptotic dimension is 1 less than the number of families: asdim $G = 1 - 1 = 0$.

By the same reasoning, any compact metric space has asymptotic dimension 0.

One may be tempted to conjecture that all metric spaces with asymptotic dimension 0 are compact. This turns out not to be correct.

Exercise 12. Construct an unbounded metric space X with asdim $X = 0$. *Hint: This can be done with a subset of $\mathbb{Z}$.*

On the other hand, if G is a finitely generated infinite group, then asdim $G \geq 1$.

Exercise 13. Let G be a finitely generated infinite group. Show that asdim $G \geq 1$. *Hint: Suppose that $r > 1$ and that $\mathcal{U}$ is a cover of G by uniformly bounded, r–disjoint sets. Fix a $U \in \mathcal{U}$ and consider $\{g \cdot s \mid g \in U, \ s \in S\}$.*

Exercise 14. In the previous exercise we don't need the group structure to get the lower bound on the asymptotic dimension. For the purposes of this problem, call a metric space *path-connected at scale k* if the space obtained by connecting every pair of points x and x' in X, with $d(x, x') = d \leq k$ with an isometric copy of $[0, d]$, yields a path-connected space. The resulting space is called a *k–thickening*.

Show that if X is a space that is path-connected at scale k for some k, then asdim $X \geq 1$. It is possible to endow this thickened space with a metric. As an example, we can see that the 1–thickening of a group G equipped with a word metric d_S is exactly the Cayley graph $\Gamma(G, S)$.

Asymptotic dimension of $\mathbb{Z}^n$. In Figure 11.5 I indicate how to construct three families of uniformly bounded, r–disjoint sets covering $\mathbb{Z}^2$. In that image, if each dot represents an element of $\mathbb{Z}^2$, then we can see that the sets are 8–bounded and 4–disjoint. By increasing the scale of the image and continuing the pattern indefinitely, we can use such a cover to prove that asdim $\mathbb{Z}^2 \leq 2$.

Exercise 15. In Figure 11.5, families are indicated showing that asdim $\mathbb{Z}^2$ is at most 2. Write down definitions of these families explicitly.

It is not difficult to imagine constructing higher-dimensional analogs of this cover to show that asdim $\mathbb{Z}^n \leq n$ for each n.

Exercise 16. Extend the sets in Figure 11.5 to three dimensions. Construct a cover of $\mathbb{Z}^3$ using four different colored blocks. How can this be extended to higher dimensions?

Asymptotic dimension of free groups. The final example is the free group on two generators, F_2. The Cayley graph is a tree in which each vertex is incident to four edges. I will show that any infinite tree has asymptotic dimension 1.

Since T contains an infinite geodesic, Exercise 14 shows that asdim $T \geq 1$. Thus, it remains to show that asdim $T \leq 1$. To this end, for each r we must find two families of uniformly bounded, r–disjoint sets whose union covers T.

Let $r > 0$ be given. Fix some vertex of the tree and call it x_0. We will use the notation $\|x\|$ to denote the distance $d(x, x_0)$ from x to the fixed vertex x_0. As a first step in the construction of the cover, for each positive integer n, let

$$A_n = \{v \in \text{Vert}(T) \mid 2r(n-1) \leq \|v\| \leq 2rn\}.$$

Although the collections $A_o = \bigcup A_{2n+1}$ and $A_e = \bigcup A_{2n}$ each consist of r–disjoint subsets, they are not uniformly bounded. See Figure 11.10, where A_o is shown in red and A_e is shown in blue. We need to subdivide these collections further.

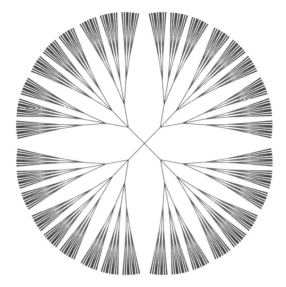

Figure 11.10 asdim $F_2 \leq 1$.

For every vertex in the tree, there is exactly one geodesic path to x_0. For two vertices x and y we define their *meet* to be the vertex, denoted by $x \wedge y$, that is common to both geodesics and farthest from x_0. Call two elements x and y in some subset A_n equivalent if $2r(n-1) \leq \|x \wedge y\|$.

Exercise 17. Prove that this is an equivalence relation. Draw equivalence classes in T for $n = 2$ and $r = 1$.

Exercise 18. Finish the proof that asdim $F_2 = 1$ by showing that the collections

$$\mathcal{U} = \{ \text{equivalence classes in } A_o \}$$
$$\mathcal{V} = \{ \text{equivalence classes in } A_e \}$$

are uniformly bounded and r–disjoint.

Nowhere in the proof do we use the fact that the rank of F_2 is 2. Indeed, the same proof shows that any tree, and hence any finitely generated free group, has asymptotic dimension at most 1.

11.7 THREE QUESTIONS

Before discussing other examples of groups and their asymptotic dimensions, let's talk about the typical questions asked about asymptotic dimension.

It is reasonable to expect that there are spaces X for which asdim $X = \infty$, but what is not immediately obvious is whether there should be any finitely generated groups that have infinite asymptotic dimension. It turns out that there are many such groups; two examples are mentioned in the next section.

The existence of finitely generated groups with infinite asymptotic dimension means the following question is interesting and often difficult to answer.

Question 19. Let G be a finitely generated group. Is asdim $G < \infty$?

This subject gained extrinsic interest following a result of Yu [267] proving that groups with finite asymptotic dimension satisfy an important conjecture on the topology of manifolds. Yu's paper sparked a great deal of interest determining which classes of groups have finite asymptotic dimension; I list some of the results of this investigation in the next section.

When the asymptotic dimension is finite, it is reasonable to ask what it equals.

Question 20. Let G be a finitely generated group for which asdim $G < \infty$. What is the exact value of asdim G?

This question is inherently more difficult, since to show asdim $X = n$ it is necessary to show that asdim $X \leq n$ and asdim $X \nleq n - 1$.

To understand this difficulty, I will give an example. In Figure 11.5 I show that asdim $\mathbb{Z}^2 \leq 2$. To show that asdim $\mathbb{Z}^2 = 2$ (which, indeed it is) is much more difficult. Using the definition, we would have to argue that it is impossible to cover $\mathbb{Z}^2$ with only two uniformly bounded r–disjoint families for some r.

Here is another related question.

Question 21. What can be said about the groups G for which asdim $G = n$?

Some progress has been made in this direction. For example, if G is finitely presented and asdim $G = 1$, we can conclude that G has a finite index subgroup isomorphic to $F_n, n \geq 1$ [138, 176].

11.8 OTHER EXAMPLES

We conclude our discussion with some other examples of groups where something is known about their asymptotic dimension. Each class of groups listed below can be found in this book.

Braid groups. Bell and Fujiwara [24] use explicit computations of asymptotic dimensions of some mapping class groups (also see the article cowritten by Fujiwara [30]) to show that the braid group B_n on $n \geq 3$ strands satisfies asdim $B_n \leq n - 2$. Braid groups are the subject of Office Hour 18.

Artin groups. They also establish that if $\mathcal{A}$ is an Artin group of finite type A_n or $B_n = C_n$, the asdim $\mathcal{A} \leq n$; if $\mathcal{A}$ is an Artin group of affine type $\tilde{A}_{n-1}$ or $\tilde{C}_{n-1}$, then asdim $\mathcal{A} = n - 1$. A particular class of Artin groups, right-angled Artin groups, is the subject of Office Hour 14.

Surface groups. Gromov [155, page 29] sketches a proof that the hyperbolic plane $\mathbb{H}^2$ satisfies asdim $\mathbb{H}^2 = 2$; combining this with the fact (see the book by de la Harpe [104, IV.24.iii], for example) that the fundamental group of a closed, orientable surface of genus at least 2 is quasi-isometric to the hyperbolic plane, we see that these surface groups have asymptotic dimension 2. The hyperbolic plane is discussed in Office Hour 5 and surface groups are discussed Office Hour 9.

Hyperbolic groups.   Gromov [155] pointed out that hyperbolic groups have finite asymptotic dimension. An explicit proof is given by Roe [227]. Hyperbolic groups are the focus of Office Hour 9.

Coxeter groups.   Dranishnikov and Januszkiewicz [117] showed that Coxeter groups have finite asymptotic dimension. Later, Dranishnikov [118] computed the dimension of right-angled Coxeter groups. Coxeter groups are introduced in Office Hour 13.

Thompson's group F.   Thompson's group F has infinite asymptotic dimension [227]. This group is the subject of Office Hour 16.

Wreath product $\mathbb{Z} \wr \mathbb{Z}$.   The wreath product $\mathbb{Z} \wr \mathbb{Z}$ also has infinite asymptotic dimension [227]. Wreath products are discussed in Office Hour 15, in particular in the context of the lamplighter group $\mathbb{Z} \wr \mathbb{Z}_2$.

FURTHER READING

For a thorough introduction to the large-scale approach to groups and metric spaces, see the book by Nowak and Yu [212] or John Roe's lecture notes [227].

   For some of the group-theoretic constructions we've discussed here, the book of Bridson and Haefliger is an outstanding resource [54].

   The classic text Hurewicz and Wallman [173] is an indispensable resource for topological dimension theory.

   A more thorough introduction to asymptotic dimension theory is given in the survey by Bell and Dranishnikov [23].

PROJECTS

Project 1.   Motivated by Theorem 11.2, we can define a variant of asymptotic dimension as follows. Say $\mathrm{asdim}_0 X \leq n$ if there are exist subsets $X_0, X_1, \ldots, X_n$ with $X = X_0 \cup X_1 \cup \cdots \cup X_n$ and $\mathrm{asdim} X_i = 0$ for all i. Is it true that $\mathrm{asdim}_0 X = \mathrm{asdim} X$?

Project 2.   Construct uniformly bounded covers by r–disjoint subsets for the groups: $\mathbb{F}_2 \times \mathbb{F}_2$, $\mathbb{Z} \times \mathbb{F}_2$, $\mathbb{Z}_2 * \mathbb{Z}$, and $\mathbb{Z}_2 * \mathbb{Z}_2$.

Project 3.   When the asymptotic dimension is infinite, we can still use the ideas of asymptotic dimension to study a metric space X as follows: for each $r > 0$, define $f(r) \in \mathbb{Z}$ to be 1 less than the minimum number of r–disjoint, uniformly bounded families required to cover X. Show that this function is nondecreasing and that $\lim f(r) = \mathrm{asdim} X$. Construct a space X that has a function f that is not eventually constant. Given a nondecreasing, nonnegative integer-valued function g, is it always possible to construct a space X so that the function g is the "asymptotic dimension function" of X? If so, how? If not, what types of restrictions need to be required?

Project 4. Give an explicit example of a finitely generated group with infinite asymptotic dimension that is not one of the examples mentioned above.

Project 5. Research the notions of random groups (see, for instance, the projects to Office Hour 9). What is the asymptotic dimension of a random group?

Project 6. Look again at Exercise 8. Improve this result to the more familiar asdim $X \times Y =$ asdim $X +$ asdim Y. Hint: Use the corresponding result for dimension and controlled maps to uniform simplices given in Exercise 9.

Project 7. Show that asdim $X \cup Y \leq \max\{$asdim $X,$ asdim $Y\}$.

Project 8. Look up Yu's property A (also called exactness). How are property A and asymptotic dimension related?

Project 9. Check out the book [173] from your library. In it, the authors define dimension inductively. What would an asymptotic inductive dimension look like? This question is answered by Dranishnikov in [116].

Project 10. Following up on the previous project, construct a dictionary translating classical dimension-theoretic results to their asymptotic analogs. What results have direct analogs? What analogous results can you prove?

Project 11. Investigate the connection between asymptotic dimension of groups and cohomological dimension of groups.

Office Hour Twelve

Growth of Groups

Eric Freden

Which group is bigger, the free abelian group $\mathbb{Z}^2$ on two generators or the free group F_2 on two generators? Of course, both groups have the same cardinality (they are countably infinite), so first we need to give some meaning to the term "bigger." To do this we will regard our groups as metric spaces and use geometric notions to compare the groups.

As usual, the closed ball of radius $r \geq 0$ centered at a point x is the set

$$B(x, r) = \{y \mid d(x, y) \leq r\}.$$

For a group with a given finite generating set, recall that the word length of a particular group element is the distance from the identity in the corresponding Cayley graph (in other words, the length of the shortest word representing the element). Thus, the ball $B(1, r)$ of radius r centered at the identity 1 is the set of group elements whose word length is less than or equal to r. The sphere $S(1, r)$ of radius r is the set of all group elements whose word length is exactly r. In this office hour we address some basic counting questions. Fix a group and some $r \geq 0$. We ask:

- How many group elements are in $B(1, r)$?
- How many group elements are in $S(1, r)$?
- How many geodesic paths in the Cayley graph start at the identity and have length r? (In general, this is not the same question as the previous one.)
- Consider all words of length n in the generators. How many of these words represent the identity?

Also:

- If we know the answers to any of the above questions, what does that tell us about the group?

A striking answer to the last question is Gromov's polynomial growth theorem. To state it requires some setup. First, we say that two nondecreasing positive functions $f(r)$ and $g(r)$ have the same growth if there is a constant $C > 0$ so that

$$f(r/C) \le g(r) \le f(Cr)$$

for all $r \ge 0$. Then we say that a function $f(r)$ has polynomial growth if $f(r)$ has the same growth as r^k for some $k \ge 0$. Finally, we say that a finitely generated group has polynomial growth if the number of group elements in the ball of radius r about the identity has polynomial growth (this makes sense because any two word metrics on a finitely generated group are quasi-isometric).

Here is Gromov's theorem.

THEOREM 12.1 (Gromov's polynomial growth theorem). *A finitely generated group has polynomial growth if and only if it has a subgroup of finite index that is nilpotent.*

In other words, the growth of a group can serve as a powerful quasi-isometric invariant. Not only can it tell us when two groups are not quasi-isometric, but it can also give us fine algebraic information about a group.

Gromov's theorem is beyond the scope of this office hour (the proof is very difficult). However, we will get our hands dirty with several specific groups. We will see that it can be a very challenging and interesting problem just to understand the answers to the above questions even for very specific groups. One class of groups we will be specifically interested in are the Baumslag–Solitar groups. We will develop some general methods for studying these groups and others.

Exercise 1. Show that "same growth" is an equivalence relation on the set of positive functions from nonnegative real numbers to positive real numbers.

Exercise 2. Show that if two groups are quasi-isometric, and one has polynomial growth, then the other does.

12.1 GROWTH SERIES

Say we have some finitely generated group G with a particular fixed finite generating set S. We want to count and keep track of how many group elements have word length r as r varies (throughout this office hour, when we say "word length," we mean the distance from 1 in the word metric). We can define a sequence of numbers to this effect: for each integer $r \ge 0$ let σ_r be the cardinality of the sphere $S(1, r)$ of radius r. In other words, σ_r is exactly the number of group elements of word length r. Of course, $\sigma_0 = 1$ and $\sigma_1 = |S \cup S^{-1}|$. The other σ_r terms depend on the relators.

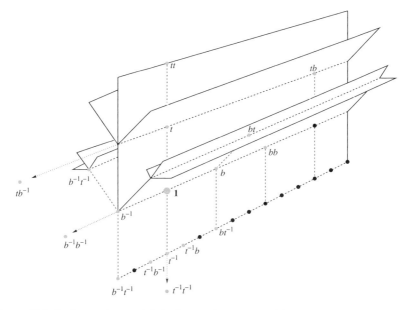

Figure 12.1 An interesting partial Cayley graph constructed from two generators. Dashed lines are graph edges, solid lines are cutaways. The green vertices comprise the ball of radius 2. Several of these vertices don't fit on the cutaway diagram and are indicated by arrows.

Given a group via a finite presentation, it is usually not easy to determine the distance from 1 of a given group element. As a starter, consider the Baumslag–Solitar group

$$BS(1,3) = \langle b, t \mid tbt^{-1} = b^3 \rangle.$$

What is the word length of $t^{-1}b^{77}$? This would be difficult to determine without a picture.

The Cayley graph for BS(1,3) is shown (in part) in Figure 12.1. This Cayley graph appears to be the direct product of a regular trivalent tree (in the vertical direction) with a line (in the horizontal direction). Consider just the main vertical sheet of this graph viewed face-on in Figure 12.2. It shows a geodesic representation for $t^{-1}b^{77}$ given by the equivalent word $t^2b^3t^{-2}b^{-1}t^{-1}b^{-1}$. This is in fact a shortest word representing this group element, and so the word length of $t^{-1}b^{77}$ is 10.

Exercise 3. Verify that the word length of the element $t^{-1}b^{77}$ with respect to the above presentation for BS(1,3) is indeed 10.

Caution! Determining the word metric turns out to be a hard problem in general. In fact, as discussed in Office Hour 8, there are groups with unsolvable word problem. For such groups it is impossible to construct the Cayley graph or do explicit growth calculations *because we cannot always determine whether the*

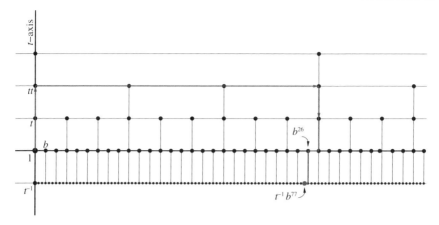

Figure 12.2 The so-called "main sheet" of the previous figure, showing a geodesic path for $t^{-1}b^{77}$.

word length of a given word is 0 or not. Only groups with solvable word problem will be considered from here on!

Generating functions and the spherical growth series. In 1730, de Moivre discovered the very useful notion of the (ordinary) *generating function* associated to a sequence of numbers. The idea is to take the rth term of the sequence, multiply it by z^r for some formal variable z, and add all these terms to make a power series. Generating functions are used extensively in combinatorics and probability. They allow the tools of algebra and analysis to be applied to discrete problems (see the book by Wilf [261] for an excellent introduction to using generating functions to solve combinatorial problems).

We define the *spherical growth series* for a given group presentation as the generating function for the sequence σ_r defined above:

$$S(z) = \sum_{r \geq 0} \sigma_r z^r \, .$$

If our group is finite, this is a polynomial and not particularly interesting. Infinite groups are another story.

Example: Spherical growth series of $\mathbb{Z}$. The group $\mathbb{Z}$ with its standard presentation has a Cayley graph that is just the usual integer number line. There are two group elements of length r for each $r > 0$, so the spherical growth series is

$$S(z) = 1 + 2z + 2z^2 + 2z^3 + \cdots = 1 + 2z \left(\frac{1}{1-z} \right) = \frac{1+z}{1-z}.$$

Example: Spherical growth series of D_∞. Recall that the infinite dihedral group D_∞ has the presentation

$$D_\infty = \langle a, b \mid a^2 = b^2 = 1 \rangle.$$

The Cayley graph for this presentation of D_∞ is very similar to the standard Cayley graph for $\mathbb{Z}$ (the Cayley graph for D_∞ has twice as many edges), and hence we obtain the same growth series. Of course, these two groups are not isomorphic, for instance, because $\mathbb{Z}$ is abelian and D_∞ is not.

A product formula for spherical growth series. Suppose G_1 and G_2 are groups with presentations $\langle S_1 \mid R_1 \rangle$ and $\langle S_2 \mid R_2 \rangle$. There is a natural presentation for the direct product $G_1 \times G_2$ where the generating set is the disjoint union of S_1 and S_2 and the set of relators consists of the disjoint union of R_1 and R_2 and all commutators of the form $[s_1, s_2]$ with $s_1 \in S_1$ and $s_2 \in S_2$ (recall that the commutator $[s_1, s_2]$ is shorthand for $s_1 s_2 s_1^{-1} s_2^{-1}$).

If the spherical growth series for G_i is $S_i(z)$, then it turns out that the spherical growth series for $G_1 \times G_2$ is the product $S_1(z) S_2(z)$. Consequently the spherical growth series for the free abelian group $\mathbb{Z} \times \mathbb{Z}$ is $\left(\frac{1+z}{1-z}\right)^2$ and more generally the free abelian group $\mathbb{Z}^d$ with standard presentation has spherical growth series $\left(\frac{1+z}{1-z}\right)^d$.

Exercise 4. Prove the above product formula for spherical growth series.

Exercise 5. If $S_1(z)$ and $S_2(z)$ are the spherical growth series corresponding to given presentations of groups G_1 and G_2, respectively, derive a formula for the spherical growth series of the free product $G_1 * G_2$ in terms of $S_1(z)$ and $S_2(z)$.

Radius of convergence and exponent of growth. The spherical growth series (and the cumulative growth series and geodesic growth series, both defined below) can be viewed as formal power series in the ring $\mathbb{Z}[[z]]$ or as analytic functions defined in some disk centered at 0 in the complex plane $\mathbb{C}$ (Wilf's text [261] is an excellent introduction to both topics). The former view is primarily combinatorial and circumvents issues of convergence. The latter view is useful for asymptotics.

Recall that the radius of convergence R for a power series can be determined by the Cauchy–Hadamard root test from calculus. In the case of general power series this involves limits superior and absolute values, but for $S(z)$ this turns out to be

$$ R = \frac{1}{\lim_{r \to \infty} \sqrt[r]{\sigma_r}} $$

(you can refer to Chapter VI of de la Harpe's text [104] for the reasons behind this simplification).

Exercise 6. Show that the radius of convergence R for $S(z)$ lies in $(0, 1]$.

When $R < 1$, the reciprocal $\frac{1}{R}$ is called the *exponent of growth* and is denoted ω. For large radii $r \gg 0$, we can make the coarse approximation

$$ \sigma_r \approx \omega^r. $$

Using more sophisticated complex analysis of the singularities of $S(z)$, one can get much better asymptotic approximations for σ_r [132]. Although $S(z)$ and R depend on the choice of presentation rather than merely on the group itself (as will be confirmed in examples later), the condition $\omega > 1$ is a group invariant. (When $R = 1$, there is a more sophisticated definition of asymptotic growth which is beyond the scope of this office hour; see [104, 150].)

Exercise 7. Compute the radius of convergence R and exponent of growth ω for all examples above. *Hint: Recall that R also represents the norm of the smallest (in absolute value) singularity of the given power series.*

Cumulative growth series.   The *cumulative growth series* for a group presentation is the generating function

$$B(z) = \sum_{r=0}^{\infty} b_r z^r,$$

where b_r is the cardinality of the ball of radius r about the identity (in other words, b_r is the number of group elements whose word length is less than or equal to r).

Exercise 8. Show that in general $B(z) = \frac{S(z)}{1-z}$, and thus the two series $S(z)$ and $B(z)$ have the same radius of convergence and are almost interchangeable.

Geodesic growth series.   There is yet a third generating function called the *geodesic growth series* $\Gamma(z) = \sum \gamma_r z^r$ where γ_r counts *all* geodesics of length r starting from the identity vertex of the Cayley graph.

There is a bijective correspondence between edge paths in the Cayley graph starting at the identity vertex and words in the generators (and their inverses). Therefore, a geodesic starting from the identity vertex can be thought of as a shortest word representative for a given group element (the one corresponding to the endpoint of the path). Such a word is called a *geodesic word*. As such, we may think of γ_r as the number of geodesic words of length r. One example of a geodesic word we already encountered was the word $t^2 b^3 t^{-2} b^{-1} t^{-1} b^{-1}$ in $BS(1,3)$.

Several results about the geodesic growth series $\Gamma(z)$ have been found recently; see [10, 50, 53, 77, 149, 198, 230]. It has been typical in the past to concentrate on $S(z)$ but computing $\Gamma(z)$ has become an increasingly popular research topic.

Computing growth series with normal forms.   The traditional method for computing $S(z)$ is to produce a preferred geodesic word for each group element g, or equivalently, a preferred geodesic path from the identity vertex of the Cayley graph to g in the Cayley graph. These geodesic words comprise what is called a set of *geodesic normal forms*. (See Office Hour 15 for examples of normal forms.) We can try to find some pattern in these normal forms that is conducive to combinatorics. Let's look at some basic examples.

Exercise 9. Consider $\mathbb{Z}$ with its presentation $\langle a, b \mid a^2 = b^3, ab = ba \rangle$. The Cayley graph for this presentation is found at the beginning of Office Hour 7. Again, geodesic normal forms are obvious from the picture. Compute $S(z)$ for this presentation.

Exercise 10. The group $\mathbb{Z}$ can also be presented as $\langle a, b \mid a^2 = b^{2k+1}, ab = ba \rangle$ for any k. Compute the spherical growth series for each such presentation.

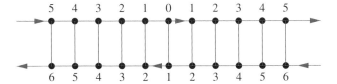

Figure 12.3 The infinite dihedral group with generators a (red) and d (blue). Vertices are labeled with their distances to the identity.

Example: Another presentation of D_∞, another growth series.   The infinite dihedral group D_∞ has another commonly seen presentation,

$$D_\infty = \langle a, d \mid a^2 = 1 = (ad)^2 \rangle,$$

obtained by the making the generator substitution (commonly referred to as a Tietze transformation) $d = ab$. We will compute $S(z)$ and $\Gamma(z)$ using normal forms.

The Cayley graph for this presentation looks like a ladder (refer to Figure 12.3). Now there are two copies of the number line oriented in opposite directions. We choose the obvious normal forms: moving horizontally from the identity uses geodesic words $d^{\pm n}$ while the other half of the ladder is composed of the words $ad^{\pm 5}$. The overall counts look like $\sigma_0 = 1$, $\sigma_1 = 3$, and $\sigma_r = 4$ for all $r > 1$. Thus

$$S(z) = 1 + 3z + 4z^2 \left(1 + z + z^2 + \cdots \right)$$

$$= 1 + 3z + \frac{4z^2}{1 - z}$$

$$= \frac{(1 + z)^2}{1 - z}.$$

We thus see that D_∞ has at least two different spherical growth series (in general, changing the presentation of a group can change its growth series).

Computing the geodesic growth series is a bit more involved. Look at the vertex on Figure 12.3 at the lower right labeled 6 that corresponds to ad^{-5}. There are six different paths of length 6 from the identity/origin to ad^{-5}, each going along a different rung of the ladder. There are another six from the origin to ad^5 and finally one each for d^6 and d^{-6}. It looks like $\gamma_0 = 1$, $\gamma_1 = 3$, and for $r > 1$, $\gamma_r = 2r + 2$. We get:

$$\Gamma(z) = 1 + 3z + \sum_{r=2}^{\infty} (2r + 2)z^r$$

$$= 1 + 3z + 2 \left(2z^2 + 3z^3 + 4z^4 + \cdots \right) + 2 \left(z^2 + z^3 + z^4 + \cdots \right)$$

$$= \frac{1 + z + z^2 - z^3}{(1 - z)^2}.$$

The plan.   The remainder of this office hour outlines a procedure to produce generating functions that represent growth series. The steps are:

1. Determine normal forms based on cone types. This is done in Section 12.2.

2. Construct a machine (a grammar) that uses rewriting rules to create the normal forms. See Sections 12.3.

3. Convert the rewriting rules into a system of equations. This follows from a famous theorem of Chomsky and Schützenberger explained in Section 12.4.

4. Solve the system of equations for the generating function to find the growth series. Section 12.4 has the details.

12.2 CONE TYPES

So far we have been able to use simple counting methods to compute several growth series. We'll need to employ more advanced combinatorics for other groups. As usual, fix a finitely generated group G and a presentation $G = \langle S \mid R \rangle$. For any geodesic word w define the *cone* based at w as

$$\Gamma_w = \left\{ w' \in F(S) \mid ww' \text{ is a geodesic word for } G \right\}.$$

Here $F(S)$ is the free group on S (we are identifying words in S with elements of $F(S)$). Equivalently, Γ_w is the set of all words that extend w in a geodesic manner.

Suppose that v is another geodesic word. We say that the cones Γ_w and Γ_v share the same *cone type* if as sets they are equal. Cone types can be used to generate recurrences.

Cone types in the free group. The standard Cayley graph of the free group F_2 is an infinite regular tree of valence 4 (refer to Office Hour 2). The cones based at a, b, a^{-1}, and b^{-1} are all illustrated on page 10 of Office Hour 2 and are labeled Γ_a, Γ_b, $\Gamma_{a^{-1}}$, and $\Gamma_{b^{-1}}$, respectively.

The same figure also shows that Γ_a and Γ_{aa} (labeled as $F_a(\Gamma_a)$) are of the same type; Γ_b and Γ_{ab} (labeled as $F_a(\Gamma_b)$) are of the same type; also $\Gamma_{b^{-1}}$ and $\Gamma_{ab^{-1}}$ (labeled as $F_a(\Gamma_{b^{-1}})$) are of the same type.

Evidently for any nontrivial geodesic word $w \in F_2$, the cone Γ_w depends only on the rightmost letter of w and agrees with one of the four types already discussed. There is a trivial fifth cone based at the identity which comprises the entire group.

We use this structure to re-derive the spherical growth series. Obviously we get $\sigma_0 = 1$ and $\sigma_1 = 4$. Observe that each nontrivial cone has one vertex leading to it from the identity and three vertices going away from the identity, with the pattern repeating ad infinitum. This defines the recursion $\sigma_r = 3\sigma_{r-1}$, which is valid for all $r > 1$. Solving this recursion we get $\sigma_r = 4 \cdot 3^r$; therefore

$$S(z) = 1 + 4z + 4 \cdot 3z^2 + 4 \cdot 3^2 z^3 + \cdots$$

$$= 1 + 4z \left(1 + 3z + 3^2 z^2 + \cdots \right)$$

$$= 1 + \frac{4z}{1 - 3z} = \frac{1 + z}{1 - 3z}.$$

We further have $\Gamma(z) = S(z)$ since geodesics are unique.

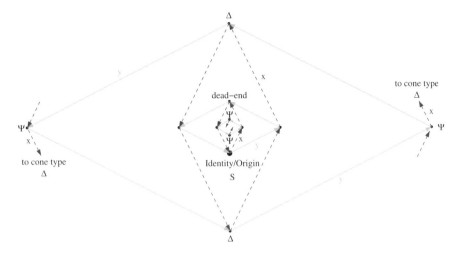

Figure 12.4 Partial Cayley graph with cone types shown.

Unlabeled cones. There is an observation to be made from the previous example. Although the cones Γ_a, Γ_b, $\Gamma_{a^{-1}}$, and $\Gamma_{b^{-1}}$ are all distinct as sets, they look identical as rooted, unlabeled subgraphs in the Cayley graph. They form a single *unlabeled cone type*, which is a weaker equivalence class on cones, namely when cones are isomorphic as rooted unlabeled graphs. We intuitively used this idea when we deduced $\sigma_r = 3\sigma_{r-1}$ above. Unless otherwise mentioned, we'll use the original (labeled) cone types in the balance of this office hour.

Example: A colorful collection of diamonds. Consider the group

$$G = \langle x, y \mid x^4 = 1 = y^4, \quad x^2 = y^2 \rangle.$$

A partial Cayley graph is drawn in Figure 12.4. From the identity vertex the generator edges are shown in their respective colors. The group is infinite since the words $(xy)^n$ are geodesic for all $n > 0$; see Exercise 11.

You can see that there is a series of shrinking diamond shapes going inward and expanding diamonds going outward. We utilize four cone types: the cone based at the identity (marked **S**), the cone based at x^2, and two others that are marked **Ψ** and **Δ**, respectively.

The cones marked **Ψ** are based at xy, xy^{-1}, and at any geodesic word of length 3 or more from the identity that ends in y or y^{-1} (a solid green edge). The cones marked **Δ** are based at yx, yx^{-1}, and at any geodesic word of length 4 or more from the identity that ends in x or x^{-1} (a dashed red edge). Notice that the vertex x^2 is a dead end, meaning that its cone is the empty set (again refer to Office Hour 15 for more on dead ends). Why is the cone based at y not labeled as **Ψ**? We'll compute the growth of this presentation later in the office hour.

Exercise 11. Verify in the previous example that $N = \langle x^2 \rangle$ is a normal subgroup and that G/N is isomorphic to D_∞.

Exercise 12. Determine and/or verify the cone types from the previous examples.

Are growth series always rational?   At this point one might think that all finitely generated groups have growth series that sum to a rational function. Lots of groups do: finite groups (polynomials are trivially rational), free groups, word hyperbolic groups, automatic groups [124], free and direct products of the previous classes, and even the non-finitely-presented wreath product $\mathbb{Z} \wr \mathbb{Z}$ [177]. It turns out that if a presentation has only finitely many cone types (as either labeled or unlabeled cones), then the growth series $S(z)$ (and hence $B(z)$ also) is a rational function; see [73].

This implies that if you can find finitely many cone types, or easy-to-compute recurrences involving geodesic normal forms for your finitely generated group, then $S(z)$ is almost surely going to be rational. This is because deriving nonrational generating functions is usually hard going. There are existence proofs showing groups with nonrational growth: groups with unsolvable word problem [36], some wreath products including lamplighter groups such as $\mathbb{Z}_2 \wr F_2$ [218], some higher Heisenberg groups [248], and groups with intermediate growth [150].

To specifically compute a nonrational growth series for a finitely generated group requires sophisticated tools, some of which are described in the next sections. We conclude this section with an interesting family of groups. Some have known growth series but most do not.

Exercise 13. Prove that a power series sums to a rational function if and only if its coefficients satisfy a linear recurrence relation with constant coefficients.

Motivating example: Baumslag–Solitar groups.   The *Baumslag–Solitar group* $BS(p, q)$ is the group with the presentation

$$BS(p, q) = \langle b, t \mid tb^p t^{-1} = b^q \rangle.$$

We'll assume $1 \le p < q$ throughout, and I already referred to $BS(1, 3)$ near the beginning of the office hour.

We'll show how to build the Cayley graph of a generic member of this family of groups. The defining relator $tb^p t^{-1} b^{-q}$ can be drawn as a rectangle referred to as a *horobrick*. The horobricks are glued end-to-end and top-to-bottom to get *sheets* (refer to Figures 12.2 and 12.5 for examples of sheets), which look like a plane made of rectangles. However, this is *not* the Euclidean plane! The horobricks are wider at the bottom than at the top: recall that each edge shown has unit length so a horobrick is p units across at the top but q units across at the base.

In our model, paths with labels t^n, $n > 0$, go up, labels $b^{\pm n}$ are horizontal, while paths labeled t^{-n} go down. We adopt standard monoid notation and use B and T in place of b^{-1} and t^{-1}, respectively (see Section 12.3 below for the definition of a monoid).

A first observation is that traveling horizontally is not the fastest way to get around (recall the example of Figure 12.2). In fact, each sheet is quasi-isometric to the hyperbolic plane, which you learned about in Office Hour 5 (think of the

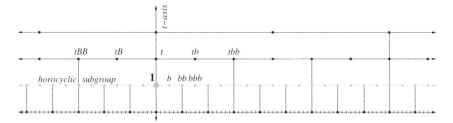

Figure 12.5 Main sheet for $BS(2, 6)$ comprised of horobricks.

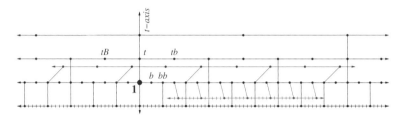

Figure 12.6 Adding a new upper (red) and lower (blue) half-sheet to the main sheet of $BS(2, 6)$.

upper half-space model), and so geodesics in a given sheet are close to semicircles centered on the x–axis.

You will notice that some of the vertices on the sheets pictured have valence 2 or 3 rather than the 4 needed for the finished Cayley graph of a two-generator group. Sheets exhibit all of the horizontal edges emanating from a particular vertex, but each vertex v for $BS(p, q)$ needs a vertical edge immediately below and above v. We need to add more vertical edges to a sheet in order to finish the Cayley graph and these edges are parts of new half-sheets (refer to Figure 12.6). In general, each horizontal line in the main sheet will need $q - 1$ new upper half-sheets glued in and $p - 1$ new lower half-sheets glued in.

The Cayley 2–complex of a presentation is obtained from the Cayley graph by filling in each basic relator and its conjugates with a topological disk (see Office Hour 8). For $BS(p, q)$ the Cayley 2–complex is homeomorphic to the product of the real line with a simplicial tree. When $1 = p < q$, all the branching is upward (as in Figure 12.1), while for $1 < p < q$ there are both upper and lower branchings (as in Figure 12.7). In the former case the group is solvable, but this is not so for the latter case.

A spherical growth series for $BS(1, q)$ has been computed and is rational [48, 88], even though there are infinitely many different cone types (this follows from Chapter 9 of Meier's text [202], although it can be done directly). Similarly $BS(p, p)$ has rational growth [121]. When $1 < p < q$, the growth series for $BS(p, q)$ are unknown, with only partial results to date [135]. In fact, a general algorithm to determine geodesic normal forms has only recently been discovered [112], and this only in the case when p divides q.

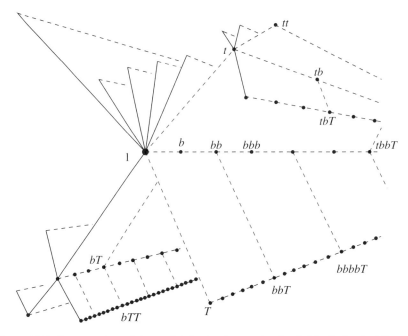

Figure 12.7 Cutaway view of the partial Cayley 2–complex for $BS(2, 6)$. Dashed lines are graph edges, solid lines are cuts through a 2–cell. Many cells omitted for clarity.

Normal forms for horocyclic subgroups of Baumslag–Solitar groups. Continuing with Baumslag–Solitar groups, consider the infinite cyclic subgroup $\langle b \rangle$, which is referred to as a *horocyclic subgroup*. Although isomorphic to $\mathbb{Z}$, it is embedded with severe distortion in the main sheet of the group (this means that distances within $\langle b \rangle$ are very different from distances in the ambient group; refer to Office Hours 8 and 9). Refer to Figure 12.5 for the horocyclic subgroup of $BS(2, 6)$ and Figure 12.9 in the case of $BS(1, 3)$.

Let's focus on the case of $BS(1, 3)$ and let's try to find a geodesic normal form for each b^n. Clearly, b, b^2, b^3, and b^4 are geodesic words, although the latter two can also be represented by tbt^{-1} and $tbt^{-1}b$, respectively. The next power, b^5, is equivalent to both $tbt^{-1}bb$ and $tbbt^{-1}b^{-1}$. But b^6 is no longer geodesic in $BS(1, 3)$ since the equivalent word $tbbt^{-1}$ is shorter. Higher powers of b have a geodesic representative of form $t^i b^j s$, where $j \in \{2, 3, 4\}$ and the suffix s always goes down (and across). Such geodesics are referred to as mesas because of their similarity to the profile of the landform of the same name (Figure 12.2 shows the general idea, and if you delete the last two blue edges you'll see a geodesic for b^{26}).

To describe the suffix portion s of such a mesa we can use a single cone type. This cone type is made by gluing together $\perp$–shapes inductively. The first three stages are shown in Figure 12.8.

Elements of the cone are finite products of the words t^{-1}, $t^{-1}b$, and $t^{-1}b^{-1}$.

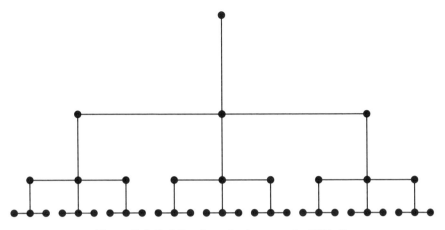

Figure 12.8 Building the main sheet cone in $BS(1, 3)$.

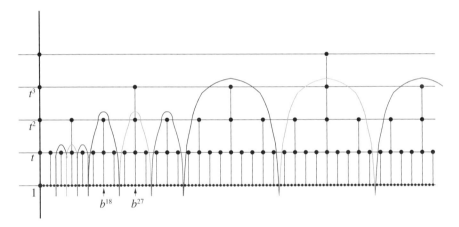

Figure 12.9 Partial sheet for $BS(1, 3)$ showing cones and horocyclic subgroup.

Figure 12.9 shows such cones in the main sheet. If we were to draw the entire main sheet, each of the cones shown would continue indefinitely downward and be isometric to the prototype cone described above. However, the horocyclic subgroup cuts the cones in Figure 12.9 after only finitely many construction stages.

The procedure for determining i, j, and s in terms of n is as follows. First identify the cone (as in the previous paragraph) containing b^n (one of the colored cones in Figure 12.9). Then choose i to be equal to the height (above $\langle b \rangle$) of this cone, and choose j so that $t^i b^j$ is the top vertex of the cone. Then s corresponds to the unique path from the top of the cone to b^n.

(This method of cones can also be used to compute the growth series for the horocyclic subgroup of $BS(2, 4)$ [135].)

12.3 FORMAL LANGUAGES AND CONTEXT-FREE GRAMMARS

In this section we will establish some of the basic algebraic objects we will use in our procedure for determining growth series.

Monoids. A *monoid* is an algebraic object that is like a group but without inverses. More precisely, a monoid is a set M together with a binary operation $M \times M \to M$ (i.e., a multiplication) that is associative and has an identity element. (A group has the added axiom that each element has a multiplicative inverse.) When we consider a finitely generated group as a monoid, we write the inverse of a generator x as X.

Let's define an *alphabet* $\mathcal{A}$ as a finite set of symbols called *letters*. The *free monoid* on an alphabet $\mathcal{A}$ is the set of all finite words formed by the letters of $\mathcal{A}$. The multiplication is concatenation. There is an identity element called the *empty word*, which is just the word of length 0, which has no letters at all. We use ε to indicate this empty word. The notation we will use for the free monoid on $\mathcal{A}$ is $\mathcal{A}^*$. The asterisk superscript is called the *Kleene star*, named after Stephen Cole Kleene (pronounced "clay-nee").[1] Whenever you have a word or formula followed by the Kleene star, it means "repeat zero or more times." So $\mathcal{A}^*$ means take any letter in alphabet $\mathcal{A}$, repeat with another letter of $\mathcal{A}$, repeat, etc.

Formal languages. A *formal language* over an alphabet $\mathcal{A}$ is a subset $\mathfrak{L} \subseteq \mathcal{A}^*$. Unlike the free monoid, $\mathfrak{L}$ is not usually closed under concatenation, and $\mathfrak{L}$ may or may not contain the empty word. Here are three basic examples.

1. A simple language: Take $\mathcal{A} = \{a, b\}$ and take $\mathfrak{L}$ to be the set of all words with any number of a's followed by any number of b's. This language is denoted by $\{a^*b^*\}$.
2. Primes in unary: Take $\mathcal{A} = \{0\}$ and define $\mathfrak{L} = \{0^p \mid p \text{ is a prime}\}$. Here, the exponent p means "repeat exactly p times." So

    $$\mathfrak{L} = \{00, \ 000, \ 00000, \ 0000000, \ 00000000000, \dots \}.$$

3. The free group: Let $\mathcal{A} = \{a, b, A, B\}$ and let $\mathfrak{L}$ be the subset of $\mathcal{A}^*$ consisting of all words that do not contain aA, Aa, bB, or Bb. There is a natural bijection between $\mathfrak{L}$ and the free group F_2.

The last example exhibits a key idea, namely, that languages can be in bijective correspondence with sets of group elements. One goal we have is to determine the language of geodesic normal forms for a group.

Let's reexamine our growth calculation for F_2 using our description of this group as a language, as in the last example. Given a word of length $r > 0$ ending in A (for instance), we can concatenate any of the letters A, b, and B to make three new freely reduced words having length $r + 1$. Thus, using the combinatorics of

---

[1] According to Kleene's son: "As far as I am aware this pronunciation is incorrect in all known languages. I believe that this novel pronunciation was invented by my father." [186].

the language, we easily recognize the recursion $\sigma_{r+1} = 3\sigma_r$ and initial conditions $\sigma_0 = 1$, $\sigma_1 = 4$ that we found using cone types.

Context-free grammars.   We would like to generalize the example of F_2 to other finitely generated groups. That is, we would like to find a geodesic word for each group element such that the set of these geodesic words forms a nice language. We can compute the growth of such a language by purely algebraic means in the next section, but now we have to wade through some more formal language theory. First let's classify what we mean by nice.

For our purposes, "nice" refers to what is known in the field of theoretical computer science as a *context-free language*. There is a huge body of literature on the subject of formal languages; an excellent exposition is found in the classic text by Hopcroft and Ullman [168] as well as its more recent editions (the original paper is by Chomsky [79]). Every context-free language can (by its definition) be generated by a *context-free grammar*, which is a rewriting system consisting of:

- variables (in bold), including a special start variable denoted by **S**
- terminals (in lower case italics), which are letters of the language to be produced
- production rules consisting of a variable, an arrow, and a word in variables and terminals; a vertical line on the right-hand side of the arrow denotes "exclusive or."

So, for example, we could have a context-free grammar where the variables are **S**, **A**, and **B**, where the language consists of some words in the terminals a, b, c, and d, built by several production rules of the form:

- **A** $\to a$**B**cd
- **S** $\to a$**X**$|b$**Y**.

The first production rule means that an instance of the variable **A** can be replaced by the string a**B**cd. The compound rule **S** $\to a$**X**$|b$**Y** is shorthand for the two productions **S** $\to a$**X** and **S** $\to b$**Y**, where we actually choose one rule or the other when creating words.

The idea is to start with **S** and compose production rules finitely many times until you run out of variables and have to stop, leaving a word made of terminals. There are usually lots of choices of which production rules to adopt. Each such set of choices will determine a terminal word. Exhausting all such choices and rules will output a formal language of words in the terminals. Such a language is exactly a context-free language.

Generating a simple language.   One of our prior examples of a language was

$$\mathcal{L} = \left\{ a^* b^* \right\}.$$

Let's show this is a context-free language by producing it with a context-free grammar.

Just as we write an element of $\mathfrak{L}$ from left to right with all the a's and then all the b's, our grammar can do the same. Let's begin with the start symbol S on the left side of our first production:

$$\mathbf{S} \to a.$$

This isn't quite what we want since there is only one possible output language, namely $\{a\}$. The right hand side of our first production needs to have a variable as well. How about:

$$\mathbf{S} \to a\mathbf{S}$$

This looks better since we can repeat the rule to get

$$\mathbf{S} \to a\mathbf{S} \to aa\mathbf{S} \to aaa\mathbf{S} \to \cdots \to a^m\mathbf{S} \to \cdots,$$

but we never get rid of the variable $\mathbf{S}$ and so we never actually produce any terminal words at all! We need another production rule. We could combine our initial guesses and try

$$\mathbf{S} \to a\mathbf{S} | \varepsilon.$$

This still allows the previous chain of productions and permits us to finally replace the dangling $\mathbf{S}$ with the empty word ε, so the output language is $\{a^*\}$. How do we attach b's? Cut to the chase and use

$$\mathbf{S} \to a\mathbf{S} | \mathbf{U} \qquad \mathbf{U} \to b\mathbf{U} | \varepsilon.$$

Observe that $\mathbf{S} \to \mathbf{U} \to \varepsilon$ makes the empty word, also written as $a^0 b^0$, whereas $\mathbf{S} \to \mathbf{U}$ followed by $\mathbf{U} \to b\mathbf{U}$ creates strings like $b^n\mathbf{U}$ and then using $\mathbf{U} \to \varepsilon$ gets rid of $\mathbf{U}$ leaving $b^n = a^0 b^n$, for $n > 1$. In the general case, starting with $\mathbf{S} \to a\mathbf{S}$ and using it several times gives $a^m\mathbf{S}$. Now use $\mathbf{S} \to b\mathbf{U}$ to get $a^m b\mathbf{U}$. Propagate b's via repeating the $\mathbf{U} \to b\mathbf{U}$ rule and finally eliminate $\mathbf{U}$ by sending it to ε. We could also have skipped the creation of b's by going from $a^m\mathbf{S}$ to $a^m\mathbf{U}$ and then to $a^m\varepsilon = a^m b^0$. Voilá! Our grammar creates

$$\{\varepsilon\} \cup \{a^m \mid m > 0\} \cup \{b^n \mid n > 0\} \cup \{a^m b^n \mid m, n > 0\} = \{a^* b^*\}.$$

It is tedious to write out compositions like $\mathbf{S} \to a\mathbf{S} \to aa\mathbf{S} \to aaa\mathbf{S} \to \cdots \to a^m\mathbf{S}$. If the composition steps are clear enough to figure out, we abbreviate the above as $\mathbf{S} \xrightarrow{*} a^m\mathbf{S}$, so this is kind of like the Kleene star operator applied to productions: "$\xrightarrow{*}$" means several (possibly different) instances of "$\to$" applied sequentially.

It is also tedious to explain all the different cases of the several production compositions. There is a diagram called a *parse tree* for each word created from a given context-free grammar. It shows the start variable as the root, terminals as leaves, and variables as interior vertices. Each downward edge in the tree represents a grammar production. You read off the leaves in a counterclockwise fashion, starting from the root, to see the terminal word that is ultimately created. A typical parse tree for the above grammar is shown in Figure 12.10.

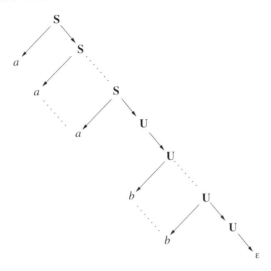

Figure 12.10 Typical parse tree producing $\{a^*b^*\}$. Variables are in bold, productions are indicated by arrows; the terminal word is written from reading off the leaves of the tree from left to right.

A grammar for the free group.   Here is a grammar for F_2. The variables are **S**, $\boldsymbol{\Gamma}_a$, $\boldsymbol{\Gamma}_b$, $\boldsymbol{\Gamma}_{a^{-1}}$, and $\boldsymbol{\Gamma}_{b^{-1}}$, the terminals are a, b, A, and B, and the production rules are:

$$
\begin{aligned}
\mathbf{S} \quad &\rightarrow \quad a\boldsymbol{\Gamma}_a \mid b\boldsymbol{\Gamma}_b \mid A\boldsymbol{\Gamma}_{a^{-1}} \mid B\boldsymbol{\Gamma}_{b^{-1}} \mid \varepsilon \\
\boldsymbol{\Gamma}_a \quad &\rightarrow \quad a\boldsymbol{\Gamma}_a \mid b\boldsymbol{\Gamma}_b \mid B\boldsymbol{\Gamma}_{b^{-1}} \mid \varepsilon \\
\boldsymbol{\Gamma}_b \quad &\rightarrow \quad a\boldsymbol{\Gamma}_a \mid b\boldsymbol{\Gamma}_b \mid A\boldsymbol{\Gamma}_{a^{-1}} \mid \varepsilon \\
\boldsymbol{\Gamma}_{a^{-1}} \quad &\rightarrow \quad A\boldsymbol{\Gamma}_{a^{-1}} \mid b\boldsymbol{\Gamma}_b \mid B\boldsymbol{\Gamma}_{b^{-1}} \mid \varepsilon \\
\boldsymbol{\Gamma}_{b^{-1}} \quad &\rightarrow \quad a\boldsymbol{\Gamma}_a \mid A\boldsymbol{\Gamma}_{a^{-1}} \mid B\boldsymbol{\Gamma}_{b^{-1}} \mid \varepsilon
\end{aligned}
$$

You may recognize the generators of the presentation as the terminals in italics and that the variables in bold represent the four cone types discussed earlier. We can derive each freely reduced word from these grammar rules in exactly one way. For instance, the word $aba^{-1}b^{-1}$, which we write as $abAB$, can be derived by $\mathbf{S} \rightarrow a\boldsymbol{\Gamma}_a$, then $\boldsymbol{\Gamma}_a \rightarrow b\boldsymbol{\Gamma}_b$ (the initial a gets to tag along), and conclude with the rules $\boldsymbol{\Gamma}_b \rightarrow A\boldsymbol{\Gamma}_{a^{-1}}$ and $\boldsymbol{\Gamma}_{b^{-1}} \rightarrow B$ (and again the italicized letters stay in place). The extended production chain for this particular word looks like

$$\mathbf{S} \rightarrow a\boldsymbol{\Gamma}_a \rightarrow ab\boldsymbol{\Gamma}_b \rightarrow abA\boldsymbol{\Gamma}_{a^{-1}} \rightarrow abAB\boldsymbol{\Gamma}_{b^{-1}} \rightarrow abAB\varepsilon = abAB.$$

A parse tree for the same is shown in Figure 12.11.

Regular languages.   Both the previous two grammars have parse trees that are similar in appearance: they form rightward trending lines with single leaves branching to the left that go directly to terminals. This is evident from the grammars: the

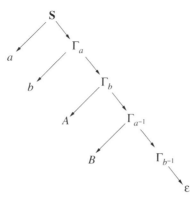

Figure 12.11 Parse tree for $abAB$.

right side of each individual grammar production has exactly one terminal followed by at most one variable, such as $\mathbf{X} \to t\mathbf{Y}$. Such a grammar is called *right-linear*.

Context-free languages produced by right-linear grammars are important enough to have their own name: they are called *regular* languages. (Actually, regular languages have an official definition that is different from this, and there is a theorem called Kleene's theorem which says that regular languages are exactly the ones produced by right-linear grammars. However, we will not need this official definition.) Thus, for example, the language of freely reduced words in the free group is regular.

We discussed cones in the Cayley graph earlier. We can generalize the concept of a cone to any language $\mathfrak{L}$. For each $w \in \mathfrak{L}$, define the cone[2] based at w as $\Gamma_w = \{h \in \mathfrak{L} \mid wh \in \mathfrak{L}\}$, and just as before say that two cones have the same type if they agree as sets. The cone Γ_w is the set of all admissible suffixes that can be concatenated after w. We have the following theorem of Myrill and Nerode.

THEOREM 12.2. *A language $\mathfrak{L}$ is regular if and only if it has only finitely many cone types.*

It follows from the theorem that a regular language has rational growth series. In fact, a right-linear grammar can be realized once all the (finitely many) cone types have been determined. (The proof of the Myhill–Nerode theorem can be found in Hopcroft and Ullman's text [168] as well as in Wikipedia.)

Grammar for the colorful collection of diamonds. Recall our group G given by

$$G = \langle x, y \mid x^4 = 1 = y^4,\, x^2 = y^2 \rangle$$

(refer to Figure 12.4). Earlier, we exhibited four cone types for G.

---

[2] Note: The people who specialize in the theory of computation don't use the word *cone*, but refer to Γ_w as the *left quotient of $\mathfrak{L}$ by w*.

As usual, let's write X, Y instead of x^{-1}, y^{-1}. We have used $\mathbf{S}$ to define the identity cone type for G as well as the start variable for any context-free grammar. This makes sense because all geodesic words belong to this cone type. The dead end cone type can be represented by the variable symbol $\boldsymbol{\phi}$, which is also appropriately used to indicate the empty set. We'll use $\boldsymbol{\Psi}$ as the variable associated with the cone type of the same name, and similarly with $\boldsymbol{\Delta}$. Our grammar starts with the production rules coming directly from Figure 12.4,

$$\mathbf{S} \to \varepsilon \mid x \mid X \mid y \mid Y \mid xx\boldsymbol{\phi} \mid XX\boldsymbol{\phi} \mid yy\boldsymbol{\phi} \mid YY\boldsymbol{\phi}$$

$$\mathbf{S} \to yx\boldsymbol{\Delta} \mid YX\boldsymbol{\Delta} \mid yX\boldsymbol{\Delta} \mid Yx\boldsymbol{\Delta} \mid xy\boldsymbol{\Psi}, \mid XY\boldsymbol{\Psi}, \mid xY\boldsymbol{\Psi} \mid Xy\boldsymbol{\Psi},$$

after which we have the obvious production rule

$$\boldsymbol{\phi} \to \varepsilon.$$

Where do the variables $\boldsymbol{\Psi}$ and $\boldsymbol{\Delta}$ lead? Sending $\boldsymbol{\Psi}$ to anything starting with y or Y represents backtracking in the Cayley graph (or going to the dead end); refer to Figure 12.4. Similarly, $\boldsymbol{\Delta}$ cannot produce anything starting with x or X. On the other hand, the Cayley graph allows

$$\boldsymbol{\Psi} \to x\boldsymbol{\Delta} \mid X\boldsymbol{\Delta},$$

and we can also allow our production chain to stop with

$$\boldsymbol{\Psi} \to \varepsilon.$$

Similarly we allow

$$\boldsymbol{\Delta} \to y\boldsymbol{\Psi} \mid Y\boldsymbol{\Psi} \mid \varepsilon,$$

and you can verify this grammar produces all geodesics in the Cayley graph.

Exercise 14. Verify that the grammar in this example produces all geodesics in the Cayley graph.

Grammar for the $BS(1, 3)$ horocycle. Earlier we showed that each $b^n \in BS(1, 3)$ with $n \geq 5$ has a geodesic representative of form $t^i b^j s$ where $j \in \{2, 3, 4\}$ and $s \in (TB \mid T \mid Tb)^i$. Note that in each geodesic the numbers of t's equals the number of T's. Recall that the suffix s belongs to a certain cone described earlier.

We could build the (infinite) cone used for the suffix s with the simple grammar rules $\mathbf{C} \to TB\mathbf{C} \mid T\mathbf{C} \mid Tb\mathbf{C}$, but this has two obvious flaws: we never eliminate the variable $\mathbf{C}$, and even if we eventually send $\mathbf{C}$ to ε we have no control over how to eventually balance t's and T's. Thus the naïve idea of building a geodesic word $t^i b^j s$ from left to right (as we would trace the geodesic with our finger) won't work. A better idea is to build such a mesa geodesic by starting with the mesa cap and simultaneously constructing the left and right sides in tandem. Let's keep the variable $\mathbf{C}$, which now represents "cap" rather than "cone," but build on both sides as per the following grammar.

$$\mathbf{S} \to b \mid b^2 \mid b^3 \mid b^4 \mid t\mathbf{C}TB \mid t\mathbf{C}T \mid t\mathbf{C}Tb$$

$$\mathbf{C} \to t\mathbf{C}TB \mid t\mathbf{C}T \mid t\mathbf{C}Tb \mid b^2 \mid b^3 \mid b^4$$

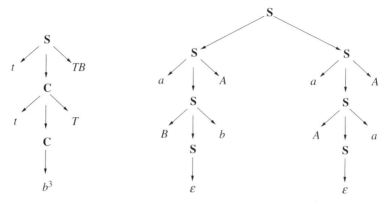

Figure 12.12 Left: parse tree for $ttbbbTTB$. Right: parse tree for $aBbAaAaA$. These are not right-linear grammars (so the languages are not regular).

A parse tree for the geodesic word $t^2b^3T^2B$ that is equivalent to b^{26} is shown in Figure 12.12 (left). The actual geodesic is found in Figure 12.2.

Grammar and the word problem. Formal languages and grammars are used in other areas of geometric group theory besides growth. One increasingly common usage is in describing the word problem. We referred to this earlier as the problem of determining whether a given word represents the identity element. The phrase "word problem" can also refer to the *set of all words in the given presentation that represent the identity*. From now on we'll stick with this newer definition which makes the word problem a formal language. The following results are due to Anīsīmov [9] and Muller and Schupp [206], respectively.

> *The word problem for a group G constitutes a regular language if and only if G is finite.*

> *The word problem for a group G constitutes a context-free language if and only if G is virtually free.*

Exercise 15. Verify that the following grammar describes the word problem for the free group F_2:

$$\mathbf{S} \to \varepsilon \mid \mathbf{SS} \mid a\mathbf{S}A \mid A\mathbf{S}a \mid b\mathbf{S}B \mid B\mathbf{S}b.$$

Is this language regular?

12.4 THE DSV METHOD

Once a context-free grammar is established for a language, it is possible to compute the growth of grammar productions by a procedure due to Delest, Schützenberger, and Viennot. The standard reference is by Chomsky and Schützenberger [78]. The simple version of this so-called DSV method is mostly mechanical: substitute all grammar symbols as per the following table. The grammar becomes a system of equations.

Replace ...	with ...
the empty word ε	the integer 1
each terminal letter	the formal variable z
the production arrow $\rightarrow$	equality $=$
each grammar variable **V**	a function $V(z)$
the exclusive or		the addition symbol $+$
concatenation	commutative multiplication

We have the following fact [78].

THEOREM 12.3. *Solving the resulting system for $S(z)$ gives the growth series for the grammar productions. Furthermore, $S(z)$ is always an algebraic function.*

Grammar equations for the language $\{a^*b^*\}$. The grammar for $\{a^*b^*\}$ described earlier transforms into the equations

$$S(z) = zS(z) + U(z) \qquad U(z) = zU(z) + 1 .$$

The second equation says $U(z) = \frac{1}{1-z}$; and substituting into the first gives

$$S(z) = \frac{1}{(1-z)^2} = 1 + 2z + 3z^2 + 4z^3 + 5z^4 + \cdots .$$

This is correct since $\{a^*b^*\} = \{\varepsilon,\ a, b,\ aa, ab, bb,\ aaa, aab, abb, bbb, \dots\}$.

Warning! The growth series of this method keeps track of the number of *grammar productions* that produce words of length $n \geq 0$, which may not be the same thing as the number of *words* of length $n \geq 0$. If you are not careful, your grammar may have several different ways to produce the same word, in which case you are counting that word several times.

Exercise 16. Verify that $\{a^*b^*\}$ is also generated by

$$\mathbf{S} \rightarrow a\mathbf{S} \,|\, \mathbf{U} \,|\, \varepsilon \qquad \text{and} \qquad \mathbf{U} \rightarrow b\mathbf{U} \,|\, \varepsilon .$$

Solve the grammar equations to get

$$S(z) = 2 + 3z + 4z^2 + 5z^3 + \cdots ,$$

which counts one extra production of each length. List the extra productions.

Ambiguous grammars. The last exercise illustrates two points. First, there are several grammars that can generate the same language. Second, you have just seen an instance of an *ambiguous* grammar, one that has multiple productions for some given word. The DSV method mandates an *unambiguous grammar* in order to calculate the growth of a language. (Unfortunately, some complicated context-free languages have only ambiguous grammars; these are *inherently ambiguous* [168]. The very interesting article by Flajolet [131] proves several languages are inherently ambiguous by showing their respective growth series are not algebraic functions.)

Grammar equations for the colorful collection of diamonds.   Earlier we derived the grammar

$$\mathbf{S} \to \varepsilon \mid x \mid X \mid y \mid Y \mid xx\boldsymbol{\phi} \mid XX\boldsymbol{\phi} \mid yy\boldsymbol{\phi} \mid YY\boldsymbol{\phi}$$

$$\mathbf{S} \to yx\boldsymbol{\Delta} \mid YX\boldsymbol{\Delta} \mid yX\boldsymbol{\Delta} \mid Yx\boldsymbol{\Delta} \mid xy\boldsymbol{\Psi} \mid XY\boldsymbol{\Psi}, \mid xY\boldsymbol{\Psi} \mid Xy\boldsymbol{\Psi}$$

$$\boldsymbol{\Psi} \to x\boldsymbol{\Delta} \mid X\boldsymbol{\Delta} \mid \varepsilon$$

$$\boldsymbol{\Delta} \to y\boldsymbol{\Psi} \mid Y\boldsymbol{\Psi} \mid \varepsilon$$

$$\boldsymbol{\phi} \to \varepsilon$$

that generates all the geodesic words of the given presentation. Let's solve the generating function for this grammar by the DSV method, starting with the third and fourth equations:

$$\Psi(z) = z\Delta(z) + z\Delta(z) + 1$$

$$\Delta(z) = z\Psi(z) + z\Psi(z) + 1.$$

We add these two equations and solve to obtain

$$\Delta(z) + \Psi(z) = \frac{2}{1 - 2z}.$$

The first, second, and fourth equations imply

$$S(z) = 1 + 4z + 4z^2 + 4z^2 \left(\Delta(z) + \Psi(z)\right).$$

We substitute to get

$$S(z) = 1 + 4z + 4z^2 + \frac{8z^2}{1 - 2z} = 1 + 4z + 12z^2 + 16z^3 + \cdots + 2^{n+1}z^n + \cdots.$$

There is a mistake in notation here! This is not the standard spherical growth series; rather, this is the geodesic growth series $\Gamma(z)$ that counts *all* geodesics. For instance, there are four geodesics that represent the group element

$$yxY = yXy = YXY = Yxy$$

instead of only one. Observe that our grammar is *not* ambiguous; there is only one way to produce each word. However, we have produced too many words!

To resolve this issue we need to determine a geodesic normal form for each group element. Fortunately this is easy in our case. The diamond-shaped relators and their conjugates are what allow multiple paths. We will restrict our normal forms to only traverse these diamonds in the *counterclockwise* direction, in other words, use only the generators x, y (with the exception of the very first letter where we allow one of X, Y). We adjust the grammar by eliminating the productions $\mathbf{S} \to XX\boldsymbol{\phi} \mid yy\boldsymbol{\phi} \mid YY\boldsymbol{\phi} \mid YX\boldsymbol{\Delta} \mid Yx\boldsymbol{\Delta} \mid XY\boldsymbol{\Psi} \mid Xy\boldsymbol{\Psi}, \boldsymbol{\Psi} \to X\boldsymbol{\Delta}$, and $\boldsymbol{\Delta} \to Y\boldsymbol{\Psi}$.

Exercise 17. Verify that, after the adjustments described in the previous paragraph, we obtain

$$S(z) = 1 + 4z + 5z^2 + \sum_{n \geq 3} 4z^n.$$

In other words, except for radii 0 and 2, there are four group elements for each radius.

Grammar equations for the free group. The grammar for the free group F_2 described earlier is unambiguous and yields the equations below:

$$S(z) = z\Gamma_a(z) + z\Gamma_b(z) + z\Gamma_{a^{-1}}(z) + z\Gamma_{b^{-1}}(z) + 1$$

$$\Gamma_a = z\Gamma_a + z\Gamma_b + \Gamma_{b^{-1}} + 1$$

$$\Gamma_b = z\Gamma_a + z\Gamma_b + \Gamma_{a^{-1}} + 1$$

$$\Gamma_{a^{-1}} = z\Gamma_{a^{-1}} + z\Gamma_b + \Gamma_{b^{-1}} + 1$$

$$\Gamma_{b^{-1}} = z\Gamma_a + z\Gamma_{a^{-1}} + \Gamma_{b^{-1}} + 1$$

Add the last four equations to get

$$\Gamma_a + \Gamma_b + \Gamma_{a^{-1}} + \Gamma_{b^{-1}} = 3(z\Gamma_a + z\Gamma_b + z\Gamma_{a^{-1}} + z\Gamma_{b^{-1}} + 1) + 1$$
$$= 3S(z) + 1.$$

On the other hand, the first equation implies

$$\frac{S(z) - 1}{z} = \Gamma_a + \Gamma_b + \Gamma_{a^{-1}} + \Gamma_{b^{-1}}.$$

Therefore

$$\frac{S(z) - 1}{z} = 3S(z) + 1$$

and so

$$S(z) = \frac{1 + z}{1 - 3z},$$

as expected. Note that the four cones based at a, b, A, and B, respectively, are pairwise disjoint, so we avoid the overcounting of the previous example.

Exercise 18. Sketch the Cayley graph, identify cone types, construct unambiguous grammars, and compute the spherical growth series for each of:

- $BS(1, -1) = \langle b, t \mid tbt^{-1} = b^{-1} \rangle$, also known as the fundamental group of the Klein bottle (see Office Hour 10),
- the free product $\mathbb{Z}_2 * \mathbb{Z}_4$ with standard presentation $\langle u, v \mid u^2 = 1 = v^4 \rangle$, and
- $H = \langle x, y \mid x^6 = 1 = y^6, \ x^3 = y^3 \rangle$.

Hints: All have only finitely many cone types. The first is almost trivial; Fibonacci numbers come into play in the second; and the third is similar to, but more complicated than, our group from the colorful collection of diamonds example.

Grammar equations for the horocyclic subgroup of $BS(1, 3)$. Earlier we described a grammar that generates a geodesic normal form for each of the words $b, b^2, b^3, \ldots$ from $BS(1, 3)$:

$$\mathbf{S} \to b \mid b^2 \mid b^3 \mid b^4 \mid t\mathbf{C}TB \mid t\mathbf{C}T \mid t\mathbf{C}Tb$$

$$\mathbf{C} \to t\mathbf{C}TB \mid t\mathbf{C}T \mid t\mathbf{C}Tb \mid b^2 \mid b^3 \mid b^4$$

(we emphasize that these are normal forms in $BS(1, 3)$, not the horocyclic subgroup, which is isomorphic to $\mathbb{Z}$).

We use the DSV method to compute

$$C(z) = \frac{z^2 + z^3 + z^4}{1 - z^2 - 2z^3}$$

from the second grammar production. The equation from the first production looks like

$$S(z) = z + z^2 + z^3 + z^4 + C(z)(z^2 + 2z^3).$$

Eliminating $C(z)$ and simplifying gives

$$S(z) = \frac{z + z^2 - z^4}{1 - z^2 - 2z^3}.$$

If we want the growth series for the entire horocyclic subgroup $\langle b \rangle$, we need to count the contribution from the identity and negative powers of b. Thus

$$1 + 2S(z) = \frac{1 + 2z + z^2 - 2z^3 - 2z^4}{1 - z^2 - 2z^3}$$

is the growth series for the horocyclic subgroup.

Growth of the word problem for F_2. We conclude this section by using the DSV method to compute the growth of the word problem for the free group F_2. As we shall see, this is a nontrivial combinatorics problem.

Unfortunately, the grammar we gave earlier for the word problem in F_2 is ambiguous. There are infinitely many different ways to generate the empty word, for instance:

$$\mathbf{S} \to \varepsilon$$

$$\mathbf{S} \to \mathbf{SS} \to \varepsilon\mathbf{S} \to \varepsilon\varepsilon = \varepsilon, \quad \text{and}$$

$$\mathbf{S} \to \mathbf{SS} \to \mathbf{SSS} \to \varepsilon\varepsilon\varepsilon = \varepsilon.$$

On the other hand, some sort of production rule involving $\mathbf{SS}$ is needed to simulate word concatenation, so it is not obvious how to avoid this.

There is an even more subtle ambiguity. Look at Figure 12.12 (right), which shows how $aBbAaAaA$ is built as a concatenation of two conjugates, $(aBbA)(aAaA)$. This word could also be built in an entirely different way as the concatenation of different conjugates, $(aBbAa A)(a A)$, or as the conjugate of yet another product, $a(BbAa)(Aa)A$.

Because of this we need an entirely different grammar that avoids all ambiguity. Creating such a grammar from scratch is a challenging exercise, but we are in luck: the word problem for F_2 is exactly the *two-sided Dyck[3] language* on two parentheses and the latter is well known. It is the free monoid on the two-sided *Dyck primes*; these are all words in (,), [, and] that are correctly nested (i.e., ([]) is allowed but ([]) is not) and cannot be written as the concatenation of two such words [26]. The language of two-sided Dyck primes is unambiguously generated by the following grammar:

$$\mathbf{P} \to \mathbf{C}_a | \mathbf{C}_b | \mathbf{C}_A | \mathbf{C}_B$$

$$\mathbf{C}_a \to a\mathbf{M}_a A \quad \mathbf{C}_b \to b\mathbf{M}_b B \qquad \mathbf{C}_A \to A\mathbf{M}_A a \quad \mathbf{C}_B \to B\mathbf{M}_B b$$

$$\mathbf{M}_a \to \varepsilon \,|\, \mathbf{C}_a\mathbf{M}_a \,|\, \mathbf{C}_b\mathbf{M}_a \,|\, \mathbf{C}_B\mathbf{M}_a \qquad \mathbf{M}_b \to \varepsilon \,|\, \mathbf{C}_a\mathbf{M}_b \,|\, \mathbf{C}_b\mathbf{M}_b \,|\, \mathbf{C}_A\mathbf{M}_b$$

$$\mathbf{M}_A \to \varepsilon \,|\, \mathbf{C}_b\mathbf{M}_A \,|\, \mathbf{C}_A\mathbf{M}_A \,|\, \mathbf{C}_B\mathbf{M}_A \qquad \mathbf{M}_B \to \varepsilon \,|\, \mathbf{C}_a\mathbf{M}_B \,|\, \mathbf{C}_A\mathbf{M}_B \,|\, \mathbf{C}_B\mathbf{M}_B$$

Each $\mathbf{C}_x$ simulates conjugation by $x \in \{a, b, A, B\}$ ("C" stands for "conjugate") while each $\mathbf{M}_x$ is a (right) multiplier of a conjugate ("M" is a mnemonic for "multiply").

We can get the simple block aA via

$$\mathbf{P} \to \mathbf{C}_a \to a\mathbf{M}_a A \to a\varepsilon A = aA,$$

and we can obtain each of Aa, bB, Bb similarly. We see that $\mathbf{P}$ generates conjugates. Showing that $\mathbf{P}$ generates *every* conjugate that represents the identity is left as an exercise. The word problem consists of products of such conjugates. To generate products of conjugates is easy; just insert these grammar productions:

$$\mathbf{S} \to \varepsilon | \mathbf{PS}.$$

Exercise 19. Show that $\mathbf{P}$ generates every conjugate that represents the identity.

Now let's compute the growth series for the word problem. The above grammar transforms into the equations below:

$$S(z) = 1 + P(z)S(z) \qquad\qquad P(z) = C_a(z) + C_b(z) + C_A(z) + C_B(z)$$

$$C_a(z) = z^2 M_a(z) \qquad\qquad C_b(z) = z^2 M_b(z)$$

$$C_A(z) = z^2 M_A(z) \qquad\qquad C_B(z) = z^2 M_B(z)$$

$$M_a(z) = 1 + (C_a + C_b + C_B)M_a(z) \quad M_b(z) = 1 + (C_a + C_b + C_A)M_b(z)$$

$$M_A(z) = 1 + (C_b + C_A + C_B)M_A(z) \quad M_B(z) = 1 + (C_a + C_A + C_B)M_B(z)$$

The first equation is easily solved:

$$S(z) = \frac{1}{1 - P(z)}.$$

So we need to solve the remaining system for the Dyck prime generating function $P(z)$.

Observe that by the symmetry of the grammar, we have:

$$C_a(z) \equiv C_b(z) \equiv C_A(z) \equiv C_B(z).$$

---

[3] *Dyck* rhymes with *Greek*

Let's just call this function $C(z)$. Similarly, all the multiplier functions are the same, so call it $M(z)$. We have

$$M(z) = \frac{C(z)}{z^2}.$$

Using the left equation for $M_a(z)$ above, we get

$$\frac{C(z)}{z^2} = 1 + (C(z) + C(z) + C(z))\frac{C(z)}{z^2}, \qquad \text{or} \qquad C(z) = \frac{1 \pm \sqrt{1 - 12z^2}}{6}$$

so

$$P(z) = 4C(z) = \frac{2 \pm 2\sqrt{1 - 12z^2}}{3}.$$

There are no Dyck primes of length 0, so use $P(0) = 0$ to choose the minus branch of the quadratic formula. Finally, solve for S to get

$$S(z) = \frac{1}{1 - P(z)}$$

$$= \frac{3}{1 + 2\sqrt{1 - 12z^2}}$$

$$= 1 + 4z^2 + 28z^4 + 232z^6 + 2092z^4 + 19864z^{10} + \cdots.$$

Exercise 20. Try various naïve counting methods to compute the number of words of length 10 (say) that represent the identity in F_2. This will emphasize the power of the formal language approach above.

FURTHER READING

You have now seen an introduction to a counting method that has been utilized in France for the last forty years but is relatively unknown elsewhere. Traditional approaches to computation of growth series exist: Chapters 9 and 11 of Meier's book [202] are geared towards an undergraduate audience; Johnson's survey [177] is also accessible, while Chapters VI and VII of de la Harpe's book [104] are more advanced. Chapter VII is devoted entirely to Grigorchuk's famous group of intermediate growth. Finding the growth series or asymptotics of the latter constitutes open research projects. Grigorchuk and Pak [150] give probably the most elementary summary of the situation. Computing growth for the several Thompson's groups (cf. Office Hour 16) and the remaining Baumslag–Solitar groups are also outstanding research problems that will probably require new methods for their respective solutions.

With the exception of the last example (word problem for F_2) our exposition above has dealt solely with rational growth series. An example of using the DSV method to compute an algebraic but nonrational growth series for a finitely generated group is given by Freden and Schofield [136]. There are several groups whose *co-word problem* (namely all the words in the generators that do *not* represent the identity) is a context-free language. One example is Thompson's group from Office Hour 16 (see [192]); other such families are described in [166].

A generalization of context-free languages (developed in the 1960s by Aho) are the *indexed languages*. The 1979 edition of Hopcroft and Ullman's text of [168] discusses these; a more modern introduction with applications to group theory has been written by Gilman [145]. The co-word problem for the Grigorchuk group has been shown to be an indexed language [167]. The DSV method has recently been shown to apply to indexed grammars [5], creating a potentially powerful new counting tool.

PROJECTS

Project 1. There are at least three unmentioned instances of exploiting the equivalence of unlabeled cones in this office hour. Find and document these instances.

Project 2. Generalize the previous exercise: find sufficient conditions in which to recognize and exploit the equivalence of unlabeled cones when using the DSV method.

Project 3. Walter Parry shows that the nonstandard lamplighter group $\mathbb{Z}_2 \wr (\mathbb{Z}_2 * \mathbb{Z}_2 * \mathbb{Z}_2)$ with the obvious presentation has algebraic but nonrational growth (see Corollary 3.4 of Parry's article [218]). Derive an unambiguous context-free grammar and use the DSV method to verify Parry's formula.

Project 4. Consider the presentation $\langle a, b, t \mid [a, b] = 1, taT = a^3, tbT = b^3 \rangle$, which defines an HNN extension of $\mathbb{Z} \times \mathbb{Z}$. This group is similar to the Baumslag–Solitar group $BS(1, 3)$ (see Figure 12.1), but instead of a 4–valent tree times a line, its Cayley 2–complex is a 10–valent tree crossed with a plane. This project computes its growth by exploiting the tree-like structure and is based on ideas from one of my papers [134]. Sánchez and Shapiro have generalized this project by showing that the horospheric subgroup has rational growth in the HNN extension of $\mathbb{Z}^m$ (for all $m > 1$) by the same cubing map I use; see [231].

- Draw some pictures of geodesics for elements of the *horospheric subgroup* $\langle a, b \rangle$. Compare the downward cones with those of the $BS(1, 3)$ horocyclic subgroup earlier in this chapter. Verify there is a single cone type whose basic

  building block looks like .
- Emulating the case of $BS(1, 3)$, build a context-free grammar that generates the horospheric subgroup. Hints: Instead of three caps there are a bunch; the basic shape above will yield nine suffixes; be careful with initial conditions (for example, $a^2 b^2$ is shorter than the mesa path $tabTAB$); finally, restricting to Quadrant I and using symmetry may be harder than just building a grammar for the entire subgroup.
- Solve the DSV system to obtain $B_0(z) = \frac{(1-z)(1+2z+2z^2)^2}{(1-2z)(1+z+2z^2)}$ as growth series for the horospheric subgroup.

- Define the *t-net-exponent-sum* for a word w as the number of occurences of t minus the number of occurrences of T. Show that in our main group, the t-net-exponent-sum of a group element g does not depend on the particular word w that represents g. Define the *level* of a horospheric coset $g \langle a, b \rangle$ as the t-net-exponent-sum of the element g. Show that this defines an equivalence relation on horospheric cosets.
- The canonical representative cosets for level n are $t^n \langle a, b \rangle$. Show that for $n \geq 0$, the coset $t^n \langle a, b \rangle$ has growth series $z^n B_0(z)$. Note that the z^n comes from the leading factor t^n, and if we use relative coordinates (measuring distance from t^n as the relative origin in the coset to element $t^n a^r b^s$ of the coset) the growth series is just $B_0(z)$.
- Show that an arbitrary horospheric coset of level $n \geq 0$ grows as $t^n \langle a, b \rangle$, provided we use relative coordinates.
- Show that $t^{-1} \langle a, b \rangle$ has growth series $z(1 + z)^2 B_0(z)$, and using relative coordinates this is just $(1 + 2z)^2 B_0(z)$. Call the latter $B_{-1}(z)$ and generalize to any level -1 horospheric coset.
- Show that the growth series (using relative coordinates) of any level $-n$ horospheric coset is $(1 + 2z)^{2n} B_0(z)$. Call this $B_{-n}(z)$.
- The above implies that horospheric cosets with nonnegative level are (with respect to relative growth series) equivalent to the horospheric subgroup. From now on reclassify all such cosets as having level 0. On the other hand, each negative level defines a unique equivalence class of horospheric cosets. Define for each $n \geq 0$ the growth series

$$X_{-n}(z) = \sum_{r \geq 0} \chi(-n, r) z^r$$

where $\chi(-n, r)$ counts the number of horospheric cosets of level $-n$ whose closest point to the group identity is r. Show that the growth series $S(z)$ of the entire group satisfies $S(z) = \sum_{n \geq 0} X_{-n}(z) B_{-n}(z)$.
- Draw a picture of the tree underlying the Cayley graph, with levels. Calculate the values of $\chi(0, r)$ and $\chi(-1, r)$ for $r = 0, 1, 2, 3, 4$. Also show $\chi(-n, r) = 0$ for $0 \leq r \leq n$.
- Validate the recurrence $\chi(-n, r) = \chi(-n - 1, r - 1) + 4\chi(-n - 1, r - 2) + 4\chi(-n - 1, r - 3)$ for all $n > 0$ and $r \geq n + 3$.
- Validate the recurrence $\chi(-n, r) = \chi(-n - 1, r + 1)$ for all $n > 0$ and $r \geq n + 3$.
- Using the new definition of level 0, validate the recurrence $\chi(0, r) = \chi(-1, r - 1) + 4\chi(-1, r - 2) + 4\chi(-1, r - 3) + \chi(0, r - 1) + 4\chi(0, r - 2) + 4\chi(0, r - 3)$ for all $r \geq 3$.
- Multiply each recurrence by z^r, sum over all suitable r, and subtract relevant initial terms to get equations relating the several coset counting generating functions. Solve these equations to get explicit rational expressions for $X_0(z)$, $X_{-1}(z)$, and a formula for the other $X_{-n}(z)$ in terms of $X_{-1}(z)$.
- Use the product structure $S(z) = \sum_{n \geq 0} X_{-n}(z) B_{-n}(z)$ to verify that

$$S(z) = \frac{(1 - z)^2 (1 + z)(1 + 2z + 2z^2)^2 (1 + 4z^2)}{(1 - 2z)^2 (1 + z + 2z^2)(1 - z - 4z^2 - 4z^3)}.$$

Project 5. Although the DSV method is powerful and elegant, it is by no means universally applicable. This project illustrates some ad hoc combinatorial methods for computing the growth series of a much studied group. We loosely follow a sketch given by Stoll in his 1996 paper [248].

- Define $\mathcal{H} = \left\{ \begin{bmatrix} 1 & a & c \\ 0 & 1 & b \\ 0 & 0 & 1 \end{bmatrix} : a, b, c \in \mathbb{Z} \right\}$ and show that ordinary matrix
  multiplication makes $\mathcal{H}$ a group (its name is the *discrete Heisenberg group*; see also Office Hour 8).

- Define
$$x = \begin{bmatrix} 1 & 1 & 0 \\ 0 & 1 & 0 \\ 0 & 0 & 1 \end{bmatrix}, \quad y = \begin{bmatrix} 1 & 0 & 0 \\ 0 & 1 & 1 \\ 0 & 0 & 1 \end{bmatrix},$$
  and $z = [x, y]$ as the commutator of x and y. Derive the matrix form for z,
  and more generally, show that $z^c y^b x^a = \begin{bmatrix} 1 & a & c \\ 0 & 1 & b \\ 0 & 0 & 1 \end{bmatrix}$. Conclude that $\mathcal{H}$ is
  generated by x and y (and their inverses).

- Show z is central, that $\langle z \rangle \lhd \mathcal{H}$, and that $\mathcal{H} / \langle z \rangle$ is isomorphic to $\mathbb{Z} \oplus \mathbb{Z}$. Conclude $\langle z \rangle$ is the full commutator subgroup of $\mathcal{H}$ and that $\mathcal{H}$ is two-step nilpotent.

- Show the exact sequence $0 \longrightarrow \langle z \rangle \longrightarrow \mathcal{H} \longrightarrow \mathbb{Z} \oplus \mathbb{Z} \longrightarrow 0$ splits and thus $\mathcal{H}$ is a semi-direct product of two free abelian groups and can be visualized as the set $\mathbb{Z} \oplus \mathbb{Z} \oplus \mathbb{Z}$ equipped with a weird product structure in the third component.

- Show that $\mathcal{H}$ has abstract presentation $\langle x, y \mid [x, [x, y]] = 1 = [y, [x, y]] \rangle$. Note: We assume this presentation for calculating the word metric and growth series below.

- You have already demonstrated that $z^c y^b x^a$ comprise a set of normal forms for $\mathcal{H}$. Show these normal forms are not generally geodesic (hint: derive $x^a y^b = z^{ab} y^b x^a$).

- Explain why the DSV method is inappropriate for the current situation. On the other hand, some combinatorial techniques are still applicable.

- Verify that $z^{ab} = [x^a, y^b] = [y^{-b}, x^a] = [y^b, x^{-a}]$ for all $a, b \geq 0$.

- Find geodesic words for z^c, $c \in \{1, 2, 3, \ldots, 20\}$ (hint: for prime c write z^c as a freely reduced product of commutators).

- Prove that for each $c \geq 0$ the word metric length of z^c is $2 \min\{r + s : r, s \geq 0 \text{ and } rs \geq c\}$. Did you correctly find a geodesic word for z^{14} in the previous step?

- Obviously z^c and z^{-c} have the same length. Determine the spherical growth series $f_{0,0}(t)$ for $\langle z \rangle$ as a rational function (hint: it is easy enough to write some code that implements the above and the next length formulas).

- Show that for $a, b, c \geq 0$ and $a \leq b$, the word metric length of $z^c y^b x^a$ is $a + b + 2 \min \{r + s : r, s \geq 0 \text{ and } (a + r)(b + s) \geq 0\}$.

- Check that the three maps $(x, y) \longmapsto (y, x)$, $(x, y) \longmapsto (x^{-1}, y)$, and $(x, y) \longmapsto (x, y^{-1})$ are isometries on $\mathcal{H}$. Show that the map $\varphi(z^c y^b x^a) = z^{ab-c} y^b x^a$ is also a length-preserving involution.

- Fix $0 < a \le b$ and consider the coset $\langle z \rangle \, y^b x^a$. Show that the elements $z^c y^b x^a$ have length $a + b$ whenever $0 \le c \le ab$ (hint: the map φ implies that $\frac{ab}{2}$ is a "midpoint" of the coset). This means that the growth series for the coset starts out as $0 + 0t + \cdots + 0t^{a+b-1} + (1 + ab)t^{a+b}$.

- For fixed $0 < a < b$ show that the growth series $f_{a,b}(t)$ for the coset $\langle z \rangle \, y^b x^a$ has tail of the form $2bt^{3b-a} + 2bt^{3b-a+2} + 2(b+1)t^{3b-a+4} + 2(b+1)t^{3b-a+6} + (b+2)t^{3b-a+8} + 2(b+2)t^{3b-a+10} + \cdots$, and conclude $f_{a,b}(t)$ is a rational function with the same denominator as $f_{0,0}(t)$.

- For fixed $0 < a = b$ show that the growth series $f_{b,b}(t)$ for the coset $\langle z \rangle \, y^b x^b$ has tail of the form $2(b+1)t^{2b+4} + 2(b+1)t^{2b+6} + 2(b+2)t^{2b+8} + 2(b+2)t^{2b+10} + 2(b+3)t^{2b+12} + 2(b+3)t^{2b+14} + \cdots$, and conclude that $f_{b,b}(t)$ is a rational function with the same denominator as $f_{0,0}(t)$.

- Find and confirm formulas for the middle terms of the coset growth series $f_{a,b}(t)$ when $0 < a < b$, and for $f_{b,b}(t)$. Derive rational expressions for $f_{a,b}(t)$ and $f_{b,b}(t)$.

- Use the isometries proved earlier to show $f_{\pm a, \pm b}(t) = f_{\pm b, \pm a}(t)$ and thus the growth series for $\mathcal{H}$ is

$$S(t) = f_{0,0}(t) + 4 \sum_{n=1}^{\infty} \left[f_{0,n}(t) + f_{n,n}(t) \right] + 8 \sum_{a=1}^{\infty} \sum_{b=a+1}^{\infty} f_{a,b}(t).$$

- Show that $S(z)$ simplifies as

$$\frac{1 + t + 4t^2 + 11t^3 + 8t^4 + 21t^5 + 6t^6 + 9t^7 + t^8}{(1 - t)^3 \left(1 - t^3\right) \left(1 + t^2\right)}.$$

Project 6. Compute the geodesic growth series for the discrete Heisenberg group.

PART 4

Examples

Office Hour Thirteen

Coxeter Groups

Adam Piggott

As this book has shown so far, in geometric group theory we study groups and geometric spaces simultaneously, linking them via group actions. Sometimes we start with a family of spaces, and we try to find groups that act on the spaces in a prescribed way. Other times we start with a family of groups, and we try to find spaces on which they act in a prescribed way. Here, I tell a story that follows first one of these approaches, and then the other. We shall see that one way to find a space on which a group acts is to build a space, readymade for the purpose, using combinatorics from the group.

13.1 GROUPS GENERATED BY REFLECTIONS

Let's begin by considering those most important spaces, the Euclidean spaces $\mathbb{R}^n$. Various subsets of Euclidean spaces are important too. After all, shapes, patterns, and solid figures are around us everyday, and everyday they sit in a space which seems Euclidean. If something is important, then we want to know about its symmetries.

The symmetries of $\mathbb{R}^n$ are isometries, that is, the maps $\mathbb{R}^n \to \mathbb{R}^n$ that preserve the Euclidean metric. As usual, $\mathrm{Isom}(\mathbb{R}^n)$ is the group of isometries. For a subset $X \subset \mathbb{R}^n$, *the symmetry group of X in $\mathbb{R}^n$* consists of those isometries of $\mathbb{R}^n$ that preserve X as a set: points in X are mapped to points in X, and points not in X are mapped to points not in X. Let's write $\Omega_{\mathbb{R}^n}(X)$ for the symmetry group, and note that it is a subgroup of $\mathrm{Isom}(\mathbb{R}^n)$.

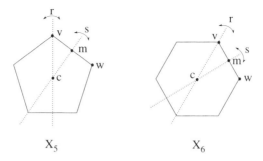

Figure 13.1 Two regular polygons in $\mathbb{R}^2$.

Let's consider some familiar subsets of Euclidean spaces and their symmetries. As we do so, we shall take particular interest in those symmetries that are reflections. A *reflection* in $\mathbb{R}^n$ is an isometry that fixes pointwise a copy of $\mathbb{R}^{n-1}$ inside $\mathbb{R}^n$, and flips the other points over the fixed set. More precisely, if p is a point not in $\mathcal{P}$, the fixed copy of $\mathbb{R}^{n-1}$, and the distance from p to $\mathcal{P}$ is d, then the reflection sends p to the unique point $p' \neq p$ that is also distance d from $\mathcal{P}$, and that lies on the line through p orthogonal to $\mathcal{P}$.

Example: The symmetries of a regular polygon in the plane. We'll start with a basic example, one that we already saw in Office Hour 1. Let m be an integer at least 3 and let X_m be a regular m–gon in $\mathbb{R}^2$. Then X_m has $2m$ symmetries because there are m possible places to send a particular corner, and once such a corner is fixed, there are two ways to arrange the two edges attached to the corner. Now there is a unique point $c \in \mathbb{R}^2$, called the *centroid* of X_m, equidistant from the corners of X_m (see Exercise 1). The setup for X_5 and X_6 is shown in Figure 13.1. Any symmetry $f \in \Omega_{\mathbb{R}^2}(X_m)$ must fix c, and send corners to corners. Let v and w be adjacent corners of X_m, with w clockwise from v; let m be the midpoint of the edge connecting v and w; let r be the reflection that fixes the line through c and v; and let s be the reflection that fixes the line through c and m. You can check: r and s are both symmetries of X_m in $\mathbb{R}^2$; the product sr, which is to be read "perform r and then perform s," spins the plane about c, and serves to spin X_m one click clockwise; the product rs spins X_m one click counterclockwise. By repeated application of sr (or rs), we can move v to any other corner of X_m. By first applying r, or not, we can arrange for the image of w to be counterclockwise or clockwise from v. Thus any symmetry of X_m in $\mathbb{R}^2$ can be achieved by a combination of r's and s's, in various orders and with repetition allowed; that is, $\Omega_{\mathbb{R}^2}(X_m)$ is generated by $\{r, s\}$.

The group $\Omega_{\mathbb{R}^2}(X_m)$ has a name: it is the *dihedral group* D_m.

Exercise 1. Prove that the centroid of the corners of a regular m–gon in $\mathbb{R}^2$ is equidistant from each of the corners. Do the same for a regular tetrahedron in $\mathbb{R}^3$.

Example: The symmetries of $\mathbb{Z}$ in $\mathbb{R}$. Here is another familiar example, from Office Hour 3. The set $\mathbb{Z}$ of integers sits on the number line $\mathbb{R}$ as equally spaced points, extending in both directions. Let $f \in \Omega_{\mathbb{R}}(\mathbb{Z})$ be an arbitrary symmetry. Then $f(0) = z$, for some integer z, and $f(1)$ is either $z + 1$ or $z - 1$. Further, f is determined by the images of 0 and 1. Write r_0 for the reflection that fixes 0 ($r_0(x) = -x$ for all real numbers x), and $r_{1/2}$ for the reflection that fixes $1/2$ ($r_{1/2}(x) = 1 - x$ for all real numbers x). We compute that:

$$(r_{1/2}r_0)^z(0) = z \text{ and } (r_{1/2}r_0)^z(1) = z + 1.$$

Further, we compute that $(r_{1/2}r_0)^z r_0(0) = z$ and $(r_{1/2}r_0)^z r_0(1) = z - 1$. Thus either $f = (r_{1/2}r_0)^z$ or $f = (r_{1/2}r_0)^z r_0$. Hence $\Omega_{\mathbb{R}}(\mathbb{Z})$ is generated by $\{r_0, r_{1/2}\}$.

The group $\Omega_{\mathbb{R}}(\mathbb{Z})$ is like a dihedral group in the sense that it is generated by two reflections. For this reason it is called the *infinite dihedral group* D_∞.

Example: The symmetries of S^1 in $\mathbb{R}^2$. As usual, let S^1 denote the unit circle in $\mathbb{R}^2$. Any symmetry $f \in \Omega_{\mathbb{R}^2}(S^1)$ is determined by the image $f(1, 0)$, and by whether $f(0, 1)$ is clockwise or counterclockwise from $f(1, 0)$. It follows that any symmetry $f \in \Omega_{\mathbb{R}^2}(S^1)$ can be achieved by a rotation of the plane to move $(1, 0)$ to the correct place on the circle, and, if $f(0, 1)$ is to be clockwise from $f(1, 0)$, a reflection that fixes the line through the origin and $f(1, 0)$. Since any rotation is the product of two reflections, $\Omega_{\mathbb{R}^2}(S^1)$ is generated by those reflections that fix the origin.

Example: The symmetries of a different type of subset of $\mathbb{R}^2$. Let X be the subset of $\mathbb{R}^2$ consisting of the point $(1, 0)$, and all the points you can reach from $(1, 0)$ by rotating about the origin through an integral number of radians, say, 1 radian, or 2 radians, etc. Then $\Omega_{\mathbb{R}^2}(X)$ is generated by just two reflections: r, which fixes the x–axis, and s, which fixes the line through the origin that makes angle $1/2$ radian with the x–axis. The group $\Omega_{\mathbb{R}^2}(X)$ is isomorphic to $\Omega_{\mathbb{R}^2}(\mathbb{Z})$; that is, as an abstract group, $\Omega_{\mathbb{R}^2}(X)$ is an infinite dihedral group.

Although $\Omega_{\mathbb{R}_2}(X)$ and $\Omega_{\mathbb{R}}(\mathbb{Z})$ are isomorphic groups, as groups of Euclidean isometries they behave quite differently. A seemingly technical difference, but a very insightful one, is the following: you cannot make a convergent sequence of distinct points just by moving a point $p \in \mathbb{R}$ using elements of $\Omega_{\mathbb{R}}(\mathbb{Z})$, but you can make such a sequence by moving a point $q \in \mathbb{R}^2$ using elements of $\Omega_{\mathbb{R}^2}(X)$. Check this! Let's say that $\Omega_{\mathbb{R}}(\mathbb{Z})$ is a *discrete subgroup* of Isom($\mathbb{R}$), while $\Omega_{\mathbb{R}^2}(X)$ is a not a discrete subgroup of Isom($\mathbb{R}^2$). Discreteness is a generalization of finiteness, in the sense that all finite subgroups of Isom($\mathbb{R}^n$) are discrete. But some infinite subgroups are discrete too. Many arguments that work for finite subgroups of Isom($\mathbb{R}^n$) can be adapted to work for infinite discrete subgroups as well, while subgroups that are not discrete usually need to be considered using different techniques. We shall restrict our attention to discrete subgroups of Isom($\mathbb{R}^n$).

Exercise 2. Here are a couple of exercises exploring the notion of discreteness.

1. Let α be an irrational number such that $0 < \alpha < 1/2$, and let $\theta = \alpha\pi$; let $r_1, r_2 \in$ Isom($\mathbb{R}^2$) be reflections that fix lines intersecting at angle θ; and let $G \subset$ Isom($\mathbb{R}^2$) be the subgroup generated by $\{r_1, r_2\}$. Prove that G is not a discrete subgroup of Isom($\mathbb{R}^2$).

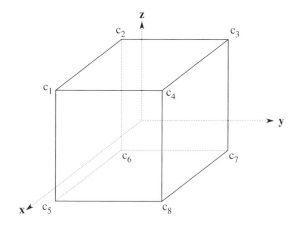

Figure 13.2 A cube in $\mathbb{R}^3$ with corners at $(\pm 1, \pm 1, \pm 1)$.

2. Find three points in $\mathbb{R}$ such that the reflections across these points together generate a subgroup of $\mathrm{Isom}(\mathbb{R})$ that is not discrete.

The product of two reflections.   Some important observations jump out from the examples we have seen.

- A product of two reflections can rotate $\mathbb{R}^n$. This happens whenever the fixed sets are distinct but not parallel, in which case the rotation is about the intersection of the fixed sets, and by twice the angle between the fixed sets. Such reflections generate a finite group if the angle between the fixed sets is a rational multiple of π, and a group that is not discrete if the angle between the fixes sets is not a rational multiple of π.
- A product of two reflections can translate, or shift, $\mathbb{R}^n$. This happens whenever the fixed sets are distinct and parallel and therefore do not intersect, in which case the translation is by twice the distance between the fixed sets. No matter the dimension, any two such reflections generate an infinite group that is isomorphic to D_∞.

Example: The symmetries of a cube in $\mathbb{R}^3$.   Let X be the cube in $\mathbb{R}^3$ with corners at $(\pm 1, \pm 1, \pm 1)$. Label the corners c_1 through c_8 as shown in Figure 13.2. In Office Hour 1 it was found that a cube has 24 symmetries. However, reflections were explicitly excluded from the analysis. The cube was treated as a rigid object, and symmetries were required to preserve the internal structure of the cube. In this office hour, our definition of a symmetry is more permissive. For example, it allows reflection across the xy–plane as a symmetry of the cube, even though the reflection permutes the corners of the cube in a way that cannot be achieved without breaking the cube. With our more permissive definition, X has 48 symmetries because there are eight possible places to send a particular corner, and once such a corner is fixed, there are six ways to arrange the three edges attached to the corner.

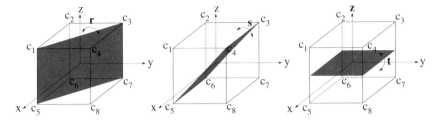

Figure 13.3 The reflections r, s, and t.

Each of the 48 symmetries of X can be achieved by combining, in various orders and with repetition allowed, the reflections r, s, and t, as shown in Figure 13.3. To see this: observe that combinations of r and s can be used to fix c_3, but rearrange the edges adjacent to c_3 in the six possible ways this can be done; the product st serves to rotate the cube by $\pi/2$ about the x–axis, and r serves to swap some corners at the back of the cube with some corners at the front; hence by rotating about the x–axis, then swapping if necessary, and then rotating about the x–axis again, we can move c_3 to any of eight corners.

Example: The symmetries of a tetrahedron in $\mathbb{R}^3$. A regular tetrahedron is made by gluing together four identical equilateral triangles.

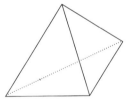

The symmetry group of a tetrahedron is a familiar group in disguise. Each symmetry of a tetrahedron must take corners to corners, and any rearrangement of the corners is possible. Further, an isometry $f \in \text{Isom}(\mathbb{R}^3)$ is completely determined once we know where the corners of a tetrahedron go. It follows that the symmetry group of a tetrahedron is isomorphic to the group of permutations of four identical objects, the four objects being the corners of the tetrahedron. This is the symmetric group S_4.

The connection with the symmetric group can help us see that, like the symmetry group of a cube, the symmetry group of a tetrahedron is generated by a set containing just three reflections. In Office Hour 1, you were invited to prove that S_4 is generated by the permutations (12), (23), and (34). As shown in Figure 13.4, label the corners of the tetrahedron c_1, c_2, c_3, and c_4, and label the centroid of the tetrahedron c_0—again, the centroid is the unique point in $\mathbb{R}^3$ equidistant from the corners (see Exercise 1). The points c_0, c_3, and c_4 lie in a plane. The reflection that fixes this plane fixes c_3 and c_4, and swaps c_1 and c_2. We can similarly use a reflection to swap c_2 and c_3, and another to swap c_3 and c_4. If you have already

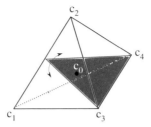

Figure 13.4 The reflection that fixes c_3 and c_4, and swaps c_1 and c_2.

shown that S_4 is generated by $\{(12), (23), (34)\}$, then you have also shown that the symmetry group of the tetrahedron is generated by just three reflections.

The symmetries of other Platonic solids.   The cube and the tetrahedron are examples of Platonic solids, a family of beautiful objects enjoyed since antiquity. The Platonic solids are the convex solids that can be made by gluing together identical regular polygons in such a way that the same number of polygons meet at each corner. There are just five Platonic solids: the cube, the tetrahedron, the octahedron, the icosahedron, and the dodecahedron. The octahedron is so closely related to the cube that the symmetry group of the octahedron is the same as that of the cube. Similarly, the symmetry group of the icosahedron is the same as that of the dodecahedron.

Exercise 3. A dodecahedron is a convex subset of $\mathbb{R}^3$ made by gluing together 12 congruent regular pentagons so that three pentagons meet at each corner. Let X be a dodecahedron in $\mathbb{R}^3$.

1. How many corners does X have? How many edges?
2. How many symmetries does X have?
3. Show that $\Omega_{\mathbb{R}^2}(X)$ is generated by just three reflections.

Exercise 4. In this exercise you will explore a certain duality inherent in the Platonic solids.

1. Explain why the symmetry group of a cube in $\mathbb{R}^3$ is the same as that of an octahedron.
2. Explain why the symmetry group of a dodecahedron in $\mathbb{R}^3$ is the same as that of an icosahedron.

Together with the above these exercises complete the argument that the symmetry group of any Platonic solid in $\mathbb{R}^3$ is generated by a set of three reflections.

The symmetric group S_n as a group generated by reflections.   Just as the symmetry group of a tetrahedron in $\mathbb{R}^3$ is the symmetric group S_4, the symmetry group of the equilateral triangle in the plane is S_3, and the symmetry group of a closed interval in the number line is S_2. In general, the symmetric group S_n can be considered a finite group generated by reflections in $\mathbb{R}^{n-1}$. Cayley's theorem tells us

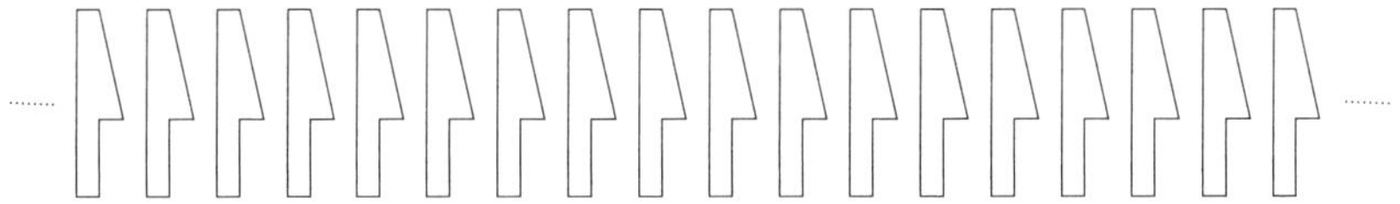

Figure 13.5 No reflection symmetries.

that every finite group is isomorphic to a subgroup of some symmetric group. Thus we can say that every finite group is isomorphic to a subgroup of some finite group generated by reflections in a Euclidean space.

A subset with no reflection symmetries.   For completeness, note that there are many subgroups of $\mathrm{Isom}(\mathbb{R}^n)$, even discrete ones, that are not generated by reflections. For example, there are infinitely many nontrivial symmetries of the planar pattern X shown in Figure 13.5, but there are no reflection symmetries. Thus $\Omega_{\mathbb{R}^2}(X)$ is not generated by reflections.

13.2 DISCRETE GROUPS GENERATED BY REFLECTIONS

We have seen that some natural, beautiful, and important subsets of Euclidean spaces have symmetry groups that are discrete, and are generated by reflections. We have also seen that some important abstract groups can be understood as discrete groups of Euclidean isometries that are generated by reflections. It is clearly worth our time to better understand the subgroups of $\mathrm{Isom}(\mathbb{R}^n)$ that are discrete and generated by reflections. In doing so, we are not necessarily interested in which isometries are in a subgroup, but in the isomorphism class of the subgroup. Another way of stating our goal is that we want to better understand those abstract groups that can act on a Euclidean space with an action that is faithful, discrete, and generated by reflections.

In the 1930s, British-born mathematician H.S.M. Coxeter set about this, and related tasks, with great success. Mathematicians before him had contributed much, but Coxeter painted a particularly clear picture by describing all of the *irreducible* discrete subgroups of $\mathrm{Isom}(\mathbb{R}^n)$ generated by reflections. These groups are the building blocks of all discrete subgroups generated by reflections because the remaining groups, which are called *reducible*, are made by combining, via the group operation of direct product, independent copies of irreducible groups. Coxeter's observations have inspired important and elegant mathematics ever since.

Let's try our hand at Coxeter's game; that is, let's see how we might understand which abstract groups arise as irreducible discrete subgroups of $\mathrm{Isom}(\mathbb{R}^n)$ generated by reflections. The key idea is to turn the algebraic problem at hand (it is algebraic because it takes the form "describe all the groups such that ...") into a geometric problem. I will describe this transformation.

Finite reflection groups.   Initially, we shall restrict our attention to irreducible finite groups generated by reflections. Let n be a positive integer, and let G be a finite group generated by reflections in $\mathbb{R}^n$.

First note that we may make some simplifying assumptions without losing generality.

- *We may assume that G fixes the origin*: Let $x \in \mathbb{R}^n$ be a point, and let $G \cdot x$ be the orbit of x. Since there are finitely many isometries in G, there are finitely many points in $G \cdot x$, and there is a unique point $c \in \mathbb{R}^n$ that is the centroid of $G \cdot x$ (see Exercise 1). If $g \in G$, then g is an isometry and g fixes setwise $G \cdot x$; since c is the unique point that possesses its defining property, we have $g(c) = c$. Up to a change in our coordinate system, we may assume that c is the origin.
- *We may assume that no other point is fixed by G*: Now suppose G fixes another point p. Then G fixes the line $\mathcal{L}$ through the origin and p. By thinking about the way in which G acts on the orthogonal complement of $\mathcal{L}$, we can see G as acting on $\mathbb{R}^{n-1}$, and no information is lost in doing so.
- *We may assume it takes at least n reflections to generate G*: Now suppose it takes only j reflections to generate G, and $j < n$. Choosing j reflections that fix the origin is equivalent to choosing j unit vectors $\vec{v}_1, \ldots, \vec{v}_j$, each orthogonal to the fixed set of one reflection. Since the vectors $\vec{v}_1, \ldots, \vec{v}_j$ cannot span $\mathbb{R}^n$, there exists a vector $\vec{w}$ that is orthogonal to each of the vectors $\vec{v}_i$. This means there is a line $\mathcal{L}$ through c that is fixed by each of the chosen reflections, and hence G fixes $\mathcal{L}$. As above, G could be thought of as acting on $\mathbb{R}^{n-1}$.

Since there is only one reflection in $\mathbb{R}$ that fixes the origin, the only finite group generated by reflections in $\mathbb{R}$ is isomorphic to $\mathbb{Z}/2\mathbb{Z}$. Another way of saying this is that any two distinct reflections in $\text{Isom}(\mathbb{R})$ together generate an infinite group. Since, for the moment, we are investigating finite groups, we may assume $n \geq 2$.

Since G fixes the origin in $\mathbb{R}^n$, it fixes setwise the unit sphere S^{n-1} in $\mathbb{R}^n$, that is, the set of points at distance 1 from the origin. In $\mathbb{R}^2$, this is the familar unit circle. The action of G on $\mathbb{R}^n$ induces an action of G on S^{n-1}. For $n \geq 2$, this new action is generated by spherical reflections in S^{n-1}; these are isometries of S^{n-1} that fix a copy of S^{n-2}. For example, a spherical reflection of S^2 is a reflection across a great circle on the sphere, such as that of swapping the northern and southern hemispheres.

Thus we have that G acts on S^{n-1} with an action generated by at least n spherical reflections in S^{n-1}. Now the fixed sets of all reflections in G—not just the generating reflections—serve to cut S^{n-1} into pieces. Let D be one of these pieces. Coxeter showed that D is a *fundamental domain* for the action of G on S^{n-1}, which means that D contains exactly one point from each G–orbit. It follows that:

- the pieces cover S^{n-1};
- the pieces are in bijective correspondence with G, so each piece is $g(D)$ for exactly one isometry $g \in G$; and
- the pieces only overlap on their boundaries.

Further, he showed that:

- whenever two copies of S^{n-2} meet in the boundary of D, they make an angle π/m for some integer $m \geq 2$;

- G is generated by those reflections R that fix a copy of S^{n-2} making the boundary of D;
- the group G is reducible if and only if the set R can be partitioned into nonempty subsets R_1 and R_2 such that the fixed set of each reflection in R_1 is orthogonal to the fixed set of each reflection in R_2.

Coxeter was able to understand which abstract finite groups arise as subgroups of Isom($\mathbb{R}^n$) because he was able to understand which shapes in S^{n-1}, bounded by copies of S^{n-2}, can be fundamental domains like D. A shape like D is called a *spherical simplex*.

Example: Finite reflection groups in Isom($\mathbb{R}^2$).   To understand the finite groups generated by at least two reflections in $\mathbb{R}^2$, we consider which connected subsets (that is, which arcs) A of the unit circle S^1 are such that: copies of A tessellate the circle; and those reflections in $\mathbb{R}^2$ that fix the origin and one of the boundary points of A can be used to move A to any other piece of the tessellation. It follows that A must have length π/m units for some integer $m \geq 2$. The group is the reducible group $\mathbb{Z}/2\mathbb{Z} \times \mathbb{Z}/2\mathbb{Z}$ if $m = 2$, and the dihedral group D_m if $m \geq 3$. Thus the finite dihedral groups are the only irreducible finite groups generated by at least two reflections in $\mathbb{R}^2$.

Example: Finite reflection groups in Isom($\mathbb{R}^3$).   To understand the finite groups generated by three reflections in $\mathbb{R}^3$, we need to understand the tessellations of the unit sphere S^2 by spherical triangles in which the angles are $\pi/p, \pi/q, \pi/r$ for some integers $p, q, r \geq 2$. Because the angle sum in a spherical triangle must exceed π, the only spherical triangles of this type have angles $(\pi/2, \pi/3, \pi/4)$, $(\pi/2, \pi/3, \pi/5)$, $(\pi/2, \pi/3, \pi/3)$, and $(\pi/2, \pi/2, \pi/r)$. The groups corresponding to $(\pi/2, \pi/2, \pi/r)$ triangles are reducible. Thus there are only three irreducible finite groups generated by at least three reflections in $\mathbb{R}^3$. These are exactly the symmetry groups of the Platonic solids.

Exercise 5. Here are two exercises about finite reflection groups in Isom(S^2).

1. On separate axes, draw the three tessellations of S^2 corresponding to the three irreducible finite subgroups of Isom(S^2) generated by three reflections.
2. On separate axes, draw three distinct tessellations of S^2 corresponding to reducible finite subgroups of Isom(S^2) generated by three reflections.

Infinite reflection groups. Now suppose G is an irreducible, infinite, discrete subgroup of Isom($\mathbb{R}^n$) that is generated by reflections. We no longer have that G fixes a point in $\mathbb{R}^n$. Nevertheless, Coxeter showed that the fundamental domain D for the action of G on $\mathbb{R}^n$ satisfies the following:

- D is a shape bounded by copies of $\mathbb{R}^{n-1}$ and of finite area called a *Euclidean simplex*;
- whenever two copies of $\mathbb{R}^{n-1}$ meet in the boundary of D, they make an angle π/m for some integer $m \geq 2$; and

- G is generated by the set R consisting of the reflections that fix a copy of $\mathbb{R}^{n-1}$ bounding D.

So, for example, to understand irreducible, infinite, discrete groups generated by three reflections in $\mathbb{R}^2$, we need to understand certain tessellations of $\mathbb{R}^2$ by triangles in which the angles are $\pi/p, \pi/q, \pi/r$ for some integers $p, q, r \geq 2$. Because the angle sum in a Euclidean triangle must equal π, the only Euclidean triangles of this type have angles $(\pi/2, \pi/3, \pi/6)$, $(\pi/2, \pi/4, \pi/4)$, and $(\pi/3, \pi/3, \pi/3)$. Thus there are only three irreducible infinite discrete groups generated by at least three reflections in $\mathbb{R}^2$.

Exercise 6. On separate sets of axes, draw the three tessellations of $\mathbb{R}^2$ corresponding to the three irreducible infinite discrete subgroups of $\mathrm{Isom}(\mathbb{R}^2)$ generated by three reflections.

Using these ideas, Coxeter described all of the irreducible discrete groups generated by reflections in all dimensions, and, as a special case, all of the irreducible finite groups generated by reflections. Pretty amazing. These lists are often described in the form of a table. I shall not list them here. To give you some sense of just how intricate Coxeter's determinations needed to be, I point out that one irreducible finite group that emerges in $\mathbb{R}^8$ has 696,729,600 elements, and it is the symmetry group of a

dischiliahectohexaconta-myriaheptachiliadiacosioctaconta-zetton.

There are other names for the object, but none as impressive as this.

13.3 RELATIONS IN FINITE GROUPS GENERATED BY REFLECTIONS

As a consequence of Coxeter's work we have an excellent understanding of the discrete groups generated by reflections and, in particular, the finite groups generated by reflections. His observations about the fundamental domains for the associated actions on Euclidean—and spherical—space lead to an abstract characterization of the finite groups generated by reflections. I shall now describe that characterization, because it is key to generalizing Coxeter's ideas.

If we want to discuss an abstract group, we need a way to label its elements. This is typically done by specifying a generating set. If each generator has order 2, then each group element can be written as a product of the generators since each generator is its own inverse.

Once we have a generating set, we need to describe the group multiplication. Performing one product of generators, and then another, will be represented by writing the second product to the left of the first (this is the opposite of the usual concatenation procedure used in other office hours). For example, performing r and then ts corresponds to tsr. But this is not enough. We'll also want to know the relations; in other words, we'll want group presentations.

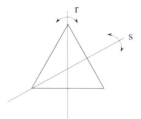

Figure 13.6 The reflections r and s generate D_3.

Example: Relations for the dihedral group D_3. Let r and s be reflections in $\mathbb{R}^2$, fixing lines that make an angle $\pi/3$ as in Figure 13.6. Together they generate D_3. Our aim is to see which equalities involving products of r's and s's are enough that we can deduce all others. The products of pairs of generators are rr, ss, rs, and sr, and their orders are recorded in the following equalities:

$$rr = ss = (rs)^3 = (sr)^3 = 1.$$

I claim that we may deduce from these equalities any other equality involving products of r's and s's; in other words, these relations suffice to give a group presentation.

To prove the claim it suffices to show that, using only these relations, an arbitrary product of r's and s's can be shown equal to one of the six products

$$1, r, rs, rsr, rsrs, rsrsr.$$

Since these products do different things in the plane (you can check this), an arbitrary product is equal to exactly one product in the list, and any equality can be shown true by showing both sides are equal to the same product in the list.

Consider, then, an arbitrary product of r's and s's. Since $rr = 1$, we may remove from the product, without changing its result, any occurrence of rr. Similarly, we may remove any occurrence of ss. Repeating these steps as necessary, we will eventually be left with an alternating product of r's and s's, perhaps even the empty alternating product. If the product has six or more terms and the leftmost letter is r, we may use the equality $(rs)^3 = 1$ to repeatedly remove the six leftmost terms, leaving a product with at most five terms. If the product has six or more terms and the leftmost letter is s, we may use the equality $(sr)^3 = 1$ to repeatedly remove the six leftmost terms, leaving a product with at most five terms. If now the leftmost letter is r, we must have one of the products in our list. If the leftmost letter is s, then we can write $(rs)^3$ on the left of the product, and repeatedly apply $rr = 1$ and $ss = 1$ to arrive at one of the products in the list.

Example: Relations for the group $\mathbb{Z}/2\mathbb{Z} \times \mathbb{Z}/2\mathbb{Z} \times \mathbb{Z}/2\mathbb{Z}$. Let r, s, and t be the reflections that fix the yz–plane, the xz–plane, and the xy–plane, respectively. Because the three fixed sets are pairwise orthogonal, the reflections operate independently and together generate the group $\mathbb{Z}/2\mathbb{Z} \times \mathbb{Z}/2\mathbb{Z} \times \mathbb{Z}/2\mathbb{Z}$. Although there are nine products of pairs of generators, we shall record only the orders of some pairs, because the orders of the other pairs can be deduced from these. We record that:

$$rr = ss = tt = (rs)^2 = (rt)^2 = (st)^2 = 1.$$

I claim that from these relations we may deduce any other relation involving products of r's, s's, and t's.

As in the previous example, it suffices to show that—using only these relations—an arbitrary product of r's, s's, and t's can be shown equal to one of the products

$$1, r, s, t, rs, rt, st, rst.$$

Consider, then, an arbitrary product of r's, s's, and t's. From the relations $(rs)^2 = rr = ss = 1$ we deduce that $rs = sr$ as follows:

$$rsrs = 1 \Rightarrow (rsrs)sr = sr$$
$$\Rightarrow rsr(ss)r = sr$$
$$\Rightarrow rs(rr) = sr$$
$$\Rightarrow rs = sr.$$

Similarly, we can deduce that $rs = sr$ and $rt = tr$. By application of these new relations we may, without changing the result of the product, replace any occurrence of sr by rs, any occurrence of tr by rt, and any occurrence of ts by st. Repeating these steps as necessary, we will eventually be left with a product in which all of the r's appear on the left, all of the s's appear next, and all of the t's appear on the right. Then, by repeated application of $rr = ss = tt = 1$, we may remove occurrences of rr, ss, and tt until the product has at most one r, at most one s, and at most one t. It follows that the given product is one of the products in our list.

Products of pairs: Coxeter's characterization. In each case we saw that we need only consider the relations that record the orders of the products of pairs of generators. Coxeter made the key observation that every discrete group generated by reflections is just as easily described. He proved:

THEOREM 13.1. *If G is a discrete, perhaps even finite, subgroup of* $\mathrm{Isom}(\mathbb{R}^n)$ *generated by reflections, then G has a presentation where each element of the generating set R is a reflection in G and each relation records the orders of products of pairs of generators.*

In Project 1 you are invited to explore the result further. The set R is, as you may suspect, the set of reflections fixing copies of $\mathbb{R}^{n-1}$ that bound a fundamental domain for the group action. Coxeter also proved the following remarkable partial converse:

THEOREM 13.2. *Let W be a finite group with a generating set S, and suppose that S has two kinds of defining relators: $s^2 = 1$ for all $s \in S$ and $(st)^k = 1$ for all distinct pairs $s, t \in S$. Then we can choose a set of reflections R in a Euclidean space such that R generates a group isomorphic to W.*

The second result means that Coxeter's abstract property captures the essence of the finite groups generated by reflections.

13.4 COXETER GROUPS

Coxeter's results inspired further investigation into groups that are as easily described as the discrete subgroups of $\mathrm{Isom}(\mathbb{R}^n)$ generated by reflections. These groups are named in recognition of his contributions.

A group W is a *Coxeter group* if it has a presentation similar to the one in Theorem 13.2. Specifically, W has a finite generating set S and the defining relations come in two types: $s^2 = 1$ for all $s \in S$, and $(st)^k = 1$ for some pairs $s, t \in S$ and some natural number k depending on s and t (it is allowed for some choices of s and t that the order of st is infinite, meaning there is no relation $(st)^k = 1$).

An immediate consequence of the definition is that if we want to tell someone which Coxeter group W we are thinking of, and be sure that there can be no misunderstanding, we need only specify a generating set S, which is presumed to consist of order 2 elements, and the orders of the products of pairs of generators. This means that Coxeter groups can be described using matrices or by labeled graphs. Notice that for all pairs $s, t \in S$, st and ts are inverses in W and so have the same order. Also, since the elements of S are involutions, st has order 1 if and only if $s = t$.

Exercise 7. Let W be a Coxeter group, with generating set $S = \{s_1, \ldots, s_n\}$ and presentation as above. Let $M = (m_{ij})$ be the $n \times n$ matrix in which the entry in the ith row and jth column is the order of the group element $s_i s_j$; M is the *Coxeter matrix* corresponding to (W, S). We allow the entry ∞. Describe those properties a matrix must possess before it is the Coxeter matrix for some Coxeter group.

Using the Coxeter matrix, a Coxeter group has a presentation of the form:

$$W = \langle S \mid (s_i s_j)^{m_{ij}} \rangle$$

(if $m_{ij} = \infty$ we interpret this to mean that there is no relation $(s_i s_j)^{m_{ij}} = 1$). This is called a *Coxeter presentation*.

Exercise 8. A Coxeter presentation is neatly encoded in a labeled graph: each generator is represented by a vertex; for $i \neq j$, the vertices representing s_i and s_j are connected by an edge with label $m_{ij} \in \{2, 3, \ldots\} \cup \{\infty\}$ if $s_i s_j$ has order m_{ij}. Note that the result is a complete graph with labeled edges. Because complete graphs are inconvenient to draw, it is conventional to omit those edges with certain labels, and to leave unlabeled certain other edges. Such omissions have the added benefit of making obvious certain decompositions of the resulting Coxeter group.

1. The classical convention is to omit those edges that would be labeled 2, and to omit the labels on those edges that would be labeled 3; the resulting labeled graph is called a *Coxeter diagram*. How does a Coxeter group decompose if it is the group determined by a disconnected Coxeter diagram? Which word, introduced in the previous section, describes exactly those Coxeter groups determined by connected Coxeter diagrams?

2. Another convention is to omit those edges that would be labeled ∞, and to omit the labels on those edges that would be labeled 2. How does a Coxeter

group decompose if, following this convention, it is the group determined by
a disconnected labeled graph?

Exercise 9. Having completed Exercise 8, you are now able to read and understand
the outcome of Coxeter's enumeration of the irreducible finite Coxeter groups.
Find a table that enumerates the Coxeter diagrams of the irreducible finite Coxeter groups. Convince yourself that, equipped with this table, you are now able to
look at any Coxeter diagram, connected or not, and determine whether or not the
corresponding group is finite. It follows from Coxeter's enumeration that there are
connected Coxeter diagrams in which every edge label in finite, yet the corresponding Coxeter group is infinite.

Exercise 10. Let W be a Coxeter group. Let's say that W is *rigid* if there is essentially only one Coxeter diagram that determines W; more precisely, W is rigid
if whenever Coxeter diagrams Γ and Δ each determine a group isomorphic to W,
there is labeled graph isomorphism $\Gamma \to \Delta$.

1. Find two nonisomorphic Coxeter diagrams that determine groups isomorphic
   to the dihedral group D_6. In doing so you will demonstrate that not all Coxeter
   groups are rigid.
2. Determine which dihedral groups are rigid.
3. Prove that if Γ and Δ are Coxeter diagrams that determine isomorphic groups,
   and every edge in Γ is labeled ∞, then every edge in Δ is labeled ∞.

Special subgroups.   Suppose that W is a Coxeter group, S is a generating set
of order 2 elements as in the definition of a Coxeter group, and T is a subset of
S. The elements of T generate a subgroup W_T of W, called a *special subgroup
of W*. Since W_T sits inside W, we know that any relation between elements of
T can be deduced from the relations that record the orders of pairs of elements
in S. What is not so obvious, but nevertheless true, is that any relation between
elements of T can be deduced from the relations that record the orders of pairs of
elements in T. More precisely, if Γ is the Coxeter diagram for (W, S), $T \subseteq S$ and
Δ is the Coxeter diagram obtained from Γ by omitting those vertices not in T and
those edges adjacent to such vertices, then W_T is isomorphic to the Coxeter group
determined by Δ.

Exercise 11. Prove this statement in the special case that W is a right-angled Coxeter group; that is, if $m_{ij} \in \{2, \infty\}$ for $i \neq j$.

Thus W_T is itself a Coxeter group. It also means that, using Coxeter's classification of the irreducible finite Coxeter groups, we can immediately identify those
subsets of S that generate finite subgroups of W—the finite subgroups they generate are called *spherical subgroups* of W. Before this office hour comes to an end,
I shall put to use our ability to recognize those subsets of S that generate spherical
subgroups.

Exercise 12. Describe the graph property that characterizes those special subgroups of right-angled Coxeter groups that are spherical.

The following exercise shows a fun connection between special subgroups and Coxeter diagrams.

Exercise 13. Let W be a Coxeter group, and let $S = \{s_1, \ldots, s_n\}$ be a generating set of order 2 elements that satisfies the definition a Coxeter group. Define a graph Γ_{odd} as follows: each generator is represented by a vertex; for $i \neq j$, the vertices representing s_i and s_j are connected by an edge if $s_i s_j$ has odd order.

1. Let $T \subset S$ be the vertices in a connected component of Γ_{odd}, and let $f \colon S \to T \cup \{1\}$ be the map

$$s_i \mapsto \begin{cases} s_i & s_i \in T \\ 1 & s_i \notin T. \end{cases}$$

   Prove that f extends to a homomorphism $\hat{f} \colon W \to W_T$, and that $\hat{f}$ restricts to the identity on W_T.

2. Prove the following, which is a special case of the result proved by Richardson [226]: For $i \neq j$, s_i and s_j are conjugate in W if and only if they lie in the same connected component of Γ_{odd}.

Constructing a geometric space for a Coxeter group.   Only a small proportion of the infinite Coxeter groups can be considered discrete groups generated by reflections. For example, there is no way to choose three reflections in a Euclidean space so that they generate a discrete group isomorphic to the Coxeter group W generated by $\{x, y, z\}$ and determined by the relations

$$(xx)^1 = (yy)^1 = (zz)^1 = (xy)^3 = (yz)^2 = 1$$

(note that, because there are no relations recording the order of xz and zx, we assume that these products have infinite order). We know there is no way to do so because you cannot choose planes $\mathcal{P}_x$, $\mathcal{P}_y$, and $\mathcal{P}_z$ in $\mathbb{R}^3$ such that $\mathcal{P}_x$ and $\mathcal{P}_z$ are parallel, $\mathcal{P}_x$ and $\mathcal{P}_y$ make angle $\pi/3$, and $\mathcal{P}_y$ and $\mathcal{P}_z$ make angle $\pi/2$. This means that the family of Coxeter groups is a genuine generalization of the family of discrete groups generated by reflections.

Whenever we have a generalization, we hope our intuition from the special case can be applied to the more general case. The challenge for us is that our intuition for the discrete groups generated by reflections is built upon our familiarity with Euclidean spaces, while groups like W are not related to Euclidean spaces. So we turn to the second approach to geometric group theory. We construct a geometric object Σ, which depends on W and the generating set we work with, that is in some ways like a Euclidean space, and with respect to which W can be seen as a group generated by reflections in Σ. This is done using the combinatorics of cosets in W. The construction we describe is due to Vinberg, Davis, and Moussong.

Example: Coset combinatorics in the dihedral group D_3.   Consider the dihedral group D_3 generated by $\{s, t\}$ and determined by the relations $(ss)^1 = (tt)^1 = (st)^3 = 1$. Each subset of $\{s, t\}$ generates a subgroup of D_3. We consider the left cosets of these subgroups. Recall that if H is a subgroup of G and $g \in G$ is an

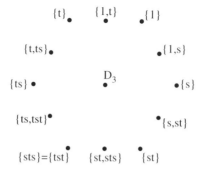

Figure 13.7 Stage I: A vertex for each left coset in D_3.

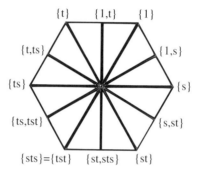

Figure 13.8 Stage II: Add edges to represent inclusion.

element, then the set $gH = \{gh \mid h \in H\}$ is the *left coset of H represented by g*.
The following table contains the results of some relevant computations.

A subset of $\{s, t\}$	The subgroup it generates	The left cosets of the subgroup
$\emptyset$	$\{1\}$	$\{1\}, \{s\}, \{t\}, \{st\}, \{ts\}, \{sts\}$
$\{s\}$	$\{1, s\}$	$\{1, s\}, \{t, ts\}, \{st, sts\}$
$\{t\}$	$\{1, t\}$	$\{1, t\}, \{s, st\}, \{ts, sts\}$
$\{s, t\}$	D_3	D_3

As can be seen, some left cosets are contained in others; this is what I meant by the
combinatorics of cosets. I create a picture of the combinatorics in three stages. At
the first stage, each left coset is represented by a vertex, as shown in Figure 13.7.

At the second stage, two vertices are connected by an edge if one left coset is
contained in the other, as shown in Figure 13.8.

Finally, if three edges form a triangle, fill the triangle with a face, as shown in
Figure 13.9. The final result is denoted by Σ.

Exercise 14. Draw the Cayley graph of D_3 with respect to the generating set consisting of two reflections and describe the relationship between the Cayley graph
and Σ.

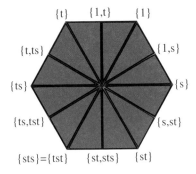

Figure 13.9 Stage III: Triangles are filled.

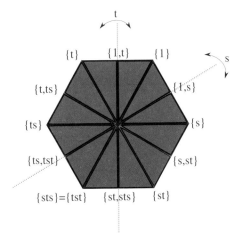

Figure 13.10 The action of D_3 on Σ.

Although skipped over in the previous paragraph, an important set of decisions was required to produce a picture of Σ. I had to choose edge lengths. Notice that I did not use equilateral triangles, so I did not make the most obvious assignment of edge lengths. Instead, I chose edge lengths so that we copied the shape and dimensions of the orbit of a point $p \in \mathbb{R}^2$ that is not fixed by any element of D_3. After choosing edge lengths, the shapes of the faces are determined.

Having constructed Σ, I can now describe how D_3 acts on Σ in a natural way. Given a subgroup H from our list, and arbitrary elements $u, w \in D_3$, the element u maps the left coset wH to the left coset uwH. For example: $s\{1\} = \{s1\} = \{s\}$; $s\{1, t\} = \{s1, st\} = \{s, st\}$; $s\{ts, tst\} = \{sts, stst\} = \{sts, ts\} = \{ts, sts\}$. Once the images of vertices are determined, there are no choices for the images of edges and triangles. As shown in Figure 13.10, s fixes a subset of Σ that looks like a line segment, and flips the other points in Σ over the fixed set. Thus s acts like a reflection on Σ. The action of t is defined similarly, and has similar properties.

A key point about this construction is that we do not need to view Σ as a subset of $\mathbb{R}^2$. Rather, we view it as an independent geometric object equipped with a group action by W. Although it is made from pieces of Euclidean spaces, it need not be thought of as sitting inside a Euclidean space. This gives us hope that we may repeat the construction for an arbitrary Coxeter group.

Davis complex. Given an infinite Coxeter group W, and a generating set S of order 2 elements, we want to build a space as we did for D_3. There are a few details to be considered.

In D_3 every subset of generators itself generates a finite group, but this is not the case in a general Coxeter group. To remedy this, we simply resolve to consider only left cosets of spherical subgroups—it is at this moment that we realize how important it is that we can easily determine which subsets of S generate finite subgroups. Since W is infinite, there are infinitely many such left cosets, so we cannot draw the entirety of Σ; we must settle for drawing it a piece at a time.

In our construction for D_3 it was never the case that four vertices were pairwise adjacent, but we must decide what to do if this occurs when considering W. If four vertices are pairwise adjacent, then from the four vertices we can choose four different groups of three vertices; each group of three corresponds to a triangle, and the four triangles will be arranged like the faces of a tetrahedron. We fill this tetrahedron with a piece of $\mathbb{R}^3$. In general, we glue in a piece of an appropriate Euclidean space whenever we have a set of pairwise adjacent vertices.

Finally, I acknowledge that our construction requires us to choose edge lengths, and not every choice will do. For example, whenever three edges are to be the boundary of a triangle, their lengths must satisfy the triangle inequality. We could avoid this problem by choosing all edge lengths equal, which would mean using pieces of the plane shaped like equilateral triangles, and pieces of $\mathbb{R}^3$ shaped like regular tetrahedra, and so on. But this may not give us a space with desirable properties. For example, if we had chosen edge lengths equal when working with D_3, our construction could not be drawn in the plane because we need 12 triangles to meet around the central point, and 12 times $\pi/3$ is not 2π. Instead, in that case, we let our edge length choices be informed by the way that D_3 acts on $\mathbb{R}^2$. Moussong showed that the analogous strategy is always a good one: even though W itself may not act on a Euclidean space, its spherical subgroups are by definition finite, and so by Theorem 13.2 they act on a Euclidean space; if we make our edge length choices based on the way in which the spherical subgroups of W act on Euclidean spaces, and we do so consistently, then we can always arrange that Σ be a CAT(0) space. Such useful spaces share some of the properties of Euclidean spaces and are prevalent through geometric group theory.

Exercise 15. Learn the definition of a CAT(0) space.

Exercise 16. This exercise is designed to illustrate the process by which we may choose edge lengths for the edges in the Davis complex of a finite Coxeter group.

1. Let $G \subset \mathrm{Isom}(\mathbb{R}^2)$ be the group generated by $\{s, t\}$, where s is the reflection fixing the line $y = x/\sqrt{3}$, and t is the reflection fixing the y–axis. Note that G is isomorphic D_3.

   (a) Let $P = (1, \sqrt{3})$. Sketch the G–orbit of P.
   (b) Let $Q = (1, 1)$. Sketch the G–orbit of Q.

(c) Compare your sketches to Figure 13.9. Describe how a choice of point, for example, P or Q, can lead naturally to a choice of edge lengths in the Davis complex for D_3.

4. Let $J \subset \text{Isom}(\mathbb{R}^3)$ be the group generated by $\{a, b, c\}$, where a is the reflection fixing the plane $y = x/\sqrt{3}$, b is the reflection fixing the yz–plane, and c is the reflection fixing the xy–plane.

   (a) Let $T = (1, \sqrt{3}, 1)$. Sketch the J–orbit of T.
   (b) Describe J as a direct product of familiar groups.

Example: Σ for an infinite Coxeter group.   Consider the infinite Coxeter group W generated by $S = \{x, y, z\}$ and determined by the relations

$$(xx)^1 = (yy)^1 = (zz)^1 = (xy)^3 = (yz)^2 = e.$$

As explained above, this group does not correspond to a discrete group generated by reflections in a Euclidean space. The following subsets of S generate finite subgroups of W: $\emptyset$ generates the subgroup $\{1\}$; $\{x\}$ generates the subgroup $\{1, x\}$; $\{y\}$ generates the subgroup $\{1, y\}$; $\{z\}$ generates the subgroup $\{1, z\}$; $\{x, y\}$, generates a copy of D_3; and $\{y, z\}$ generates a copy of $\mathbb{Z}/2\mathbb{Z} \times \mathbb{Z}/2\mathbb{Z}$. We choose the edge lengths in our construction so that Σ is made from copies of regular hexagons corresponding to D_3, and squares corresponding to $\mathbb{Z}/2\mathbb{Z} \times \mathbb{Z}/2\mathbb{Z}$. Part of the result is shown in Figure 13.11. The full construction Σ cannot fit into $\mathbb{R}^2$ because the arms quickly start running into each other. This is not a problem though, as the construction is not necessarily thought of as sitting inside any space at all.

The elements x, y and z act like reflections, as shown in the figure.

Explore on your own.   That each Coxeter group acts on a CAT(0) space, and the action is generated by what passes for reflections in that space, means that the study of Coxeter groups allows for intuitive arguments based on geometry. There are other, equally productive, ways to approach the study of Coxeter groups, but this way appeals to the taste of many. I conclude with two exercises guiding you through some of the ideas constructed above.

Exercise 17. A *triangle group* is a Coxeter group determined by a Coxeter diagram with exactly three vertices, and no edges labeled infinity. Each such group corresponds to a tessellation of the sphere, the Euclidean plane, or the hyperbolic plane, using congruent triangles. Office Hour 5 explores the hyperbolic plane for those not familiar with it. The angles in the triangle correspond directly to the labels on the edges in the Coxeter diagram.

1. Let W be the triangle group corresponding to the diagram in which all three edges are labeled 3. Describe the Davis complex for W.
2. Let V be the triangle group corresponding to the diagram in which two edges are labeled 3 and one is labeled 4. This group corresponds to a tessellation of the hyperbolic plane. What is the relationship between the tessellation of hyperbolic plane and the Davis Complex for V?

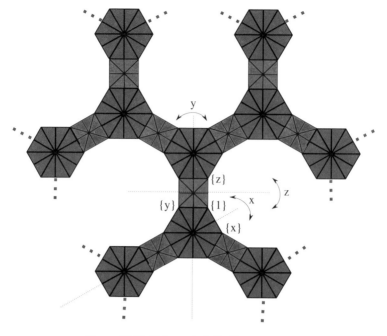

Figure 13.11 Σ for an infinite Coxeter group.

Exercise 18. Exercise 17.2 exhibited a Coxeter group V that acts on the hyperbolic plane as a discrete group generated by reflections, and such that the fundamental domain is a triangle. The triangles in question are bounded, and they are what is called compact. The group V is an example of a compact hyperbolic Coxeter group. As you know from working with improper integrals in calculus, there are shapes in the Euclidean plane that have finite area, even though they seem to stretch on forever; such shapes are said to be non-compact shapes of finite area. In the Euclidean plane, non-compact shapes of finite area cannot have boundaries made from finitely many straight lines. Hyperbolic geometry is different. In the hyperbolic plane, one can use finitely many straight lines to make a shape that has finite area, even though it seems to stretch on forever. Such a shape is called a non-compact simplex of finite area.

1. Give a simple argument that the sphere does not contain a shape with infinite area.
2. Give a simple argument that the Euclidean plane does not contain a non-compact simplex of finite area.
3. Let W be the Coxeter group generated by $S = \{x, y, z\}$ and determined by the relations

$$(xx)^1 = (yy)^1 = (zz)^1 = (xy)^3 = (yz)^2 = 1.$$

This is the group considered in the text above, and with Davis complex as shown in Figure 13.11. In your favorite model of the hyperbolic plane, find

a shape bounded by straight lines, and that is a fundamental domain for an action of W on $\mathbb{H}^2$. The existence of a such a shape shows that W is a non-compact
hyperbolic Coxeter group.

4. Determine whether or not the Davis complex for W, as shown in (13.11), can be drawn in the hyperbolic plane.

5. Research the classification (due to Lannér) of compact and non-compact hyperbolic groups.

Exercise 19. Describe, in general, the relationship between the Cayley graph (with respect to our preferred set of generators) of a Coxeter group and the corresponding Davis complex.

Exercise 20. For each of the Coxeter diagrams you found in Exercise 10.1, construct the corresponding Davis complex.

FURTHER READING

Coxeter's papers [93, 94] are must-reads if you are interested in the topic of Coxeter groups. They are the primary source for the material covered in the first three sections of this office hour. For a modern treatment of these ideas, and to read about the Davis complex in particular, you may consult one of the excellent books of more recent vintage. The two I found most helpful in the preparation of this office hour are those by Bahls [14] and Davis [102]. The first is a great place to start reading more on the topic, and in particular to get a more detailed description of the Davis complex, while the second is the most comprehensive book written from the point of view of geometric group theory. For a different point of view on the topic, the text by Humphreys [170] is a classic.

PROJECTS

Project 1. In classifying the finite groups generated by reflections, Coxeter showed that the fixed sets of the reflections in such a group serve to cut a sphere, of whatever dimension is appropriate, into pieces, each of which is a spherical simplex. This sphere, cut into these pieces, is an example of a simply connected simplicial complex. Similarly, a Euclidean space of the appropriate dimension, cut into the pieces corresponding to the action of an infinite discrete group generated by reflections, is an example of a simply connected simplicial complex. Brown described how to extract from this situation a finite presentation for the group at hand [62]. Using the internet or a library, learn the language of simplicial complexes and CW–complexes, read Brown's paper, and use his result to prove the following result: If G is a discrete, perhaps even finite, subgroup of $\mathrm{Isom}(\mathbb{R}^n)$ generated by the reflections R in the faces of a Euclidean or spherical simplex, then any relation between generators can be deduced from the relations that record the orders of pairs of generators.

Project 2. A Coxeter presentation is called a *right-angled Coxeter presentation*, and the corresponding group is called a *right-angled Coxeter group* if $m_{ij} \in \{2, \infty\}$ whenever $i \neq j$. It follows from Exercise 10.3 that if W is a right-angled Coxeter group, then all Coxeter presentations of W are right-angled. When depicting right-angled Coxeter presentations, it is usually best to follow the non-classical convention described in Exercise 8.2; the result is an unlabeled graph, a *right-angled Coxeter graph*, in which $m_{ij} = 2$ if the corresponding vertices are adjacent, and $m_{ij} = \infty$ otherwise. It is known that each right-angled Coxeter group is rigid; this result is a special case of a result by Radcliffe [225].

1. In what sense are right-angled Coxeter groups "right-angled"?
2. Explain the following (you may need to look up some of the terms used): If W is a right-angled Coxeter group, and $\mathcal{C}(W, S)$ is the Cayley graph of W with respect to S, then the Davis complex for W is the barycentric subdivision of the cubical completion of $\mathcal{C}(W, S)$.
3. Prove that if W is a right-angled Coxeter group, then the Davis complex is CAT(0).

Project 3. The deletion condition is a statement about working with words written in the generators of a Coxeter group. You can tell that it must be explanatory, because it provides a characterization of Coxeter groups in a certain sense (see [14, Theorem 1.3]). The deletion condition is clearly described, and a quite beautiful proof is outlined, in Chapter 1 of Bahls [14]. Read this, then complete the following.

1. Write down a detailed proof of the deletion condition in the case that W is a right-angled Coxeter group (see project 2). Can the conclusion of the deletion condition be strengthened in this special case.
2. *Use the deletion condition* to prove that right-angled Coxeter groups satisfy a quadratic isoperimetric inequality (you might need to look this up, but you can get the idea from Office Hour 8).

Project 4. Brink and Howlett [58] proved that Coxeter groups are automatic. Read about automatic groups in [124]. Read Brink and Howlett's article. Demonstrate your understanding by writing down an explicit automatic structure for an arbitrary right-angled Coxeter group W. Using the Deletion Condition, or otherwise, give an alternate proof of Brink and Howlett's result in the special case that W is a right-angled Coxeter group.

Project 5. Right-angled Artin groups are cool! See Office Hour 14. Davis and Januszkiewicz [103] prove that an isomorphic copy of every right-angled Artin group lives as a finite-index subgroup in some right-angled Coxeter group. One of the exercises in Office Hour 14 invites you to consider how such a copy is found. Read Davis and Januszkiewicz's paper. Demonstrate your understanding by describing three finite simple graphs Γ_i for $i = 1, 2, 3$, each with at least five vertices and 12 edges, and three finite simple graphs Δ_i for $i = 1, 2, 3$ such that an isomorphic copy of the right-angled Artin group determined by Γ_i lives as a finite-index subgroup in the right-angled Coxeter group determined by Δ_i.

Office Hour Fourteen

Right-Angled Artin Groups

Robert W. Bell and Matt Clay

In this book you've heard a lot about free groups and free abelian groups. The goal of this office hour is to introduce you to a wide class of groups called right-angled Artin groups. As we will see, this is a broad spectrum of groups, with free groups on one end, free abelian groups on the other end, and lots of other interesting groups in between. Recently, right-angled Artin groups have taken center stage in geometric group theory, in large part due to their involvement in the solution to one of the main open questions in the topology of 3-manifolds. See the end of this office hour for more on that.

A *right-angled Artin group* is a group $G(\Gamma)$ defined in terms of a graph Γ. So let Γ be a simple, loopless graph with a finite vertex set $V(\Gamma)$ and with a finite edge set $E(\Gamma)$; we think of an element of $E(\Gamma)$ as an unordered pair $\{v, w\}$, where v and w are distinct elements of $V(\Gamma)$. Recall that simple means that there is at most one edge connecting a given pair of vertices and loopless means that there is no edge connecting a vertex to itself. All graphs we consider will be simple and loopless.

The right-angled Artin group (sometimes abbreviated RAAG) defined by Γ is the group with presentation

$$G(\Gamma) = \langle V(\Gamma) \mid vw = wv \text{ for each } \{v, w\} \in E(\Gamma) \rangle.$$

What this means is that we take as generators for $G(\Gamma)$ the vertices of Γ and declare that two vertices commute if they are adjacent in Γ. That's it!

The first examples to consider are the extremes. If Γ is the complete graph with n vertices, meaning that every vertex is adjacent to every other vertex, then $G(\Gamma)$ is isomorphic to $\mathbb{Z}^n$, the free abelian group of rank n. On the other hand, if Γ consists

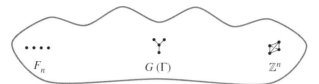

Figure 14.1 The world of right-angled Artin groups.

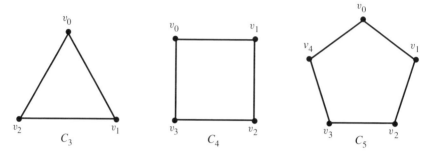

Figure 14.2 Cycle graphs, C_n.

of n vertices and no edges, then $G(\Gamma)$ is isomorphic to F_n, the free group of rank n. So, as we already advertised, right-angled Artin groups interpolate between free abelian groups and free groups, exhibiting features of each class. See Figure 14.1.

For a more typical example, consider the *cycle graph* C_n: this graph has vertex set $V(C_n) = \{v_i \mid i \in \mathbb{Z}/n\mathbb{Z}\}$ and edges $E(C_n) = \{\{v_i, v_{i+1}\} \mid i \in \mathbb{Z}/n\mathbb{Z}\}$. See Figure 14.2. We find:

$$G(C_3) = \langle v_0, v_1, v_2 \mid v_0 v_1 = v_1 v_0, \ v_1 v_2 = v_2 v_1, \ v_2 v_0 = v_0 v_2 \rangle$$
$$\cong \mathbb{Z}^3, \quad \text{and}$$
$$G(C_4) = \langle v_0, w_0, v_1, w_1 \mid v_i w_j = w_j v_i \text{ for } i, j \in \{0, 1\} \rangle$$
$$\cong F_2 \times F_2.$$

For $n \geq 5$, $G(C_n)$ does not admit an alternative description in terms of free groups and free abelian groups.

In this office hour, we explore some of the other possibilities between the two extremes, investigate examples related to the other office hours, and establish some foundational results in the theory of right-angled Artin groups. An interesting challenge in the study of right-angled Artin groups—and one to keep in mind while you are reading—is to determine if and how graph properties of Γ influence the algebraic or geometric structure of $G(\Gamma)$. Here are some sample theorems on this theme:

1. $G(\Gamma)$ is abelian if and only if Γ is complete.
2. $G(\Gamma)$ is isomorphic to a direct product $A \times B$ if and only if Γ is a *join*.[1]

---

[1] A graph Γ is a join if $V(\Gamma) = V_1 \bigsqcup V_2$ and for each $v_1 \in V_1$ and $v_2 \in V_2$, there is an edge $\{v_1, v_2\} \in E(\Gamma)$.

Figure 14.3 Path graphs, P_n.

3. $G(\Gamma)$ is isomorphic to a free product $A * B$ if and only if Γ is not connected.
4. $G(\Gamma)$ has infinitely many ends if and only if Γ is not connected.

An unresolved research question in this direction is that of characterizing when two graphs define quasi-isometric right-angled Artin groups. By the end of this office hour, you should have a good idea about how to start proving the four theorems listed and gain some appreciation for this open problem.

14.1 RIGHT-ANGLED ARTIN GROUPS AS SUBGROUPS

Before we start looking at more examples, we need to recall a property of group presentations. Namely, if $G = \langle S \mid R \rangle$ and H is another group, then a function $f \colon S \to H$ extends to a homomorphism $f \colon G \to H$ if and only if $f(r) = 1$ for each $r \in R$.

In our case, this means that to define a homomorphism from $G(\Gamma)$ to some other group H, we only need to specify the image of each generator and then verify that the images of commuting generators of $G(\Gamma)$ commute in H.

For example, if $G(\Gamma)$ is generated by $v_1, \ldots, v_n$, then we can choose n integers $m_1, \ldots, m_n \in \mathbb{Z}$ and define a homomorphism $f \colon G(\Gamma) \to \mathbb{Z}$ by $f(v_i) = m_i$. This always defines a homomorphism as any two elements in $\mathbb{Z}$ commute.

Exercise 1. Show that the abelianization of $G(\Gamma)$ is $\mathbb{Z}^n$, where $n = \#|V(\Gamma)|$. The abelianization of a group is introduced in Office Hour 16, in case you have not seen it previously.

Subgroups from subgraphs.   A graph Δ is a *subgraph* of Γ if $V(\Delta) \subseteq V(\Gamma)$ and $E(\Delta) \subseteq E(\Gamma)$. We say that Δ is an *induced subgraph* if, in addition, for every $v, w \in V(\Delta)$,

$$\{v, w\} \in E(\Gamma) \implies \{v, w\} \in E(\Delta).$$

That is, if two vertices are adjacent in Γ, then they are adjacent in Δ.

If $X \subseteq V(\Gamma)$, by $\Gamma(X)$ we denote the induced subgraph of Γ with vertex set X. Thus the edges of $\Gamma(X)$ are all of the edges in Γ whose endpoints are in X.

Here are some examples to consider. Let n be a positive integer. The *path graph*, denoted by P_n, has vertex set $V(P_n) = \{v_i \mid i \in \mathbb{Z}/n\mathbb{Z}\}$ and edges $E(P_n) = \{\{v_i, v_{i+1}\} \mid i \in \mathbb{Z}/n\mathbb{Z}, i \neq n - 1\}$. See Figure 14.3. Then P_n is a subgraph of C_n, but not an induced subgraph, as v_0 and v_{n-1} are adjacent in C_n but not P_n. The graph P_{n-1} is an induced subgraph of P_n. The graph C_{n-1} is not a subgraph of C_n.

Exercise 2. Provide an alternative description of $G(P_3)$ in terms of F_2 and $\mathbb{Z}$ as we did for $G(C_4)$. Can you provide an alternative description of $G(P_4)$ in terms of three copies of $\mathbb{Z}^2$? You might want to consult Office Hour 3 for the latter.

A *graph map* $f\colon \Delta \to \Gamma$ is a function $f\colon V(\Delta) \to V(\Gamma)$ that if $\{v, w\} \in E(\Delta)$, then $\{f(v), f(w)\} \in E(\Gamma)$ (this agrees with the definition in Office Hour 4 in the case of simple, loopless graphs). That is, the images of adjacent vertices are adjacent. Hence f also defines a map $f\colon E(\Delta) \to E(\Gamma)$.

The image of a graph map f is a subgraph of Γ with vertex set $f(V(\Delta)) \subseteq V(\Gamma)$ and edge set $f(E(\Delta)) \subseteq E(\Gamma)$. If f is a bijection and $f^{-1}\colon \Gamma \to \Delta$ is a graph map, then we say that f is an *isomorphism* and that Δ and Γ are isomorphic. For example, in P_n, both the induced subgraph on $\{v_0, \ldots, v_{n-2}\} \subset V(P_n)$ and the induced subgraph on $\{v_1, \ldots, v_{n-1}\} \subset V(P_n)$ are isomorphic to P_{n-1}. Our interest in graph maps stems from the following exercise.

Exercise 3. Show that a graph map $f\colon \Delta \to \Gamma$ induces a group homomorphism $f\colon G(\Delta) \to G(\Gamma)$.

The interesting case is when Δ is an induced subgraph of Γ.

THEOREM 14.1. *Let Δ be an induced subgraph of a graph Γ and let $i\colon \Delta \to \Gamma$ be the inclusion map. Then i induces an injective homomorphism $i\colon G(\Delta) \to G(\Gamma)$.*

Proof. We define a homomorphism $\pi\colon G(\Gamma) \to G(\Delta)$ as follows. As usual, it will be convenient to ignore the distinction between vertices of Γ and Δ with the corresponding group elements of $G(\Gamma)$ and $G(\Delta)$. We define π on generators of $G(\Gamma)$ by

$$\pi(v) = \begin{cases} v & v \in V(\Delta) \\ 1 & v \notin V(\Delta) \end{cases}.$$

As above, it is straightforward to see that this defines a homomorphism (images of commuting generators automatically commute).

Notice that $\pi \circ i\colon G(\Delta) \to G(\Delta)$ is the identity homomorphism (in other words π is a left inverse to i). Thus $\pi \circ i$ is injective and therefore so is i. $\square$

Theorem 14.1 is useful in that it provides an easy way to construct known subgroups of $G(\Gamma)$: if $X \subset V(\Gamma)$; then the subgroup generated by X is canonically isomorphic to the right-angled Artin group defined by the induced subgraph $\Gamma(X)$.

Exercise 4. Let $v, w \in G(\Gamma)$ be distinct generators. Show that if v and w are not adjacent, then $\langle v, w \rangle \cong F_2$. Conclude item (1) from the beginning of this office hour: $G(\Gamma)$ is abelian if and only if Γ is complete.

Exercise 5. Prove the "if" part of item (2) from the beginning of this office hour. Specifically, if Γ is a join with respect to the decomposition $V(\Gamma) = V_1 \coprod V_2$, then $G(\Gamma) = G(\Gamma(V_1)) \times G(\Gamma(V_2))$.

Subgroups of GL(n, $\mathbb{R}$).   Consider the following three matrices:

$$A = \begin{bmatrix} 1 & 0 & 0 \\ 0 & 1 & 2 \\ 0 & 0 & 1 \end{bmatrix} \qquad B = \begin{bmatrix} 2 & 0 & 0 \\ 0 & 1 & 0 \\ 0 & 0 & 1 \end{bmatrix} \qquad C = \begin{bmatrix} 1 & 0 & 0 \\ 0 & 1 & 0 \\ 0 & 2 & 1 \end{bmatrix}.$$

You can readily check that $AB = BA$, $BC = CB$, and $AC \neq CA$. Is $G = \langle A, B, C \rangle \subset \mathrm{GL}(3, \mathbb{R})$ a right-angled Artin group? We have that $G = HK$, where $H = \langle B \rangle$ and $K = \langle A, C \rangle$. These computations show that there is a homomorphism $G(P_3) \rightarrow G$ defined by sending $v_0 \rightarrow A$, $v_1 \rightarrow B$ and $v_2 \rightarrow C$. In order for this to be injective, we would need A and C to satisfy no relations, i.e., generate a free subgroup.

Exercise 6. Prove that $\langle A, C \rangle \cong F_2$. *Hint: Look at Office Hour 5!*

By the exercise, we have $K \cong F_2$, and since $H \cap K$ is the trivial group (can you prove this?), we have that $G \cong F_2 \times \mathbb{Z}$. So, indeed, $G \cong G(P_3)$ and is a right-angled Artin group.

Exercise 7. Find a subgroup of $\mathrm{GL}(4, \mathbb{R})$ isomorphic to $G(C_4)$.

14.2 CONNECTIONS WITH OTHER CLASSES OF GROUPS

Did Theorem 14.1 seem easy? Seem obvious? Not so fast. In general, given a group and subset X of the group, it is very difficult to characterize the subgroup generated by X even if we assume that X is a subset of the defining generators. For instance, Magnus proved the very difficult theorem (affectionately known as the Freiheitssatz) that says that if we take a one-relator group $G = \langle s_1, \ldots, s_n \mid r = 1 \rangle$ and we take some proper subset of the generators, then this subset generates a free subgroup of G. Try to prove that one! (Or see [197].)

Of course, in Magnus's theorem the group G itself is generally not free. An example of such a group is $G = \langle x, y, z \mid x^2 y^2 z^2 = 1 \rangle$. Any pair of generators of G generates a free group, but G is not free.

Therefore, we should not be too surprised if recognizing an arbitrary finitely generated subgroup of $G(\Gamma)$ or recognizing a right-angled Artin group as a subgroup of a known group is a challenging problem. The examples that follow illustrate these challenges.

Subgroups of the braid group.   Let B_n be the braid group on n–strings. Office Hour 18 discusses these groups in detail. The group B_n has the presentation:

$$B_n = \langle \sigma_1, \ldots, \sigma_{n-1} \mid \sigma_i \sigma_j = \sigma_j \sigma_i \text{ if } |i - j| > 1,$$
$$\sigma_i \sigma_{i+1} \sigma_i = \sigma_{i+1} \sigma_i \sigma_{i+1} \text{ for } 1 \leq i \leq n - 2 \rangle.$$

The element σ_i is represented by a left-to-right diagram of n horizontal strands numbered from top to bottom by $1, 2, \ldots, n$; the half-twist braids the ith strand over the $(i + 1)$st strand and leaves the other strands unaltered. See Figure 18.2(c) for a picture of $\sigma_3 \in B_7$.

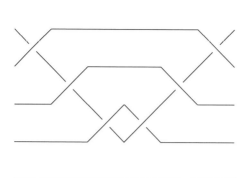

Figure 14.4 The generator $A_{2,5}$ in the pure braid group PB_6. The braid is drawn from left to right with the first string on bottom.

Consider the case of $n = 4$; the group B_4 is generated by the three elements σ_1, σ_2, and σ_3. The first and third generators commute (refer to the presentation), but neither of these commutes with σ_2. However, σ_1 and σ_2 do not generate a free group since

$$\sigma_1\sigma_2\sigma_1\sigma_2^{-1}\sigma_1^{-1}\sigma_2^{-1} = 1$$

(again this is one of the defining relators). Therefore, the homomorphism $G(P_3) \to B_4$ mapping v_i to σ_{i+1} is not injective.

Exercise 8. Do Exercise 9 in Office Hour 18 and use it to conclude that B_n is a right-angled Artin group if and only if $n = 2$, in which case $B_2 \cong G(P_1) \cong \mathbb{Z}$.

The pure braid group on three strings. We will also make use of Office Hour 18 in this example. Let PB_n be the pure braid group on n strings. The group consists of all braids on n strings such that the ith string terminates at the ith node for each i. The pure braid group is generated by $\binom{n}{2}$ elements $\{A_{ij} \mid 1 \le i < j \le n\}$, where A_{ij} is the element that passes the jth strand behind the strands $j - 1, \ldots, i$, loops it over strand i, and then passes behind strands $i + 1, \ldots, j - 1$ as in Figure 14.4. As a subgroup of B_n, we have that $A_{i,i+1} = \sigma_i^2$ and, more generally,

$$A_{ij} = (\sigma_{j-1}\sigma_{j-1}\cdots\sigma_{i+1})\sigma_i^2(\sigma_{j-1}\sigma_{j-1}\cdots\sigma_{i+1})^{-1}.$$

In PB_3 there are three generators, A_{12}, A_{13}, and A_{23}, and each pair of generators generates a free group. The following exercise gives an outline for how to prove this.

Exercise 9. Let $a_1 = \begin{bmatrix} -t & 0 \\ -1 & 1 \end{bmatrix}$ and $a_2 = \begin{bmatrix} 1 & -t \\ 0 & -t \end{bmatrix}$, where t is a formal variable with formal inverse t^{-1}.

1. Verify that $\sigma_i \mapsto a_i$ defines a homomorphism from B_3 to the group of 2×2 matrices with coefficients in $\mathbb{Z}[t, t^{-1}]$ that have nonzero determinant. This map is known as the *(reduced) Burau representation*.

2. Prove that the image of $\langle \sigma_1^2, \sigma_2^2 \rangle$ under the Burau representation is a free group of rank 2. *Hint: Modify a proof that $A = \begin{bmatrix} 1 & 0 \\ 2 & 1 \end{bmatrix}$ and $B = \begin{bmatrix} 1 & 2 \\ 0 & 1 \end{bmatrix}$ generate a free subgroup of* SL$(2, \mathbb{Z})$; *see Office Hour 5.*

3. Use the fact that a surjection $F_2 \to F_2$ is in fact an isomorphism to prove that $\langle \sigma_1^2, \sigma_2^2 \rangle \cong F_2$.

4. Deduce that A_{ij} and A_{jk}, where $i \neq k$, generate a free group of rank 2.

But PB_3 is not a free group! The element

$$z = A_{13}A_{23}A_{12}$$

commutes with every element of the group. The following exercise walks you through a proof.

Exercise 10. We'll actually show something stronger than what was just claimed: the element $A_{13}A_{23}A_{12}$ actually commutes with every element of B_3. Here is the idea. First (using the above formulas) rewrite A_{12} as $\sigma_1\sigma_1$, A_{23} as $\sigma_2\sigma_2$, and A_{13} as $\sigma_2\sigma_1\sigma_1\sigma_2^{-1}$. And then use the braid relations (from the presentation for B_n) to directly show that

$$\sigma_i\sigma_2\sigma_1\sigma_1\sigma_2^{-1}\sigma_2\sigma_2\sigma_1\sigma_1 = \sigma_2\sigma_1\sigma_1\sigma_2^{-1}\sigma_2\sigma_2\sigma_1\sigma_1\sigma_i$$

for i either 1 or 2.

It turns out that we have found all of the relations in PB_3! It has the presentation

$$PB_3 \cong \langle A_{12}, A_{13}, z \mid [A_{12}, z] = [A_{23}, z] = 1 \rangle.$$

In other words, PB_3 is isomorphic to $F_2 \times \mathbb{Z}$, a right-angled Artin group!

By what we have already shown, there is an obvious homomorphism from $F_2 \times \mathbb{Z}$ to PB_3; the generators for F_2 map to A_{12} and A_{13} and the generator for $\mathbb{Z}$ maps to $z = A_{13}A_{23}A_{12}$. It is easy to see that this homomorphism is surjective. To check the injectivity, we'll first check it on each factor. We already said that A_{12} and A_{23} generate a free group, so that gives injectivity on that factor. The injectivity on the $\mathbb{Z}$–factor comes for free since the braid group is torsion free (see Office Hour 18). The only thing you have left to show is that the images of these two factors have trivial intersection. That does it!

Exercise 11. Verify the above claim that the images of the F_2–factor and the $\mathbb{Z}$–factor have trivial intersection.

Subgroups of the pure braid group. Let SQ_n be the subgroup of the n braid group B_n generated by the squares of the standard generators. This is equal to the subgroup of PB_n generated by $A_{12}, \ldots, A_{n-1,n}$. Since $A_{i,i+1}$ and $A_{j,j+1}$ commute if $|i - j| > 1$ and $A_{i,i+1}$ and $A_{i+1,i+2}$ generate a free group of rank 2, could it be that this subgroup is a right-angled Artin group? Remarkably, Collins [89] and Humphries [172] showed that the answer is yes.

THEOREM 14.2. *The squares of the standard generators of the braid group generate a subgroup isomorphic to $G(P_{n-1})$.*

Coxeter groups and right-angled Artin groups. There are many interesting connections between Coxeter groups (see Office Hour 13) and right-angled Artin groups. Given a graph Γ, the *right-angled Coxeter group* $W(\Gamma)$ has generators corresponding to the vertices of Γ and has as defining relations the commuting relations corresponding to edges plus the relations that every generator has order 2. Right-angled Coxeter groups seem smaller than right-angled Artin groups, because of the added relations. So you might be surprised to read the following theorem of Davis and Januszkiewicz [103].

THEOREM 14.3. *Every right-angled Artin group is isomorphic to a subgroup of finite index in some right-angled Coxeter group.*

The construction by Davis and Januszkiewicz is explicit. A key part is figuring out where to send commuting generators (remember, that's all we need to worry about to define a homomorphism!). The next exercise will help you figure this step out.

Exercise 12. Prove that $G(P_2) \cong \mathbb{Z}^2$ is isomorphic to a subgroup of the right-angled Coxeter group

$$W(C_4) = \langle a_0, b_0, a_1, b_1 \mid a_i^2 = b_i^2 = 1, a_i b_j = b_j a_i \text{ for } i, j \in \{0, 1\}\rangle$$
$$\cong (\mathbb{Z}/2\mathbb{Z} * \mathbb{Z}/2\mathbb{Z}) \times (\mathbb{Z}/2\mathbb{Z} * \mathbb{Z}/2\mathbb{Z})$$

via the map $v_0 \mapsto a_0 a_1$, $v_1 \mapsto b_0 b_1$. Generalize this construction to exhibit $G(P_n)$ as a subgroup of a right-angled Coxeter group.

Imitating the method of the exercise will exhibit your favorite right-angled Artin group as a subgroup of a right-angled Coxeter group, but to realize it as a subgroup of finite index requires more care.

As every Coxeter group $W(\Gamma)$ is isomorphic to a group of matrices, i.e., a subgroup of $GL(n, \mathbb{C})$, the same holds for every right-angled Artin group. As in Office Hour 5, we say that such groups are linear.

14.3 SUBGROUPS OF RIGHT-ANGLED ARTIN GROUPS

A subgroup of a free group is itself a free group, this appears in both Office Hours 3 and 4. The same is true for free abelian groups: their subgroups are themselves free abelian groups. This is far from being true for all right-angled Artin groups. We'll present a couple of examples illustrating just the tip of the iceberg of this fascinating phenomenon.

Kernels of homomorphisms $G(\Gamma) \to \mathbb{Z}$. Choosing an integer for each generator of $G(\Gamma)$ defines a homomorphism $G(\Gamma) \to \mathbb{Z}$ (see the discussion at the beginning of the office hour). One obvious choice is to send every generator to $1 \in \mathbb{Z}$. We'll use $\phi \colon G(\Gamma) \to \mathbb{Z}$ to denote this homomorphism. The kernels of these homomorphisms are subgroups with some very interesting properties.

For instance, the group $\ker(\phi\colon F_2 \to \mathbb{Z})$ is not finitely generated (see Theorem 4.9). So even though it is a free group, it is not a right-angled Artin group.

As a more exotic example, the kernel of $\phi\colon F_2 \times F_2 \to \mathbb{Z}$ is finitely generated, but not finitely presented and therefore not a right-angled Artin group.

Exercise 13. Show that $\ker(\phi\colon F_2 \times F_2 \to \mathbb{Z})$ is generated by the three elements $s_0 = v_0 w_0^{-1}$, $s_1 = v_1 w_0^{-1}$, and $t_0 = v_0 w_1^{-1}$ where we have identified $F_2 \times F_2 \cong G(C_4)$ and used the presentation of $G(C_4)$ as in the beginning of this office hour. Let $t_1 = s_1 s_0^{-1} t_0 = v_1 w_1^{-1}$. Show that the relation $s_0^n s_1^{-n} = t_0^n t_1^{-n}$ holds for all $n \in \mathbb{Z}$. Can you find other infinite families of relations?

Next, the kernel of $\phi\colon F_2 \times F_2 \times F_2 \to \mathbb{Z}$ is finitely presented.

Exercise 14. Find a finite generating set for $\ker(\phi\colon F_2 \times F_2 \times F_2 \to \mathbb{Z})$ and show that the analogous infinite families of relations you found in the previous exercise are the consequence of finitely many, hinting at the fact that this group is finitely presented.

These groups form the beginning of a sequence of groups $\ker(\phi\colon (F_2)^n \to \mathbb{Z})$ studied by Robert Bieri [32] and more generally by Mladen Bestvina and Noel Brady [29]. As the examples might suggest, there are certain finiteness properties (generalizations of finite generation and finite presentability) that are achieved only after n gets large enough.

Surface subgroups of right-angled Artin groups.   Next we will show how to find surface groups inside right-angled Artin groups. Surface groups are introduced in Office Hour 9. Given a surface Σ, the corresponding surface group is the group of homotopy classes of based loops in Σ and the group operation is concatenation.

We will be content with considering only the closed oriented surface with genus 2, Σ_2. This group has presentation (Theorem 9.1):

$$\pi_1(\Sigma_2) = \langle a_1, b_1, a_2, b_2 \mid [a_1, b_1][a_2, b_2] = 1 \rangle.$$

We will describe an injective homomorphism of $\pi_1(\Sigma_2) \to G(C_6)$ using a technique discovered by Crisp and Wiest [97].

A homomorphism $\phi\colon \pi_1(\Sigma_2) \to G(C_6)$ is described by Figure 14.5 as follows. In the figure are shown six black curves in Σ_2, each labeled with a vertex of C_6, and each given a transverse orientation (that's the arrow orthogonal to the curve). The key feature is that curves that intersect are labeled with adjacent vertices. Check this property in the figure.

Now given a loop $\gamma \subset \Sigma_2$ based at $x_0 \in \Sigma_2$, we define $\phi(\gamma) \in G(C_6)$ to be the element obtained by recording the labels of curves γ crosses, using v_i^{-1} in the case that γ crosses the curve opposite to the given transverse orientation. For the loop γ shown in the figure we have $\phi(\gamma) = v_0^{-1} v_1 v_2^{-1} v_4 v_3 v_2^{-1}$.

Why is ϕ well defined? We need to show that altering the loop γ by small perturbations doesn't change the element $\phi(\gamma) \in G(C_6)$ (indeed, any deformation can be broken into a sequence of small perturbations). This is shown in Figure 14.6

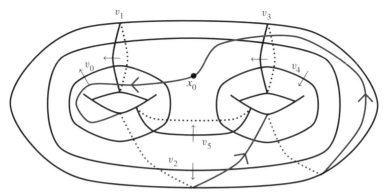

Figure 14.5 An injective homomorphism $\phi \colon \pi_1(\Sigma_2) \to G(C_6)$.

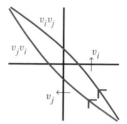

Figure 14.6 Perturbing γ doesn't affect $\phi(\gamma)$ as v_i and v_j are adjacent.

where we show a small portion of some loop. The red curve can be perturbed to the blue curve. Since the vertices v_i and v_j are adjacent by construction, we have $v_i v_j = v_j v_i$ in $G(C_6)$ and so reading off the labels along either the red curve or blue curve gives the same element!

Exercise 15. Determine all other "events" that can happen during a small perturbation of a loop in Σ_2 and convince yourself that none of these affect the image of ϕ.

That ϕ is a homomorphism is clear, since the group operation is concatenation. For ϕ to be injective, it is necessary, but not sufficient, that the six curves in Σ_2 cut it into a collection of disks: otherwise some curve lying in the non-disk subsurface would map to the identity. To show that our specific ϕ is really injective requires further analysis; see Project 1.

Exercise 16. Find the images of $a_1, b_1, a_2, b_2 \in \pi_1(\Sigma_2)$ in $G(C_6)$. Verify that $[\phi(a_1), \phi(b_1)][\phi(a_2), \phi(b_2)] = 1$.

Droms, Servatius, and Servatius [239] proved that for each $n \geq 4$, $\pi_1(S_g)$ is a subgroup of $G(C_n)$ when $g = 1 + (n - 4)2^{n-3}$. On the other hand, the problem of which right-angled Artin groups contain a subgroup isomorphic to $\pi_1(S_g)$ for some $g \geq 2$ is unresolved. Kim's article [184] is a good starting point for learning more about this problem.

Exercise 17. Describe a homomorphism $\phi: \pi_1(\Sigma_2) \to G(C_5)$ by finding a collection of curves on Σ_2 labeled by $V(C_5)$ where two curves intersect if and only if their labels correspond to adjacent vertices in C_5. *Hint: The collection of curves will need more than five curves and so some of the curves will have the same label.*

14.4 THE WORD PROBLEM FOR RIGHT-ANGLED ARTIN GROUPS

We now begin a systematic study of properties of right-angled Artin groups. Our primary goal is to solve the word problem. This is discussed for general groups in Office Hour 8. Specifically, we will describe an algorithm that accepts any word w in the generators of $G(\Gamma)$ and returns as output an answer to the question, "Does w represent the identity element?"

The word problem in $G(C_5)$. Let's consider an example to start. Take the word

$$w = v_0 v_1^{-1} v_4 v_0^{-1} v_3 v_4^{-1} v_2 v_3^{-1} v_1 v_2^{-1}$$

in the generators for $G(C_5)$. Does w represent the identity? First of all, notice that each v_i and v_i^{-1} appear the same number of times in w. This is a necessary condition. Do you see why? (Consider the homomorphism $\pi: G(\Gamma) \to \mathbb{Z}$ from Theorem 14.1, which is the left inverse to the inclusion $i: \langle v_i \rangle \to G(\Gamma)$.)

If you play around with the commuting relations, you will probably be able to show that w does represent the identity. Here is one such calculation:

$$
\begin{aligned}
w &= v_0 v_1^{-1} v_4 v_0^{-1} v_3 v_4^{-1} v_2 v_3^{-1} v_1 v_2^{-1} \\
&= v_1^{-1} v_4 v_0 v_0^{-1} v_4^{-1} v_2 v_3 v_3^{-1} v_1 v_2^{-1} && \text{as } v_0 \text{ commutes with } v_1 \text{ and } v_4 \text{ and} \\
& && v_3 \text{ commutes with } v_4 \text{ and } v_2 \\
&= v_1^{-1} v_4 v_4^{-1} v_2 v_1 v_2^{-1} && \text{canceling } v_0 v_0^{-1} \text{ and } v_3 v_3^{-1} \\
&= v_1^{-1} v_2 v_1 v_2^{-1} && \text{canceling } v_4 v_4^{-1} \\
&= v_1^{-1} v_1 v_2 v_2^{-1} && \text{as } v_1 \text{ commutes with } v_2 \\
&= 1 && \text{canceling } v_1^{-1} v_1 \text{ and } v_2 v_2^{-1}
\end{aligned}
$$

That was great, but what about other words? Well, for instance, we know that $v_0 v_2 v_0^{-1} v_2^{-1} \neq 1$ since $\langle v_0, v_2 \rangle$ is a free group of rank 2 and so v_0 and v_2 do not commute. But what about the word $w = v_0 v_1 v_2 v_3 v_0^{-1} v_1^{-1} v_2^{-1} v_3^{-1} \in G(C_5)$? Playing around with commuting relations, it seems like we cannot move the v_0 past the v_2 or v_3 term to cancel with the v_0^{-1}. Similarly with the other generators: it seems like there is always another generator blocking it from its inverse. This sure seems to suggest that this word is not the identity; but this is not a proof. How do we know that the calculation showing that $w = 1$ doesn't need to make w longer, replacing v_i by $v_{i+1} v_i v_{i+1}^{-1}$, for example?

Normal forms for right-angled Artin groups. We will describe a set of normal forms for an element in right-angled Artin groups that will solve the word problem for us. You have seen this idea before in Office Hour 8 when you solved the word problem for $\mathbb{Z}^2$; any element can be put into the normal form $a^m b^n$ for some integers m and n, and an element is trivial if and only if $m = n = 0$.

Back to right-angled Artin groups. Suppose that $w = x_1^{e_1} \cdots x_k^{e_k}$ is a word in the generators $x_i \in V(\Gamma)$, where $e_i \in \mathbb{Z}$. We call each $x_i^{e_i}$ a *syllable* of w. For example, $w = v_0^3 v_1^5 v_0^{-2} v_1^{-5} v_2^{-1}$ has five syllables: v_0^3, v_1^5, v_0^{-2}, v_1^{-5}, and v_2^{-1}.

We consider three moves:

1. Remove a syllable $x_i^{e_i}$ if $e_i = 0$.
2. If $x_i = x_{i+1}$, then replace consecutive syllables $x_i^{e_i} x_{i+1}^{e_{i+1}}$ by $x_i^{e_i + e_{i+1}}$.
3. If $[x_i, x_{i+1}] = 1$, then replace $x_i^{e_i} x_{i+1}^{e_{i+1}}$ with $x_{i+1}^{e_{i+1}} x_i^{e_i}$.

These three moves do not change the element in $G(\Gamma)$ represented by the word. Applying these moves to our example $w = v_0^3 v_1^5 v_0^{-2} v_2^{-1} \in G(C_5)$ we find:

$$v_0^3 v_1^5 v_0^{-2} v_1^{-5} v_2^{-1} \underset{3}{=} v_0^3 v_0^{-2} v_1^5 v_1^{-5} v_2^{-1} \underset{2}{=} v_0 v_1^0 v_2^{-1} \underset{1}{=} v_0 v_2^{-1}$$

For an element $g \in G(\Gamma)$, let $\mathrm{Min}(g)$ denote the set of words representing g that have the fewest number of syllables. So $w = 1$ if and only if $\mathrm{Min}(g) = \{1\}$. The following theorem is not explictly written down but is implict in the work of Baudisch [16], Green [148], and Hermiller–Meier [165].

THEOREM 14.4. *Any word representing $g \in G(\Gamma)$ can be transformed into any element of* $\mathrm{Min}(g)$ *by applying a sequence of the moves above. In particular, in any such sequence, the number of syllables and the length do not increase.*

Why does this solve the word problem? It shows us that given a word w representing g, we can effectively construct $\mathrm{Min}(g)$. The key point is that we only need to consider words with length less than or equal to the length of w as none of the three moves increases length. Since there are only finitely many words of length at most the length of w (can you write down how many there are?) we can find all the words in $\mathrm{Min}(g)$. To say it another way, starting with w, we can simply list all possible sequences of moves, as there are finitely many such sequences.

For example, using the three moves we can show that for the element represented by $w = v_0 v_1 v_2 v_3 v_0^{-1} v_1^{-1} v_2^{-1} v_3^{-1} \in G(C_5)$, w is a minimal length word representing it. Therefore $w \neq 1$.

Notice that this strategy is a direct generalization of the strategy used for free abelian groups.

Pilings. While we have this nice representative set $\mathrm{Min}(g)$, we still don't have our normal forms. There is a visual way to describe normal forms for elements of right-angled Artin groups called *pilings*. This builds on the observation we made earlier in our example for $G(C_5)$, where certain generators seem to be blocking certain other generators. This construction was introduced by Crisp, Godelle, and Wiest to study not only the word problem in right-angled Artin groups but also to study the conjugacy problem [96].[2]

---

[2] The conjugacy problem is the problem that takes two elements of a given group and decides whether or not they are conjugate. This is a generalization of the word problem, since an element is conjugate to the identity if and only if it is equal to the identity.

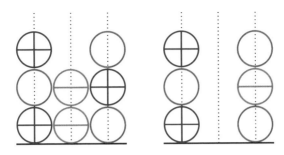

Figure 14.7 Two pilings: $(+0+, --, 0+0)$ and $(+0+, \varepsilon, 0-0)$.

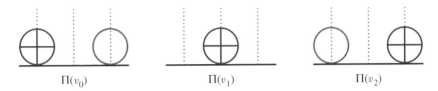

$\Pi(v_0)$ $\Pi(v_1)$ $\Pi(v_2)$

Figure 14.8 Pilings $\Pi(v_0)$, $\Pi(v_1)$, and $\Pi(v_2)$ in $G(P_3)$.

A piling is an n–tuple of finite strings using the symbols $\{+, -, 0\}$. We allow the empty string ε. For example, $(+0+, --, 0+0)$ and $(+0+, \varepsilon, 0-0)$ are pilings ($n = 3$). Pilings can be visualized using a diagram as in Figure 14.7. Each string is displayed vertically and is read from the bottom to the top, as if you were to drop a sequence of balls one on top of the the other: first a ball with a plus, then a ball with no symbol (zero), and finally another ball with a plus.

What do pilings have to do with elements in a right-angled Artin group $G(\Gamma)$? To each word w in the generators $V(\Gamma)$ we can associate a piling $\Pi(w)$ in the following manner. The pilings will be n–tuples where $n = \#|V(\Gamma)|$. They are defined recursively.

Naturally, $\Pi(1)$ is the empty piling $(\varepsilon, \dots, \varepsilon)$.

Next, the pilings $\Pi(v_i)$ have ith column $+$ and jth column 0 for all j where v_j is not adjacent to v_i. The remaining columns are ε. In Figure 14.8 we show the pilings for $\Pi(v_0)$, $\Pi(v_1)$, and $\Pi(v_2)$ for $G(P_3)$. The pilings $\Pi(v_i^{-1})$ are similar except that we use $-$ instead of $+$.

Now suppose that w is a word and that we have defined $\Pi(w)$ already. We will define the piling $\Pi(wv_i)$. If the ith column of $\Pi(w)$ is ε or has 0 or $+$ on top, then combine $\Pi(w)$ with $\Pi(v_i)$. That is, drop balls according to $\Pi(v_i)$ onto the piling $\Pi(w)$. Otherwise, the ith column has $-$ on top and we remove this ball and the top ball of the jth column for all j where v_j is not adjacent to v_i.

The piling $\Pi(wv_i^{-1})$ is defined similarly, interchanging the roles of $+$ and $-$.

In Figure 14.9 we show how to build the piling for $w = v_0 v_2^{-1} v_1 v_2 v_0 \in G(P_3)$.

THEOREM 14.5 (Crisp–Godelle–Wiest [96]). *A word w in the generators $V(\Gamma)$ represents the identity in $G(\Gamma)$ if and only if $\Pi(w) = (\varepsilon, \dots, \varepsilon)$. Moreover, two words w, w' in the generators $V(\Gamma)$ represent the same element in $G(\Gamma)$ if and only if $\Pi(w) = \Pi(w')$.*

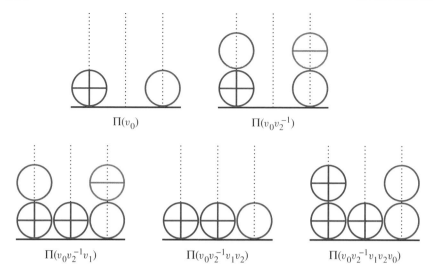

Figure 14.9 Constructing the piling $\Pi(v_0 v_2^{-1} v_1 v_2 v_0)$ in $G(P_3)$.

Exercise 18. Build the pilings for the words $v_0 v_1^{-1} v_4 v_0^{-1} v_3 v_4^{-1} v_2 v_3^{-1} v_1 v_2^{-1}$ and $v_0 v_1 v_2 v_3 v_0^{-1} v_1^{-1} v_2^{-1} v_3^{-1}$ in $G(C_5)$.

Exercise 19. Describe the piling for $\Pi(w^{-1})$ using $\Pi(w)$.

Exercise 20. Experiment with some examples and hypothesize how to construct $\mathrm{Min}(g)$ using $\Pi(w)$ where w is a word representing g. Can you prove your hypothesis?

Applications of normal forms. What do normal forms give us? They give us a canonical way to represent elements of right-angled Artin groups that can be used to understand their algebraic structure. We already explained how this allows us to solve the world problem. Here is a series of exercises that give other applications of normal forms.

Exercise 21. Prove that if $g \in G(\Gamma)$, $g \neq 1$, then g has infinite order. *Hint: Every element is conjugate to one that can be represented by a cyclically reduced word.*

Exercise 22. Suppose that H is a subgroup of $G(\Gamma)$ of index $n < \infty$. Let L be the subgroup of $G(\Gamma)$ generated by $\{v_i^N \mid v_i \in V(\Gamma)\}$, where $N = n!$. Prove that $L < H$ and $L \cong G(\Gamma)$.

Exercise 23. Fix a generator $v_i \in V(\Gamma)$ and $N \geq 2$ and define a homomorphism $\phi \colon G(\Gamma) \to \mathbb{Z}/N\mathbb{Z}$ by $\phi(v_i) = 1$ and $\phi(v_j) = 0$ if $j \neq i$. Show that $\ker \phi$ is a right-angled Artin group $G(\Gamma_\phi)$ and describe the graph Γ_ϕ. You might start by looking at the examples P_n and C_n for small values of n.

Exercise 24. Show that if $n \geq 4$, then $G(P_n)$ is isomorphic to a subgroup of $G(P_4)$. *Hint: Show that the map $G(P_5) \to G(P_4)$ defined by*

$$v_0 \mapsto v_0$$
$$v_1 \mapsto v_1$$
$$v_2 \mapsto v_2$$
$$v_3 \mapsto v_3 v_1 v_3^{-1}$$
$$v_4 \mapsto v_3 v_0 v_3^{-1}$$

is injective.

FURTHER READING

Many of the articles listed as references in this office hour can be read and understood with some persistence and a willingness to seek out help from your peers and mentors. In particular, both the article by Crisp and Wiest [97] and the article by Kim [184] fall into this category. These articles will further introduce you to a fascinating research problem: which right-angled Artin groups contain hyperbolic surface subgroups?

You can also read more about solutions to the word and conjugacy problems in right-angled Artin groups. Elisabeth Green's Ph.D. thesis [148] gives a very careful solution to the word problem and can serve as a good introduction to some of the foundational concepts in combinatorial group theory. Crisp, Godelle, and Wiest's article [96] is much more challenging to read, but it is here that the visually appealing pilings are introduced.

The interplay between the geometry of cube complexes and the structure of right-angled Artin groups is very rich. A fun place to look for inspiration is in first few chapters of Dani Wise's book [262]. But, this is strictly for inspirational reading: to work through this book in earnest is a task for someone with graduate level training in geometric group theory.

We also mention the survey article by Charney [76]. This has a much more advanced tone than this office hour, but it's the place to look if you want to get a better sense of how right-angled Artin groups arise naturally in various contexts. You will need to be willing to skip around the various sections since some (e.g., K$(\pi, 1)$ conjecture) assume training in algebraic topology.

As we alluded to at the beginning, right-angled Artin groups play an important role in the recent resolution of an important open question in the topology of three-dimensional manifolds. As in Office Hour 9, we have known the complete list of surfaces, or two-dimensional manifolds, for more than 100 years. Of course the world of three-dimensional manifolds is much more complicated. In 2003, Grigori Perelman resolved the famous Poincaré conjecture (every simply connected three-dimensional manifold is homeomorphic to the 3–sphere). At the same time he proved something much more general, conjectured by Thurston: three-dimensional manifolds can be broken up along spheres and tori and the remaining pieces are either Seifert fibered spaces, which are totally classified, and hyperbolic manifolds, which form sort of a jungle. The virtual Haken conjecture, which was made by Waldhausen in 1968 and was proven in 2012 by Dani Wise and Ian Agol, gives structure to the set of hyperbolic three-dimensional manifolds. Specifically, they all come from the following construction: start with a surface, cross it with an interval, and glue the ends together. The most amazing thing is that the proof goes through geometric group theory, and the theory of right-angled Artin groups in particular. Read Mladen Bestvina's excellent survey on the subject to see how this works [28].

PROJECTS

Project 1. Show that $\phi: \pi_1(\Sigma_2) \to G(C_6)$ is injective (see Exercise 16). You can work through Crisp and Wiest's article [97] to put together a solution (part of which they have left as an exercise), but first try tackling this problem on your own. You might want to look at Office Hour 17, in particular the concept of a bigon. Crisp and Wiest give a geometric solution, but it would be interesting to see an algebraic solution.

Project 2. Write a proof of Theorem 14.4 and the solution to the word problem by following this outline. Let $G(\Gamma)$ be a right-angled Artin group with geneerating set S. Let $L = S \cup S^{-1}$. The set of words over S is denoted by L^*. Elements of L are denoted by a or b; elements of L^* are denoted by U or V. A reduction is a move that carries a word containing aa^{-1} to the word that results from the deletion of aa^{-1}. A shuffle is a move that carries a word containing ab to the word obtained by interchanging these letters, provided the corresponding letters in S are adjacent in Γ. After finitely many reduction and shuffle moves, the length of a given word will reach a minimum and we say such a word is reduced. The claim is that two reduced words represent the same element of $G(\Gamma)$ if and only if there is a sequence of shuffles carrying one to the other. Recursively define a function $\pi : L^* \to L^*$ that maps each word to a reduced word as follows: $\pi(1) = 1$, $\pi(a) = a$, and $\pi(Ua) = \pi(U)a$ if there is no sequence of shuffles resulting in a being deleted by a reduction move. The remaining case in the definition of π occurs precisely when $Ua = Va^{-1}W$ for some $W \in L^*$ each of whose corresponding letters in S are adjacent to that of a. In this case, define $\pi(Va^{-1}Wa) = \pi(VW)$.

1. Prove $\pi(U)$ is reduced, $\pi(\pi(U)) = U$, and $\pi(UV) = \pi(\pi(U)V)$.
2. Prove that if U is obtained from V by inserting aa^{-1}, then $\pi(U)$ and $\pi(V)$ are related by a sequence of shuffles.
3. Suppose that a and b have adjacent corresponding letters in S. Prove that $\pi(UabV)$ can be shuffled to $\pi(UbaV)$.
4. Prove that U and $\pi(U)$ define the same element of $G(\Gamma)$. Prove the same is true for $\pi(U)\pi(V)$ and $\pi(UV)$.
5. Show $\pi(U)$ and $\pi(V)$ define the same element in $G(\Gamma)$ if and only if they are related by shuffles. The stated solution to the word problem follows from this.

Project 3. Learn how to solve the conjugacy problem in right-angled Artin groups using pilings. When are two elements of $G(C_5)$ conjugate?

Project 4. Learn about graph braid groups (see [97]) and their relation to right-angled Artin groups and braid groups. In this same article, you will learn about one approach to proving Theorem 14.4 on the solution to the word problem. Crisp and Wiest indicate that their approach can be extended to graph products of groups. Can you carry out this extension?

Project 5. A research problem is to characterize which right-angled Artin groups are subgroups of a fixed right-angled Artin group. Ideally, such a characterization could be stated in terms of the defining graphs. Read Kim and Koberda's article [185] for an introduction to this problem.

Project 6. A right-angled Coxeter group $W(\Gamma)$ is the quotient of $G(\Gamma)$ obtained by adding a relation $s^2 = 1$ for each defining generator $s \in S$. As stated in Theorem 14.3, each right-angled Artin group is a finite index subgroup of a right-angled Coxeter group. A much more challenging problem is to determine when a right-angled Artin group is quasi-isometric (see Office Hour 7) to a right-angled Coxeter group. Show that $W(C_n)$ is quasi-isometric to a right-angled Artin group if and only if $n = 4$. Can you identify other very specialized families of graphs in which this problem has an easy to state solution?

Project 7. Write a computer program that takes as input a defining graph Γ and a word ω over the generating set and that returns as output a normal form. Refer to Hermiller and Meier's article [165] for how you can resolve the ambiguity of a word possibly having several distinct normal forms.

Project 8. As in the previous project, write a computer program that returns a normal form, but this time your program should show the step-by-step process in terms of pilings. You could also make this more interactive, and let the user choose which moves to perform on the pilings.

Project 9. What kinds of automorphisms does a right-angled Artin group admit? See Office Hour 6 where the extreme cases $\mathbb{Z}^n$ and F_n are discussed. When you think you've identified a collections of automorphisms that seem to generate the entire automorphism group, look at Herman Servatius's article [238]. Then use MathSciNet to learn about how this problem was ultimately solved by Laurence. There has been a lot of recent interest in automorphisms of right-angled Artin groups with the hope of developing aspects of the theory of automorphisms of free groups in a more general context. Look for articles by R. Charney, M. Day, and K. Vogtmann, among others.

Project 10. Can you give a direct proof that right-angled Artin groups are residually finite? See Office Hour 4, where this is shown for free groups. A group G is residually finite if each nontrivial element g has a nontrivial image in a finite quotient of G. That right-angled Artin groups are residually finite follows from Theorem 14.3 together with the fact that Coxeter groups are linear groups and that finitely generated linear groups are residually finite.

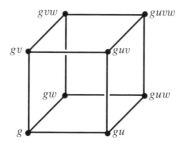

Figure 14.10 A three-dimensional cube.

Project 11. Let σ_r be the number of elements of $G(\Gamma)$ whose length is exactly r. Let

$$S(z) = \sum_{r=0}^{\infty} \sigma_r z^r$$

be a formal power series, called the spherical growth series of $G(\Gamma)$. Determine spherical growth series for various right-angled Artin groups. For example, prove that $G(P_1)$ has $S(z) = 1 + \sum_{r=1}^{\infty} 2z^r = \frac{1+z}{1-z}$ and $G(P_2)$ has $S(z) = \frac{(1+z)^2}{(1-z)^2}$. What is the growth series of $G(P_n)$? For much more on growth functions, refer to Office Hour 12.

Project 12. There is a nice geometric space associated to a right-angled Artin group. The group acts on the space freely, properly, and cocompactly by isometries, and so by the Milnor–Schwarz lemma (Office Hour 7) the space is quasi-isometric to the group. Here is the construction. Let $G = G(\Gamma)$ and let $C^{(1)} = \Gamma(G, S)$ be the Cayley graph of G with respect to the generating set S. The Cayley graph can be filled in by squares in the following way: for each $g \in G$ and each pair of distinct commuting generators $v, w \in S$, glue in a square along the 4–cycle consisting of the edges (g, gv), (gv, gvw), (g, gw), and (gw, gwv), where $gvw = gwv$. This is just like what was done in Office Hour 8 and is called the Cayley 2–complex.

At this point, the space is quasi-isometric to G, but for a right-angled Artin group $G(\Gamma)$, more can be filled in as well and makes the space look more like our familar Euclidean space $\mathbb{R}^n$.

For each $g \in G$ and each triple of distinct mutually commuting generators $u, v, w \in S$, a cube is added along the squares as in Figure 14.10. The resulting space is denoted by $C^{(3)}$.

Inductively, one can attach k–cubes, by which we mean copies of the space $[-1, 1]^k \subset \mathbb{R}^k$, one for each g and each collection of k distinct mutually commuting generators in S. When no more cells of higher dimension can be attached, the resulting space will denoted simply by $C = C(\Gamma)$. The dimension of C is the largest $k \geq 0$ such that C has a k–cube. In terms of the graph Γ, what is the dimension of $C(\Gamma)$?

Let $T \subset S$ be a subset of mutually commuting generators and let G_T be the subgroup generated by T. Prove that the vertices of C which belong to a left coset

gG_T are the vertices of an isometrically embedded copy of $|T|$–dimensional Euclidean space.

Project 13. This quotient of the space described in the previous project is called the Salvetti complex. It has an alternative description in terms of k–dimensional tori:

$$\mathbb{T}^k = \underbrace{S^1 \times \cdots \times S^1}_{k \text{ times}}.$$

Specifically, we start with a disjoint collection of circles $S^1_{v_1} \coprod \cdots \coprod S^1_{v_n}$ labeled by the vertices of Γ and glue them altogether at one point. Call this space X. Next, to every complete subgraph $\Delta \subseteq \Gamma$ with vertices $v_{i_1}, \ldots, v_{i_k}$ we attach $\mathbb{T}^k$ by identifying the sets:

$$\bigcup_{j=1}^{k} S^1_{v_{i_j}} \subseteq X \longleftrightarrow \bigcup_{j=1}^{k} \{0\} \times \cdots \times \underbrace{S^1}_{jth \text{ coordinate}} \times \cdots \times \{0\} \subset \mathbb{T}^k.$$

Reconcile these two descriptions.

Office Hour Fifteen

Lamplighter Groups

Jennifer Taback

Imagine a quaint rural town, with an infinite main street lined with evenly spaced lampposts. A lamplighter walks up and down the street, lighting some of the light bulbs, then ends his walk leaning against one of the lampposts.

If you can envision that, then you understand what an element of the lamplighter group L_2 looks like: each element corresponds to a finite set of illuminated bulbs, and a final position, or displacement, of the lamplighter.

To make this mathematically precise, we envision the infinite main street as a copy of the integer points on the number line, with a bulb placed at every integer. An element of the lamplighter group L_2 consists of two parts:

1. a finite set of integers, corresponding to the bulbs illuminated by the lamplighter, and
2. a single integer denoting the position of the lamplighter.

So $(\{-3, -1, 5, 6, 18\}, 44)$ would correspond to bulbs illuminated in the lamps placed at the integers $-3, -1, 5, 6,$ and 18, with the lamplighter finishing his walk at the integer 44. This interpretation of an element of the lamplighter group L_2 was originally given by Jim Cannon, and is called the lamplighter picture associated to the element. The infinite string of light bulbs placed at integer points is called the lampstand.

In this office hour, I'll first describe several ways of understanding a single element of the lamplighter group L_2, which will lead us to a presentation for the entire group. This is not the way one usually meets a new group, but elements of L_2 are very special!

Figure 15.1 A whimsical picture of an element of the lamplighter group L_2 with bulbs illuminated (that is, darkened) in positions -6, -1, 4, 5, and 6 and the lamplighter at position -2. The vertical bar denotes $0 \in \mathbb{Z}$ and the arrow represents the position of the lamplighter. Bulbs not drawn in the figure are understood not to be illuminated.

Figure 15.2 The standard representation of the element of the lamplighter group given in Figure 15.1, with bulbs illuminated (that is, darkened) in positions -6, -1, 4, 5, and 6, and the lamplighter at position $-2 \in \mathbb{Z}$. The vertical bar denotes $0 \in \mathbb{Z}$ and the arrow represents the position of the lamplighter. Bulbs not drawn in the figure are understood to be not illuminated.

A mathematical view of the group elements. There is another way to think of an element of the lamplighter group L_2, which better reflects the fact that any given light bulb is either on or off. The cyclic group $\mathbb{Z}/2\mathbb{Z}$ has two elements: 0 and 1. We use 0 to denote a bulb that is not illuminated, and 1 to denote a bulb that is. Instead of listing a sequence of integers, such as $\{-3, -1, 5, 6, 18\}$ in the above example, where the bulbs are illuminated, we represent the same information with an element of the group denoted by $\bigoplus_{i=-\infty}^{\infty} (\mathbb{Z}/2\mathbb{Z})_i$; you read this as: the infinite direct sum of copies of $\mathbb{Z}/2\mathbb{Z}$. Elements of this group are infinite-tuples, whose entries are indexed by integers, and each entry is an element of $\mathbb{Z}/2\mathbb{Z}$ (i.e., either 0 or 1) with the condition that there can be only finitely many nonzero entries. As the lamplighter can only light a finite number of bulbs, we use the element 1 in the copies of $\mathbb{Z}/2\mathbb{Z}$ indexed by the positions of the illuminated bulbs. So in our example, we would have nonzero entries only in the copies of $\mathbb{Z}/2\mathbb{Z}$ in $\bigoplus_{i=-\infty}^{\infty} (\mathbb{Z}/2\mathbb{Z})_i$ indexed by -3, -1, 5, 6, and 18.

If you now think that the lamplighter group L_2 is made from a combination of $\mathbb{Z}$ and this new (large) group $\bigoplus_{i=-\infty}^{\infty} (\mathbb{Z}/2\mathbb{Z})_i$, you would be correct. We use an element of $\mathbb{Z}$ to denote the position of the lamplighter, and an element of $\bigoplus_{i=-\infty}^{\infty} (\mathbb{Z}/2\mathbb{Z})_i$ to encode which bulbs are illuminated. We will see the precise mathematical construction used here later in this office hour. First we should arrive at a presentation for this group.

15.1 GENERATORS AND RELATORS

Our first major goal is to give a presentation for the lamplighter group L_2. We will start by discussing generators and then proceed to relations.

Group generators. To create an element of the group, there are two basic "actions" that we will take as our generators. Namely, the lamplighter must be able to move one unit in either direction along the integers, and the lamplighter must be able to change the state of a bulb in the position he is at, that is, turn it on if it is off, or vice versa.

1. Define a generator t of L_2 that corresponds to changing the position of the lamplighter by moving him one unit to the right along the integers (so that t^{-1} moves him one unit to the left).

2. Define a generator a of L_2 that corresponds to changing the state of the bulb at the current position of the lamplighter. Since a bulb is mathematically modeled by the group $\mathbb{Z}/2\mathbb{Z}$, we can think of a as the generator 1 of $\mathbb{Z}/2\mathbb{Z}$, and then addition of 1 and reduction modulo 2 always changes the state of the bulb.

Let's see how a string of generators constructs the lamplighter picture of an element by acting on the lampstand. Let $h = t^2 a t^4 a$, and the lamplighter begins at position $0 \in \mathbb{Z}$, with no bulbs illuminated on the lampstand. Think of the set of instructions corresponding to each generator in h.

1. First, t^2 moves the lamplighter two units to the right, namely to the integer 2.
2. Then, a changes the state of the bulb in position 2, that is, turns on that bulb.
3. Next, t^4 moves the lamplighter to position $6 \in \mathbb{Z}$. Finally, a changes the state of the bulb in position 6, turning it on.

Now we have constructed the lamplighter picture for h, with illuminated bulbs in positions 2 and 6, and the lamplighter at position 6.

Group multiplication. To multiply two group elements, we again use the action of the generators a and t described above on the lampstand. For example, consider the element g in Figure 15.2. Suppose we want to multiply g by the element $h = t^2 a t^4 a$. Beginning with the lamplighter picture for g, follow the instructions corresponding to the generators in h, as listed above—only now the lamplighter begins in position -2, as in Figure 15.2. Thus the product gh has illuminated bulbs in positions $-6, -1, 0, 5$, and 6, with the lamplighter at the integer 4.

An equivalent way to construct this product is to first construct the lamplighter pictures for g and h, and stack these diagrams so that the origin of the lampstand in the picture for h is placed at the final position of the lamplighter in the picture for g. In the lamplighter picture for the product gh, add the entries in overlapping bulbs, reducing mod 2. In this way it is clear that the positions of the lamplighter in g and h are added together when these group elements are multiplied.

A simple example: We can immediately see the change in the lamplighter picture for g after multiplication by either generator. To obtain the lamplighter picture for ga from that for g, we simply change the state of the bulb at the position of the lamplighter. To obtain the lamplighter picture for gt from that for g, we increase the position of the lamplighter by 1.

Exercise 1. Above, we described how to multiply two elements of L_2 using the interpretation of elements as strings of illuminated bulbs, along with a position of the lamplighter. Describe how to multiply group elements when they are represented as a pair $(k, \mathbf{x})$ where $k \in \mathbb{Z}$ and $\mathbf{x} \in \bigoplus(\mathbb{Z}/2\mathbb{Z})_i$, that is, give a formula for the product $(k, \mathbf{x}) \cdot (\ell, \mathbf{y})$.

Group presentation. Clearly the generators given above can be combined to create any particular configuration of illuminated bulbs and position of the lamplighter. However, to give a presentation for the group L_2 we must provide the relations as well. An immediate relation is that a^2 equals the identity in $\mathbb{Z}/2\mathbb{Z}$, as a is the generator of $\mathbb{Z}/2\mathbb{Z}$. In the lamplighter picture of an element, this means that if the lamplighter changes the state of the same bulb twice, it reverts to its original state, whether on or off.

The presentation is completed with an infinite collection of relations, which one can view as follows: imagine the lamplighter beginning at the origin (that is, $0 \in \mathbb{Z}$) with no bulbs illuminated, moving k units to the right, illuminating the bulb in position $k \in \mathbb{Z}$, and moving back to the origin of $\mathbb{Z}$. Next, the lamplighter moves from the origin to position $j \neq k \in \mathbb{Z}$, illuminates the bulb in that position, and walks back to the origin. If you think about the final picture, with two bulbs illuminated in positions k and j, and the lamplighter at the origin, you will realize that it doesn't matter in which order the lamplighter makes these two separate walks, as the end result is the same. But this is exactly what it means for $t^k a t^{-k}$ and $t^j a t^{-j}$ to commute, which we write as $[t^k a t^{-k}, t^j a t^{-j}] = 1$. (Recall that $[a, b]$ denotes $aba^{-1}b^{-1}$ and is called the commutator of a and b.) As this is true for any $k \neq j \in \mathbb{Z}$ (and trivially true for $j = k$), we have described an infinite collection of relators for this group. As always, products and conjugates of these relators are again equal to the identity in the group. Note that in this group the identity element has a lamplighter picture in which the lamplighter is at position $0 \in \mathbb{Z}$ and no bulbs are illuminated on the lampstand.

There is no way to present this group using only a finite number of these relations (see [18]); hence L_2 is an example of an infinitely presented group that has no finite presentation.

Combining these two types of relations, we obtain the presentation

$$L_2 = \langle a, t \mid a^2, \ [t^k a t^{-k}, t^j a t^{-j}] \text{ for all } j, k \in \mathbb{Z} \rangle.$$

In the notation L_2, the index 2 is used to indicate that the light bulbs are modeled by the group $\mathbb{Z}/2\mathbb{Z}$, that is, they are either on or off.

L_2 as a group of matrices. We will now give a different point of view on the lamplighter group—we will describe it as a group of matrices! We will also use this (independently interesting) idea to describe a Cayley graph for the lamplighter group later on.

We have seen already that an element of L_2 corresponds uniquely to a pair $h \in \bigoplus_{i=-\infty}^{\infty}(\mathbb{Z}/2\mathbb{Z})_i$ and $k \in \mathbb{Z}$. We can encode the information given by the element h as a polynomial P in the variables t and t^{-1} with coefficients in $\mathbb{Z}/2\mathbb{Z}$, where the only nonzero coefficients correspond to the only nonzero entries in h. That is, we use the entries of h as the coefficients in our polynomial. So the element $(\ldots, 0, 0, 1, 0, 1, 1, 1, 0, 0, \ldots)$, where the first 1 in the tuple corresponds to the copy of $\mathbb{Z}/2\mathbb{Z}$ indexed by -2 and the ... denotes an infinite string of zeros, yields the polynomial $t^{-2} + 1 + t + t^2$.

Now we can think of an element of L_2 as a pair (P, k), where $P = \sum_{i \in \mathbb{Z}} c_i t^i$ is a polynomial in the variables t and t^{-1} with only finitely many nonzero coefficients, all $c_i \in \mathbb{Z}/2\mathbb{Z}$, and k is an integer. We will encode this information in a matrix, using t as a formal variable:

$g \in L_2$ corresponds to the pair (P, k) which corresponds to the matrix

$$\begin{pmatrix} t^k & P \\ 0 & 1 \end{pmatrix}.$$

One can prove the following fact:

> *The lamplighter group L_2 is isomorphic to the group of matrices of the form $\begin{pmatrix} t^k & P \\ 0 & 1 \end{pmatrix}$, where $k \in \mathbb{Z}$ and P ranges over all polynomials in t and t^{-1} with finitely many nonzero coefficients in $\mathbb{Z}/2\mathbb{Z}$.*

The generator t of L_2 is identified with the matrix $\begin{pmatrix} t & 0 \\ 0 & 1 \end{pmatrix}$ and a with $\begin{pmatrix} 1 & 1 \\ 0 & 1 \end{pmatrix}$.

Exercise 2. Write down the identification between L_2 and this group of matrices and prove it is an isomorophism.

Let's understand how multiplying by $\begin{pmatrix} t & 0 \\ 0 & 1 \end{pmatrix}$ (corresponding to the generator t) or by $\begin{pmatrix} 1 & 1 \\ 0 & 1 \end{pmatrix}$ (corresponding to a) changes an arbitrary $g = \begin{pmatrix} t^k & P \\ 0 & 1 \end{pmatrix}$ in L_2.

Suppose $g = \begin{pmatrix} t^k & P \\ 0 & 1 \end{pmatrix}$ represents some element of L_2, with the lamplighter at position k and the coefficients of the polynomial P describing a set of illuminated bulbs. In the lamplighter picture of this element, we know that multiplication by the generator t simply moves the lamplighter one unit to the right without changing any illuminated bulbs. Notice what happens if we multiply the matrix for g by the matrix corresponding to the generator t:

$$\underbrace{\begin{pmatrix} t^k & P \\ 0 & 1 \end{pmatrix}}_{g} \underbrace{\begin{pmatrix} t & 0 \\ 0 & 1 \end{pmatrix}}_{t} = \underbrace{\begin{pmatrix} t^{k+1} & P \\ 0 & 1 \end{pmatrix}}_{gt}.$$

The resulting matrix has the same collection of illuminated bulbs, since P has not changed, and the exponent of t in the upper left entry has increased by 1—the lamplighter is now at position $k + 1$. This is exactly what we expected to happen.

We know that multiplication by the generator a alters the lamplighter picture of $g \in L_2$ by changing the state of the bulb at the position of the lamplighter. So if we begin with the element $\begin{pmatrix} t^k & P \\ 0 & 1 \end{pmatrix}$, in which the lamplighter is at position k, and multiply by $\begin{pmatrix} 1 & 1 \\ 0 & 1 \end{pmatrix}$ (corresponding to the generator a), we expect only the lamp in position k to change state. In terms of the polynomial P, this multiplication should only change the coefficient of the t^k term, to which we should add the generator 1 of $\mathbb{Z}/2\mathbb{Z}$. This has the effect of turning the bulb off, if it was already on, or vice versa.

Let us verify that this matrix multiplication has the desired effect:

$$\underbrace{\begin{pmatrix} t^k & P \\ 0 & 1 \end{pmatrix}}_{g} \underbrace{\begin{pmatrix} 1 & 1 \\ 0 & 1 \end{pmatrix}}_{a} = \underbrace{\begin{pmatrix} t^k & t^k + P \\ 0 & 1 \end{pmatrix}}_{ga}.$$

Thus we add 1 to the coefficient of t^k in P and reduce modulo 2, as required.

A family of groups. The above presentation of the lamplighter group L_2 easily generalizes to an infinite family of groups. Imagine that the street lamps had their light bulbs replaced with three-way bulbs, that is, bulbs which can be turned on low, high, or turned off. Such a bulb can be modeled using the elements of $\mathbb{Z}_3$ under addition modulo 3, as adding the generator 1 of $\mathbb{Z}_3$ changes the state of the light bulb accordingly. If the light bulb is off, then adding 1 turns it to low, which we denote by 1; if the bulb is on low, adding 1 turns it to high, which we denote by 2; adding 1 again (modulo 3) turns the bulb to 0, that is, off. Then the set of all possible configurations of illuminated bulbs is given by $\bigoplus_{i=-\infty}^{\infty} (\mathbb{Z}_3)_i$, and an element of this new lamplighter group, which we call L_3, is uniquely determined by an element of $\bigoplus_{i=-\infty}^{\infty} (\mathbb{Z}_3)_i$ along with an integer representing the position of the lamplighter.

This model can be extended to bulbs with n states, which we represent by $\mathbb{Z}/n\mathbb{Z}$. This yields a family of lamplighter groups L_n with presentation

$$L_n = \langle a, t \mid a^n = 1, \ [t^k a t^{-k}, t^j a t^{-j}] = 1 \text{ for all } j, k \in \mathbb{Z} \rangle.$$

For the remainder of the office hour, we will study L_2 in detail, with some analogous results stated for the entire family of groups. You will see that these groups have interesting Cayley graphs (with respect to the correct generating set). Lamplighter groups have connections to finite state automata and automata groups, random walks, language-theoretic problems, graphs and other topological problems. Something which appears in so many different contexts must surely be interesting! We begin by understanding these groups algebraically.

15.2 COMPUTING WORD LENGTH

The next big goal is to give a formula for the word lengths of elements of the lamplighter group L_2 with respect to the generating set $\{t, a\}$. Remember that the word length of an element is the minimum number of generators we need to multiply to obtain that element. We can think of such a minimal-length word in L_2 as describing an efficient way for the lamplighter to create a particular configuration of the lampstand with the lamplighter ending at a designated position.

Efficient descriptions of group elements. Suppose $g \in L_2$ has illuminated bulbs in positions $-3, 4$, and 5, and the lamplighter is at position $2 \in \mathbb{Z}$. Without thinking too hard, you might come up with the following way of constructing this element using the generating set $\{t, a\}$ acting on the lampstand:

- Begin with the lamplighter at $0 \in \mathbb{Z}$ and no bulbs illuminated on the lamp-stand.
- Move the lamplighter to position $4 \in \mathbb{Z}$ and turn on the light bulb.
- Move forward one unit to $5 \in \mathbb{Z}$ and turn on that light bulb.
- Move back to $-3 \in \mathbb{Z}$ and turn on that light bulb.
- Finally, move the lamplighter to position $2 \in \mathbb{Z}$, where he remains.

It is intuitive that it is inefficient to first move the lamplighter to position 5, turn on the bulb, then to position -3, turn on the bulb, and finally back to position 4 to turn on the remaining bulb before moving to the final position 2, even though this would create the correct picture of the element.

An efficient path of the lamplighter analogous to the example above can easily be constructed for any group element according to the following rules:

(1) Have the lamplighter move to the smallest nonnegative index where a bulb must be illuminated and turn on that bulb.
(2) Now move to the right, illuminating the desired bulbs as you come to them.
(3) When all desired bulbs with nonnegative indices are illuminated, have the lamplighter move back to position 0.
(4) Now have the lamplighter move to the left, illuminating the desired bulbs as he comes to them.
(5) When all desired bulbs with negative indices are illuminated, have the lamplighter move to his final position.

We denote a path constructed for a group element g according to rules (1)–(5) above by $\gamma(g)$, and below introduce some notation to describe it. There is an analogous efficient path which first illuminates bulbs with negative indices. While these paths may not be geodesic (that is, of minimal length) either can be used to compute the word length of g with respect to the generating set $\{t, a\}$, as shown below.

Exercise 3. In the efficient path $\gamma(g)$ described above for $g \in L_2$, notice that the bulbs with nonnegative indices are illuminated first.

(a) Write down instructions for an analogous efficient path $\gamma'(g)$ in which bulbs with negative indices are illuminated first.
(b) Describe conditions under which both $\gamma(g)$ and $\gamma'(g)$ are geodesic, that is, of minimal length.
(c) Find conditions under which only $\gamma(g)$ is geodesic.

Writing down the efficient paths. Let $a_k = t^k a t^{-k}$ denote the conjugate of a by t^k. Beginning with the bi-infinite string of light bulbs which are all turned off and the lamplighter at $0 \in \mathbb{Z}$, we see that a_k moves the lamplighter to the bulb in position k, turns it on, and returns the lamplighter to the origin. It is clear that the a_k all commute, as discussed above. Occurrences of a_k in a word $w = \Pi_{i=1}^{m} a_i$ cancel in pairs, corresponding to turning on a bulb, and later turning it off.

Notice that when two of these conjugates are multiplied together, with indices in increasing order, say a_2a_7, the cancellation between adjacent powers of t ensures that the lamplighter moves to position 2, illuminates the bulb, then moves to position 7, illuminates the bulb, and finally moves back to the origin:

$$a_2a_7 = t^2at^{-2}t^7at^{-7} = t^2at^5at^{-7}.$$

Thus, if a group element g has illuminated bulbs in nonnegative positions $i_1 < i_2 \cdots < i_k$ and negative positions $-j_1 > -j_2 > \cdots > -j_l$ (with $j_l > 0$) with the final position of the lamplighter at the integer m, the path $\gamma(g)$ can be written as

$$\gamma(g) = a_{i_1}a_{i_2} \ldots a_{i_k}a_{-j_1}a_{-j_2} \ldots a_{-j_l}t^m.$$

To find the analogous efficient paths describing elements of the lamplighter groups L_n for $n \geq 3$, we simply allow powers of the generator a in the above expression for $\gamma(g)$.

Exercise 4. Examples of efficient paths. First draw the lamplighter pictures for two elements $g, h \in L_2$. Then write down the expressions $\gamma(g)$ and $\gamma(h)$.

Computing word length of group elements. There is a straightforward algorithm to compute the word length of $g \in L_2$ with respect to the generating set $\{t, a\}$ from either type of efficient path described above. Given $g \in L_2$, first write down the path $\gamma(g)$ using rules (1)–(5) and the notation $a_k = t^kat^{-k}$. Then compute the numerical quantity that we call $D(g)$ from this path. If $\gamma(g) = a_{i_1}a_{i_2} \ldots a_{i_k}a_{-j_1}a_{-j_2} \ldots a_{-j_l}t^m$, then

$$D(g) = k + l + \min\{2j_l + i_k + |m - i_k|, 2i_k + j_l + |m + j_l|\}.$$

If you look closely at the indices in the formula, you will see that geometrically, $D(g)$ is the sum of several quantities related to the lamplighter picture of g:

1. the number of bulbs that are illuminated,
2. twice the distance of the furthest illuminated bulb from the origin in one direction (that is, the direction that minimizes the two quantities in $D(g)$),
3. the distance of the furthest illuminated bulb from the origin in the other direction, and
4. the distance of the lamplighter from the furthest illuminated bulb in the second direction.

Sean Cleary and I proved [83] that $D(g)$ is exactly the word length of g with respect to the generating set $\{t, a\}$!

Exercise 5. Prove that $D(g)$ is exactly the word length of $g \in L_2$ with respect to the generating set $\{t, a\}$ in two steps:

(a) Show that any word in the generators t and a representing g has length at least $D(g)$.
(b) Show that there is a word in the generators t and a representing g with length exactly $D(g)$.

Hint: Think about why taking the minimum of the two quantities in the expression for $D(g)$ is important.

Example: Efficient paths and word length.   Using the group element given in Figure 15.2, we write the efficient path $\gamma(g) = a_4 a_5 a_6 a_{-1} a_{-6} t^{-2}$ and compute the word length with respect to $\{t, a\}$ as

$$D(g) = 3 + 2 + \min\{18 + |-2 - 6|, 18 + |-2 + 6|\} = 27.$$

Exercise 6. Compute the word length in L_2 using the generating set $\{t, a\}$ of

1. $g = a_3 a_5 a_6 a_8 a_{-1} a_{-2} a_{-7} t^3$
2. $g = a_{10} a_{15} a_{25} a_{-100} t^{17}$

15.3 DEAD END ELEMENTS

Let us now summarize what we know thus far about the lamplighter group L_2:

1. We understand the lamplighter picture associated to each element.
2. We know what the generators and relations are for the group presentation, and how multiplication by a generator alters the lamplighter picture of a specific element.
3. We understand an efficient path $\gamma(g)$ (although not always of minimal length) for the lamplighter to follow to construct each group element $g \in L_2$, how this path can be written as a sequence of generators, and how reading each generator in the sequence alters the lamplighter picture.
4. Using this efficient path we can compute the word length of an element with respect to the generating set $\{t, a\}$.

This is already quite a lot of information! In this section and the next, we consider some geometric properties of the Cayley graph of L_2 with respect to the generating set $\{t, a\}$ that can be deduced from the above four pieces of information.

Bumping into walls.   Suppose you were out taking a walk away from your house, and suddenly you reached a point where all paths took you closer to home. Perhaps your path runs into a big stone wall, and to get around the wall you first need to backtrack a bit.

Now imagine taking a walk in a Cayley graph, where you walk along the edges of the graph away from the identity vertex. Suppose that each edge takes you one step further away from the identity. But then you come to a vertex where all edges either take you closer to the identity, or maintain your distance from the identity. Can that happen? Yes it can, and such a vertex is called a dead end element because any geodesic path that reaches that vertex cannot be extended and remain geodesic.

Example: A dead end element in $\mathbb{Z}$.   Since the idea of a dead end element can be a little counterintuitive the first time you encounter it, here is an example of a dead end element in the Cayley graph of $\mathbb{Z}$ with respect to the generating set $\{2, 3\}$; see Figure 11.4. This example is not related to lamplighter groups, but it will illustrate what a dead end element is.

We will use $d(a, b)$ to denote the distance in this Cayley graph between two integers a and b. Notice that $d(0, 2) = d(0, 3) = 1$ and $d(0, 1) = 2$ since $1 = 3 - 2$. We will show that 1 is a dead end element in this Cayley graph, that is, all edges emanating from 1 lead to vertices which are at most a distance 2 from the origin 0. We check all four such edges, which terminate at the vertices 3, −1, 4, and −2.

1. $d(0, 3) = 1$ since 3 is a generator.
2. $d(0, −1) = 2$ since $−1 = 2 − 3$, and 1 is not a generator.
3. $d(0, 4) = 2$ since $4 = 2 + 2$ and 4 is not a generator.
4. $d(0, −2) = 1$ since 2 is a generator.

Thus $1 \in \mathbb{Z}$ is connected by a single edge only to vertices at distance at most 2 from the identity in $\Gamma(\mathbb{Z}, \{2, 3\})$, and hence is a dead end element in this Cayley graph.

Exercise 7. Consider different possible generating sets for your favorite finite group, for example, the dihedral group of order 16, and identify the dead end elements with respect to each generating set.

Backtracking from dead end elements. A good analogy for dead end elements in Cayley graphs is to think about an incorrect path through a maze. A dead end element is analogous to the end of such a path, where you cannot continue through the maze. In this case, you have to retrace your steps for a certain distance through the maze before you are back on a more promising path. In a Cayley graph, this amount of backtracking is called the depth of the dead end element.

More precisely, suppose an element $g \in G$ is a dead end element and the word length of g with respect to a particular generating set S for G is n, that is, $|g|_S = n$. We say that the depth of g is k if the shortest path from g to any group element h with $|h|_S = n + 1$ has length $k + 1$. Intuitively, $k + 1$ is the minimum number of edges you must traverse in the Cayley graph $\Gamma(G, S)$ in order to reach a vertex farther from the identity than g. If a group has dead end elements of arbitrary depth it means that there is an infinite sequence of dead end elements in the group whose depths grow without bound.

The depth of $1 \in \mathbb{Z}$, which is shown above to be a dead end element in $\Gamma(\mathbb{Z}, \{2, 3\})$, is easily seen to be 1, as there is a path of length 2 from the vertex 1 to the vertex 7 (follow two edges labeled 3) and 7 has word length 3 in this generating set.

Exercise 8. Show that the group $\mathbb{Z}$ has dead end elements with respect to the presentation $\mathbb{Z} = \langle a, b \mid a^{12} = b, ab = ba \rangle$. What are the possible depths of dead end elements with respect to this generating set?

Dead end elements in L_2. Are there dead end elements in the Cayley graph $\Gamma(L_2, \{t, a\})$? Next we will discuss the proof of the following theorem, due to Sean Cleary and me:

THEOREM 15.1. *The lamplighter group L_2 contains dead end elements of arbitrary depth with respect to the generating set $\{t, a\}$.*

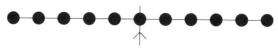

Figure 15.3 The lamplighter picture of d_5, which is a dead end element in the Cayley graph $\Gamma(L_2, \{t, a\})$. A solid circle represents a bulb that has been illuminated. All bulbs not shown in the figure are assumed not to be illuminated.

To prove this theorem we must first produce dead end elements in this Cayley graph. We will leave the proof that the dead end elements we find have arbitrary depth to the projects at the end of this office hour.

Consider the element d_m of L_2 in which all bulbs at positions between $-m$ and m are illuminated, and the lamplighter is at $0 \in \mathbb{Z}$. Using the notation $a_k = t^k a t^{-k}$ introduced above, an efficient path producing these elements is

$$d_m = a_0 a_1 a_2 \ldots a_m a_{-1} a_{-2} \ldots a_{-m}.$$

The lamplighter picture of the element d_5 is shown in Figure 15.3. We will show that d_m is a dead end element in $\Gamma(L_2, \{t, a\})$.

The word length of d_m is $D(d_m) = (2m + 1) + 4m = 6m + 1$. Notice that our efficient path to d_m has exactly this length—it is a geodesic path from the identity to this element in the Cayley graph.

In order for d_m to be a dead end element, we must show that the word lengths of $d_m a$, $d_m t$, and $d_m t^{-1}$ are at most the word length of d_m. We consider these three cases separately.

1. The difference in the lamplighter pictures for d_m and $d_m a$ is that the bulb at the origin is illuminated in d_m and is not illuminated in $d_m a$. Thus the efficient path $\gamma(d_m a)$ needs one less term than $\gamma(d_m)$, and hence $D(d_m a)$ is 1 less than $D(d_m)$.
2. The difference in the lamplighter pictures for d_m and $d_m t$ is in the position of the lamplighter: both have the same illuminated bulbs, but in $d_m t$ the position of the lamplighter is 1 rather than the 0 in d_m. Thus there is an efficient path for $d_m t$ that ends in t and it is straightforward to compute that $D(d_m t) = D(d_m) - 1$.
3. As an exercise, show that $D(d_m t^{-1}) = D(d_m) - 1$.

Together these three cases show that d_m is a dead end element in $\Gamma(L_2, \{a, t\})$.

Exercise 9. It remains to show that the dead end element d_m introduced above has depth at least m. Prove this by following the two steps below, and conclude that there is no upper bound on the depth of dead end elements in L_2.

(a) Let H_n be the set of all elements of L_2 in which all illuminated bulbs have indices i with $-n \leq i \leq n$ and the lamplighter is at position $l \in \mathbb{Z}$ with $-n \leq l \leq n$. Show that d_n has the maximum word length of any element in H_n.
(b) What is the shortest path from d_n to a vertex outside of H_n that is farther from the identity than d_n?

15.4 GEOMETRY OF THE CAYLEY GRAPH

You may have noticed that in the above discussion of the lamplighter picture of an element of L_2, and in our proof that the Cayley graph $\Gamma(L_2, \{t, a\})$ has dead end elements, we have never actually described the Cayley graph $\Gamma(L_2, \{t, a\})$!

In order to obtain a beautiful Cayley graph for this group we must change our generating set slightly. Consider the generating set $\{t, at\}$ for L_2 (and $\{t, at, a^2t, \cdots, a^{n-1}t\}$ for L_n when $n \geq 3$). This is sometimes called the electricity-free generating set, as moving and turning on a bulb can be accomplished by a single generator, that is, the electricity (i.e., turning on the bulb) does not require its own generator. It is not hard to see that the efficient paths for elements and computation of the word metric can easily be adapted to work for this generating set as well.

Exercise 10. Consider an element $g \in L_2$.

(a) Describe an algorithm to construct efficient paths $\eta(g)$ and $\eta'(g)$ using the electricity-free generating set $\{t, at\}$, analogously to $\gamma(g)$ and $\gamma'(g)$.

(b) Can you use these new paths $\eta(g)$ and $\eta'(g)$ to compute the word length of $g \in L_2$ with respect to $\{t, at\}$? Can you redefine the quantity $D(g)$ and mimic the analogous proof for the generating set $\{t, a\}$?

Diestel–Leader graphs. The Cayley graph of L_2 with respect to the $\{t, at\}$ generating set will be shown below to be a very special type of graph, called a Diestel–Leader graph. These graphs were introduced in order to answer the question: *Is every "nice" infinite graph quasi-isometric to the Cayley graph of some finitely generated group?* Here, "nice" involves some technical conditions, for instance, the graph must be connected and have the same degree at each vertex, among other conditions.

Notice that we usually consider this question in reverse, namely: *What is the Cayley graph corresponding to a group I am interested in?* Diestel and Leader conjectured that some of the graphs in the family they defined (called Diestel–Leader graphs) were *not* quasi-isometric to the Cayley graph of any finitely generated group, and this was later proven to be true [125]. Diestel–Leader graphs are constructed below from a pair of trees, but this construction holds with a greater number of trees as well. These larger graphs and the groups whose Cayley graphs are Diestel–Leader graphs with respect to a carefully chosen generating set have been studied in detail [15, 246].

To define a Diestel–Leader graph, consider two infinite trees of fixed valence $d + 1$, which we denote by T_1 and T_2. We orient the trees so that each vertex has one incoming edge and d outgoing edges. See, for example, the trees in Figures 15.4 and 15.5 (only a portion of each tree is shown). Fix a height function $h_i : T_i \to \mathbb{Z}$ on each tree, for $i = 1, 2$. This function fixes a level in the tree which is thought of as height 0; the outgoing edges from a vertex at height 0 terminate at a vertex of height 1 and an edge that ends at a vertex at height 0 begins at height -1. Continuing in this way, we have identified each level of vertices in T_i with an integer. See Figure 15.4 below for an example of a height function on

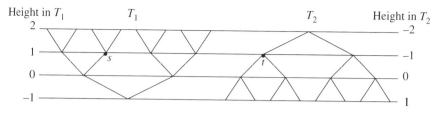

Figure 15.4 The product of trees $T_1 \times T_2$ with height functions fixed on each tree. The vertex set of the Diestel–Leader graph $\mathrm{DL}_2(2)$ is the subset of $T_1 \times T_2$ in which both coordinates lie on a horizontal line in the picture. For example, the pair (s, t) is a vertex of $\mathrm{DL}_2(2)$. Remember that this picture shows only a small part of these infinite trees!

a tree. If you had picked a different height function on the trees, you would obtain an isomorphic Diestel–Leader graph.

Let $T_1 \times T_2$ denote the product of the two trees T_1 and T_2. The points in $T_1 \times T_2$ are ordered pairs of the form (t_1, t_2) where $t_1 \in T_1$ and $t_2 \in T_2$. The Diestel–Leader graph $\mathrm{DL}_2(d)$ is a special subset of this product of trees. To identify this graph (as with any graph) we describe its vertices and edges.

Vertices in the Diestel–Leader graph $\mathrm{DL}_2(d)$. The vertex set of the Diestel–Leader graph $\mathrm{DL}_2(d)$ is the subset of the product of trees $T_1 \times T_2$ for which the sum of the heights of the two coordinates is 0:

$$\{(t_1, t_2) \mid t_i \in T_i \text{ and } h_1(t_1) + h_2(t_2) = 0\}.$$

One way to visualize vertices of $\mathrm{DL}_2(d)$ is to orient T_1 and T_2 in opposite directions, so that if you drew them on the same page, the height function on T_1 would increase as you moved toward the top of the page, but the height function on T_2 would decrease in that direction. Draw the trees in this way so that height k in T_1 is at the same level on your paper as height $-k$ in T_2. Now the vertices of the Diestel–Leader graph $\mathrm{DL}_2(d)$ are simply those pairs of vertices, one from each tree, which lie on the same horizontal line on your page. See Figure 15.4 for this representation of the pair of trees, with the height in each tree labeled on the sides of the diagram. Using the labels in Figure 15.4 we see that $(s, t) \in \mathrm{DL}_2(2)$. Remember that in Figure 15.4 you are only seeing a small part of these infinite trees!

Edges in the Diestel–Leader graph $\mathrm{DL}_2(d)$. The edges of the graph $\mathrm{DL}_2(d)$ are defined as follows, using the notation in Figure 15.5: vertices (s_0, t_0) and (s_1, t_1) are connected by an edge if the vertices s_0 and s_1 are connected by an edge in T_1, and vertices t_0 and t_1 are connected by an edge in T_2. As the sum of the height functions of the two coordinates must stay at 0, to find the vertices in $\mathrm{DL}_2(d)$ connected to (s_0, t_0) by a single edge we can only do one of two things:

1. increase height by 1 in T_1 while decreasing height by 1 in T_2, or
2. decrease height by 1 in T_1 while increasing height by 1 in T_2.

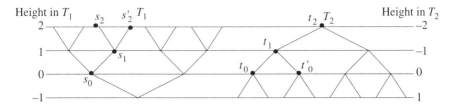

Figure 15.5 Using the vertices labeled in each tree, we see that in $DL_2(2)$ the vertices (s_0, t_0), (s_0, t_0'), (s_2, t_2), and (s_2', t_2) are all connected to the vertex (s_1, t_1) by a single edge.

As there are d edges in T_1 (resp. T_2) which connect s_1 (resp. t_1) to a vertex where the height has increased by one and a single edge connecting s_1 (resp. t_1) to a vertex in T_1 (resp. T_2) where the height has decreased by 1, we see that there are $2d$ vertices in $DL_2(d)$ connected to (s_1, t_1) by a single edge. In Figure 15.5 we see that the vertices (s_1, t_1) and (s_0, t_0) are connected by an edge in $DL_2(2)$. The three remaining vertices connected by a single edge to (s_1, t_1) are (s_0, t_0'), (s_2, t_2), and (s_2', t_2); see Figure 15.5 below.

Important Reminder. When looking at Figures 15.4 and 15.5 we must remember that the edges drawn in each tree do not represent edges which are present in the graph $DL_2(2)$. Drawing the product of trees as in Figures 15.4 and 15.5 gives us an easy way of visualizing the vertices of $DL_2(2)$ or, equivalently—as we will show below—the vertices of the Cayley graph $\Gamma(L_2, \{t, at\})$. The edges we described above are edges we know must exist in the Diestel–Leader graph, but are not drawn in either Figure 15.4 or 15.5. For an example of an edge drawn between vertices of $DL_2(2)$, see Figure 15.8.

Vertices in $DL_2(2)$ as elements of L_2. We now show that the Cayley graph $\Gamma(L_2, \{t, at\})$ is the Diestel–Leader graph $DL_2(2)$. We must show how to associate a vertex in $DL_2(2)$ with an element of L_2 in a unique way that captures the group structure. This involves an alternate method of identifying a vertex in a tree.

1. *Labeling edges in a tree.* Consider a single infinite tree T of valence 3 with a fixed height function. Each vertex of T has two outgoing edges, and we label the left one with a 0 and the right one with a 1.
2. *Identifying a vertex in a tree I.* Choose a vertex v at height k in the tree T whose edges are labeled with 0s and 1s as described above. From v, there is a unique downward path through T along which height is decreased by 1 at every step. Reading the labels along this path one obtains a string of 0s and 1s which is eventually always 0. We denote this string by $\mathcal{A} = a_1, a_2, a_3, a_4, \ldots$, where $a_i \in \{0, 1\}$ and $a_j = 0$ for j greater than some constant.
3. *Identifying a vertex in a tree II.* Conversely, given a string $\mathcal{A} = a_1, a_2, a_3, a_4, \ldots$, where $a_i \in \{0, 1\}$ and $a_j = 0$ for j greater than some constant and an integer $k \in \mathbb{Z}$, one can find a unique vertex w at height k in T for which the unique downward path from w is labeled by $\mathcal{A}$.

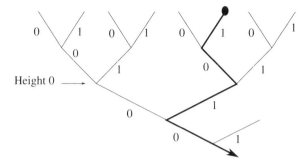

Figure 15.6 The string $1, 0, 1, 0, 0, 0, \ldots$ and height $k = 2$ uniquely describe a vertex in the tree T.

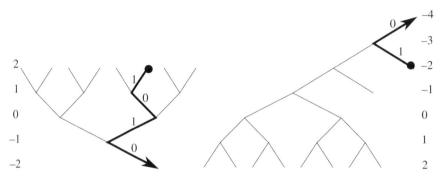

Figure 15.7 The element of $DL_2(2)$ uniquely described by the triple $(\mathcal{A}_1, \mathcal{A}_2, 2)$, where $\mathcal{A}_1 = 1, 0, 1, 0, 0, \ldots$ and $\mathcal{A}_2 = 1, 0, 0, 0, 0, \ldots$. The sequences are assumed to consist entirely of 0s after the given terms.

For example, in Figure 15.6 the vertex of T corresponding to the string $1, 0, 1, 0, 0, 0, \ldots$ (with all subsequent terms equal to 0) and $k = 2$ is shown.

4. *Describing a vertex in* $DL_2(2)$. To describe a vertex in $DL_2(2)$, we need two strings $\mathcal{A}_1$ and $\mathcal{A}_2$ of elements of $\{0, 1\}$ whose entries are eventually all 0s and of a single height k: the first string describes a vertex at height k in T_1 and the second a vertex at height $-k$ in T_2. For example, if $\mathcal{A}_1 = 1, 0, 1, 0, 0, \ldots$ and $\mathcal{A}_2 = 1, 0, 0, 0, 0, \ldots$ (where both strings are forever 0 after the given terms) and $k = 2$, the corresponding vertex of $DL_2(2)$ is given in Figure 15.7.

Exercise 11. Identify the four vertices of the infinite tree of valence 3 depicted in Figure 15.7 corresponding to the string $\mathcal{A} = 1, 0, 1, 0, 0, \ldots$ which we assume to consist entirely of 0s after the given terms, and with heights 3, 1, 0, and -2, respectively. You might have to draw in more of the trees to find one of these vertices!

$DL_2(2)$ as a Cayley graph of L_2. We are now ready to exhibit a bijection between the vertices of the Diestel–Leader graph $DL_2(2)$ and the elements of L_2. We have described above how to view the elements of $DL_2(2)$ as tuples of the form $(\mathcal{A}_1, \mathcal{A}_2, k)$. Consider an element g of L_2 given by the pair $g = (P, k)$, where P is a polynomial in the variables t and t^{-1} whose coefficients lie in $\mathbb{Z}/2\mathbb{Z}$, only finitely many of which are nonzero, and $k \in \mathbb{Z}$. Let $\{a_i\}$ be the sequence of coefficients of P where a_i is the coefficient of t^i. Divide $\{a_i\}$ into two sequences: $\mathcal{A}_1 = \{a_i\}_{i<k}$ and $\mathcal{A}_2 = \{a_i\}_{i\geq k}$, where the former sequence is listed in reverse order: $a_{k-1}, a_{k-2}, \ldots$. Notice that both sequences must eventually consist entirely of 0s.

Thus we view g as the tuple $(\mathcal{A}_1, \mathcal{A}_2, k)$ and the correspondence to a vertex of $DL_2(2)$ is clear. For example, the vertex of $DL_2(2)$ described in Figure 15.7 corresponds to the matrix

$$\begin{pmatrix} t^2 & t^{-1} + t^1 + t^2 \\ 0 & 1 \end{pmatrix}.$$

The proof that $DL_2(2)$ is the Cayley graph of L_2 with respect to the generating set $\{t, at\}$ is completed by showing that this identification extends to a graph isomorphism between the two graphs.

Exercise 12. Show that the identification between vertices of $DL_2(2)$ and elements of L_2 extends to a graph isomorphism between the two graphs, proving that $DL_2(2)$ is the Cayley graph of L_2 with respect to the generating set $\{t, at\}$.

Exercise 13. Using Figure 15.7, choose a vertex of $DL_2(2)$. Identify this vertex as:

1. a matrix in L_2;
2. a triple $(\mathcal{A}_1, \mathcal{A}_2, k)$.

Moving around in this Cayley graph. We would now like to understand how to move around in this Cayley graph, in the same way that we understand how to move around in the Cayley graph of $\mathbb{Z} \times \mathbb{Z}$, for example, with the standard generators. We know that from any vertex g in $\Gamma(\mathbb{Z} \times \mathbb{Z}, \{(1,0), (0,1)\})$ (where we use g both for the vertex and the group element associated to that vertex) there is an edge labeled $(1, 0)$ emanating from g that terminates at a vertex one unit away in the x direction. Similarly, there is an edge labeled $(0, 1)$ emanating from g that terminates at a vertex one unit away in the y direction. There are analogous edges labeled by the inverses of the generators as well. Traversing such an edge labeled by s emanating from g yields the vertex corresponding to gs.

For $DL_2(2) = \Gamma(L_2, \{t, at\})$ we would like the same intuitive sense of how vertices connected by a single edge differ in their $(\mathcal{A}_1, \mathcal{A}_2, k)$ coordinates. Namely, if we are at a vertex $g = (\mathcal{A}_1, \mathcal{A}_2, k) \in DL_2(2)$ and traverse an edge labeled by the generator t (equivalently, we view g as an element of L_2 and multiply g by the generator t), what are the coordinates of the vertex we wind up at?

To answer this question, write the generator t as the matrix $\begin{pmatrix} t & 0 \\ 0 & 1 \end{pmatrix}$, and let $(\mathcal{A}_1, \mathcal{A}_2, k)$ correspond to the element $g \in L_2$ represented by the matrix $\begin{pmatrix} t^k & P \\ 0 & 1 \end{pmatrix}$. In particular, if we write the infinite sequence of polynomial coefficients as $\{a_i\}$, then $\mathcal{A}_1 = \{a_i\}_{i<k}$ and $\mathcal{A}_2 = \{a_i\}_{i\geq k}$.

As shown above,

$$gt = \begin{pmatrix} t^{k+1} & P \\ 0 & 1 \end{pmatrix}.$$

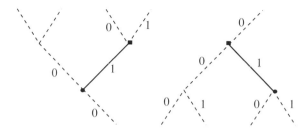

Figure 15.8 The pair of solid edges represents the edge in $DL_2(2)$ labeled by the generator t between the vertex of $DL_2(2)$ identified by the solid circles and the vertex of $DL_2(2)$ identified by the solid squares. The dotted lines represent the edges of the individual trees and are not edges in $DL_2(2)$.

How does this change the 3-tuple describing the corresponding vertex of $DL_2(2)$? It is clear that the integer coordinate changes from k to $k + 1$. Since the polynomial P was unchanged, the coefficients of P are exactly the same in gt as in t. What has changed from g to gt is the point at which we divide the coefficients $\{a_i\}$ into two subsequences: from gt we obtain $\mathcal{B}_1 = \{a_i\}_{i<k+1}$ and $\mathcal{B}_2 = \{a_i\}_{i\geq k+1}$, where the former has a_k as its initial entry. Notice that to create $\mathcal{B}_1$, the first entry of $\mathcal{A}_2$ was removed, and inserted as the first entry of $\mathcal{A}_1$ to create $\mathcal{B}_1$. The sequence $\mathcal{B}_2$ is simply the sequence $\mathcal{A}_2$ with the initial entry removed.

Now that we see the precise difference between the coordinates of g and gt in $DL_2(2)$, we can locate gt in $DL_2(2)$ relative to g as follows. In T_1, proceed along the edge labeled a_k to a new vertex that is at height $k + 1$. In T_2, simply remove the initial edge in the path, so that the truncated path begins at a vertex at height $-(k + 1)$. This new vertex corresponds to $(\mathcal{B}_1, \mathcal{B}_2, k + 1)$ and is connected to $(\mathcal{A}_1, \mathcal{A}_2, k)$ by a single edge.

To understand how to traverse an edge labeled by at emanating from the vertex g, we again multiply the two relevant matrices to obtain

$$g(at) = \begin{pmatrix} t^{k+1} & t^k + P \\ 0 & 1 \end{pmatrix}.$$

The polynomial entry in gat differs from P in only one coefficient: the coefficient of t^k has increased by 1, and as all coefficients are taken in $\mathbb{Z}/2\mathbb{Z}$, this means that it has changed from 0 to 1 or vice versa. Since the integer coordinate is now $k + 1$, when we separate the coefficients of $t^k + P$ into two sequences, we obtain $\mathcal{B}_1$, whose initial entry is $a_k + 1$ (mod 2) and whose subsequent entries are the sequence $\mathcal{A}_1$, and $\mathcal{B}_2 = \{a_i\}_{i\geq k+1}$. The vertex gat has the same T_2 coordinate as gt, and the T_1 coordinates share a common parent vertex.

Looking at the pair of trees in Figure 15.8, where the vertex g is identified by the pair of solid circles, the edge of $DL_2(2)$ emanating from g and labeled t is drawn with a solid line. The vertex g is identified by the 3-tuple $((0, 0, 0, \dots),$ $(1, 0, 0, \dots), k)$ for some $k \in \mathbb{Z}$, where both sequences continue with all entries equal to 0. Hence the edge labeled t terminates at the vertex described by the 3-tuple $((1, 0, 0, 0, \dots), (0, 0, \dots), k + 1)$, as illustrated in the figure.

Exercise 14. Let $g \in L_2$ correspond to the vertex depicted in Figure 15.8 by a solid dot. Find the vertices corresponding to the elements gt, gt^{-1}, $g(at)$, $g(at)^{-1}$, and ga. (Recall that a is not a generator in the presentation which yields the Cayley graph $DL_2(2)$.)

A stack of cards. Tim Riley came up with a unique way of viewing vertices of the Diestel–Leader graph $DL_2(2)$, and hence the Cayley graph of L_2, as two stacks of cards with an integer counter that stores the value k. Imagine the sequence $\mathcal{A}_i$ as an infinite stack of cards, so that each entry in the sequence corresponds to a card, with a_{k-1} on top and the indices decreasing when $i = 1$, and a_k on top with the indices increasing when $i = 2$.

Consider a vertex $g = (\mathcal{A}_1, \mathcal{A}_2, k)$ in $DL_2(2)$. There is an edge labeled by t connecting g to a vertex h; to create the stacks of cards representing h we remove the top card from the stack $\mathcal{A}_2$ and place it on top of the first stack $\mathcal{A}_1$, while increasing the counter by 1. This encodes the changes in the matrix entries between g and gt as described above. Similarly, there is an edge labeled by at connecting g to a vertex v; to create the stacks of cards representing v we remove the top card from the stack $\mathcal{A}_2$, add 1 and reduce modulo 2, and place the card with the new value on top of the first stack $\mathcal{A}_1$, while increasing the counter by 1. This encodes the changes in the matrix entries between g and gat.

15.5 GENERALIZATIONS

The lamplighter group L_2 is an example of an algebraic construction called a wreath product. Any product construction is designed to build a larger group out of two components. As we said at the beginning of this office hour, it seemed like L_2 was built from the groups $\bigoplus_{i=-\infty}^{\infty} (\mathbb{Z}/2\mathbb{Z})_i$ and $\mathbb{Z}$. We write $L_2 = \mathbb{Z}/2\mathbb{Z} \wr \mathbb{Z}$ to denote this wreath product, and in general, $L_n = \mathbb{Z}/n\mathbb{Z} \wr \mathbb{Z}$. We say this as: L_2 is the wreath product of $\mathbb{Z}/2\mathbb{Z}$ and $\mathbb{Z}$ or L_2 is $\mathbb{Z}/2\mathbb{Z}$ wreath $\mathbb{Z}$.

The wreath product is a very general construction. To begin with, let F be a finite group and G any finitely generated group. To understand elements of the wreath product $F \wr G$, consider a city whose map is described by the Cayley graph of G (with respect to a particular generating set S) and at each element of G there is a copy of the Cayley graph of F (also with respect to a particular generating set) attached. Instead of turning on a lightbulb (that is, choosing an element of $\mathbb{Z}/nZ$), we have the choice of selecting a particular element of F from each copy of F. An element of $F \wr G$ then consists of a finite collection of elements of F from these copies of the Cayley graph of F placed at the vertices of $\Gamma(G, S)$, as well as a single element of G, analogous to the position of the lamplighter. We can also replace F with a finitely generated group to obtain wreath products such as $\mathbb{Z} \wr \mathbb{Z}$.

As with the lamplighter groups L_n, an element of $F \wr G$ also consists of two parts: an element of $\bigoplus_{i \in I}(F)_i$ and a single element of G. We adjust the index set I depending on the group G.

Lamplighters and traveling salesmen. The wreath product construction described above is quite general, and does not depend in any way on the groups

$\mathbb{Z}/n\mathbb{Z}$ and $\mathbb{Z}$ which have been discussed in this office hour as part of the wreath product $L_n = \mathbb{Z}/n\mathbb{Z} \wr \mathbb{Z}$. To end, we consider a slight generalization of the lamplighter group, that is, the wreath product $\mathbb{Z}/2\mathbb{Z} \wr (\mathbb{Z} \times \mathbb{Z})$. The picture of an element of this group can be constructed analogously to L_n: imagine a lamplighter walking around a city whose street map is the standard Cayley graph of $\mathbb{Z} \times \mathbb{Z}$ where there is a light bulb at each vertex. The lamplighter illuminates a finite number of bulbs and then walks to an ending vertex. We take as our generators the set of elements which correspond to

- moving one unit in the x direction,
- moving one unit in the y direction, and
- turning on the bulb in the position of the lamplighter.

In this group, the word problem is solvable, as it is in $\mathbb{Z}/n\mathbb{Z} \wr \mathbb{Z}$; that is, one can determine whether a particular string of generators reduces to the identity. However, consider the problem of finding a minimal length word representing a given element. Suppose that the element $g \in \mathbb{Z}/2\mathbb{Z} \wr (\mathbb{Z} \times \mathbb{Z})$ has bulbs illuminated at a collection of vertices $\{(x_1, y_1), (x_2, y_2), \ldots, (x_r, y_r)\}$ in $\mathbb{Z} \times \mathbb{Z}$, and the lamplighter has position $(x, y) \in \mathbb{Z} \times \mathbb{Z}$. To find a minimal length word in these generators representing w, we must find a minimal path which visits the $r + 1$ vertices of $\mathbb{Z} \times \mathbb{Z}$ listed above. This problem is equivalent to the traveling salesman problem in the plane. This is a very hard problem!

Traveling salesman problem. *Given a list of cities and a map (so that you know how far apart the cities are), determine a minimal path for a salesman who must visit each city exactly once.*

A quick Google search will turn up a wide variety of results related to this problem. So we see that although the word problem for $\mathbb{Z}/2\mathbb{Z} \wr (\mathbb{Z} \times \mathbb{Z})$ is straightforward, the question of determining minimal-length representatives is very hard! Notice that this problem does not arise for $L_n = \mathbb{Z}/n\mathbb{Z} \wr \mathbb{Z}$, as the problem of determining the shortest path visiting a collection of integers is easily solved.

FURTHER READING

For an accessible introduction to lamplighter groups with a slightly different viewpoint than the one given here, I enthusiastically refer you to the chapter on lamplighter groups in John Meier's *Groups, Graphs and Trees* [202]. For a more detailed introduction to these groups, as well as Diestel–Leader graphs, one can read "What is a horocyclic product, and how is it related to lamplighters?" by Wolfgang Woess [263]; this paper also discusses broader questions related to quasi-isometries, and becomes quite technical for a beginning reader.

References for some of the results mentioned in this chapter are found in "Dead end words in lamplighter groups and other wreath products" [83] by me and Sean Cleary, as well as "Cone types and geodesic languages for lamplighter groups and Thompson's group F" [80] by Sean Cleary, Murray Elder, and me.

If you are interested in understanding lamplighter groups and their higher-dimensional generalizations, a good overview is given by Riley and Amchislavska in "Lamplighters, metabelian groups, and horocyclic products of trees" [8]. The canonical reference for these more general groups is "Horocyclic products of trees" [15], by Laurent Bartholdi, Markus Neuhauser, and Wolfgang Woess. Stein and I also present a detailed introduction to the higher rank lamplighter groups (also called Diestel–Leader groups) in "Metric properties of Diestel–Leader groups" [246], and we consider generalizations of the properties mentioned in this office hour for those groups.

There are other contexts in which lamplighter groups often appear. Here are two examples. Many mathematicians are interested in random walks on Diestel–Leader graphs such as the Cayley graph of the lamplighter group with respect to the generating set given in this office hour. A reference for beginning to understand these random walks is "Lamplighters, Diestel–Leader graphs, random walks, and harmonic functions" [264], by Wolfgang Woess. Some people like to study the lamplighter group in the context of automata groups. If this appeals to you, you should read "The lamplighter group as a group generated by a 2-state automaton, and its spectrum" [152], by Grigorchuk and Zuk, and "Metric properties of the lamplighter group as an automata group" [84], by Sean Cleary and me. So many people find different things to study about lamplighter groups that if you do a quick search on the *arxiv.org*, you will find over 100 papers—that should give you plenty of reading ideas!

PROJECTS

Project 1. Draw the balls of radius 3 and 4 in the Cayley graph $DL_2(2)$, representing elements via either the generators t and at, or matrices, or triples $(k, \mathcal{A}_1, \mathcal{A}_1)$. Which of these balls do you expect to be a tree, that is, a graph with no closed loops? Why? Could it be that both balls are trees?

Project 2. Draw the balls of radii 3, 4, and 5 in the Cayley graph of $\mathbb{Z}/2\mathbb{Z} \wr (\mathbb{Z} \times \mathbb{Z})$ with respect to the generating set given above.

Project 3. Extend the efficient paths $\gamma(g)$ and $\gamma'(g)$ to elements $g \in L_n$ for $n \geq 3$. Make an analogous definition of $D(g)$ for $g \in L_n$ with $n \geq 3$. Following the outline for $n = 2$, show that $D(g)$ is the word length of $g \in L_n$ with respect to the generating set $\{t, a\}$.

Project 4. This project is about comparing word lengths.

1. Find an example of an element of L_2 whose word length is shorter with respect to $\{t, at\}$ than it is with respect to $\{t, a\}$. Can you generalize your example to a family of examples?
2. Describe all elements $g \in L_2$ with $|g|_{\{t,at\}} < |g|_{\{t,a\}}$. What about $|g|_{\{t,at\}} \leq |g|_{\{t,a\}}$?
3. Can you find an example of an element of L_2 whose word length is shorter with respect to $\{t, a\}$ than it is with respect to $\{t, at\}$? Why or why not?

4. Can you quantify how much shorter or longer the word length of $g \in L_2$ can become when changing from one generating set to another? That is, find a constant K so that

$$\frac{1}{K}|g|_{\{t,a\}} \leq |g|_{\{t,at\}} \leq K|g|_{\{t,a\}}.$$

Project 5. Write a computer program to compute the word length of $g \in L_2$ with respect to the generating set $\{t, a\}$.

Project 6. Write a computer program to compute the word length of $g \in L_2$ with respect to the generating set $\{t, at\}$.

Project 7. Look up the definition of a semidirect product in Office Hour 10.

1. Figure out how to write $H \wr K$ as a semidirect product of K and a group related to H. Try this first assuming that H and K are finite groups.
2. Write L_n as a semidirect product.

Project 8. Is the dead end element d_m still a dead end element with respect to the generating set $\{t, at\}$? Can you find a family of dead end elements with respect to this generating set? Perhaps a family related to the elements d_m?

Project 9. Write a computer program to decide whether $g \in L_2$ is a dead end element with respect to generating set either $\{t, a\}$ or $\{t, at\}$. Use your program to analyze the following problems.

1. Find other families of dead end elements in L_2 with respect to either generating set.
2. Can you find any elements which are dead end elements with respect to *both* generating sets?
3. Use the programs you wrote to compute word length with respect to either generating set to list the elements in the ball of radius n, for as many n as feasible. What proportion of these elements are dead end elements? Can you make a guess as to the percent of elements in the n-ball that are dead end elements?

Office Hour Sixteen

Thompson's Group

Sean Cleary

In this office hour we will investigate one of the most well-studied groups in geometric group theory, Thompson's group F. This group seems to be ubiquitous. It was first defined in 1965 by Richard Thompson, who was studying the word problem for groups.[1] Since then, the group F—along with its cousins T and V— has arisen in the contexts of logic, homotopy theory, category theory, shape theory, mapping class groups, and data storage. Thompson's enigmatic group F has a number of seemingly paradoxical behaviors. For instance:

1. F is finitely presented and it contains a copy of $F \times F$,
2. F is (again) finitely presented and is an HNN extension[2] of itself, and
3. F has exponential growth but contains no free groups of rank 2.

The first item implies that F contains the direct sum of infinitely many copies of F, and the second item implies that we can write F as an HNN extension of an HNN extension of an HNN extension …of an HNN extension of F. You might guess that any group with either of these properties would not be finitely generated, let alone finitely presented!

Finally, we know that free groups have exponential growth (meaning that the number of vertices of the Cayley graph in the ball of radius r grows exponentially in r; see Office Hour 12); but it is hard to imagine a group that has exponential growth where the growth does not come from free groups. Thompson's group F was the first known example without torsion.

---

[1] The name F comes from the notion that F is somewhat "free" in the sense of some universal properties, but that is somewhat confusing as F contains no actual nonabelian free subgroups.

[2] See the Projects section in Office Hour 3 for the definition of an HNN extension.

The group F is also the subject of one of the most famous questions in geometric group theory:

  Is F amenable?

We will discuss this question at the end of the office hour. It is related to item (3) above and also the Banach–Tarski paradox. We will also explain why either answer to the question leads to more paradoxes!

To begin, I will give three different descriptions of Thompson's group F:

- analytically, as a group of piecewise-linear homeomorphisms of [0, 1],
- geometrically, as pairs of rooted binary trees, and
- combinatorially, in terms of generators and relations.

There are other descriptions; some of these will be mentioned later in the office hour.

The main theorem we will prove about F is the following.

THEOREM 16.1. *F is finitely generated.*

After we see the first description of the group F, I think you will agree that this fact is quite surprising. We will then discuss some of the finer algebraic properties of F, some of which I already touched on above. Finally we will touch upon Blake Fordham's ingenious method for computing distances in the Cayley graph for F.

Cannon, Floyd, and Parry wrote an excellent survey on Thompson's group, from which we borrow heavily. We highly recommend the motivated reader to read their article after this office hour [72].

16.1 ANALYTIC DEFINITION AND BASIC PROPERTIES

A homeomorphism of the unit interval [0, 1] is a bijective, continuous function [0, 1] → [0, 1] with continuous inverse. The graph of such a function lies in [0, 1] × [0, 1] and has the property that each horizontal line and each vertical line intersects the graph in a single point.

Next, a piecewise-linear homeomorphism of [0, 1] is a homeomorphism where the graph is a finite union of line segments in [0, 1] × [0, 1]. The slopes of these line segments must either be all positive or all negative (why?). We will only consider the orientation-preserving piecewise-linear homeomorphisms of [0, 1], namely, the ones where all of the slopes are positive (equivalently, a piecewise linear homeomorphism of [0, 1] preserves the orientation of [0, 1] if it fixes both endpoints). These homeomorphisms preserve the left-to-right orientation of [0, 1].

The set of all orientation-preserving piecewise-linear homeomorphisms of [0, 1] naturally forms a group where the multiplication is given by function composition: the identity in the group is the identity homeomorphism, inverses exist by definition (these are the usual inverses from calculus class), and associativity is automatic (again something you discussed in calculus).

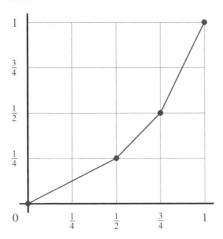

Figure 16.1 Interpolating between the two subdivisions of the unit interval $\{0, \frac{1}{2}, \frac{3}{4}, 1\}$ and $\{0, \frac{1}{4}, \frac{1}{2}, 1\}$ to make a piecewise-linear homeomorphism.

Here is one way to get an orientation-preserving piecewise-linear homeomorphism of $[0, 1]$. Fix a positive integer n and choose two sets of n points in $[0, 1]$, with the first and last points in each set being 0 and 1, say $0 = p_1 < p_2 < \cdots < p_n = 1$ and $0 = q_1 < q_2 < \cdots < q_n = 1$. Then draw the points $(p_1, q_1), \ldots,$ (p_n, q_n) in the square $[0, 1] \times [0, 1]$. Finally, connect the dots (in order!) by line segments. For any choice of p_i's and q_i's, this is the graph of an orientation-preserving piecewise-linear homeomorphism of $[0, 1]$; see Figure 16.1 for an example.

What is more, every orientation-preserving piecewise-linear homeomorphism of $[0, 1]$ arises from the connect-the-dots method of the previous paragraph. We refer to the (p_i, q_i) as the *break points* and the numbers $(q_{i+1} - q_i)/(p_{i+1} - p_i)$ as the *slopes* of the homeomorphism (the former are of course the endpoints of the various line segments in the graph and the latter are the slopes of the line segments).

From the connect-the-dots description, we immediately see that there are uncountably many orientation-preserving piecewise-linear homeomorphisms of $[0,1]$. So the group we have so far is uncountable. We will add more restrictions in order to obtain a countable group. But first we need a definition.

A dyadic rational number is a rational number of the form $\frac{k}{2^n}$ for some integers k and n. Within the group of orientation-preserving piecewise-linear homeomorphisms of $[0, 1]$ we will consider those with the following properties:

- there are finitely many break points, each of which is a pair of dyadic rational numbers and
- the slopes are all powers of 2.

The resulting subgroup of the group of orientation-preserving piecewise-linear homeomorphisms of $[0, 1]$ is (finally) Thompson's group F.

Exercise 1. Check that F is a group.

An example. The piecewise-linear orientation-preserving homeomorphism in Figure 16.1 satisfies the last two conditions and so is itself an element of F. We can write it down as the piecewise function:

$$x_0(t) = \begin{cases} \frac{t}{2} & \text{for } 0 \le t \le \frac{1}{2} \\ t - \frac{1}{4} & \text{for } \frac{1}{2} \le t \le \frac{3}{4} \\ 2t - 1 & \text{for } \frac{3}{4} \le t \le 1 \end{cases}$$

Indeed the graph has slope $\frac{1}{2}$ until the point $(\frac{1}{2}, \frac{1}{4})$ is reached, at which point it has slope 1 until reaching $(\frac{3}{4}, \frac{1}{2})$, and then has slope 2 until it reaches $(1, 1)$.

It is straightforward to compute x_0^2:

$$x_0^2(t) = \begin{cases} \frac{t}{4} & \text{for } 0 \le t \le \frac{1}{2} \\ \frac{t}{2} - \frac{1}{8} & \text{for } \frac{1}{2} \le t \le \frac{3}{4} \\ 2t - \frac{5}{4} & \text{for } \frac{3}{4} \le t \le \frac{7}{8} \\ 4t - 3 & \text{for } \frac{7}{8} \le t \le 1 \end{cases}$$

Another example. The piecewise-linear orientation-preserving homeomorphism corresponding to the two subdivisions $\{0, \frac{1}{2}, \frac{3}{4}, \frac{7}{8}, 1\}$ and $\{0, \frac{1}{2}, \frac{5}{8}, \frac{3}{4}, 1\}$ is:

$$x_1(t) = \begin{cases} t & \text{for } 0 \le t \le \frac{1}{2} \\ \frac{t}{2} + \frac{1}{4} & \text{for } \frac{1}{2} \le t \le \frac{3}{4} \\ t - \frac{1}{8} & \text{for } \frac{3}{4} \le t \le \frac{7}{8} \\ 2t - 1 & \text{for } \frac{7}{8} \le t \le 1 \end{cases}$$

Exercise 2. Draw the graph of x_1 and compare it to the graph of x_0.

Exercise 3. Compute the homeomorphism $x_0^{-1}x_1x_0$, draw its graph, and compare it to the graphs of x_0 and x_1.

These two homeomorphisms play an essential role for F; indeed they generate F!

Two basic facts. Composing piecewise functions in this naive way is tedious and we will soon have more efficient ways to calculate with elements of F. Nevertheless, there are a number of properties of F that are easy to see from this analytic characterization in terms of piecewise-linear functions.

Fact 1. F is torsion free (that is, every nontrivial element has infinite order).

This first fact is easy to prove: given any nontrivial element of F, consider the smallest x-value where the (right-sided) slope is not equal to 1. This is 0 for x_0 and $\frac{1}{2}$ for x_1. Any nontrivial power of this element will have slope not equal to 1

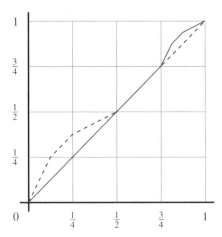

Figure 16.2 The dashed red element has support in the interval $[0, \frac{1}{2}]$ and the solid blue element has support in $[\frac{3}{4}, 1]$, so the two elements commute.

at the same x-value. Hence this nontrivial power is a nontrivial element of the group—just as we wanted!

Exercise 4. Fill in the details in our proof of Fact 1 (actually, there is only one detail to check!).

Exercise 5. Show any nontrivial group of piecewise-linear homeomorphisms of an interval is torsion free.

Fact 2. F contains a subgroup isomorphic to $\mathbb{Z}^k$ for each $k \geq 0$.

For any element of F, the support is the complement of the fixed set. It is not hard to see that elements of F with disjoint supports will necessarily commute. Therefore, given any k, it is enough to construct k functions with disjoint supports. (Note that we are using Fact 1 here!) See Figure 16.2 for an example of a pair of elements with disjoint supports.

A particularly useful kind of function here is a bump function: an element that coincides with the identity for most of the interval, but deviates from the identity in a short interval. For these bump functions, the support is the short interval. In this short interval, the graph should have one line segment with slope greater than 1 and another line segment with slope less than 1. We leave it as an exercise to show that for any k we can find k bump functions with disjoint supports.

Exercise 6. Complete the proof of Fact 2 by constructing the required bump functions.

Figure 16.3 A caret.

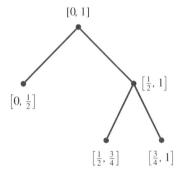

Figure 16.4 The rooted binary tree for the subdivision $\{0, \frac{1}{2}, \frac{3}{4}, 1\}$.

16.2 COMBINATORIAL DEFINITION

I'll now give a completely different way to think about the elements of F. The first definition should remind you of calculus class; this second definition should remind you of combinatorics or graph theory.

In this new definition, an element of F will be represented by a pair of rooted binary trees. Think of a rooted binary tree as being made of a collection of binary *carets*, each of which has a parent node at the top, two downward edges, and two child nodes at the bottom; see Figure 16.3. Start with a *root caret* and iteratively attach more carets at child nodes to obtain a rooted binary tree.

A *tree pair diagram* is two such trees with the same number of carets; look ahead to Figure 16.5 for examples.

From tree pairs to homeomorphisms. Let's see how we can get a piecewise-linear orientation-preserving homeomorphism of $[0, 1]$ from a tree pair diagram. Each tree can be regarded as instructions for subdividing the unit interval $[0, 1]$ by a successive halving procedure. We begin with the root node (the parent node of the root caret) and the unit interval $[0, 1]$ with just its two endpoints. The root caret corresponds to dividing the entire interval in half, giving a new dyadic subdivision of the interval $[0, 1]$ along the points $\{0, \frac{1}{2}, 1\}$.

Each added caret specifies an interval to be halved, introducing a new point into the subdivision. For example, if the the right child of the root node has a caret attached, then the interval $[\frac{1}{2}, 1]$ is subdivided, introducing $\frac{3}{4}$ into the subdivision to get the new subdivision along the points $\{0, \frac{1}{2}, \frac{3}{4}, 1\}$; see Figure 16.4.

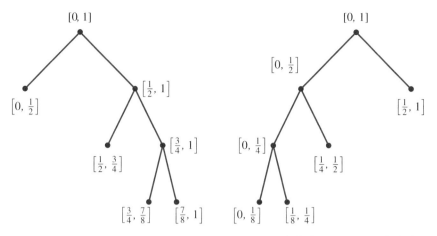

Figure 16.5 A pair of trees giving subdivisions $\{0, \frac{1}{2}, \frac{3}{4}, \frac{7}{8}, 1\}$ and $\{0, \frac{1}{8}, \frac{1}{4}, \frac{1}{2}, 1\}$ respectively.

In this way, each node of the tree corresponds to a subinterval of $[0, 1]$ whose endpoints are dyadic rationals, and if one node is a "descendant" of another, that means that the interval corresponding to the first is contained in the interval corresponding to the second.

Thus, to each rooted binary tree with n leaf nodes (a leaf node is a node with no children, i.e., a valence 1 vertex), we associate a subdivision of the unit interval into segments each of length 2^{-k} (k is the depth, or distance from the root of the corresponding node), containing a total of $n + 1$ points in the subdivision.

Given a pair of such trees with the same number of carets, we associate the piecewise-linear interpolation between the subdivisions, thus giving an element of F. The trees illustrated in Figure 16.5 give the subdivisions used for the element x_0^2.

Exercise 7. Find tree pair diagrams representing the elements x_0, x_1, and $x_0^{-1}x_1x_0$.

We often regard a tree pair diagram (S, T) as having the *source tree S* and the *target tree T* as the subdivisions happen in the domain and range of the associated piecewise-linear homeomorphism.

Exercise 8. If the tree pair diagram (S, T) reperesents f, can you describe a tree pair diagram that represents f^{-1}?

Exercise 9. How can you read the slope of f on the kth subinterval of the subdivision from a tree pair diagram representing f?

Exercise 10. Find tree pair diagram representatives of three elements that generate a subgroup of F isomorphic to $\mathbb{Z} \times \mathbb{Z} \times \mathbb{Z}$.

Reduced tree pair diagrams. There are many tree pair diagrams corresponding to an element f of F. For example, if S and T are the same tree then (S, T) represents the identity homeomorphism (so by "many" I really meant infinitely many).

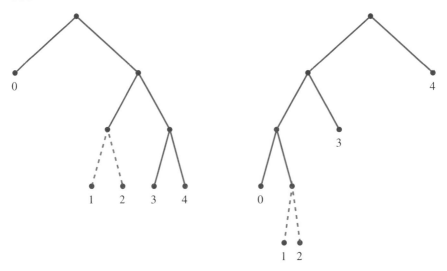

Figure 16.6 The left tree gives the subdivision $\{0, \frac{1}{2}, \frac{5}{8}, \frac{3}{4}, \frac{7}{8}, 1\}$ and the right tree gives the subdivision $\{0, \frac{1}{16}, \frac{1}{8}, \frac{1}{4}, \frac{1}{2}, 1\}$ and the piecewise-linear interpolation gives x_0^2, as did the tree pair in Figure 16.5.

Given a tree pair representative of f, we can create additional representatives by introducing subdivisions simultaneously on corresponding leaf nodes in the two trees to obtain a different tree pair diagram representing f. The piecewise-linear orientation-preserving homeomorphism corresponding to the new tree pair diagram will be exactly the same as that for the old pair when we play connect-the-dots to form the homeomorphism; there is now just one additional dot in the middle of some line segment.

An example is illustrated in Figure 16.6, where I've added the green dashed caret to corresponding nodes in each tree of Figure 16.5. As the pictures will get more complex, I will no longer label the nodes with the subintervals of [0, 1]. To keep track of which leaf nodes correspond to each other, we can label the leaf nodes of each tree in the tree pair from 0 to n in a left-to-right order.

This leads us to the notion of a reduced tree pair diagram. A tree pair diagram is *unreduced* if there is an i so that in both trees the ith and $(i + 1)$st leaf nodes are children of the same parent node. An unreduced tree pair diagram admits a reduction, whereby we remove the pair of redundant carets (and renumber the nodes in the resulting trees). The tree pair in Figure 16.6 is of course unreduced: leaf nodes 1 and 2 are children of the same parent in both trees.

Let us then say that two tree pair diagrams are equivalent if they have a common reduction; this makes sense because the corresponding elements of F are the same. There is a unique reduced (meaning: not unreduced) tree pair diagram in each such an equivalence class.

Exercise 11. Show carefully that the reduced tree pair diagram representing an element of F is unique.

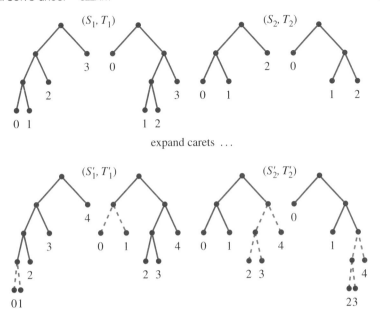

expand carets ...

Figure 16.7   To multiply the elements given by tree pairs (S_1, T_1) and (S_2, T_2) on the top, we create unreduced representatives (S_1', T_1') and (S_2', T_2') of the same group elements by adding the dashed carets and renumbering. For the new pairs T_1' and S_2' are identical, allowing us to form the product $(S_2', T_2') \cdot (S_1', T_1') = (S_1', T_2')$. There may be a possible reduction in the resulting tree pair, although that does not happen in this example.

Multiplying trees.   From the perspective of tree pair diagrams, group multiplication in F proceeds as follows. To multiply two tree pair diagrams (S_1, T_1) and (S_2, T_2),

$$(S_2, T_2) \cdot (S_1, T_1)$$

(remember the multiplication is function composition) we compare T_1 and S_2. If they are identical, then we immediately have the composition as (S_1, T_2). But probably they are not identical, in which case we create unreduced representatives of the two equivalence classes of tree pair diagrams so that the middle two trees coincide. That is, we find equivalent representatives $(S_1', T_1') \sim (S_1, T_1)$ and $(S_2', T_2') \sim (S_2, T_2)$, where the trees T_1' and S_2' are identical. We can do this by successively expanding the leaf nodes, that is, by adding carets and creating new leaf nodes. Specifically, we expand leaf nodes of T_1 that are not leaf nodes in S_2, and then similarly expand leaf nodes of S_2 that are not leaf nodes in T_1, all the while expanding the appropriate leaf nodes in S_1 and T_2 respectiviely so as not to change the elements. If we think of the trees T_1 and S_2 as living inside the infinite rooted binary tree, then we have $T_1' = T_1 \cup S_2 = S_2'$. This expansion process is illustrated in Figure 16.7. Once we have the appropriate representatives, we have $(S_2', T_2') \cdot (S_1', T_1') = (S_1', T_2')$

Exercise 12. Use the description of multiplication of tree pairs to compute x_0^2 and $x_0^{-1}x_1x_0$. Check that your answers agree with the piecewise functions from earlier.

Exercise 13. Write down piecewise functions for the tree pairs (S_1, T_1) and (S_2, T_2) in Figure 16.7, compose these functions, and check that you get the function represented by (S_1', T_2').

Exercise 14. Describe the tree pair representing $x_0^{-2}x_1x_0^2$.

16.3 PRESENTATIONS

I will give two group presentations for F, first an infinite one that has the advantage of being very symmetric and then a finite one, which (of course) has the advantage of being finite.

An infinite presentation. There is a standard infinite presentation of F with infinitely many generators and relators. There is one generator x_i for each integer $i \geq 0$, and one relation for each $0 \leq i < j$, as given by the presentation:

$$F \cong \langle x_0, x_1, \ldots \mid x_i^{-1}x_jx_i = x_{j+1}, \ i < j \rangle.$$

How does this match up with our earlier description of F? Well, as you might have guessed, x_0 and x_1 line up exactly with the piecewise-linear orientation-preserving homeomorphisms x_0 and x_1 that we've already looked at. As for the rest, the presentation inductively tell us what these should be; for $i \geq 2$, we want $x_{i+1} = x_0^{-1}x_ix_0$. In particular, this presentation claims that F is generated just by the two elements x_0 and x_1 as we stated earlier. We'll prove this surprising fact in a bit.

Exercise 15. Write $x_2 = x_0^{-1}x_1x_0$ and $x_3 = x_0^{-1}x_2x_0$ as piecewise functions and draw their graphs. Also give tree pair diagrams representing each. Exercise 14 is useful here.

Exercise 16. Describe a tree pair diagram for each x_i (again Exercise 14 is useful here). What do the graphs of the x_i look like?

Exercise 17. Use tree pair diagrams to verify the relation $x_0^{-1}x_2x_0 = x_1^{-1}x_2x_1$. Feeling confident? Try to verify the relations $x_i^{-1}x_jx_i = x_{j+1}$ for $j > i$. *Hint: First understand how x_i acts on the support of x_j when $j > i$.*

The relations all say that a lower-index generator conjugates a higher-index generator to a next generator. If we write an element of F as a word in the x_i, these relations give an easy way to move generators past each other. For example, we have

$$x_4x_0 = x_0x_5.$$

So moving the x_0 to the left of the x_4 changes the 4 to a 5. We can rewrite the last equation as

$$x_0^{-1}x_4 = x_5 x_0^{-1}.$$

In this case, moving x_0^{-1} to the right of x_4 again changes the 4 to a 5.

Exercise 18. Find tree pair diagrams for three elements of F that generate a subgroup isomorphic to $F \times \mathbb{Z}$.

A normal form.   In Office Hour 14, normal forms for right-angled Artin groups are used to solve the word problem in those groups. The idea is that if we want to tell if two words in the generators represent the same element, then we should convert the words to their normal forms and compare those. If there is a unique normal form for each group element, then this solves the word problem.

Back to F. If we have an element of F written as a word in the x_i, we can use the above defining relations to push all of the lower-index generators with positive exponents toward the beginning, and the lower-index generators with negative exponents toward the end. If we do this, we obtain a normal form for elements of F: a word in the x_i is in normal form if it starts with positive powers of the x_i, with indices in increasing order, and ends with negative powers of the x_i, with indices in decreasing order; in other words, it is a word of the form

$$x_{i_1}^{r_1} x_{i_2}^{r_2} \ldots x_{i_k}^{r_k} x_{j_l}^{-s_l} \ldots x_{j_2}^{-s_2} x_{j_1}^{-s_1}$$

where all of the exponents satisfy $r_i, s_i > 0$, and the indices are arranged so that $i_1 < i_2 \ldots < i_k$ and $j_1 < j_2 \ldots < j_l$.

Again, we are happiest when normal forms are unique. Unfortunately, this one is not unique in general. There is a "deconjugation" operation that gives equivalent words in normal form. For example, $x_0 x_5 x_0^{-1}$ is in normal form, but we can see from the relations that it is equal to x_4, which is also in normal form.

We can fix this by adding the restriction that when both x_i and x_i^{-1} occur in the word, so does at least one of x_{i+1} or x_{i+1}^{-1}. This eliminates the deconjugation possibility and the resulting normal forms are unique. This solves the word problem for F, albeit with respect to the infinite generating set $\{x_i\}$.

Exercise 19. Find normal forms for the elements:

1. $x_2 x_1 x_0$

2. $x_0^{-1} x_1^{-1} x_2^{-1} x_0$

3. $x_3^{-1} x_2 x_1^{-1} x_0^{-1} x_1 x_2 x_3$

The normal form we just gave was first described by Brown and Geoghegan [63]. From now on, when we refer to the normal form for an element of F, we will mean this unique normal form.

Exercise 20. Carefully prove that this normal form is unique.

Multiplying normal forms.   Multiplying two elements in normal form consists of setting the words next to each other, and then moving the positive generators from the second word past the negative generators of the first word, combining, reordering, and canceling as needed. For example, to multiply the words $x_0^2 x_3 x_2^{-1} x_1^{-1} x_0^{-1}$ and $x_0^2 x_1 x_3 x_2^{-1} x_1^{-1} x_0^{-1}$ we proceed by moving positive generators left from the beginning of the second word and work until we have a word in normal form:

$$(x_0^2 x_3 x_2^{-1} x_1^{-1} x_0^{-1})(x_0^2 x_1 x_3 x_2^{-1} x_1^{-1} x_0^{-1}) = (x_0^3 x_4 x_3^{-1} x_2^{-1})(x_1 x_3 x_2^{-1} x_1^{-1} x_0^{-1})$$
$$= (x_0^3 x_1 x_5 x_4^{-1} x_3^{-1})(x_3 x_2^{-1} x_1^{-1} x_0^{-1})$$
$$= x_0^3 x_1 x_5 x_4^{-1} x_2^{-1} x_1^{-1} x_0^{-1}.$$

Essentially, the process is to move the generators that are out of place (the negative ones at the end of the first word and the positive ones at the beginning of the second) into their proper places, while during the movement the indices of the generators passed may change or cancel. There are many methods for proceeding and with the uniqueness of the normal forms all methods will give the same final result.

A finite presentation.   As we've observed, the infinite generating set used above is excessive: it turns out that F is generated by x_0 and x_1. Indeed, the third generator x_2 is nothing else than $x_2 = x_0^{-1} x_1 x_0$. By the same logic, we can write x_3 in terms of x_1 and x_2, and hence in terms of x_0 and x_1. More explicitly, for $i > 1$:

$$x_i = x_0^{-i+1} x_1 x_0^{i-1}.$$

The infinite family of defining relations clearly has some redundancies as well. For example, consider the relation

$$x_2^{-1} x_5 x_2 = x_6.$$

If we conjugate the entire expression by x_0^{-1}, we increase all of the indices by 1, and so we obtain as an immediate consequence that

$$x_3^{-1} x_6 x_3 = x_7.$$

These two equations appeared as two distinct relations in the infinite family of relations given above. In fact, all of those infinitely many relations are consequences of the first two nontrivial ones.

Exercise 21. Write a proof of this last claim. As a hint, here are the steps to showing that $x_1^{-1} x_4 x_1 = x_5$ is a consequence of $x_1^{-1} x_2 x_1 = x_3$ and $x_1^{-1} x_3 x_1 = x_4$.

1. First show that $x_2^{-1} x_3 x_2 = x_4$ and $x_3^{-1} x_4 x_3 = x_5$ are consequences of the first relation by conjugating by x_0^{-1}, as in the example.
2. Then compute:

$$x_1^{-1} x_4 x_1 = x_1^{-1} x_2^{-1} x_3 x_2 x_1 = x_3^{-1} x_1^{-1} x_3 x_1 x_3 = x_3^{-1} x_4 x_3 = x_5.$$

From these observations, we obtain a finite presentation for F:

$$F \cong \langle x_0, x_1 \mid x_1^{-1} x_2 x_1 = x_3, \ \ x_1^{-1} x_3 x_1 = x_4 \rangle.$$

The relations given in this presentation are expressed in terms of x_2, x_3, and x_4, which are not permitted generators, but of course those can be each expressed in terms of x_0 and x_1 as above. Doing so gives the two relations as

$$x_1 x_0^{-1} x_1 x_0 x_1 = x_0^{-2} x_1 x_0^2$$

and

$$x_1^{-1} x_0^{-2} x_1 x_0^2 x_1 = x_0^{-3} x_1 x_0^3.$$

These give relators of length 10 and 14, respectively.

Proof of the presentation.   We have one important job remaining, namely, to prove that the purported presentations really are presentations of F. We already discussed why the two presentations give isomorphic groups, and so it remains to examine why the infinite presentation really is a presentation for F. For this, we will think of the elements of F as tree pair diagrams (although, as discussed, it is easy to convert between tree pair diagrams and piecewise-linear homeomorphisms).

Our main task is to construct a bijection between reduced tree pair diagrams and normal form words in the x_i. The key lies in the notion of a leaf node exponent, which I now define.

First, we say an edge in a rooted binary tree is a *left edge* if it connects a parent node to its left child. Similarly we have *right edges*. The right side of the tree is the root node and all edges and nodes connected to the root via paths consisting entirely of right edges. For a single tree T, the leaf node exponent e_i is the length of the longest path with the following properties:

1. the path starts at the leaf node labeled i,
2. the path consists entirely of left edges, and
3. the path does not reach the right side of the tree.

(The length of a path is the number of edges.) Necessarily the leaf node exponent of any leaf node on a right edge is 0. Similarly, any leaf node that is a child of a node on the right side of the tree is 0.

Consider the tree shown in Figure 16.8. The leaf node exponent of leaf node 0 is 2, since the third left edge starting at leaf node 0 touches the root, which is part of the right side of the tree. The leaf node exponents for leaf nodes 1 and 2 are both 1, as the second edges up from them are right edges, not left edges. The leaf node exponent for leaf node 3 is 0, as it is a right child and has no path consisting entirely of left edges. Similarly, the leaf node exponents of 6, 9, 12, 13, and 14 are all 0. The other nonzero leaf node exponents are those of 4, 5, 7, and 8, which are all 1; and finally there is leaf node 11, which has leaf node exponent 2.

Given a tree pair diagram (S, T), we associate to it a normal form word in the x_i via leaf node exponents. Specifically, the word is

$$x_0^{i_0} x_1^{i_1} \ldots x_n^{i_n} x_n^{-j_n} \ldots x_1^{-j_1} x_0^{-j_0},$$

where the exponents i_k of the positive part of the word are formed from the leaf node exponents for leaf nodes in the tree S, and the exponents j_k of the negative part of the word are formed from those of the tree T. For example, the leaf node

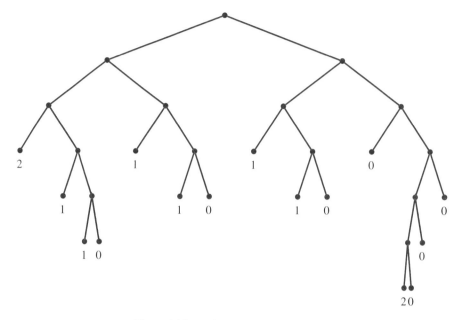

Figure 16.8 Leaf node exponents of a tree.

exponent i_3 is the leaf node exponent of the third leaf node in S and determines the power of x_3 in the normal form. Many of these leaf node exponents may be 0 and we omit those when writing the normal form.

If the example tree from Figure 16.8 were the first tree in a reduced tree pair, the positive part of the associated normal form would be $x_0^2 x_1 x_2 x_4 x_5 x_7 x_8 x_{11}^2$.

Exercise 22. Using the reduced tree pair diagram for x_i that you found earlier, verify that this process returns x_i.

Exercise 23. Find the normal forms for the relevant elements used for the multiplication pictured in Figure 16.7 and for the resulting product.

Exercise 24. It is possible to construct a tree pair diagram given a normal form. Take the example of multiplication $(x_0^2 x_3 x_2^{-1} x_1^{-1} x_0^{-1})(x_0^2 x_1 x_3 x_2^{-1} x_1^{-1} x_0^{-1})$ and construct tree pair diagrams for the factors; multiply and then confirm that the leaf node exponents for the resulting product are the same as those from the normal form.

We note that while we generally use this process on reduced tree pair diagrams to obtain unique normal forms, we can also apply the process to unreduced tree pair diagrams, which will result in normal forms which do not satisfy the condition that if x_i and x_i^{-1} are both present, so must one of $x_{i+1}^{\pm 1}$. Performing the reduction on the tree pair diagram is analogous to the deconjugation operation on the normal form. That observation helps ensure that indeed these two descriptions are of the same group.

Exercise 25. Prove that the process just given really gives a bijection between reduced tree pair diagrams and normal forms for words in the x_i. Use this to prove that the purported infinite presentation for F really is a presentation for F.

At the same time we prove that the purported presentations of F are really presentations of F, we also prove the important fact that F is finitely generated (Theorem 16.1), as one of the presentations has finitely many generators.

Other descriptions of F.   There are many other descriptions of F and in fact none of the ones given here are the original description of F. Thompson [254] originally described F as a group of reassociations of expressions in finitely many variables, which can be seen to be equivalent to the tree pair description. Thompson's group can also be described as a diagram group, using essentially duals to the tree pair descriptions and where the group multiplication operation becomes the concatenation of diagrams.

16.4 ALGEBRAIC STRUCTURE

We have already seen that F is torsion free, it contains large abelian subgroups, it is finitely generated, and (further) it is finitely presented. In general, we can understand the algebraic structure of a group by understanding (1) its subgroups, (2) its quotients, (3) its endomorphisms, and (4) its group actions. In this section we explore each of these four aspects of Thompson's group F.

Subgroups.   We've already seen that we can use commuting bump functions to show that F contains subgroups isomorphic to $\mathbb{Z}$, $\mathbb{Z} \times \mathbb{Z}$, ..., and in fact free abelian subgroups of countable rank.

Even more remarkable, Thompson's group contains multitudes of copies of itself, sitting inside it in many different ways. The unit interval is naturally dydically isomorphic to the interval $[0, \frac{1}{2}]$, so if we consider all elements of F whose support lies in $[0, \frac{1}{2}]$, we obtain a subgroup isomorphic to F. Furthermore, if we do the same thing for elements whose support lies in $[\frac{1}{2}, 1]$, we obtain another copy of F that commutes with the first. Thus F has a subgroup isomorphic to $F \times F$.

Once we have a subgroup of F isomorphic to $F \times F$, we continue to find many subgroups of the form $F \times F \times \cdots \times F$. In fact, we can find a subgroup isomorphic to the direct product of F with itself countably many times, merely by taking copies of F in successive dyadic intervals $[0, \frac{1}{2}], [\frac{1}{2}, \frac{3}{4}], \ldots [\frac{2^n-2}{2^n}, \frac{2^n-1}{2^n}], \ldots$. So:

F contains a subgroup isomorphic to $F \oplus F \oplus F \oplus \cdots$.[3]

---

[3] There is a slight technical issue here: the direct sum $\oplus$ versus the direct product $\times$. These are related objects; the direct product consists of tuples of elements, the direct sum is the subgroup where only finitely many of the elements of each tuple are not the identity. When there are only finitely many factors, there is no difference. Since elements in F have only finitely many break points, we need the direct sum here.

This gives a natural self-similar structure to F, and is one reason that F turns up in so many different settings. On one hand, F has this natural self-similar structure which is a useful way of copying phenomena in F into its factors, yet somehow F is finitely presented. This is part of the miraculous tension that makes F so interesting to study.

Thompson's group F also has wreath products as subgroups, sitting in a natural way that make them a little easier to see once we have a better understanding of the actions of the generators on tree pair diagrams, so we will look at that below.

Homomorphisms to abelian groups.   There are natural homomorphisms from F to $\mathbb{Z}$ that come from the description of F in terms of piecewise-linear homeomorphisms. Given $f \in F$, we can consider the slope of f near 0; this makes sense since there are only finitely many break points. This slope is a power of 2. So the map

$$\psi_0 : F \to \mathbb{Z}$$

that records the base 2 logarithm of the slope near 0 is well defined. In fact it is a homomorphism: if two elements of F have slopes 2^j and 2^k near 0 then their composition has slope 2^{j+k}.

Exercise 26. Check the last statement.

Similarly, there is a homomorphism $\psi_1 : F \to \mathbb{Z}$ that records the slope near 1. These homomorphisms are easy to define and understand. When we look for more homomorphisms from F to $\mathbb{Z}$—or to any other abelian group—we actually won't find any! Let's explain this.

First, the abelianization G^{ab} of a group G is the group obtained from G by declaring that all elements of G commute. If we have a presentation for G, we obtain a presentation for the abelianization by adding relations that say that all pairs of generators commute. There is a natural homomorphism from G to its abelianization. This homomorphism is called the abelianization map.

Exercise 27. Check that the kernel of the abelianization map for G is the commutator subgroup of G.

Exercise 28. Show that the abelianization of the free group F_n is $\mathbb{Z}^n$. Show that the abelianization of any abelian group is isomorphic to the original group. Find an example of a nontrivial group whose abelianization is trivial.

The reason that the abelianization of a group is so important is because it has the following universal property:

> If A is any abelian group and $\phi : G \to A$ is any homomorphism then ϕ factors through the abelianization, that is, ϕ is equal to a composition
>
> $$G \to G^{ab} \to A.$$

By the first isomorphism theorem, this universal property immediately gives that A is a quotient of G^{ab}. So, for example, if G^{ab} is trivial, then G has no homomorphism onto a nontrivial abelian group. And if G^{ab} is finite then G has no surjective homomorphisms onto an infinite abelian group. In other words, computing the single group G^{ab} tells us about arbitrary homomorphisms from G to abelian groups. That is pretty powerful!

Exercise 29. Prove the universal property of the abelianization.

Exercise 30. Show that there is no surjective homomorphism $F_2 \to \mathbb{Z}^3$.

Earlier I said that there are no homomorphisms $F \to \mathbb{Z}$ besides the ones ψ_0 and ψ_1 already given. In our new language, what this means is that the abelianization of F is $\mathbb{Z}^2$ and the abelianization map is nothing other than:

$$\psi_0 \times \psi_1 : F \to \mathbb{Z} \times \mathbb{Z}.$$

This is not so hard to prove. First let us check that $\psi_0 \times \psi_1$ is surjective. This follows immediately from the computations:

$$\psi_0(x_0) = -1 \qquad \psi_1(x_0) = 1$$
$$\psi_0(x_1) = 0 \qquad \psi_1(x_1) = 1.$$

Exercise 31. Compute the images under ψ_0 and ψ_1 of each of the x_i in F. Find an element $f \in F$ that has $\psi_0(f) = 0$ and $\psi_1(f) = 1$.

Since $\psi_0 \times \psi_1$ is a surjective homomorphism from F to $\mathbb{Z}^2$, we know that the abelianization of F has $\mathbb{Z}^2$ as a quotient. On the other hand, we know that F has a generating set consisting of two elements and so the abelianization of F is an abelian group that is generated by two elements. But any such group is a quotient of $\mathbb{Z}^2$. Combining these two statements: the abelianization of F is a quotient of $\mathbb{Z}^2$ and has $\mathbb{Z}^2$ as a quotient. It follows that the abelianization is $\mathbb{Z}^2$!

Exercise 32. Check the last statement.

Let's do a reality check. We might be thinking that we've missed some obvious homomorphisms from F to $\mathbb{Z}$. For instance, since the relations of F can be written as conjugation relations, we have the exponent sum homomorphisms: write an element of F as a word in the x_i and add the exponents on some particular x_i. But those homomorphisms are just variations on ψ_0 and ψ_1. The exponent sum for x_0 is merely $-\psi_0$ and the exponent sum homomorphism for x_1 is $\psi_1 + \psi_0$.

We can use this to help understand the kernel of $\psi_0 \times \psi_1$, namely, the commutator subgroup of F. These are elements that coincide with the identity on neighborhoods of both 0 and 1. The exponent sums on both x_0 and x_1 will necessarily both be 0, and with some manipulation we can write any such element as a product of commutators.

Exercise 33. Show that the exponent sum homomorphism for x_1 is indeed given by $\psi_1 + \psi_0$.

Exercise 34. Find the normal form for the group element given by the tree pair diagram in Figure 16.5. Calculate the slopes at 0 and 1 for the element and check that they are the same as when using the exponent sum homomorphism. Do the same for the element whose graph is shown in Figure 16.9.

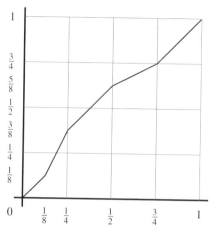

Figure 16.9 An element in the commutator subgroup and the kernel of the slope homomorphism, as the slopes at the left and right endpoints are both 1. Any element
with such "flat ends" lies in the kernel of the slope map and the commutator
subgroup.

Exercise 35. Find a recipe for writing an element of the the kernel of $\psi_0 \times \psi_1$ as
a product of commutators.

We can see the slopeat 0 map ψ_0 in terms of the tree pair diagram representation
easily. All we need to do is look at the depth of leaf node 0 in the element given
by (S, T) in each tree S and T. If the depths are the same, the slope at 0 is 1. If
the depths differ by i, then the slope at 0 is either 2^i or 2^{-i}, depending upon which
leaf node 0 is deeper in the tree. Similarly, by looking at the depths of leaf node n
in each tree, we can obtain the slope at 1 from the tree pair diagram.

Endomorphisms: The shift map.   Inspecting the relations of F, we notice that
the relations merely depend upon the relative sizes of the subindices, and not on
their absolute size. Thus if we uniformly increase the subindices of the relations,
nothing really changes. This observation gives rise to an important endomorphism
from F into itself, called the *shift map* ϕ, which is given by

$$\phi(x_i) = x_{i+1}.$$

It is immediate that the shift map is a homomorphism, and it is straightforward to
see that it is injective.

Exercise 36. Check that the shift map ϕ is injective.

The shift map fails to be surjective as it misses words whose normal form has
an $x_0^{\pm 1}$.

Geometrically, the shift map takes an element given by a tree pair diagram
(S, T), and returns an element whose tree pair diagram is (S', T') where the primed
versions of the trees are the original trees with a single caret added on top of the

root, with the former roots of S and T now the right children of the roots of S' and T'.

Analytically, the shift map just takes an element f of F thought of as piecewise-linear isomorphism from $[0, 1]$ to itself and shifts it to another piecewise-linear homeomorphism, which is the identity on $[0, 1/2]$ and then is a shrunken version of f on the interval $[1/2, 1]$. This can be seen by noting that the new tree pair diagram is guaranteed to have leaf node 0 be the left child of the root in both trees, making the analytic map the identity on the interval $[0, 1/2]$. Then, the right child of the root contains the previous element, necessarily scaled to fit the interval $[1/2, 1]$.

If we recall our standard examples of $F \times F$ subgroups of F, we see that the shift map ϕ is the map $F \to F$ taking each element of F to the second factor of $F \times F$.

The shift map is almost a conjugation by x_0. The generator x_0 conjugates all higher index generators to be one index higher. So on the set of words that do not involve an x_0, the shift map ϕ and conjugation by x_0 coincide. One way to make sure that a word does not involve an x_0 is to hit it with the shift map first, so we see that $\phi^2(w) = x_0 \phi(w) x_0^{-1}$. This looks quite similar to the idempotent relation $\phi^2 = \phi$. The shift map is not a genuine idempotent, but it is an idempotent up to conjugacy. In fact, Freyd and Heller [137] discovered F as the universal example of such a group conjugacy idempotent.

Action on the Cantor set.   Since we use trees to represent elements of F, it is tempting to look for a nice action of F on a tree. For better or worse, though, there is no such action. Instead of acting on a tree itself, elements of F act on the set of ends of a tree, or equivalently, the Cantor set. Let us start by making sense of this last statement.

You probably learned the definition of the Cantor set somewhere (for instance, Office Hour 10): you start with the unit interval $[0, 1]$ and successively delete "middle thirds" infinitely many times and what is left over is the Cantor set. You probably also learned that the Cantor set is both uncountable and has measure (or total length) zero: another good example of tension in mathematics.

What you might not have thought about as much is the topology on the Cantor set. Since the Cantor set is a subset of $[0, 1]$, it makes sense to say that a sequence of elements of the Cantor set is approaching some specific element of the Cantor set. To say it another way: a small neighborhood of an element of the Cantor set is the intersection with the Cantor set of a small open interval in $[0, 1]$ around that element (why are these essentially the same?).

What does this have to do with trees? Well, one way to specify an element of the Cantor set is to say which "thirds" it lives in. For instance, in the construction of the Cantor set, we first delete $(1/3, 2/3)$. So any given element of the Cantor set lies in the left interval $[0, 1/3]$ or the right interval $[2/3, 1]$. But we can keep going. Within one of these intervals, our chosen element lives in either the left or the right third, etc., etc. If we specify an infinite number of lefts and rights, we eventually pin down a specific element of the Cantor set. In other words, elements of the Cantor set correspond to infinite paths in a rooted binary tree! (Compare the description here with the description in terms of L-R sequences in the beginning of Office Hour 10.)

Exercise 37. Here we have identified the Cantor set with the set of ends of a rooted binary tree. In Office Hour 10 the Cantor set is identified with the set of ends of a regular 4-valent tree. Make sense of this.

Yet another way to see the connection between the ends of the tree and the elements of the Cantor set as it is usually presented is via ternary (base 3) representation of real numbers between 0 and 1. Deleting the closed middle third from 1/3 to 2/3 is equivalent to removing real numbers whose ternary representative has the digit 1 in the first decimal place (the thirds place).[4] The next stage of deleting the middle thirds [1/9, 2/9] and [7/9, 8/9] removes those with a 1 digit in the second (ninths) place of the ternary representation. Each further middle third deletion will forbid a 1 digit in another ternary place. Thus the Cantor set can be represented as those real numbers between 0 and 1 whose ternary representatives have only 0's and 2's occurring in their expansions. Since ends of the rooted binary tree can be described as infinite strings of 0's and 1's via a natural addressing scheme where 0 indicates left and 1 indicates right, there is a natural bijection sending 0 to 0 and 2 to 1 in these two descriptions of the Cantor set. Pictorially, we can think about an end of the tree telling us which direction to choose at each stage to dodge being in the middle third that gets chopped out.

General group actions on the Cantor set have the potential to be mind-bogglingly complicated, but there are certain kinds of actions that are tractable. For instance, we will show in a moment that F acts on the set of ends of the rooted infinite binary tree (= Cantor set), but these actions are all order-preserving and hence describable in some finite way.

We begin by thinking about how x_0 acts on the Cantor set, realized as the set of ends of the infinite rooted binary tree. The reduced tree pair diagram for x_0 has two carets in each tree. To each leaf node of each tree we can tack on an infinite rooted binary tree. In both (now infinite) trees, we label these added subtrees from left to right as A, C, and E. We think of x_0 as rearranging the first tree by promoting A from a grandchild to a child, switching the parent of C, and demoting E from a child to a grandchild, as shown in Figure 16.10.

This action is sometimes called "right rotation at the root" of a binary tree, despite the distracting fact that it is not a continuous motion as we typically expect from something called a rotation. The action forcibly rips the tree apart, and nodes that were joined before (such as B and C) are no longer adjacent. However, the ends of the tree are preserved in a manner that respects a left-to-right order. (These rotations also turn up as manipulations on binary trees considered as data structures.) The generator x_1 has a similar action, but instead of acting at the root, it acts in the same way as x_0 but at the right child of the root.

---

[4] In base 10, some real numbers such as 1 have two decimal representations, since $0.99999\cdots = 1.0000000\ldots$. We can obtain unique decimal representatives by forbidding an infinite string of 0's. Similarly, in base 3, there are some real numbers that have two representatives, such as $1/3 = 0.1 = 0.0222\ldots$. For uniqueness, we can again forbid representations which end in an infinite string of 0's.

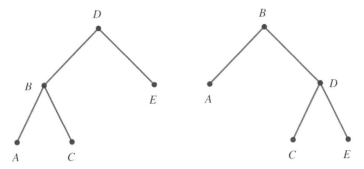

Figure 16.10 A right rotation of a binary tree at the root. The nodes marked A, C, and E could represent leaf nodes or subtrees.

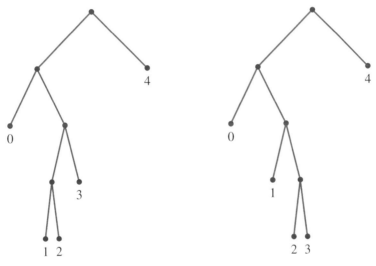

Figure 16.11 The element $x_0 x_1^2 x_2^{-1} x_1^{-1} x_0^{-1}$ which generates a copy of $\mathbb{Z} \wr \mathbb{Z}$ in F together with x_0.

A wreath product subgroup. I now explain how to find a subgroup of F isomorphic to $\mathbb{Z} \wr \mathbb{Z}$. In Office Hour 15 it was shown that the lamplighter group is isomorphic to the wreath product $\mathbb{Z}_2 \wr \mathbb{Z}$. The group $\mathbb{Z} \wr \mathbb{Z}$ has a similar geometric model, with the binary lamps (each with an on or off state) replaced by counters (each with countably many states).

To represent the states of the zeroth "lamp," take an element of F with small support such as $x_0 x_1^2 x_2^{-1} x_1^{-1} x_0^{-1}$, shown in Figure 16.11. Its powers generate an infinite cyclic subgroup that can serve as a counter. The conjugates of this element by powers of x_0 all have disjoint support and commute, so we have a copy of $\mathbb{Z} \wr \mathbb{Z}$ as a subgroup of F. The generator for the first $\mathbb{Z}$ in the wreath product is $x_0 x_1^2 x_2^{-1} x_1^{-1} x_0^{-1}$ and the generator for the second $\mathbb{Z}$ in the wreath product is x_0 (this corresponds to the movement of the lamplighter).

Exercise 38. Show that the generators mentioned in Figure 16.11 actually satisfy the relations of $\mathbb{Z} \wr \mathbb{Z}$; that is, $\langle a, t \mid [t^{-i}at^i, t^{-j}at^j] \rangle$ for all integers i, j.

16.5 GEOMETRIC PROPERTIES

Let's end by discussing several geometric properties of the group F: word length, distortion, dead ends, almost convexity, growth, and amenability.

Word length.   We of course expect a complicated element of F to have large word length, but we need to be careful about what we mean by complicated. A piecewise-linear homeomorphism of $[0, 1]$ with many break points seems complicated, and it is, but there are elements with just a few break points that are actually quite large with respect to word length. For instance, x_{100} has word length 199 ($x_{100} = x_0^{-99}x_1x_0^{99}$) but it has only three break points, which happens to be the same number of break points as its conjugate x_1.

We can measure word length much more effectively from the perspective of tree pair diagrams. In this picture, an element looks complicated if its reduced tree pair diagram has many carets. Blake Fordham surprised the geometric group theory community in 1995 by giving a simple and elegant formula for the word length of an element of F in terms of the reduced tree pair diagram. Not only do his methods show that word length is roughly proportional to the number of carets in the reduced tree pair diagram, but, even better, he gave an easy-to-follow recipe to compute the exact word length by examining the tree pair diagram! Check his paper [133] for the details.

Distortion.   Let G be a finitely generated group and let H be a finitely generated subgroup. Choose finite generating sets for both. Recall from Office Hour 8 that the distortion function for the inclusion $H \to G$ is the function $\mathbb{Z}_{\geq 0} \to \mathbb{Z}_{\geq 0}$ that measures the largest word length in the H-metric of any element of H whose word length in the G-metric is at most n:

$$D_H^G(n) = \max\{|h|_H \mid h \in H, \ |h|_G \leq n\}.$$

Also recall from Office Hour 8 that this function is well defined up to a certain equivalence.

Why care about distortion? Well, for example, if an infinite cyclic subgroup is undistorted, meaning that the distortion function is equivalent to the identity function, we can think of the elements of the cyclic subgroup as tracing out a path that is close to being a geodesic. A first step to understanding a metric space is to understand its geodesics, so this fits in quite well with the main goal of geometric group theory. Understanding the distortion of a more complicated subgroup should be thought of as a generalization, and if such a subgroup is undistorted, that should remind us of totally geodesic (or convex) subspaces.

Using Fordham's method, we can deduce that the $\mathbb{Z}^n$ subgroups, the $\mathbb{Z} \wr \mathbb{Z}$ subgroups, and the $F \times F$ subgroups of F constructed earlier in the office hour are all undistorted.

Exercise 39. One consequence of Fordham's method is that the word length of an element of F is proportional to the number of carets in the reduced tree pair diagram representing it. Use that to prove the facts stated in the last paragraph.

Dead ends. Let us work with the finite generating set $\{x_0, x_1\}$ and the resulting Cayley graph and word metric on F. As in Office Hour 15, a *dead end element* is an element with word length n that is not adjacent in the Cayley graph to any elements of word length $n + 1$. Recall that dead end elements have varying degrees of depth, measuring how far one needs to move from a dead end element in order to reach an element with larger word length. There are arbitrarily deep dead ends in wreath products with $\mathbb{Z}$ (for instance, in the lamplighter group). The group F does have dead ends, but they are all of depth 2; that is, they have elements at distance three that have larger word length.

Almost convexity. We say that a group with a fixed finite generating set is minimally almost convex if every two elements with word length n have distance strictly less that $2n$ from each other within the ball of radius n centered around the identity (so we are looking for the shortest path that connects the two points and stays within the ball of radius n). Of course the distance is always at most $2n$, as we can see by concatenating two paths of length n to the identity. So minimal almost convexity is the smallest possible improvement over this bound.

It is known that F is not minimally almost convex [21, 81]. This means that if you compute the entire ball of radius n you will find two elements that appear to be very far apart, but then you find that there is a very short path between them that travels outside the ball of radius n. In other words, even if we write down the entire ball of radius n, we still might not have a good idea of the geometry of the ball of radius $n + 1$, let alone the whole group.

Groups that do not have a presentation with finitely many relators are never minimally almost convex (why?). But there it makes sense because we will never understand all of the infinitely many relations by looking in a finite ball. The group F is finitely related and yet is also not minimally almost convex, another of the many tensions found in this group.

Amenability. A group is *amenable*[5] if there is a finitely additive left-invariant probability measure on the set of all subsets of F. A combinatorial characterization of amenability is the existence of *Følner families*, which are increasingly large sets where the fraction of elements in the boundary decreases to zero.

A more amusing definition (due to Gromov) goes as follows. We say that a group is non-amenable if it admits a Ponzi scheme. More precisely, imagine that there is a person at every vertex of the Cayley graph for the group and each person has one dollar. Fix some number D. If all at once each person can give his or her dollar to someone else within distance D so that every person has more than one dollar at the end, then the group has a Ponzi scheme.

---

[5] The term "amenable" was coined by Mahlon M. Day in 1949. It is thought to be a pun, as a group is amenable (pronounced: a-mean-able) if it has a specific type of measure, or mean.

Exercise 40. Show that all non-abelian free groups admit a Ponzi scheme.

By the previous exercise, non-abelian free groups are not amenable, and groups that contain non-abelian free groups are not amenable. Brin and Squier [57] proved that F contains no non-abelian free group and so we cannot determine the amenability of F on that basis.

Finite and abelian groups are amenable, and groups built from amenable groups via taking subgroups, forming quotients, forming extensions, and by directed unions are called elementary amenable. The group F cannot be constructed in this way (see Cannon, Floyd, and Parry's article [72, Theorem 4.10]), so we cannot determine the amenability of F on that basis either.

A great deal of effort has gone into determining the amenability of F, so far without success. Either possibility turns out to be interesting: if amenable, F is a finitely presented amenable group that is not elementarily amenable; if non-amenable, F is a finitely presented non-amenable group that does not contain a non-abelian free group. Though by now we have other examples of both these phenomena, those examples are not in a group that occurs naturally, such as F.

Growth. The growth of F, à la Office Hour 12, is also a source of interesting questions. The exact growth function (the numbers of elements in balls size n) of F is known to be larger than an exponential function, but the exact growth function is not known. Even the rate of growth is not known, and it is not known if the growth series is rational or of a reasonably tame nature. Nevertheless, the first 1,500 terms in the growth series are known [122] and the growth appears to be very close to the upper bounds constructed by Guba [157]. Computing this growth on the nose is a tantalizing open question.

Hopefully I have convinced you that Thompson's group F is an odd, and in some ways wild, group, with a number of perplexing properties. Despite many years of study, it has shown strong resistance to understanding many of its geometric aspects. I hope you are inspired to attack some of these questions.

FURTHER READING

As we already mentioned, the classic introduction to Thompson's group F and its cousins (T and V) is the survey by Cannon, Floyd, and Parry [72]. Brown and Geoghegan [63] showed that F is infinite-dimensional (read the paper to find out what that means) and that article includes important fundamental properties of the presentations for F. Guba and Sapir [159] describe the diagram group approach to understanding F. Progress on some of the fine-scale metric properties of F were made possible after Fordham [133] developed his method for computing word length of F exactly, including some convexity results of Cleary and Taback [81] and Belk and Bux [21], as well as the description of dead end elements in F by Cleary and Taback [82]. Farley constructs an infinite-dimensional nonpositiviely curved cube complex [128] for F. Moore and Lodha [193] describe interesting groups of piecewise projective homeomorphisms that contain F.

PROJECTS

Project 1. Thompson's group F can also be regarded as the group of piecewise-linear homeomorphisms of the real line, satisfying certain conditions. Develop those conditions precisely, by conjugation with the piecewise-linear map from $[0, 1]$ to the real line that sends $\frac{1}{2}$ to 0, $\frac{3}{4}$ to 1, $\frac{7}{8}$ to 2, $\frac{1}{4}$ to -1, and so on. Find the restrictions on such maps needed to give another characterization for F, and describe the map that corresponds to the slope-at-the-endpoints homomorphism $\psi_0 \times \psi_1$.

Project 2. Thompson's group F can also be described as the group of piecewise-integral projective homeomorphisms of the unit interval. This can be seen from using the tree pair description. Instead of viewing rooted trees as describing subintervals according to dyadic subdivision, the subintervals come from the Farey subdivision. Construct examples to show that in this description, the break points are now potentially any rational number in the unit interval and the slopes are potentially any positive rational number. Investigate the role of Minkowski's question mark function $?(x)$ in connecting this description to the analytic description.

Project 3. Thompson's group T is analogous to Thompson's group F except that we consider piecewise-linear homeomorphisms of the unit circle rather than the unit interval. We regard the circle as the quotient of the real line by the relation of two numbers being equivalent if they differ by an integer, so there are representatives in $[0, 1)$. So, for example, there is torsion in T: the map that sends t to $t + \frac{1}{2}$ has order 2 and in fact this element t, together with x_0 and x_1, forms a generating set for T. It is easy to construct similar elements of orders 2^k. Begin by constructing an element of order 3 in T. These maps can be described by tree pair diagrams where the leaf nodes are numbered cyclically from 0 to n, not necessarily beginning with 0. Find the element obtained by conjugating $x_0 x_1 x_3 x_2^{-1}$ by the generator t. Begin to explore relations in T obtained by multiplying the tree pair diagrams from generators. Note that the slope-at-the-endpoint homomorphism is no longer present as there are no endpoints. In fact, T has no nontrivial homomorphisms and was an early example of a finitely presented infinite simple group, constructed by Thompson [254].

Project 4. It may seem as though the piecewise-linearity is essential in the description of F. That is, due to the resulting many possible changes in slope, it appears as though it may be difficult to realize F as a group of differentiable maps rather than merely continuous albeit piecewise-linear ones. If you understand French, investigate the description of F as a group of not just differentiable but actually C^∞ diffeomorphisms of the unit circle given by Ghys and Sergiescu [144].

Project 5. Transducers are simple computers that take an infinite string of 0's and 1's as input, and output an infinite string of 0's and 1's. Transducers can be synchronous, meaning that at every intermediate stage the length of the input string processed is the same as the output produced at that point. Or they can be asynchronous, without that restriction. There is a natural operation of composition of transducers, and the identity transducer is a synchronous transducer that merely copies the input to the output. Some transducers are invertible, meaning merely that there is another transducer so that the composition is the identity transducer, and such invertible ones form a group under composition. Groups generated by even just a few invertible transducers have proven to be potentially interesting examples of a number of unusual phenomena, including the construction by Grigorchuk [153] of a finitely generated group with subexponential growth. Each invertible transducer can be regarded as a bijection of the Cantor set via the addresses of the ends of the tree. Thompson's group F can be described as a group generated by asynchronous transducers. Learn more about transducers and construct transducer descriptions for the generators x_0 and x_1 as well as for a more complicated element such as $x_2 x_3^2 x_1^{-2} x_0^{-2}$.

Project 6. Thompson's group V is analogous to Thompson's groups F and T except that we consider right-continuous piecewise-linear bijections of the unit circle rather than the unit interval. As with T, we regard the circle as the quotient of the real line. Elements of T necessarily preserve a cyclic order of points on the circle, but elements of V need not do so. Read enough of Cannon, Floyd, and Parry's paper [72] to see how to describe elements of V via labeled tree pair diagrams and how to compose them, and then find subgroups of V that are isomorphic to finite permutation groups. Since V is finitely presentable, we now have a finitely presented group that contains all possible finite groups as subgroups.

Project 7. The description of F via dyadic rationals and slopes being powers of 2, with binary trees, has analogs where we replace 2 by 3 to get what is known as $F(3)$, where break points lie in the triadics $\mathbb{Z}[\frac{1}{3}]$ instead of the dyadics $\mathbb{Z}[\frac{1}{2}]$, and the slopes, when defined, are powers of 3. The tree description of this group is similar, replacing binary trees with ternary trees. Similarly, there are groups $F(n)$ for the analogous versions with $\mathbb{Z}[\frac{1}{n}]$ break points and slopes powers of n, or via n-ary trees. Even though the dyadic rationals inside the unit interval do not intersect the triadic rationals inside the unit interval, the dyadic version of F (the original F) occurs as a subgroup of the triadic version $F(3)$! Read Burillo, Cleary, and Stein's paper [66] and note that there are many ways that these groups sit inside of each other.

Project 8. There are descriptions of F as pairs of forest diagrams, similar to the tree description where we cut off the backbone along the right side of the tree, leaving a forest of trees whose roots correspond to the right side of the tree. Read Belk and Brown's paper [22] to understand this description and develop the ability to convert back and forth from forest diagrams to the tree pair description. Investigate how this description can give a manageable way of measuring word length in F with respect to the standard generators of F.

Project 9. Cannon, Floyd, and Parry [71] describe F via "pipe diagrams." Read the short "What is Thompson's group F?" article and see how to convert elements from that description to any of the others described above and vice versa.

Project 10. There are braided versions of Thompson's groups, described by Brin [56] and Dehornoy [108], which replace the permutations associated with elements of Thompson's group V with braids. Read enough of those or the paper by Burillo and Cleary [67] to understand how to compose elements, and how there is a natural factorization of such elements into tree-braid-tree triples.

Project 11. There are higher-dimensional versions of Thompson's group V, described by Brin [55], where elements act on products of Cantor sets and can be visualized as bijections between squares or cubes of the appropriate dimension. Read enough of Brin's paper or Burillo and Cleary's [68] to see how to describe and compose elements of these groups, and understand some of the differences between the one-dimensional and higher-dimensional versions.

Project 12. Investigate the dynamical approach to understanding elements of Thompson's group V by the notion of "revealing pairs" described by Brin [55]. These not-necessarily-reduced pairs capture information helpful for understanding how points behave under repeated iterations of applications of a given element of V.

Office Hour Seventeen

Mapping Class Groups

Tara Brendle, Leah Childers, and Dan Margalit

An overarching theme in mathematics is that we can learn about an object by studying its group of symmetries. For example, in abstract algebra we study two fundamental objects in mathematics—a finite set and a regular polygon—via the symmetric group and the dihedral group, respectively.

The goal of this chapter is to introduce the *mapping class group*, that is, the group of symmetries of a fundamental object in topology: a surface. We will acquaint you with a few of its basic properties and give a brief glimpse of some active research related to this group.

Our main goal is to find a nice generating set for the mapping class group. We will introduce elements called Dehn twists, symmetries of surfaces obtained by twisting an annulus. At the end we will sketch a proof of the following theorem of Max Dehn.

THEOREM 17.1. *The mapping class group of a compact orientable surface is generated by Dehn twists.*

In Section 17.1 we give an introduction to surfaces and explain the concept of a homeomorphism, our working notion of isomorphism for surfaces. In Section 17.2 we give examples of homeomorphisms and in Section 17.3 we define the mapping class group as a certain quotient of the group of homeomorphisms of a surface. In Section 17.4, we discuss Dehn twists and some of the relations they satisfy. We prove Theorem 17.1 in Section 17.5. Finally, we conclude with some projects and open problems.

The mapping class group is connected to many areas of mathematics, including complex analysis, dynamics, algebraic geometry, algebraic topology, geometric topology (particularly in the study of three- and four-dimensional spaces), and group theory. Within geometric group theory, the close relationships between mapping class groups and groups such as braid groups, Artin groups, Coxeter groups, matrix groups, and automorphism groups of free groups have proved to be a fascinating and rich area of study. We refer you to Farb and Margalit's book *A Primer on Mapping Class Groups* [127] for further details and references on many topics mentioned in this office hour. Although their text is aimed at graduate students and researchers, large portions of it are accessible to undergraduates.

17.1 A BRIEF USER'S GUIDE TO SURFACES

You already heard a little bit about surfaces and their classification in Office Hour 9. Here we will do a little bit more, starting with the formal definition of a surface and going into more detail about the classification.

The word "surface" comes from the French for "on the face." Indeed, we all have an intuitive notion of a surface as the outermost layer of an object, as when we speak of resurfacing a road, or of the boundary between two substances, such as the surface of the sea. Each of these kinds of surfaces is inherently two-dimensional in nature, and mathematicians think of surfaces in similar terms.

A *homeomorphism* between surfaces (or any two topological spaces) is a continuous function with continuous inverse; equivalently, it is an invertible function f so that f and f^{-1} preserve open sets (this agrees with the definition of a homeomorphism of the interval, as in Office Hour 16). We should think of a homeomorphism as a function that stretches and bends, but does not break or glue. For example, a circle is homeomorphic to a square since you can bend one into the other. But a circle is not homeomorphic to a line segment, since you would have to break the circle to turn it into a segment.

Exercise 1. Convince yourself that the following subsets of $\mathbb{R}^2$ are all homeomorphic: a circle, an ellipse, and a rectangle (or any polygon!). In other words, find explicit homeomorphisms between these spaces.

The official definition of a surface is: a space so that every point has an open set around it that is homeomorphic to either (1) an open set in the plane or (2) an open set in the half-plane $\{(x, y) \mid y \geq 0\}$ (technically, the space has to be second countable and Hausdorff, but we'll ignore this). The points that don't have open neighborhoods homeomorphic to open sets in the plane make up the *boundary* of the surface. To state the definition informally: a surface is a space where every point has an open set around it that looks planar (or a stretched, bent version of the plane). Let us give some examples.

The plane and the disk.   Hopefully it is not hard to convince yourself that the plane is a surface: every point definitely has an open neighborhood around it that

Figure 17.1 A list of surfaces.

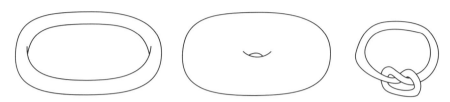

Figure 17.2 Three tori.

is homeomorphic to the plane, namely, the whole plane itself! Similarly, with the half-plane $\{(x, y) \mid y \geq 0\}$; the points on the x–axis make up the boundary of this surface. The closed unit disk is another surface with boundary. The next exercise asks you to prove this.

Exercise 2. Prove that the closed unit disk $\{(x, y) \mid x^2 + y^2 \leq 1\}$ is a surface. What is its boundary?

The sphere and the torus.   The leftmost surface in Figure 17.1 is familiar to us as the *sphere* S^2. We can think of S^2 as the set of points in $\mathbb{R}^3$ that are distance 1 from the origin. The next surface in the figure is the *torus* T^2, which may be familiar from calculus as a surface of revolution. For example, you can obtain a torus by taking the circle of radius 1 in the xy–plane centered at the point $(2, 0) \in \mathbb{R}^2$, and revolving it around the y–axis in $\mathbb{R}^3$. The sphere and the torus have empty boundary.

Higher genus.   The torus T^2 is often described as the frosting on a doughnut, without the doughnut. The doughnut illustration is useful for obtaining another infinite family of examples, by imagining (the frosting of) a doughnut with any number of holes, as shown in Figure 17.1. The number of holes is the *genus* of the surface. The genus can also be thought of as the number of handles on the surface.

  The list of Figure 17.1 depicts surfaces (without boundary) increasing in genus. The sphere S^2 has genus 0. The next surface, with genus 1, is the torus T^2. The torus T^2 is followed by surfaces of genus 2 and higher.

Three different tori?   Consider the three subsets of $\mathbb{R}^3$ shown in Figure 17.2. The surface on the left is much skinnier than the torus in the middle, yet we can still recognize its basic donut shape. If we inflate the leftmost surface until it looks like the surface in the middle, we obtain a homeomorphism from the first to the second; the inverse map is obtained by deflating (notice that we can convince ourselves of this without writing down an explicit map—if you can get comfortable with this,

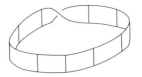

Figure 17.3 A Möbius band.

you are becoming a topologist!). So these two surfaces, which look a little different, are really homeomorphic. So to a topologist, they are the *same*.

But what about the other surface in Figure 17.2? We claim that it is also homeomorphic to the other two. For a moment, we imagine the first torus as a flexible hollow tube. We cut the tube, tie it in a knot, and then reglue the tube so that every point on one side of the cut is matched up exactly as before to the points on the other side. We claim that this process gives a homeomorphism from the first torus to the third. But wait! Didn't we say no cutting or gluing? Yes, that's true, but the official definition of a homeomorphism that we gave just means that homeomorphisms must preserve open sets. Certainly all open sets away from the cut were not disturbed by this process. And by our careful regluing, we have not changed any of the open sets around points along the circle where we cut. So it's okay to cut and glue, just as long as there is no evidence of this when you are done! We will return to this issue in the next section when we talk about homeomorphisms known as Dehn twists.

The classification of surfaces. We can think of lots of other surfaces: paraboloids, a sphere with a few points deleted, an icosahedron, a Möbius strip, the unit disk, etc. But amazingly there is a way to list them all! Let us restrict ourselves to the special case of compact, orientable surfaces.

A surface is *compact* if every infinite sequence has a convergent subsequence (the surfaces shown in Figure 17.1 are compact, but the plane is not compact and a sphere or torus with finitely many points deleted is not compact).

Next, *orientable* means that we can tell the difference between clockwise and counterclockwise. A Möbius band is not orientable, because if you take a small counterclockwise loop and push it around the Möbius band, it turns into a clockwise loop. Actually, a surface is nonorientable if and only if it contains a Möbius band.

For example, a *Klein bottle*, shown in Figure 17.4, is a nonorientable surface.[1] You should try to find a Möbius strip in this surface.

The classification of surfaces is the following amazing fact.

THEOREM 17.2 (Classification of Surfaces I). *Every compact orientable surface without boundary is homeomorphic to one of the surfaces shown in Figure 17.1.*

In other words, two compact, orientable surfaces without boundary are homeomorphic if and only if they have the same genus *g*.

---

[1] We're fans of the Acme Klein Bottle company; check them out at kleinbottle.com.

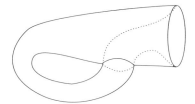

Figure 17.4 A Klein bottle.

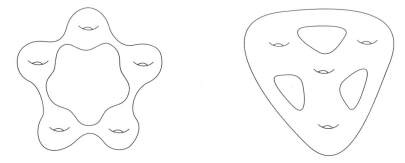

Figure 17.5 Two surfaces.

From this, we can easily deduce a stronger version of the classification of surfaces:

THEOREM 17.3 (Classification of Surfaces II). *Every compact orientable surface is homeomorphic to a surface obtained from one of the surfaces shown in Figure 17.1 by deleting the interiors of finitely many disjoint closed disks.*

On a first pass through this office hour, the student might want to ignore the case of surfaces with nonempty boundary as much as possible.

Exercise 3. Determine the genus of each of the two surfaces shown in Figure 17.5.

Exercise 4. Take a compact orientable surface S of genus 1 with one boundary component. The boundary of $S \times [0, 1]$ is a compact surface without boundary. Which one is it? What if we start with a surface of genus g with b boundary components?

Euler characteristic.   If we decompose a surface S into polygons (this means that we obtain the surface from a disjoint union of polygons by gluing edges in pairs), then the Euler characteristic $\chi(S)$ is $V - E + F$, where V, E, and F are the numbers of vertices, edges, and faces (= polygons) in the decomposition. Notice that some edges and vertices get identified in the gluing and so you need to keep track of all of this. It is an amazing fact that $\chi(S)$ does not depend on the decomposition into polygons! The Euler characteristic of a compact orientable surface of genus g with b boundary components is $2 - 2g - b$. It follows that a compact orientable

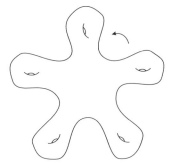

Figure 17.6 Rotation by $2\pi/g$ about the center of the surface pictured is a homeomorphism.

surface without boundary is determined up to homeomorphism by any two of the three numbers χ, g, and b.

Exercise 5. Prove the last statement by finding polygonal decompositions of the compact, orientable surfaces (you may assume that the Euler characteristic does not depend on the decomposition).

17.2 HOMEOMORPHISMS OF SURFACES

So far we have been living in the world of topology, but the notion of homeomorphism of a surface immediately leads us to groups and group theory. Let Homeo(S) denote the set of homeomorphisms of S. The set Homeo(S) is closed under the operation of function composition. Composition is associative and by definition every homeomorphism has an inverse. We therefore see that Homeo(S) is a group with the identity homeomorphism as the identity element of the group.

When we first encountered homeomorphisms, we said that homeomorphic surfaces should be thought of as the same surface. In the same vein, a self-homeomorphism of a surface is precisely what we should think of as a symmetry of a surface. Normally, when we think of symmetries, we think of rigid motions, as in the dihedral group. But here in the world of topology, our symmetries are allowed to stretch and bend, but never break or glue.

The mapping class group will be defined as a quotient of (a certain subgroup of) Homeo(S). Before we say more about that, we introduce several important examples of elements in the group Homeo(S). Even though we just said that homeomorphisms are usually not rigid symmetries, the first few examples of homeomorphisms we give will in fact be rigid symmetries. These are the simplest ones to visualize.

Rotation. If we arrange our surface of genus g as in Figure 17.6, we can rotate it by $2\pi/g$, or by one "click," to obtain an element of order g in the group Homeo(S).

Exercise 6. Explain how the above example of a rotation of a surface of genus g gives a homeomorphism of the surface of genus g as depicted in Figure 17.1

Figure 17.7 Rotation by π about the indicated axis is a hyperelliptic involution.

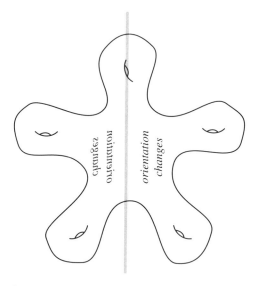

Figure 17.8 The surface is reflected across the vertical plane indicated. This homeomorphism reverses the orientation.

Hyperelliptic involution. Another example of a rotation is the *hyperelliptic involution* given by skewering the surface about the axis indicated in Figure 17.7 and rotating it by π.

Reflections. Reflections of $\mathbb{R}^3$ can also give rise to homeomorphisms of a surface. As in Figure 17.8, we can just reflect across a plane that slices the surface in half. This homeomorphism is fundamentally different from the others we have discussed so far because the orientation of the surface has been reversed. If you think of writing a word on the surface, then after the reflection the words will be reversed in the same way that words look backward in a mirror. More precisely, an orientation-reversing homeomorphism is one that takes small counterclockwise loops to small clockwise loops (we need to be in an orientable surface for this to even make sense).

Dehn twists. We come now to some homeomorphisms of surfaces that cannot be realized by rigid motions, namely, Dehn twists. Remember, one of our main goals is to convince you that these generate the mapping class group.

First, a *simple closed curve* in a surface S is the image of a circle in the surface under a continuous, injective function; three examples are shown in Figure 17.9.

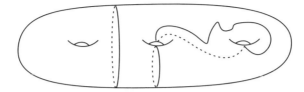

Figure 17.9 Three examples of simple closed curves in a surface.

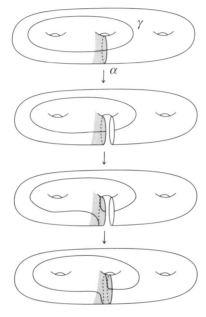

Figure 17.10 A Dehn twist seen as cut along α, twist, and reglue. The simple closed curve γ intersecting α acquires an extra twist about α.

We can picture a simple closed curve in a surface S as a loop in the surface that does not intersect itself.

Any simple closed curve in a surface S gives rise to an important example of an element of the group Homeo(S). Imagine cutting a surface along a simple closed curve α, twisting one of the two resulting boundary components by a full 360–degree twist to the right, and then carefully regluing, as shown in Figure 17.10.

This is really continuous! A map is continuous when it takes nearby points to nearby points, and nearby points that are separated in the cutting are carefully reunited again when regluing (look back at the knotted torus example we discussed at the end of Section 17.1, which also involved cutting and regluing). The inverse is also continuous, since it is obtained by just twisting the other way. Thus we have a homeomorphism, known as a *Dehn twist about* α, denoted by T_α.

Notice that twisting to the right makes sense as long as we have an orientation on the surface and this does not depend on any orientation of the curve we are twisting around. If we approach the curve we are twisting around from either direction, the

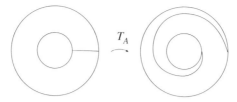

Figure 17.11 A Dehn twist on an annulus.

Figure 17.12 The core of an annulus.

surface gets stretched to the right. See Figure 17.11: the small horizontal arc in the left-hand picture gets twisted to the right no matter which side it approaches the annulus from.

Dehn twists via annuli.   We can make the definition of a Dehn twist more precise as follows. Using polar coordinates (r, θ) for points in the plane $\mathbb{R}^2$, we consider the annulus A made up of those points with $1 \leq r \leq 2$. Then we can define a map $T_A : A \to A$ by

$$(r, \theta) \mapsto (r, \theta - 2\pi r).$$

But this discussion was supposed to be about simple closed curves, not annuli. The key realization is that every simple closed curve α in S is the core[2] of some annulus A, as in Figure 17.12 (we are only considering orientable surfaces). We can identify (via homeomorphism) this annulus A in the surface with our annulus A in the plane. So our formula can be thought of as describing a homeomorphism of the annulus in the surface.

The important thing to notice about our formula is that *each point on the boundary of the annulus A is fixed by the map T_A.* This means that once we do our twisting on the annulus A in S, we can obtain an element of Homeo(S) by extending by the identity, that is, by fixing every other point of S outside of A. The point is that our twist on the annulus A and the identity map on $S \backslash A$ agree where they meet, on the boundary of the annulus A.

Exercise 7. Find a simple closed curve in the Klein bottle that is not the core of an annulus.

In fact, the map T_A we have just defined is really just T_α, a Dehn twist about the curve α as defined above. To see this, look again at Figure 17.10. We can

---

[2] In our previous discussion, the *core* of A in the plane $\mathbb{R}^2$ is the set of points with $r = \frac{3}{2}$.

Figure 17.13 A Dehn twist preserves open sets.

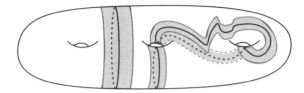

Figure 17.14 Three simple closed curves with their corresponding annuli.

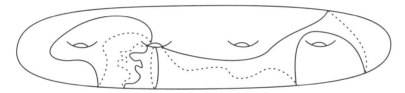

Figure 17.15 The simple closed curve on the left is not homotopic to the other three curves, which are all pairwise homotopic.

understand this map by seeing what happens to a simple closed curve γ that crosses α: away from A, nothing happens to γ, but as it nears α, the simple closed curve γ suddenly turns right and traces α before continuing on its way.

17.3 MAPPING CLASS GROUPS

We are trying to understand the symmetries of a surface, and we already said that homeomorphisms gave the right notion of symmetry. However, the group Homeo(S) is somehow much too large. We would like to lump together homeomorphisms that are in some sense the same, and declare them to *be* the same. In other words, we are going to introduce an equivalence relation, called homotopy, on the set Homeo(S). The goal is to distill Homeo(S) into a more manageable group that still incorporates all the essential features of Homeo(S).

Homotopy. We like to think of homotopy as the technical tool that allows us to get away with not being very good artists when drawing simple closed curves in surfaces—a bump here or a wiggle there does not matter; drawing objects to scale is unimportant. Informally, we say two simple closed curves in a surface are *homotopic* if one can be deformed to the other; see Figure 17.15 for examples and non-examples. One way to think of this is to imagine that the simple closed curve in the surface is made of a rubber band. If you stretch the rubber band and move it around, you will get a new curve homotopic to the original.

(We used homotopy in Office Hour 9 in the definition of the fundamental group. The idea is the same but the context is slightly different. There the loops were all based at some base point, and the loops there were not required to be simple.)

More precisely, a homotopy is a continuous deformation of one simple closed curve to another. Even more precisely, if we think of a simple closed curve in S as the image of a continuous map $S^1 \to S$, then two curves are homotopic if there is a continuous map $S^1 \times [0, 1] \to S$ so that the image of $S^1 \times \{0\}$ is the first curve and the image of $S^1 \times \{1\}$ is the second curve. We use t as the parameter for the $[0, 1]$ factor because we often think of a homotopy as a movie where at time $t = 0$ we see the first curve and then we watch the curve slowly being deformed so that by the time $t = 1$ we have arrived at the second curve.

Next we discuss homotopy for homeomorphisms. Two elements f and g in Homeo(S) are *homotopic* if one can be deformed to the other. More precisely, f and g are homotopic if there is a continuous map $F : S \times [0, 1] \to S$ so that the restrictions of F to $S \times \{0\}$ and $S \times \{1\}$ are equal to f and g, respectively. Again, we can think of the homotopy F as a movie going from one homeomorphism to the other.

Exercise 8. Think of a rotation of the circle S^1 as an element of Homeo(S^1), and convince yourself that it is homotopic to the identity. (Even better, convince yourself that every element of Homeo(S^1) is homotopic to either the identity or reflection about the x–axis.)

Exercise 9. Find elements of Homeo(T^2) that are homotopic to the identity. Find some that are not!

Exercise 10. For any S, describe a nontrivial element of Homeo(S) that is homotopic to the identity.

Exercise 11. Show that "is homotopic to" is an equivalence relation on the set Homeo(S).

A homotopy of surface homeomorphisms is a little harder to draw and visualize than a homotopy of curves. However, it turns out that we can understand the former in terms of the latter.

It is not too hard to see that if two homeomorphisms f and h of S are homotopic, then for all simple closed curves α, the curves $f(\alpha)$ and $h(\alpha)$ are homotopic. For a closed, orientable surface S of genus at least 3, the converse is true:

> If $f(\alpha)$ is homotopic to $h(\alpha)$ for all simple closed curves α, then f is homotopic to h.

This is a useful tool for showing that two homeomorphisms are homotopic as it reduces a problem about surfaces to a problem about curves. A priori, there are infinitely many such curves to check, but it turns out you can get away with only checking finitely many. Can you guess such a finite set of curves that determines a homeomorphism of a closed orientable surface of genus 3?

If the genus is 1 or 2, the above statement is almost true: if $f(\alpha)$ is homotopic to $h(\alpha)$ for all α, then f is homotopic to either h or the product of h with the hyperelliptic involution. For other surfaces—for instance, surfaces with boundary—some version of the statement is true. Usually it is enough to consider curves and arcs instead of just curves. We will discuss the details of this as necessary.

Mapping class groups.   We now go about the business of defining the mapping class group formally. The basic idea is that it is the group of homotopy classes of homeomorphisms of a surface, but the actual definition is slightly more technical.

Let S be a compact, orientable surface and let $\text{Homeo}^+(S, \partial S)$ denote the group of homeomorphisms of S that preserve the orientation of S and that restrict to the identity map on each component of the boundary ∂S.

Exercise 12. Show that if S is a surface with nonempty boundary, then every homeomorphism of S that restricts to the identity on ∂S must also preserve the orientation of S.

If $h \in \text{Homeo}^+(S, \partial S)$, we let $[h]$ denote the set of all homeomorphisms from S to S that are homotopic to h; here we insist that our homotopies do not move any points on the boundary of S. We say that $[h]$ is the *mapping class* of the homeomorphism h. Alternatively, we say that the homeomorphism h represents the mapping class $[h]$.

The set of all mapping classes of a surface S is denoted by $\text{Mod}(S)$ and is called the *mapping class group* of S. Since the elements of $\text{Mod}(S)$ are classes of homeomorphisms, we will use composition of homeomorphisms to define a group operation on $\text{Mod}(S)$. If f and g are elements of $\text{Homeo}^+(S, \partial S)$ and if $[f]$ and $[g]$ in $\text{Mod}(S)$ are their respective mapping classes, then we can define an operation on $\text{Mod}(S)$ as follows:

$$[f] \cdot [g] = [f \circ g].$$

It is not too hard to show that this operation is well defined and associative, that the mapping class of the identity function on S is the identity element of $\text{Mod}(S)$, and that $[f]^{-1} = [f^{-1}]$ for any mapping class $[f] \in \text{Mod}(S)$. Thus $\text{Mod}(S)$ together with this operation is truly a group.

Exercise 13. Show that the mapping class group $\text{Mod}(S)$ is the same as the quotient of $\text{Homeo}^+(S, \partial S)$ by the subgroup consisting of all homeomorphisms homotopic to the identity (again, homotopies must fix the boundary pointwise).

The notation $\text{Mod}(S)$ is short for *Teichmüller modular group*, an alternative name sometimes used for this group.

Dehn twists as mapping classes.   Returning to our example of Dehn twists, let us consider a simple closed curve α in a surface S. Recall that in defining a Dehn twist corresponding to α, the resulting homeomorphism depended heavily on our choice of annulus A (with core α) and our parametrization of A. We seem to have a serious problem: in the context of homeomorphisms, it makes no sense to talk

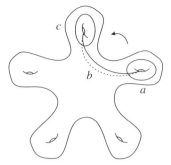

Figure 17.16 The homotopy classes a, b, and c.

about "the" Dehn twist about the simple closed curve α. Rather, we obtained an uncountably infinite number of different Dehn twists about α!

However, *the homotopy class of the resulting homeomorphism is independent of the choices*, although it is a challenging exercise to prove this carefully. In other words, while it does not make sense to talk about "the" Dehn twist T_α in the context of $\text{Homeo}(S)$, it *does* make sense in the context of $\text{Mod}(S)$.

Even better, it turns out that if α' is another simple closed curve in the surface S that is homotopic to α, then the corresponding Dehn twists are also homotopic! So not only can we choose whatever annulus and whatever parametrization we like, we are also free to choose any simple closed curve that is homotopic to α. Thus if a is a homotopy class of simple closed curves, it makes sense to write T_a as a well-defined element of $\text{Mod}(S)$.

Notice that—as in the case of homeomorphisms—the inverse of a Dehn twist is simply the Dehn twist about the same curve in the other direction.

We have finally arrived at the correct notion of the group of symmetries of a surface S: it is the mapping class group $\text{Mod}(S)$.

17.4 DEHN TWISTS IN THE MAPPING CLASS GROUP

Recall that Dehn's Theorem 17.1 says that the mapping class group of a compact, orientable surface is generated by Dehn twists. To get some appreciation for this theorem let's consider the counterclockwise rotation of order 5 of the surface in Figure 17.16. By Dehn's theorem there should be a product of Dehn twists achieving this rotation. It is not at all obvious how to do this!

The rotation takes the homotopy class of curves a to the homotopy class c. As a warm-up, we might simply want to find at least some product of Dehn twists that takes a to c.

First, a few words about homotopy classes of simple closed curves versus simple closed curves. This is important because the mapping class group does not act on the set of simple closed curves in a surface but does act naturally on the set of homotopy classes of simple closed curves. For two homotopy classes of simple closed curves a and b, we define the *geometric intersection number* $i(a, b)$ to be the minimum of $|\alpha \cap \beta|$ over all representatives α of a and β of b.

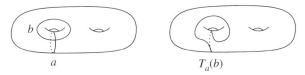

Figure 17.17 The homotopy classes a, b, and $T_a^{-1}(b) = T_b(a)$.

In order to find a product of Dehn twists taking a to c in Figure 17.16, we first observe that there is a homotopy class of curves b with $i(a, b) = i(b, c) = 1$. This is good, because we claim the following.

Claim. If a and b are the homotopy classes of two simple closed curves that intersect in one point, then $T_a T_b(a) = b$.

Using this claim, we can take a to c by first taking a to b and then taking b to c.

Proof of the claim. On the left-hand side of Figure 17.17 we have drawn two curves that intersect in one point; we have denoted the homotopy classes by a and b. If we multiply the desired equality $T_a T_b(a) = b$ by T_a^{-1} on both sides, we obtain the equivalent equality $T_b(a) = T_a^{-1}(b)$. This is straightforward to check; see the right-hand side of Figure 17.17.

That proves the claim! Except for one thing: it might seem like cheating that we have only checked what happens for a single pair of curves, while in fact there are infinitely many pairs of curves in a surface that intersect exactly once, even up to homotopy. The crucial point is that for every such pair of curves, there is a homeomorphism of the surface taking that pair to our pair. This homeomorphism (or rather the inverse) takes our calculation to the required calculation for the other pair. In other words, our calculation actually does *all* of the calculations! This is known as the *change of coordinates principle*; it is similar to the principle of changing basis in linear algebra.

Exercise 14. Prove that any two nonseparating simple closed curves in a surface differ by a homeomorphism of the surface (a simple closed curve is nonseparating if it does not divide the surface into two pieces). *Hint: What surfaces can you get when you cut a given compact, orientable surface along a nonseparating curve?*

Exercise 15. Prove the assertion that any two pairs of simple closed curves that intersect once differ by a homeomorphism of the surface.

The problem of finding a product of Dehn twists taking one simple closed curve to another should remind you of the problem of solving a Rubik's cube using the finitely many possible twists of the Rubik's cube. In fact, there is a really fun computer game called Teruaki, written by Kazushi Ahara of Meiji University, which realizes this idea (at the time of this writing, it is available for free from his website [6]).

The braid relation. We can rephrase the claim that $T_a T_b(a) = b$ as a relation between Dehn twists in the mapping class group, namely: if $i(a, b) = 1$ then we have the relation

$$T_a T_b T_a = T_b T_a T_b.$$

This relation is called the *braid relation*.

To prove that the braid relation holds we will need the following useful fact.

Fact. For any $f \in \mathrm{Mod}(S)$ and any homotopy class a of simple closed curves in S we have

$$T_{f(a)} = f T_a f^{-1}.$$

A bit of thought will convince you that this equation does not really require proof: following a homeomorphism to another copy of S, doing the Dehn twist there, and then going back again is the same as doing the Dehn twist about the image of your curve under the very same homeomorphism.

Using this fact, it is easy to prove the braid relation. Indeed, the relation

$$T_a T_b T_a = T_b T_a T_b$$

is the same as

$$(T_a T_b) T_a (T_a T_b)^{-1} = T_b,$$

and by our fact this is the same as

$$T_{T_a T_b(a)} = T_b.$$

Now there is another fact that $T_c = T_d$ if and only if the homotopy classes c and d are the same (this is believable, but not obvious!), and so that last equality is equivalent to

$$T_a T_b(a) = b.$$

But our above claim says that this holds when $i(a, b) = 1$ so we are done!

The term "braid relation" comes from the theory of braid groups. Indeed this relation is directly connected to an analogous relation in the braid group; see Office Hour 18 for more explanation.

Groups generated by two Dehn twists.   It turns out that we can completely characterize the subgroup of $\mathrm{Mod}(S)$ generated by two Dehn twists T_a and T_b in terms of $i(a, b)$. Here are the groups we get:

$i(a, b)$	$\langle T_a, T_b \rangle$
0	$\langle T_a, T_b \mid T_a T_b = T_b T_a \rangle$
1	$\langle T_a, T_b \mid T_a T_b T_a = T_b T_a T_b \rangle$
≥ 2	$\langle T_a, T_b \mid \ \rangle$

The first group is isomorphic to $\mathbb{Z}^2$, the second to the braid group B_3, and the third to the free group F_2. What is more, the last isomorphism can be proved by applying the ping-pong lemma (cf. Office Hour 5) to the action of $\langle T_a, T_b \rangle$ on the

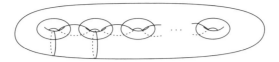

Figure 17.18 Dehn twists about these $2g + 1$ simple closed curves generate the mapping
class group.

set of homotopy classes of simple closed curves in the surface. The two sets in
the ping-pong lemma are the sets of homotopy classes of simple closed curves c
with $i(a, c) > i(b, c)$, and vice versa. See the book by Farb and Margalit [127,
Chapter 3] for the proofs.

17.5 GENERATING THE MAPPING CLASS GROUP BY DEHN TWISTS

Let's dive right in now and and prove Theorem 17.1:

> *For any compact, orientable surface S, the mapping class group*
> $\mathrm{Mod}(S)$ *is generated by Dehn twists.*

Even better, Stephen Humphries [171] showed that for a closed, orientable surface
S of genus g, the group $\mathrm{Mod}(S)$ is generated by Dehn twists about the $2g + 1$
simple closed curves in Figure 17.18.

As a warm-up for the proof of Theorem 17.1 we will convince ourselves that it
is true for the compact, orientable surfaces of genus 0 with zero, one, two, or three
boundary components, namely, the sphere, the disk, the annulus, and the pair of
pants. We will later use these examples as base cases for our inductive proof of
Theorem 17.1.

Reminder. Throughout, it is important to remember that as part of our definition
of the mapping class group, all homeomorphisms are required to fix each point of
the boundary.

The disk.   The first particular surface we will discuss is the disk D^2: the compact,
orientable surface of genus 0 with one boundary component. We claim that any
homeomorphism h of the disk that fixes the boundary pointwise is homotopic to
the identity. Here is the homotopy: at time t, "do" h on the sub-disk of radius
$1 - t$ and act as the identity everywhere else (here we are taking D^2 to be the
disk of radius 1 and t to vary from 0 to 1). At time 1 we just take the identity
map of D^2. For each t in $[0, 1]$ this gives a homeomorphism precisely because h
fixes the boundary of D^2 pointwise. Further, no matter what h is, this rule defines
a homotopy from h to the identity (this clever homotopy is called the Alexander
trick). In particular, $\mathrm{Mod}(D^2)$ is trivial. So it is true in thus case that the mapping
class group is generated by Dehn twists!

Exercise 16.   Write down a precise formula in terms of h for the homotopy in the
last example and verify that h is continuous.

The sphere.   It is intuitively clear that any two simple closed curves in the sphere S^2 are homotopic—sketching and staring at a few pictures should convince you, although writing down a careful proof is a nontrivial exercise. It follows from this that any homeomorphism of S^2 can be modified by homotopy so that it fixes the equator pointwise (this is again intuitively clear but nontrivial to prove; there is a theorem in differential topology called the isotopy extension theorem that does the trick if you assume all of the maps in question are smooth). Any homeomorphism of S^2 that fixes the equator and preserves orientation must send the northern and southern hemispheres to themselves. But each hemisphere is just a disk, and so using the fact that $\mathrm{Mod}(D^2)$ is trivial, we conclude that our homeomorphism of S^2 is homotopic to the identity. It follows that $\mathrm{Mod}(S^2)$ is again just the trivial group.

The annulus.   Let A denote the annulus $S^1 \times [0, 1]$. We will argue that

$$\mathrm{Mod}(A) \cong \mathbb{Z}$$

and further that $\mathrm{Mod}(A)$ is generated by the Dehn twist about the core curve of A. The key claim is the following.

Claim. An arc connecting two given points on different boundary components of A is completely determined up to homotopy by how many times it winds around the circle direction of A.

(For this claim to work we need to require that a homotopy of an arc keep the endpoints fixed throughout the homotopy.) What exactly do we mean by the number of times that an arc winds around the circle direction of A? One way to make this precise is to choose some oriented arc δ in A (fixed once and for all) that connects the two boundary components. Given any other arc α connecting the two boundary components, we orient α so that it connects the boundary components of A in the same order as δ, and then we count all of the intersections of α with δ; if α crosses δ from left to right we count a $+1$ and if it crosses from right to left we count a -1. The sum of these numbers is the desired winding number.

Let's prove the claim. Say that α and β are two arcs that have the same endpoints and that both wind k times (in the same direction) around the circle direction of A. We would like to show that β is homotopic to α by a homotopy that fixes the endpoints of β. Actually, by applying the kth power of a Dehn twist about the core curve of A (or rather the inverse), we can assume without loss of generality that $k = 0$. Moreover, it doesn't hurt to assume that α is the arc δ we used in the previous paragraph to define the winding of an arc around A.

The assumption that $k = 0$ then means that β has just as many positive intersections with α as negative intersections. Therefore, if we follow along β we will find somewhere two consecutive intersections with opposite sign. The picture looks like the one in Figure 17.19; specifically, this sub arc of β, together with the arc of α connecting the endpoints of the sub-arc of β, bounds a disk in A (the fact that this disk does not surround the hole of the annulus is exactly because the signs of intersection disagree). There may be other arcs of β intruding on this disk, but no matter: just push (or homotope) the sub-arc of β through the disk. We have reduced the number of intersections of β with α, and continuing this process inductively we

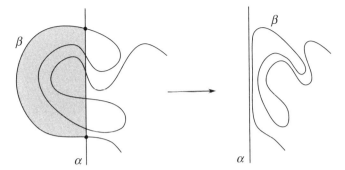

Figure 17.19 Consecutive intersection points of β with α and the disk (shaded) we use to push the arc of β.

remove all intersections of α with β. The only points of intersection that remain at the end are the two points of intersection at the endpoints of α and β. But then $\alpha \cup \beta$ form a simple closed curve in A, and hence bound a disk which can then be used to homotope β onto α.

Exercise 17. We secretly used the Jordan–Schönflies theorem—that every simple closed curve in the plane bounds a disk—twice in the proof of the claim. Find both instances of this and explain why it is valid to apply this theorem to the annulus instead of the plane.

To complete the proof of the claim, you should check that arcs that wind different numbers of times around A are not homotopic. This is the easier direction, and the essential ideas are in the previous paragraph.

Exercise 18. Prove that homotopic arcs in A wind around A the same number of times.

Now, back to showing that $\mathrm{Mod}(A)$ is isomorphic to $\mathbb{Z}$. Let α be an arc that winds zero times around the circle direction. We can define a homomorphism to $\mathrm{Mod}(A) \to \mathbb{Z}$ whereby $f \in \mathrm{Mod}(A)$ maps to the number of times (with sign) that $f(\alpha)$ wraps around the circle direction of A. It is easy to show that this map is a surjective homomorphism: the nth power of the Dehn twist about the core of the annulus maps to n. Using the same logic we used for the sphere, we can argue it is injective. Indeed, if a homeomorphism fixes α up to homotopy, then we can modify the homeomorphism by homotopy so that it fixes α pointwise. Then there is an induced homeomorphism of the disk obtained by cutting A along α. This homeomorphism of the disk is homotopic to the identity since $\mathrm{Mod}(D^2)$ is trivial. This homotopy gives us a homotopy of the original homeomorphism of A to the identity, as desired.

The pair of pants. A *pair of pants* P is a compact, orientable surface of genus 0 with three boundary components (in other words, a sphere with three disks removed). We would like to show that

$$\mathrm{Mod}(P) \cong \mathbb{Z}^3$$

and moreover that $\text{Mod}(P)$ is the free abelian group generated by the Dehn twists about the three boundary components of P (really, this means we take curves parallel to the boundary). Since every compact, orientable surface is made by pasting together some number of spheres, disks, annuli, and pairs of pants, this is the final ingredient we will need for our proof of Theorem 17.1.

We can take a similar tack to the one used for the annulus. Let $f \in \text{Mod}(P)$. We take a simple arc α in P connecting two distinct boundary components. The key point is that an arc connecting two specific boundary components is determined up to homotopy by the number of times it winds around each boundary component. Then we can use a similar argument to the one used for the last two cases: we can modify f by Dehn twists about the two boundary components at the end points of α so that f fixes α up to homotopy; then we can cut along α to obtain an annulus, whose mapping class group we already understand.

Exercise 19. Prove that an arc in a pair of pants connecting a pair of particular points on different boundary components is determined by the number of times it winds around the boundary components at either end. You may do this by modifying the argument for the analogous claim for the annulus. First you should make sense of these winding numbers. (*Hint: Count points of intersection with arcs connecting the endpoints of the given arc to the component of the boundary of the pair of pants not containing an endpoint of the given arc.*)

Proving Theorem 17.1. We are now ready to prove that the mapping class group of any compact, orientable surface is generated by Dehn twists. Our exposition follows closely a set of notes written by Au, Luo, and Yang [13]. The key is the following.

Main Lemma. If c and d are simple closed curves in a compact, orientable surface S, then there is a product h of Dehn twists so that

$$i(c, h(d)) \le 2.$$

Proof. We will show that if $i(c, d) \ge 3$, then there is a simple closed curve b so that $i(c, T_b(d)) < i(c, d)$.

The idea is to look at the pattern of intersections along c. We can draw c as a vertical arc on the page (imagine that this is a small piece of c) and draw the intersections of d with c, so that d looks like a collection of horizontal arcs. All of these arcs are connected up somewhere outside the picture, but we do not need to worry about exactly how they are connected.

We also orient c and d arbitrarily. All we care about is whether signs of intersection agree or disagree, and this does not depend at all on how we orient the two curves.

If the intersection number $i(c, d)$ is at least 3, then along the vertical arc in our picture we either have to see two consecutive intersections of the same sign or three consecutive intersections with alternating signs. In either case we can find a simple closed curve b with the desired property. The curve b is indicated in Figure 17.20.

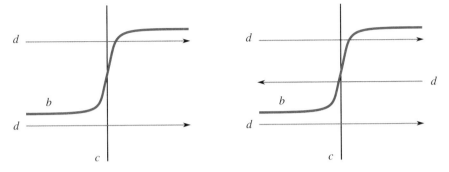

Figure 17.20 The two cases for the main lemma.

Figure 17.21 Checking the first case of the main lemma.

We have only shown a small part of b in the picture—after it leaves the page, b just follows along d.

It is a straightforward computation to check that b has the desired property. We show the computation for the first case in Figure 17.21. The first of the five pictures in the figure just shows c, d, and b, as in Figure 17.20. The second picture shows $T_b(d)$; again, this curve follows d outside the picture. We see a portion of this curve (at the bottom right) that looks like it is backtracking, so we can simplify the picture by pushing this portion to the right as in the third picture. In the fourth picture we have pushed around d until we arrive at the top left of the picture. If we keep pushing, we get the fifth picture, at which point we see that $T_b(d)$ has (at least) one fewer point of intersection with c than d does, as desired. We leave the computation for the second case as an exercise. This completes the proof of the lemma.

Exercise 20. Check in the second case of the last proof that $i(c, T_b(d)) < i(c, d)$.

Proof of Theorem 17.1. We proceed by induction on the genus g of our (compact, orientable) surface S. The base case is $g = 0$, and to prove Theorem 17.1 in this case we use induction on the number n of boundary components (induction inside induction!).

If n is 0, 1, 2, or 3, then we have a sphere, a disk, an annulus, or a pair of pants, and we already verified the theorem in those cases. Now suppose $n \geq 4$ and let $f \in \mathrm{Mod}(S)$. Let c be a curve that cuts off a pair of pants in S (necessarily on the other side of c we have a surface of genus 0 with $n - 1$ boundary components).

By the main lemma there is a product of Dehn twists h so that $i(c, h \circ f(c)) \leq 2$. We claim that this implies that $h \circ f(c) = c$! The reason is simple. First of all, since c is a separating curve (as are all curves in a surface of genus 0), we may only have even numbers of intersection. Next, we can check that if d is a curve in S and $i(c, d)$ is equal to 0 or 2 and $c \neq d$, then c and d surround different sets of boundary components (to see this, draw your favorite pictures of curves that intersect zero or two times and then argue that all pairs of curves with those intersection numbers look like this). Since f and h act as the identity on the boundary of S, we know that c and $h \circ f(c)$ surround the same boundary components. It follows that $h \circ f(c) = c$, as desired.

Exercise 21. Verify the last sentence.

So what we have now is that there is a product of Dehn twists h so that $h \circ f(c) = c$. In other words, we can assume without loss of generality that f fixes c. Now we would like to argue by induction that f is a product of Dehn twists.

When we say that f fixes c we really mean that the homotopy class of homeomorphisms f fixes the homotopy class of curves c. It follows that we can choose a representative homeomorphism of f that fixes pointwise a representative curve of c (isotopy extension again). Then if we cut our surface along c we get two surfaces (a pair of pants and a surface of genus 0 with $n - 1$ boundary components, as above) and our representative of f induces a homeomorphism of each of these surfaces. By induction, the corresponding mapping classes are both equal to products of Dehn twists. Then it follows that the original mapping class f is a product of the same Dehn twists!

Now let $g \geq 1$. We assume by induction that every surface of genus $g - 1$ (with any number of boundary components) satisfies the theorem. Let c be any nonseparating curve. The key point is that if we cut our surface along c then we get a surface of genus $g - 1$ with two additional boundary components. We can prove this using three facts: (1) the Euler characteristic of a surface of genus g with n boundary components is $2 - 2g - n$, (2) when we cut we create two additional boundary components, and (3) when we cut we do not change the Euler characteristic. If you do not believe this argument, you can instead take c to be a specific nonseparating curve and just check for that curve.

Again, given any mapping class f, we can apply the main lemma to say that, without loss of generality, we have that $i(c, f(c))$ is 0, 1, or 2. In any of these three cases, we claim that we can find a curve b so that $i(c, b) = i(b, f(c)) = 1$. But we showed in Section 17.4 that if $i(c, b) = 1$ then there is a product of Dehn twists taking b to c (in the case $i(c, f(c)) = 1$, such a b exists but we clearly don't need it). Therefore we can modify f by a product of Dehn twists so that $f(c)$ is in fact equal to c. As in the genus 0 case, this gives us a mapping class of the surface of genus $g - 1$ obtained by cutting along c. By induction on genus, that mapping class is equal to a product of Dehn twists, and it follows that f is itself a product of Dehn twists, as desired. That does it!

Exercise 22. Verify the claim that if a and c are nonseparating curves with $i(a, c)$ equal to 0 or 2 then there is a curve b with $i(a, b) = i(b, c) = 1$. *Hint: There are three cases: $i(a, c) = 0$, $i(a, c) = 2$ with same signs of intersection, and $i(a, c) = 2$ with opposite signs of intersection. Draw your favorite configuration for each case and then argue that the general case looks like your configuration.*

Figure 17.22 Two simple closed curves in a torus.

The torus. We end with a discussion of the torus. While it follows from The-orem 17.1 (which we just proved) that $\mathrm{Mod}(T^2)$ is generated by Dehn twists, we will see that this mapping class group carries the extra structure of a linear group:

$$\mathrm{Mod}(T^2) \cong \mathrm{SL}(2, \mathbb{Z}).$$

Recall from Office Hour 9 that the fundamental group of T^2 is isomorphic to $\mathbb{Z}^2$. The idea of the isomorphism $\mathrm{Mod}(T^2) \cong \mathrm{SL}(2, \mathbb{Z})$ is that there is a homomorphism from $\mathrm{Mod}(T^2)$ to the group of automorphisms of the fundamental group of the torus:

$$\mathrm{Mod}(T^2) \to \mathrm{Aut}(\pi_1(T^2)).$$

Given an element of $\mathrm{Mod}(T^2)$, we can choose a representative that fixes our base point for $\pi_1(T^2)$; hence the action on $\pi_1(T^2)$. It turns out that for the torus this action is well defined, and independent of the choice of representative homeomor-phism (what special property of $\pi_1(T^2)$ are we using here?).

Since $\mathrm{Aut}(\pi_1(T^2)) \cong \mathrm{GL}(2, \mathbb{Z})$, the entire game now is to show that the map $\mathrm{Mod}(T^2) \to \mathrm{Aut}(\pi_1(T^2))$ is injective and that the image is exactly the subgroup of index 2 corresponding to matrices in $\mathrm{GL}(2, \mathbb{Z})$ with positive determinant. The first statement is proven using arguments similar to the ones already given (if a homeomorphism induces the trivial action then it fixes a curve, which we then cut along ...). The second statement is proven by showing that the image matrix has positive determinant if and only if the homeomorphism preserves orientation.

Consider the two simple closed curves in the torus of Figure 17.22.

The Dehn twist about each of these simple closed curves is nontrivial and in fact has infinite order in $\mathrm{Mod}(T^2)$. It turns out that these two Dehn twists generate $\mathrm{Mod}(T^2)$. Indeed, it is classically known that $\mathrm{SL}(2, \mathbb{Z})$ is generated by the matrices

$$\begin{pmatrix} 1 & 1 \\ 0 & 1 \end{pmatrix} \text{ and } \begin{pmatrix} 1 & 0 \\ -1 & 1 \end{pmatrix}.$$

And these matrices exactly correspond to the two given Dehn twists.

Exercise 23. Complete the proof that the mapping class group of the torus is isomorphic to $\mathrm{SL}(2, \mathbb{Z})$.

FURTHER READING

The seminal text *Braids, Links and Mapping Class Groups* by Joan Birman [41] stimulated much of the early work in this area and has become a true classic, with many researchers in the field today continuing to learn basics about mapping class groups and related topics from this source. The problem list contained therein has generated new and exciting insights into the structure of mapping class groups and poses several as yet unanswered questions.

Many helpful references have been published since; as we have mentioned a few times, the text *A Primer on Mapping Class Groups* [127] is one place to find out more about mapping class groups, and much of it is accessible to undergraduates. Among the many excellent references in its bibliography, we would like to highlight the volume *Problems on Mapping Class Groups and Related Topics* [126], a survey of open questions that still gives a fairly accurate sense of the most active areas of research in the field. Even though some of the questions have been solved, the answers have often just led to more questions, in keeping with the Haitian proverb quoted at the start of the book: "Behind the mountains, more mountains"—a useful proverb to keep in mind when doing mathematics research!

PROJECTS AND OPEN QUESTIONS

The study of $\mathrm{Mod}(S)$ is a vast area of current research involving many branches of mathematics. We will end here with a collection of projects for further exploration, interspersed with a few open questions in the field that highlight areas of active research. We begin with a few projects designed to tease out more of the basic structure of $\mathrm{Mod}(S)$.

Project 1. Download the software Teruaki [6] and play around with it. Explain Dehn twists to your local math club, and then organize a Teruaki tournament.

Project 2. Explore other generating sets for $\mathrm{Mod}(S)$. For example, we have mentioned that the mapping class group can be generated by $2g + 1$ Dehn twists. Can we generate $\mathrm{Mod}(S)$ by a smaller set of elements? It turns out we can't do any better using Dehn twists (see Project 10 below), but, perhaps counterintuitively, we can find smaller generating sets consisting of finite order elements. Compile a list of known "small" generating sets. Two useful starting points are Wajnryb's short paper showing that the mapping class group is generated by two elements of finite order [259], and Brendle–Farb's paper describing generating sets consisting of *involutions*, or elements of order 2 [49]. The latter explains why $\mathrm{Mod}(S)$ cannot be generated by two involutions; give a detailed proof of this fact.

Open Question. *Can $\mathrm{Mod}(S)$ be generated by three involutions?*

Project 3. For a closed, orientable surface of genus at least 3, use the famous *lantern relation* in mapping class groups (see [127], Chapter 5, for example) to prove that the corresponding mapping class groups are *perfect*, that is, their abelianizations are trivial. Make sure you can explain why the genus needs to be at least 3 for this to work. In the remaining cases, find finite presentations for $\mathrm{Mod}(S)$ and use these to calculate the corresponding abelianizations. (A paper of Labruère and Paris [189] handles the case of surfaces with punctures and provides a good survey of the literature on presentations as well.)

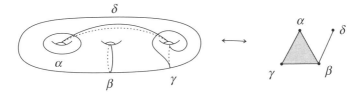

Figure 17.23 Four simple closed curves in a surface and their corresponding span in the
complex of curves.

By the universal property of abelianization (see Office Hour 16), the fact that
mapping class groups are perfect (for closed surfaces of genus at least 3) tells us
that there are no homomorphisms from these mapping class groups onto any nontrivial abelian groups, such as the integers $\mathbb{Z}$. However, this does not rule out the
possibility that some very large (finite index) subgroup of a mapping class may
admit such a homomorphism.

Perhaps surprisingly, the existence of homomorphisms of finite index subgroups
of a group onto $\mathbb{Z}$ can have deep implications for the geometry and topology of related structures. For example, this question is related to Kazhdan's Property (T) and
to a conjecture of Thurston about aspherical 3–manifolds. Gonciulea has proven
that infinite Coxeter groups have this property [147]. Gonciulea's paper is one of
many excellent starting points for references on these topics. Intriguingly, the fact
mentioned above, that mapping class groups are generated by involutions, implies
that mapping class groups are quotients of infinite Coxeter groups, but we do not
currently know if mapping class groups have this property.

Open Question. *Do any finite index subgroups of mapping class groups admit a
surjective homomorphism onto the integers?*

A recurrent theme in this book is that we learn a great deal about groups by
studying actions of groups on "nice" spaces, such as graphs or other simplicial
complexes. The next three projects explore various complexes on which Mod(S)
acts naturally.

Project 4. Learn about the complex of curves associated to an orientable surface
S. We give a brief introduction to this topic here; for a more detailed survey of this
topic, see Chapter 4 of Farb and Margalit's *Primer* [127]. For any surface S, we can
define an infinite graph, which we denote by $\mathcal{C}^1(S)$ and refer to as the *curve graph*.
The curve graph $\mathcal{C}^1(S)$ has one vertex for each homotopy class of simple closed
curves in S, and we join two vertices by an edge if (and only if) we can find a pair
of simple closed curves, one representing each vertex, that are disjoint. The curve
graph $\mathcal{C}^1(S)$ is a one-dimensional simplicial complex. We can add dimensions by
turning $\mathcal{C}^1(S)$ into a special kind of simplicial complex called a *flag complex*. This
just means that we glue in a "filled-in" triangle each time we see the outline of a
triangle in our graph, and then we fill in a solid tetrahedron every time we see the
boundary of one, and so on in higher dimensions. The resulting simplicial complex
is the *complex of curves* associated to S. Figure 17.23 shows a subcomplex of the
complex of curves of a surface of genus 3.

Use what you have learned about Dehn twists to give a simple proof that the curve graph (and hence the complex of curves) is always infinite, and for most surfaces is also locally infinite (that is, each vertex has infinitely many edges emanating from it). Learn about *pants decompositions* of a surface, and use these to explain why the dimension of the complex of curves for a closed surface of genus $g \geq 2$ is $3g - 4$. Using ideas similar to those in the proof of Theorem 17.1, prove that $\mathcal{C}^1(S)$ is connected. Along the way you will find that there are some exceptional surfaces S for which this statement is not in fact true. What are these exceptional surfaces?

Project 5. View the graph $\mathcal{C}^1(S)$ as a metric space by assigning length 1 to each edge. Give examples of pairs of curves in the surface S that represent vertices at distance N from each other in $\mathcal{C}^1(S)$, for $N = 0, 1, 2, \ldots$. Give rigorous proofs that your examples indeed realize the various distances: "I can't figure out how to do it in fewer than N steps" doesn't count. How high can you get before you get stuck? Use the MICC software written by Glenn–Menasco–Morrell–Morse [146] to find examples after you get stuck. Compare known algorithms for computing distance in the curve graph; the introduction of a recent paper of Birman–Margalit–Menasco gives an overview [39].

Project 6. Explore other simplicial complexes associated to surfaces. There are many: the separating curve complex, the nonseparating curve complex, the pants complex, the arc complex, and the cut system complex, to name just a few (see, for example, [127], [199], and [162]). Find out why each of these other complexes was introduced: what properties of the group Mod(S) have been deduced by studying these?

Now, when encountering a new group, the first question we often ask is: is this group familiar? Have we ever encountered an isomorphic copy of it in some other totally different context?

Project 7. We have seen that the mapping class group of the torus is an example of a *linear group*, that is, it is isomorphic to a multiplicative group of matrices $\mathrm{GL}(n, \mathbb{C})$, or one of its subgroups, for some natural number n. Find out how braid groups (the subject of Office Hour 18) can be viewed as a type of mapping class group, and how Bigelow–Budney used representations of braid groups to deduce that the mapping class group of a genus 2 surface is also a linear group [34]. See also Project 10 in Office Hour 18.

Open Question. *Is Mod(S) a linear group for a surface S of genus greater than 2?*

The next trio of projects provides a snapshot of the fascinating subgroup structure of Mod(S). One subgroup discussed here arises from viewing a surface in the context of an ambient 3–manifold, and the second arises from a natural linear representation of Mod(S).

Project 8. Consider the surface S as the boundary of a *handlebody* in $\mathbb{R}^3$ (that is, "fill in" the doughnut). Prove that the set of all elements of Mod(S) that extend to the entire handlebody forms a proper subgroup of Mod(S), known as the *handlebody subgroup*. Which of the homeomorphisms that we have studied in this office hour lie in this subgroup, and which do not? In particular, give an example of a Dehn twist that extends to a homeomorphism of the entire handlebody, and also give an example of a Dehn twist that does not have this property. Use your results to give a necessary condition for a Dehn twist in Mod(S) to extend to a homeomorphism of the entire handlebody. Find a complete set of generators for the handlebody subgroup; results on this topic by Suzuki [252] and Wajnryb [260] will be a helpful starting point.

Project 9. This project and the next require a tiny bit of algebraic topology. The mapping class group Mod(S) acts naturally on the first homology group $H_1(S; \mathbb{Z})$ (for a closed surface of genus g, the group $H_1(S; \mathbb{Z})$ is isomorphic to $\mathbb{Z}^{2g}$, and we think of each copy of $\mathbb{Z}$ as being generated by an oriented simple closed curve). Find examples of Dehn twists, or products of Dehn twists, that lie in the kernel of this action; this kernel is known as the *Torelli subgroup* of Mod(S). Early work in this direction was done by Birman [40], Powell [224], and Johnson [179]. Among many groundbreaking results on Torelli groups, Johnson showed that the Torelli group of a closed surface is finitely generated when the genus is at least 3 [178]. For closed surfaces of genus 3, Johnson gave a minimal generating set for the Torelli group consisting of 35 elements, where each generator is a product of two Dehn twists[178]. Give an explicit description of Johnson's generating set for the genus 3 Torelli group, ideally including pictures of a representative sample of the 35 pairs of curves involved.

Open Question.   *Is the Torelli group finitely presented for surfaces of genus at least 3?*

Project 10. The representation described in the previous project turns out to give a surjective map from Mod(S) to the group of symplectic matrices with integer entries Sp($2g, \mathbb{Z}$). Understand why this map is surjective, and why it is also injective in genus 1. Explore properties of the group Sp($2g, \mathbb{Z}$), including its generation by elements known as *transvections*. Find analogies between Dehn twists and transvections, and use what you learn to prove that Mod(S) cannot be generated by fewer than $2g + 1$ Dehn twists, a fact due to Humphries [171].

Last, but not least, we revisit non-orientable surfaces.

Project 11. In this office hour, we restricted ourselves to the case of orientable surfaces, yet one can easily define mapping class groups for non-orientable surfaces by simply forgetting the requirement that homeomorphisms be orientation-preserving. Read Stukow's work on mapping class groups of non-orientable surfaces (for example, see [249] and [250]) in order to write down an analog of Theorem 17.1 for non-orientable surfaces. In particular, describe the mapping class group of a Klein bottle and compare it to that of the torus.

Office Hour Eighteen

Braids

Aaron Abrams

Clear your coil of kinkings
Into perfect plaiting,
Locking loops and linkings
Interpenetrating.

James Clerk Maxwell, from a poem sent to Peter Guthrie Tait, 1877

This office hour is about braids. No doubt you already have some idea of what a braid is; two familiar examples are shown in Figure 18.1.

Braids are among the oldest topics discussed in this book. Their history begins thousands of years ago,[1] and they entered the realm of mathematics at least several centuries ago. The modern mathematical study of braids involves lots of beautiful geometry and group theory, some of which I will introduce in this office hour. In addition to giving several different ways to think about mathematical braids and some of the basic theorems about them, I will also indicate some of the ways that braids relate to other parts of mathematics and science.

18.1 GETTING STARTED

Just as mathematical points, lines, and surfaces are idealized abstractions of physical objects, so mathematical braids are an idealized abstraction of the familiar hair

---

[1] Look up *quipu* to learn about the early history of braids.

Figure 18.1 Two braided things: Thalia, challah.

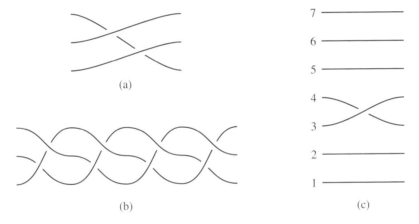

(a)

(b)   (c)

Figure 18.2 *(a)* A 3–string braid with two crossings. *(b)* The "usual" 3–string braid pattern.
(c) String 3 crosses over string 4.

and bread braids of Figure 18.1. One important difference is what happens at the
ends: in a mathematical braid, the "strings" remain separate at the ends rather than
being fused together. The other main difference is that mathematical braids can
have any number of braided strings and the braiding can occur in any pattern.

Here are some examples of mathematical braids:

When you braid hair or bread, usually you break it up into thirds, and then repeatedly pick up an outside piece and cross it over the middle piece. To make the
usual braid pattern, you alternate picking up the two outside pieces, resulting in a
braid whose mathematical form looks like Figure 18.2(*b*).

Exercise 1. Draw the 4–string braid you get by taking the top string and crossing it
over the middle two strings, and then taking the bottom string and crossing it over

Figure 18.3 A product of two braids.

Figure 18.4 The identity in B_3.

Figure 18.5 Let's make these equivalent.

the middle two strings, and then repeating. What happens if you braid your hair this way?

From a mathematical point of view, there are a lot of interesting things about braids. One of the most interesting is the raison d'être for this office hour: for each fixed n, the set B_n of all n–string braids has a natural group structure.

The way you multiply two braids is by sticking them together, as in Figure 18.3.

What would it mean to have the inverse of a braid? For example, what's the inverse of the braid α shown in Figure 18.2(b)? Well, first you need to know what the identity is. For $n = 3$, the identity braid is shown in Figure 18.4.

So, how could you extend α to make it look like the identity? Literally, it's impossible: α has crossings in the picture, and drawing more stuff doesn't change the fact that you have crossings in the picture. What you need is a way to talk about canceling crossings, so that, for instance, the two braids in Figure 18.5 would be considered equivalent.

The idea is that you should be able to move the braids around as you can in physical space—so, without passing any string through another—and arrive at an equivalent braid. To define equivalence, we should also say that when you move the braids around, the endpoints have to stay fixed—otherwise any braid would be equivalent to the identity! It may help you to imagine that the left endpoints of each braid are fastened to the vertical wall $x = 0$ in 3–space, and the right end-points are fastened to the wall $x = 1$. Now, two braids are *equivalent* if you can move one braid, keeping it between the walls and keeping the endpoints fixed, and

Figure 18.6 Two equivalent braids.

make it look like the other.[2] With this definition, the two braids in Figure 18.5 are equivalent. So are the two braids in Figure 18.6.

There is a technical point to make now. If we fix the bounding walls for the braid to be $x = 0$ and $x = 1$, then we should be a bit more careful about what it means to multiply braids. Precisely, to form β times β', we need to shift β' by one unit in the x–direction, take the union of β with the shifted version of β', and then scale the whole thing in the x–direction by a factor of $1/2$. This makes the product of two braids into a braid.

Exercise 2. *(a)* Convince yourself that the braid in Figure 18.4 is actually an identity element with respect to the product operation.
(b) Find a braid β such that α (the braid from Figure 18.2*(b)*) times β is equal (i.e., equivalent) to the identity.
(c) Describe a general procedure for constructing the inverse of a braid. *(Hint: Use a mirror.)*

It hasn't been said yet, but there is one more important feature of a braid: if you follow any individual string from the left endpoint to the right endpoint, you have to move monotonically from left to right. The strings can't ever "backtrack." So in particular no individual string can ever have a knot in it. Without this restriction, we would be able to make braids that don't have inverses!

There are no other rules. You now know the complete definition of a braid.

Exercise 3. Show that for fixed n, the (equivalence classes of) n–string braids form a group.

This group is called the n–string *braid group.* It's denoted by B_n and it's the primary subject of this office hour.

18.2 SOME GROUP THEORY

If you are trying to draw a braid, you can always use equivalence to avoid having three strings cross at the same point in the drawing. You can also make it so that no crossing occurs directly above another in the picture as in Figure 18.7.

Because of this, you can chop up any braid using vertical lines so that each chunk only contains one crossing, as in Figure 18.8.

---

[2] This is very similar to the notion of *isotopy* that you may have seen in knot theory.

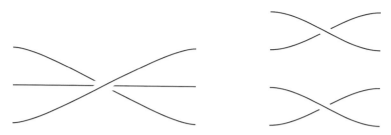

Figure 18.7 Drawings like these are avoidable.

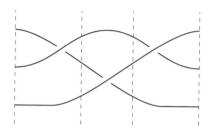

Figure 18.8 A braid as a product of crossings.

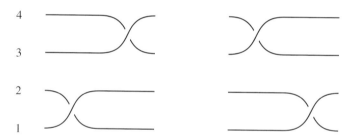

Figure 18.9 $\sigma_1\sigma_3$ is equivalent to $\sigma_3\sigma_1$.

In other words, B_n is generated by the set of all n–string braids that have exactly one crossing. The braid in which string i crosses over string $i + 1$ is commonly called σ_i, so, for example, Figure 18.2(c) shows the braid σ_3 in B_7. Think about what its inverse is. You can now see that B_n is generated by the set $\{\sigma_1, \ldots, \sigma_{n-1}\}$; this is the braid group version of Exercise 1 from Office Hour 1. (Note that the notation σ_i doesn't tell you which braid group you're in, except that it must be at least B_{i+1}. The context usually clarifies this.)

Exercise 4. Write down each of the braids we've seen before as products of the σ_i and their inverses.

Let's think about how the different σ_i relate to each other. Notice first that if σ_i and σ_j involve completely different strings, then they commute; see Figure 18.9.

Exercise 5. Show that neighboring σ_i's do not commute: $\sigma_i\sigma_{i+1} \neq \sigma_{i+1}\sigma_i$.

The neighboring σ_i's do satisfy a relation, however. This is a famous equation called the *braid relation*:

$$\sigma_i \sigma_{i+1} \sigma_i = \sigma_{i+1} \sigma_i \sigma_{i+1}.$$

(A similar braid relation in the mapping class group is discussed in Office Hour 17.)

Exercise 6. Draw a picture to convince yourself that the braid relation holds. Have you seen a picture like this before?

It turns out that these two types of relations, the commuting relations and the braid relations, imply all other relations that hold among the σ_i. In other words:

THEOREM 18.1. *The braid group has the presentation*

$$B_n = \langle \sigma_1, \ldots, \sigma_{n-1} \mid \sigma_i \sigma_j = \sigma_j \sigma_i \text{ for } |i - j| > 1,$$
$$\sigma_i \sigma_{i+1} \sigma_i = \sigma_{i+1} \sigma_i \sigma_{i+1} \text{ for } 1 \leq i \leq n - 2 \rangle.$$

We'll discuss how this theorem is proved later in this office hour.

Exercise 7. *(a)* Using Theorem 18.1, show that $B_2 \cong \mathbb{Z}$. Does this agree with your intuition?
(b) Using Theorem 18.1, write down a presentation for B_3.
(c) Show that $B_3 \cong \langle x, y \mid x^3 = y^2 \rangle$. *(Hint: Call the latter group G, and define a map from G to B_3 by sending x to $\sigma_1 \sigma_2$ and y to $\sigma_1 \sigma_2 \sigma_1$. Show that this is well defined and an isomorphism.)*

Exercise 8. *(a)* Using Theorem 18.1, show that the abelianization of B_n is isomorphic to $\mathbb{Z}$ when $n \geq 2$.
(b) Describe the abelianization map from B_n to $\mathbb{Z}$. For example what number is the image of $\sigma_1 \sigma_2 \sigma_1^{-1}$?
(c) Define a function ℓ, called the *length homomorphism*, from B_n to the integers as follows. If w is a word in the generators σ_i and their inverses, set $\ell(w)$ to be the sum of the exponents of the σ_i in w. In other words, each σ_i counts for 1 and each σ_i^{-1} counts for -1. Show that ℓ is well-defined and a homomorphism.
(d) Show that ℓ is exactly the same function as the abelianization map.

Exercise 9. *(a)* Show that the generators σ_i of B_n are all conjugate to each other.
(b) Let $\delta = \sigma_1 \sigma_2 \sigma_1$ and let $\gamma = \sigma_2 \sigma_1^{-1}$. Show that $\delta \gamma \delta^{-1} = \gamma^{-1}$. Thus the braid group contains an element γ that is conjugate to its inverse.
(c) Show that if γ is any braid that is conjugate to its inverse, then $\ell(\gamma) = 0$.

Permutations.   Look back at the picture of a permutation in Section 1.1 in Office Hour 1. It looks a lot like the pictures of braids we are drawing, doesn't it? The only difference is that with a braid, you keep track of over and under crossings, whereas with a permutation there's just a crossing.

What this means is that there is a map $\pi : B_n \to S_n$ where you take a braid, ignore the overs and unders, and just read the corresponding permutation. (Recall that S_n stands for the *symmetric group*.)

Exercise 10. Show that $\pi : B_n \to S_n$ is a homomorphism.

By the way, how did you solve Exercise 5? Do you have any tools to show that two braids (such as $\sigma_1\sigma_2$ and $\sigma_2\sigma_1$) are different? Well, now you do: look at their images under the map π!

Let's use the symbol τ_i for the image $\pi(\sigma_i)$. When we ignore the crossings of a braid and turn it into a permutation, we are effectively saying that the image of σ_i is the same as the image of σ_i^{-1}. That is, $\tau_i = \tau_i^{-1}$, or equivalently $\tau_i^2 = 1$.

Exercise 11. Consider the following presentation, which resembles the presentation in Theorem 18.1:

$$G = \langle \tau_1, \ldots, \tau_{n-1} \mid \tau_i\tau_j = \tau_j\tau_i \text{ for } |i - j| > 1,$$
$$\tau_i\tau_{i+1}\tau_i = \tau_{i+1}\tau_i\tau_{i+1} \text{ for } 1 \leq i \leq n - 2,$$
$$\tau_i^2 = 1 \text{ for } 1 \leq i \leq n - 1 \rangle.$$

(a) Find elements $\tau_i \in S_n$ that satisfy all these relations. *(Hint: Use the preceding discussion.)*

(b) Use part *(a)* to show that the map taking $\tau_i \in G$ to $\tau_i \in S_n$ extends to a well-defined surjective homomorphism from G to S_n.

(c) Show that this homomorphism is also injective. Thus the above is a presentation of S_n.

Exercise 12. Write down the permutations associated to all the braids we've drawn so far.

In a way, this says that you can think of a braid as being like a permutation, except that when you describe it in terms of successive transpositions $(i \ i + 1)$, you have to specify not just what i is but also *how* you are swapping i with $i + 1$: does i go over $i + 1$ or does it go under?

Pure braids. The kernel of the map $\pi : B_n \to S_n$ is an important subgroup of B_n. It is called the *pure braid group* PB_n, and it consists of all braids in B_n such that the strings line up in the same order on the left as they do on the right. (Being in the kernel of π means that they map to the identity permutation.) We will see a nice interpretation of PB_n in a little while.

You could also explore different subgroups of B_n that are related to π, but that aren't just the kernel. For instance, you could define an "even" braid to be an element of $\pi^{-1}(A_n)$, where A_n is the alternating group (consisting of the even permutations).

Exercise 13. Find an easy way to tell whether a braid is even.

Exercise 14. *(a)* What is the index of PB_n inside B_n?

(b) Recall that $B_2 \cong \mathbb{Z}$. Describe the subgroup PB_2.

Coming up with generators for the pure braid group is a bit trickier than for the full braid group, because you can't always chop up a pure braid into obvious chunks that are pure. Can you come up with a candidate generating set? For example, does the set $\{\sigma_i^2\}$ generate PB_n?

In fact there is a "standard" generating set for the pure braid groups that was defined first by Artin (see [11, 12]). There are $\binom{n}{2}$ generators called A_{ij}; there is one for each i and j with $1 \leq i < j \leq n$. These generators are defined in terms of the standard generators for B_n by

$$A_{ij} = (\sigma_{j-1}\sigma_{j-2}\cdots\sigma_{i+1})\sigma_i^2(\sigma_{j-1}\sigma_{j-2}\cdots\sigma_{i+1})^{-1}.$$

Exercise 15. Draw A_{ij} and describe it in words.

Can you see any relations between the A_{ij}'s? Some of them obviously commute, right? In fact Artin gave a complete presentation of PB_n using these generators, but his relations are a bit complicated. It turns out that these generators can be used in a presentation that has only *commuting relations*, i.e., relations that say that two elements commute. Mind you, these elements are not always just generators.[3] For example, the element A_{23} commutes with the element $A_{12}A_{13}$.

Exercise 16. Verify the last claim.

Project 1 at the end of this office hour invites you to study these and other presentations of the pure braid groups.

Combing and the word problem. Artin used the generators A_{ij} to describe a process he called *combing* for a pure braid. The idea behind combing is to take a pure braid and put it into a standard form, so that if you start with two pure braids that are equivalent (even if they don't look equivalent!) then you will end up with the same picture after combing. The normal forms we find here are in the same spirit as the normal forms found in Office Hours 14 and 16 for right-angled Artin groups and Thompson's group.

The combing process is iterative: roughly speaking, starting with a braid β in PB_n, you first remove the nth string to produce a braid called β_{n-1}. Then remove string $n-1$ to get the braid β_{n-2}. Eventually you get down to just a single string. Draw this as a straight line. Now reinstate the second string, creating β_2, but don't change how you've drawn the first string. (So the first string stays straight.) Then add in the third string, without changing the first two, being careful to *put any crossings involving the new string to the left of all previous crossings*. Continue like this. What you end up with is a picture of β that naturally decomposes into $n-1$ "blocks," where each block contains only one string that isn't straight. An example of a combed braid is shown in Figure 18.10, in which the vertical dotted lines delimit the blocks.

For a more detailed and truly excellent demonstration of braid combing, watch the animated video [99] produced by Ester Dalvit as part of her Ph.D. thesis. It is available at http://matematita.science.unitn.it/braids/index.html or on YouTube.[4]

---

[3] In particular, if $n \geq 4$ then PB_n is *not* a right-angled Artin group, although this is not so easy to prove. The subgroup of B_n generated by the elements σ_i^2, on the other hand, *is* a right-angled Artin group! See Office Hour 14 for more details.

[4] I wonder which will live longer: this book or YouTube?

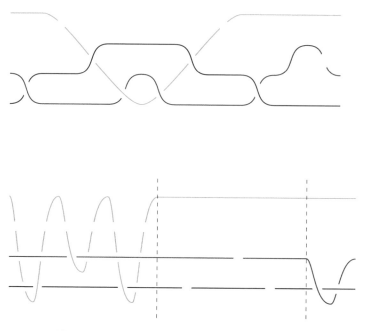

Figure 18.10 The result of combing. It's tricky; try it!

The part on combing starts about five minutes into Chapter 2, and lasts just a few minutes. In fact, I stole Figure 18.10 from her video: this is exactly the example she works out!

There are some things to point out about the combed braid. For one, the result is obviously not in its minimal form. However, this form makes it easy to write the braid in terms of the A_{ij}. Also, note that even after you remove the obvious cancellations, there are still a lot more crossings in the combed version than in the original! The fact that the combing is nevertheless useful illustrates the tenet that, depending on your goals, the "simplest" form of an object is not always the most convenient form to use.

Exercise 17. (*a*) Write the top braid in Figure 18.10 in terms of the generators σ_i.
(*b*) Write the bottom braid in Figure 18.10 in terms of the generators A_{ij}.
(*c*) Show that the two braids are equivalent.

The idea of combing helps us solve the word problem in B_n (refer to Office Hour 8 for a discussion of the word problem). Why? Well, Artin proved the following theorem about combed braids. See Artin's paper [11] for a proof.

THEOREM 18.2. *If two pure braids are equivalent, then they have the same combing.*

In other words, the combed version of a braid is unique; two braids might not initially look the same, but if they are, then once they're combed they'll be identical. Because of this the combed version is called a "normal form" for the braid.

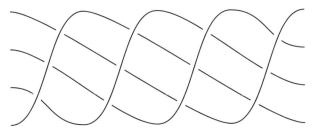

Figure 18.11 The full twist, Δ^2.

Exercise 18. Comb the braids $A_{23} \cdot (A_{12}A_{13})$ and $(A_{12}A_{13}) \cdot A_{23}$.

So what does this have to do with the word problem? Well, the problem after all is to determine for any given braid whether or not it is equivalent to the identity. So, given a braid, step one is to check whether or not it is pure. If not, it's not the identity! Thus we can suppose it's pure. Now comb it. By Artin's theorem, if it's equivalent to the identity, then the combed version will look exactly like the combed version of the identity, namely, the identity. So that's a solution to the word problem: as soon as we comb the (pure) braid, we know whether or not it's equivalent to the identity.

THEOREM 18.3. *The braid groups have solvable word problem.*

This is just one of many known ways to solve the word problem in the braid groups, and in fact it's a relatively inefficient one: it takes exponential time as a function of the number of crossings in the braid. (Note, in particular, that the 10–crossing braid in Figure 18.10 has a combed form with 28 crossings! This gives a hint, though not a proof, that the combing process might not be particularly efficient.) There are now quadratic-time algorithms for this problem, meaning that the number of steps required to come up with the answer is a quadratic function of the length of the braid word. One of the projects at the end of the office hour is to learn about the many different solutions to the word problem for braid groups.

The twist.   There is a braid called a *full twist* that you can imagine as follows: take the identity braid and spin the entire right wall by 360 degrees. When the endpoints return to their original positions, the braid created in the strings is the full twist. (Remember, you can't show the full twist is equivalent to the identity by unspinning the wall, because braid equivalence requires the endpoints to stay still throughout the process.) The full twist is denoted by Δ^2. See Figure 18.11.

Exercise 19. *(a)* Write down the full twist Δ^2 in B_n as a word in the braid generators $\sigma_i^{\pm 1}$.
(b) Show that Δ^2 commutes with every other braid in B_n.
(c) Show that, as the notation suggests, Δ^2 is indeed the square of a braid in B_n. (Note: Your argument should work regardless of whether n is even or odd.)

(d) The braid you found in part *(c)* is called the *half twist*, denoted by Δ. Does Δ also commute with everything in B_n?

(e) Show that Δ^2 has both a square root and an nth root.

Another way to say the conclusion of part *(b)* of Exercise 19 is that the full twist is in the *center* of B_n. In fact, Artin proved the following theorem in [11].

THEOREM 18.4. *If $n \geq 3$, then the center of B_n is an infinite cyclic subgroup generated by the full twist.*

In other words, any braid that commutes with all other braids must be a power of the full twist. (Why is this not true for $n = 2$?)

Exercise 20. Without using the fact that the center is generated by the full twist, show that the center of B_n is contained in PB_n.

Exercise 21. Prove Theorem 18.4 in the case of $n = 3$: the center of B_3 is generated by the full twist. *Hint: Use Exercise 7(c).*

Understanding the center is helpful for various reasons, foremost among them that the center (of any group) is a *characteristic* subgroup. This means that every automorphism (i.e., isomorphism) from a group G to itself takes the center to the center. In the case of braid groups we can use this, for example, to prove the following theorem, also due to Artin.

THEOREM 18.5. *If $B_m \cong B_n$, then $m = n$.*

This is a reassuring statement, isn't it? To prove this theorem, Artin used the abelianization map, or equivalently the length homomorphism ℓ as defined in Exercise 8. Remind yourself now what those are.

Exercise 22. Show that $\ell(\Delta^2) = n(n-1)$ if Δ^2 is the full twist in B_n.

Here is Artin's proof of Theorem 18.5. First note that we may assume $m, n \geq 3$, since $B_2 \cong \mathbb{Z}$ and B_n is non-abelian if $n \geq 3$, so B_2 cannot be isomorphic to B_n for any $n \geq 3$. Now, suppose $B_m \cong B_n$ with $m, n \geq 3$, and suppose $\phi : B_m \to B_n$ is an isomorphism. To minimize confusion, let's use Δ_i^2 to refer to the full twist in the group B_i. The center of B_i is infinite cyclic and generated by Δ_i^2, so because the center is characteristic, the isomorphism $\phi : B_m \to B_n$ must take Δ_m^2 to $\Delta_n^{\pm 2}$. Now, the map ϕ induces a map ϕ' on the abelianizations, and ϕ' is an isomorphism too. (See Exercise 23 below.) The map ϕ' takes the image of Δ_m^2 (in the abelianization of B_m) to the image of Δ_n^2 (in the abelianization of B_n). As ϕ' goes from $\mathbb{Z}$ to $\mathbb{Z}$, it must be multiplication by ± 1. Thus, by Exercise 22, we see that $m(m-1) = \pm n(n-1)$. Since m and n are positive, this implies $m = n$.

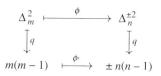

Exercise 23. If $\phi : A \to B$ is an isomorphism of groups, show that ϕ induces an isomorphism ϕ' from the abelianization of A to the abelianization of B.

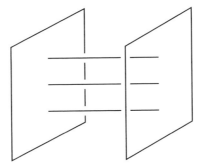

Figure 18.12 The identity braid is a boring movie: the particles sit still as the "time" coordinate x goes from 0 to 1.

18.3 SOME TOPOLOGY: CONFIGURATION SPACES

There are numerous important connections between braid groups and topology, and in this section we discuss one of the most basic ones. The central concept is that of a *configuration space*.

The phrase "configuration space" generally refers to a space that encapsulates all the possible states of a complex system. Configuration spaces are also sometimes called state spaces or parameter spaces. For example, configuration spaces can be used to model the collective motions of several objects, such as cars on city streets or packets in a network or molecules in solution or robot workers in a factory. For a perhaps more concrete example, consider the three joints in your arm: the shoulder, elbow, and wrist. Your shoulder and wrist each have two degrees of rotational freedom, and your elbow has one, for a total of five dimensions of configurations. A robotic arm modeled on yours has to navigate through a five-dimensional configuration space in order to take advantage of all the flexibility built into this joint structure. To illustrate this, let's imagine you don't have a hand, but rather just a rigid platform attached to your wrist. Suppose you want to lift a full glass of water from below your waist to above your shoulder. Can you simply rotate your shoulder joint? This is what babies do, and they spill the water every time! Eventually, they learn how to coordinate the shoulder rotation with the requisite elbow and wrist motions to keep the glass upright. Once you're good at it it seems easy, but the five-dimensional configuration space takes some getting used to.

So, what is the connection with braids? We have been talking about individual braids as topological objects, but it is also true that the braid *groups* are topological objects, in the sense that each braid group describes, in a natural way, the different types of loops that exist in a certain configuration space. The space in question models the collective motion of n distinct particles in the plane that are not allowed to collide.

Let's get specific. Begin with n distinct particles in the plane. Stand the plane up so that it forms the left wall of a braid. Now slide the plane from left to right, and imagine that each particle leaves a trail behind it in space. What happens?

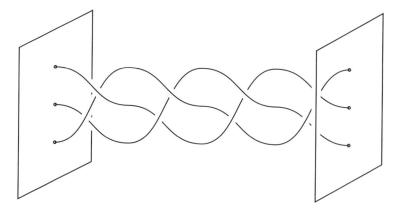

Figure 18.13 The braid of Figure 18.2(b) makes for a more interesting movie.

If the particles sit still (within the moving plane), then when you're done you see n parallel trails: the identity braid, as in Figure 18.12. But if the particles are moving around in the plane while the plane is sliding from left to right, then what happens? As long as no two particles ever collide, you see a braid! And clearly every braid can be made this way. See Figure 18.13. The point is that we are now interpreting the x direction as a *time* axis.

Exercise 24. *(a)* Place three coins on a table and move them in such a way that the resulting movie would look as in Figure 18.13.
(b) For jugglers: consider any juggling pattern that keeps all balls in a (vertical) plane, such as the standard 3–ball pattern. Translate the pattern into a braid, using either a picture or the σ_i notation.[5]
(c) For contra dancers (or square dancers): translate your favorite dance pattern into a braid.
(d) Watch the first few minutes of [101] (Chapter 4) and see what kinds of braids are produced by maypole dances.

Exercise 25. Look back at the two equivalent braids of Figure 18.5, and think about the equivalence itself, i.e., the motion that takes one braid to the other. If the braids are movies of motions of particles in the plane, describe:
(a) the pair of motions yielding the first braid,
(b) the pair of motions yielding the second braid, and
(c) what happens to the pair of motions during the equivalence between the two braids.

In order to form products of braids using this idea, you should make sure that collectively the particles are sitting at the same place at the end of the braid as at the beginning. That way you can concatenate two braids without getting a discontinuity in the middle.

---

[5] For more on connections between braids and juggling, see the book by Polster [222].

The starting position is called a *configuration* of n particles. The set of all possible configurations is the *configuration space*

$$C_n(\mathbb{R}^2) = \{(p_1, \ldots, p_n) \in (\mathbb{R}^2)^n \mid p_i \neq p_j \text{ for } i \neq j\},$$

where the condition $p_i \neq p_j$ expresses the requirement that the particles not collide.

Exercise 26. (*a*) Show that $C_2(\mathbb{R}^2)$ is homeomorphic to $\mathbb{R}^2 \times (\mathbb{R}^2 - \{(0,0)\})$. *Hint: Place the particles down one at a time.*
(*b*) Can you prove something similar about $C_3(\mathbb{R}^2)$?

Actually what we've just defined is the set of *ordered* or *labeled* configurations, meaning that a point of $C_n(\mathbb{R}^2)$ is an ordered n–tuple of distinct points in $\mathbb{R}^2$. For the purpose of describing braids, we would like to ignore the ordering and just pay attention to the set of particles—for example, we want to consider (p_1, p_2) to be the same as (p_2, p_1), and we want to think of it as simply $\{p_1, p_2\}$. For this we define the unordered version

$$UC_n(\mathbb{R}^2) = \{\{p_1, \ldots, p_n\} \subset \mathbb{R}^2 \mid p_i \neq p_j \text{ for } i \neq j\},$$

or, in words, the set of (unlabeled) n–point subsets of the plane.

Exercise 27. (*a*) If you are familiar with group actions, show that the symmetric group S_n acts on $C_n(\mathbb{R}^2)$ by permuting coordinates. In other words, if ρ is a permutation in S_n, then ρ acts on $C_n(\mathbb{R}^2)$ via $\rho(p_1, \ldots, p_n) = (p_{\rho(1)}, \ldots, p_{\rho(n)})$.
(*b*) Show that if ρ is not the identity permutation then ρ acts without fixed points on $C_n(\mathbb{R}^2)$.
(*c*) Show that two points $(p_1, \ldots, p_n)$ and $(q_1, \ldots, q_n)$ are in the same orbit if and only if $\{p_1, \ldots, p_n\} = \{q_1, \ldots, q_n\}$.
(*d*) Therefore the quotient $C_n(\mathbb{R}^2)/S_n$ is equal to $UC_n(\mathbb{R}^2)$.

Now, look back at Figure 18.13. The left wall represents a point of $UC_3(\mathbb{R}^2)$, and so does the right wall. In fact, the two walls represent the same point. (They do *not* represent the same point of $C_3(\mathbb{R}^2)$. Do you see why?) Not only that, but when you slide the left wall to the right in order to create the braid, at each instant in time you have a point of $UC_3(\mathbb{R}^2)$. This is because each string is monotonic. In other words, the braid can be viewed as a *loop* in $UC_3(\mathbb{R}^2)$. To reiterate, you can think of left-to-right as a time axis, and then the whole three-dimensional braid is like a movie of three particles dancing around in the two-dimensional plane (without colliding) and then returning to where they started. Equivalently, the braid is the graph of a function $[0, 1] \to UC_3(\mathbb{R}^2)$, whose starting and ending values agree.
The moral is that you can think of braids as loops of configurations. In fact, this is essentially a one-to-one correspondence.

THEOREM 18.6. *The fundamental group of $UC_n(\mathbb{R}^2)$ is isomorphic to the braid group B_n.*

Recall from Office Hour 9 that the *fundamental group* keeps track of the different based loops in a topological space. The notion of equivalence for loops—namely, *homotopy*—translates directly to our notion of equivalence of braids. (Think about this!) Thus this theorem states that not only can braids be viewed as loops in the configuration space, but also you can get every loop in the configuration space this way. This theorem was probably understood by A. Hurwitz [174] around 1890 while he was studying the way the (complex) roots of a polynomial move around when you change the coefficients of the polynomial. A source for this is [123].

Exercise 28. *(a)* Explain how to define a natural function from B_n to B_{n+1} for each n. Is it a homomorphism?
(b) Can you similarly define a map from B_{n+1} to B_n? Is it a homomorphism?
(c) Using the preceding theorem, show that the fundamental group of $C_n(\mathbb{R}^2)$ is isomorphic to the pure braid group PB_n. (This is easier if you know about covering spaces.)
(d) Show that there is a natural homomorphism from PB_{n+1} to PB_n defined by erasing the last string.
(e) Describe the kernel of this map in terms of combings (see Section 18.2).
(f) Show that PB_n can be expressed as an iterated semidirect product of free groups:
$PB_n \cong F_{n-1} \rtimes (F_{n-2} \rtimes (\cdots \rtimes (F_2 \rtimes \mathbb{Z})) \cdots)$.

In addition to giving us a fun new way to think about braids, this topological perspective is useful both for proving things about braids and for coming up with interesting generalizations. For example, it is possible to prove Theorem 18.1 using these ideas; see Project 3. For another example, note that the notation itself, $C_n(\mathbb{R}^2)$, suggests an obvious generalization: namely, we could replace $\mathbb{R}^2$ with another space X, and see what happens. You could try letting X be a surface, or a graph, or other manifolds or spaces. The set of (equivalence classes of) loops in a configuration space of n particles in a space X is called a *braid group* associated to X. If the configuration space is ordered, the braid group is *pure*, and we write $PB_n(X)$; if not, we write $B_n(X)$. So the usual braid groups PB_n and B_n are the same as $PB_n(\mathbb{R}^2)$ and $B_n(\mathbb{R}^2)$. We will look at the case where X is a sphere in the next section, and the case where X is a graph a little later.

Exercise 29. *(a)* Define and describe the ordered and unordered configuration spaces of two particles in a line. How many connected components do they have?
(b) How many connected components do $C_3(\mathbb{R})$ and $UC_3(\mathbb{R})$ have?
(c) What about $C_n(\mathbb{R})$ and $UC_n(\mathbb{R})$?
(d) Repeat for a circle instead of $\mathbb{R}$.

Spherical braids.   Suppose X is a 2–sphere. How do "spherical braids" differ from "planar braids"? You can imagine that the left and right walls in Figure 18.13 are each a small patch of a huge sphere; so each (usual) braid can be viewed as a spherical braid. (In fact, this idea can be applied to braids on any surface, not just a sphere.) This gives a map $f : B_n \to B_n(S^2)$. Can you get every spherical braid this way? Is f a 1–1 correspondence? The answer is: yes and no (see the following exercise).

Exercise 30. *(a)* Show that f is surjective.
(b) Show that the square of the full twist is in the kernel of f!

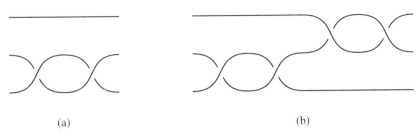

Figure 18.14 Trivial in $B_3(S^2)$?

(c) Look up and read about the "belt trick," and then demonstrate and explain it to your friends.

Part *(b)* might be tricky; try drawing a picture of the image of $(\Delta^2)^2$ inside $S^2 \times [0, 1]$. Another braid that's in the kernel of f is this one: $\gamma = \sigma_1\sigma_2 \cdots \sigma_{n-1}\sigma_{n-1}\sigma_{n-2} \cdots \sigma_1$. (Note that there are two σ_{n-1}'s in the middle.)

Exercise 31. *(a)* Draw the braid γ.
(b) Convince yourself that $f(\gamma)$ is the identity braid. *Hint: $f(\gamma)$ lives in $B_n(S^2)$, so strings can "wrap around the back."*

It turns out that the kernel of f is normally generated by γ, meaning that anything in the kernel of f is a product of conjugates of γ or γ^{-1}. Another way to say this, using the first isomorphism theorem, is that if you add the relation $\gamma = 1$ to the standard braid group presentation, then you will get a presentation for $B_n(S^2)$.

THEOREM 18.7. *The spherical braid group has the presentation*

$$B_n(S^2) = \langle \sigma_1, \ldots, \sigma_{n-1} \mid \sigma_i\sigma_j = \sigma_j\sigma_i \text{ for } |i - j| > 1,$$
$$\sigma_i\sigma_{i+1}\sigma_i = \sigma_{i+1}\sigma_i\sigma_{i+1} \text{ for } 1 \leq i \leq n - 2,$$
$$\sigma_1\sigma_2 \cdots \sigma_{n-2}\sigma_{n-1}^2\sigma_{n-2} \cdots \sigma_1 = 1\rangle.$$

Exercise 32. Show algebraically that $(\Delta^2)^2$ is in the normal subgroup of B_n generated by γ.

Exercise 33. *(a)* Using Theorem 18.7, show that $B_2(S^2)$ is a finite group of order 2.
(b) Using Theorem 18.7, show that $B_3(S^2)$ is a finite group of order 12, and write down (representatives of) the 12 elements.
(c) How many groups of order 12 do you know? Can you identify $B_3(S^2)$ from among the ones you know?
(d) Show that $PB_3(S^2)$, the pure spherical 3–string braid group, is cyclic of order 2, and find a (the) nontrivial element.
(e) Determine whether each of the braids in Figure 18.14 is trivial, when thought of as elements of $B_3(S^2)$.

Torsion.   Back to planar braids now. Here is a basic question: could you have a nontrivial braid β that has finite order (meaning $\beta^n = 1$ for some $n > 1$)? Intuitively it seems like this couldn't happen: if you have a braid, how could you unravel it by braiding it again? I am suggesting that the braid groups should be *torsion free*.

But this is not obvious: in fact, we saw in Exercise 33 that some small spherical braid groups aren't torsion free! In fact, for all $n \geq 2$, $B_n(S^2)$ contains finite-order elements. Remember the full twist Δ^2? In view of Exercises 19*(e)* and 30*(b)*, the roots of Δ^2 have finite order in $B_n(S^2)$. By the next exercise, they are nontrivial, so this makes them torsion elements of $B_n(S^2)$.

Exercise 34. Show that the square root of Δ^2 and the nth root of Δ^2 that you constructed in Exercise 19 represent nontrivial braids in $B_n(S^2)$. Is Δ^2 also nontrivial in $B_n(S^2)$?

None of the finite order braids we found in $B_n(S^2)$ have finite order when thought of as (planar) braids in B_n: you always need the last ("spherical") relation to reduce the braids to the identity. Could it be that B_n is actually torsion free?

We will resolve this question in the next section. In fact, planar braid groups really are torsion free. This is pretty hard to prove using just group theory, but there are several different geometric arguments that do the trick. If you know some algebraic topology, here is a quick outline of one such proof. You can prove by induction on n that the space $UC_n(\mathbb{R}^2)$ is *aspherical*, which means that for every $k > 1$, any continuous map from a k–sphere into the space can be continuously shrunk to a point. It is a basic result of algebraic topology that fundamental groups of aspherical spaces are always torsion free, so that does it. The proof we will give is more elementary, using pictures called *curve diagrams*.

18.4 MORE TOPOLOGY: PUNCTURED DISKS

Look back at Figure 18.13 for a moment. We will now describe a completely different way to interpret this picture. Let's pretend that the braid is made of stiff wire, and that the plane is actually just a finite square-shaped part of a plane (as it is actually drawn), which is to say, a topological disk, except that this disk has holes poked in it that line up with the left side of the braid. Now imagine that the interior of the disk is made of very flexible fabric and that the boundary square is a rigid wire frame (sort of like those cloth frisbees people use as dog toys, only square (not that the shape matters here)). Picture the process of taking this punctured disk and pushing it across the rigid wire braid until it gets to the right wall. The braid hasn't moved, but the disk itself has gotten all twisted up. The braid has produced a change in the surface.

The specific thing that has happened to the punctured disk is that the braid has implemented a homeomorphism from the surface to itself. The outer boundary stays fixed, because it is rigid, and the punctures get permuted according to the usual permutation associated to the braid. Maps like this, up to homotopy, form the

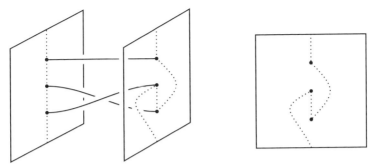

Figure 18.15 The curve diagram associated to σ_1 is drawn on the right wall. The braid σ_1 is left veering.

mapping class group of D_n, a disk with n punctures:

$$\text{Mod}(D_n) = \{f : D_n \to D_n \mid f \text{ is a homeomorphism,}$$

$$f|_{\partial D_n} = \text{Id}\}/ \text{Homotopy}$$

The group operation is composition of functions: two homeomorphisms f and g can be composed to give a homeomorphism $g \circ f$ (or $f \circ g$, which is usually different). Visit Office Hour 17 for more details.

We have described a map ψ from B_n to $\text{Mod}(D_n)$, namely, given a braid, slide the disk across the braid and consider the resulting homeomorphism.

Exercise 35. Show that ψ is a homomorphism.

In fact, the following theorem shows that this is yet another way to think about the braid group.

THEOREM 18.8. *The map ψ is an isomorphism. So, B_n is isomorphic to $\text{Mod}(D_n)$.*

Curve diagrams. Let's go a little deeper into this perspective on braid groups. Look once again at Figure 18.13, focusing now on the square disks at either end of the picture. Imagine that there is a logo on the disk on the left shaped like a vertical dotted line going through all the punctures. Where will the logo be after the disk has been pushed across the braid to the right?

Figure 18.15 shows a simpler example. The braid is simply σ_1, and both the original logo and the twisted result are drawn. The dotted curve on the right wall is called the *curve diagram* induced by the braid (σ_1, in this case).

Let's call the original logo the "axis" of the disk. It consists of $n + 1$ dotted segments $a_0, \ldots, a_n$, numbered in order from bottom to top. Then, in general, the *curve diagram* associated to a braid β is the union of the arcs $\psi(\beta)(a_i)$, or, in other words, you think of β as a homeomorphism of the disk, and the curve diagram is where the axis goes. We will call the arcs of the curve diagram c_i, and the whole diagram c. Note that since β fixes the boundary of the disk, the c_i's (i.e., the curve diagram) fit together end to end, in order, to form the single arc c that starts at the

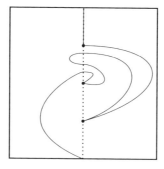

Figure 18.16 This curve diagram reduces to the one in Figure 18.15. (The axis is made of dots.)

bottom center, never crosses itself, visits every puncture exactly once, and ends at the top center. See again Figure 18.15.

Exercise 36. Draw the curve diagrams for $\sigma_1\sigma_2\sigma_1$ and $\sigma_2\sigma_1\sigma_2$ in D_3. What do you notice?

Curve diagrams can be extremely complicated, as you would expect if you start with a very long braid. For example, the curve diagram for the braid in Figure 18.13 is pretty hard to draw (but it's fun to try!). One preliminary way to simplify a curve diagram is to make sure it is *reduced*, in the following sense.

Start with a curve diagram on a punctured disk, and also draw the axis on the same disk. The diagram is *reduced* if there are no *bigons* in the picture. A bigon is a region whose boundary consists of one sub-arc of a single a_i and one sub-arc of a single c_i. These bigons can easily be eliminated, so every curve diagram can be put into reduced form. The curve diagram in Figure 18.16 has two bigons. Do you see them?

One last definition: two curve diagrams are *equivalent* if they have the same reduced form.

Now, it is time to make good on our promise to prove that the braid group is torsion free. We will do this using curve diagrams.

Given a braid β, look at its reduced curve diagram c. Start at the bottom and look for the first c_i that isn't equal to the corresponding axis arc a_i. Let's say that β is *right veering* if this first c_i lies to the right of a_i, and say that β is *left veering* if c_i is to the left of a_i. If there is no such i, meaning that $c_i = a_i$ for all i, then β must be the identity braid. (Technically, we are using Theorem 18.8 here!) The braid in Figure 18.15 is left veering because starting at the bottom, the arc c_0 veers to the left.

Exercise 37. (a) Look back at some of the braids we've seen, and determine whether they're right veering or left veering (or neither). For example, try the braid from Figure 18.13.
(b) If β is right veering, show that β^{-1} is left veering.
(c) Can you tell by looking at a word in the σ_i's whether the corresponding braid is right veering or left veering?

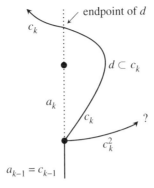

Figure 18.17 Illustrating the proof of the lemma.

Now, remember that we are trying to prove that B_n is torsion free. This is a consequence of the following lemma.

LEMMA 18.9. *If β is right veering, then β^n is right veering for all $n > 0$.*

THEOREM 18.10. *The braid group B_n is torsion free.*

Exercise 38. Why does the theorem follow from the lemma?

Here is a proof of the lemma. Assume β is right veering, and look at the reduced curve diagram c for β. Find the first c_i that differs from the corresponding a_i, and call this arc c_k. Next let c^2 denote the curve diagram for β^2, and let's call the arcs of this curve diagram c_i^2. Imagine two separate disks, one in which the a_i and c_i are drawn and a second in which the c_i and the c_i^2 are drawn. Observe that the second picture is obtained from the first by applying the braid β (thought of as a homeomorphism of course!). Thus, since c_k is to the right of a_k, it follows that c_k^2 is to the right of c_k. Also, since there are no bigons in the first picture, there are no bigons in the second picture either.

What we want to show is that the braid β^2 is right veering, i.e., that c_k^2 is to the right of a_k. We know that β doesn't affect the arcs a_i for $i < k$, and so neither does β^2. We also know that in the pictures, c_k is to the right of a_k and c_k^2 is to the right of c_k. This sounds pretty good. Are we missing anything?

The only missing piece is that the curve diagram for β^2 might not be reduced. We know there are no bigons between the axis and c, nor are there any between c and c^2, but if c^2 and the axis make a bigon, then c^2 needs to be reduced and it might not end up right veering.

Fortunately this can't happen. To see this, let d be the initial segment of c_k that goes until the first time c_k touches the axis, so that d is contained in one half (the right half) of the disk. (It is possible that d is all of c_k, but in general c_k may wander and weave for a long time before reaching a puncture.) Now the only way for c_k^2 to be to the left of (or equal to) the axis arc a_k is by forming a bigon with a_k. But this would force c_k^2 to cross d, thereby creating a bigon between c and c^2, which we know doesn't exist. See Figure 18.17.

Thus β^2 is indeed right veering. By repeating this argument we can now easily show that β^n is right veering for all $n > 0$, as desired. □

Thus the braid groups B_n are torsion free. One lesson to learn from this set of ideas, and from this book in general, is that a topological or geometric approach can often provide a lot of insight into algebraic results. To be fair, there are some details missing from this proof. For example, homeomorphisms can be complicated: for one thing, the arc c_0 might intersect a_0 infinitely many times. Then how do you reduce it? This and other details can be dealt with in a completely rigorous way, but once people get the hang of these types of technicalities, they tend to find it more palatable to omit them than to include them.

The word problem revisited. We are now primed to see several more approaches to the word problem. The first is an exercise.

Exercise 39. Using Theorem 18.8, describe a solution to the word problem in B_n using curve diagrams.

To delve further, see Projects 2, 4, and 5 at the end of the office hour.

18.5 CONNECTION: KNOT THEORY

Having proved some basic theorems about braid groups, we will now discuss some further connections between braids and other areas of mathematics. The first of these is knot theory.

If you have ever studied knot theory, then you have probably already noticed some similarities between knots—or more generally links—and braids (a link is just like a knot, except that the number of components is allowed to be greater than 1). For instance, they are both defined in terms of pictures of strings in 3–space, and they both have a notion of *equivalence* (usually called *isotopy*) that describes when pictures are equivalent. Of course there are also some differences, most notably that braids have endpoints that aren't allowed to move during isotopies. Another difference is that each string of a braid goes monotonically from left to right, whereas there is no such restriction with links.

In fact, one of the early motivations for studying braids was to help understand the theory of (knots and) links.[6] The connection is that every braid can be "closed up" to form a link. It was hoped that the group structure on braids would lead to some sort of group structure on the set of links. It turns out that that doesn't work, but there are nevertheless other ways to use braids to help with the study of links.

Here is how you close up a braid: just connect the left endpoints to the right endpoints without introducing any new crossings. Suddenly you have a link!

---

[6] The quote that opens this office hour is part of the poem, *(Cats) Cradle Song*, written by Maxwell in response to Tait's manuscript *On Knots*, the first mathematical treatise on knot theory. According to Silver [241], the word "coil" refers to what we now call a braid.

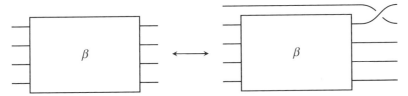

Figure 18.18 The Markov move: here β stands for an arbitrary braid (with any number of strings). The Markov move adds (or subtracts) a string and a single crossing.

Exercise 40. *(a)* Show that when you close up the 3–string braid $(\sigma_1\sigma_2^{-1})^2$, you get the Figure 8 knot.
(b) Show that the Borromean rings can be realized as a closed braid (look this up if you don't know what it is).
(c) Show that every link can be realized as a closed braid.

Part *(c)* of the preceding exercise is sometimes called Alexander's theorem. If you are stuck, Chapter 3 of the movie [100] illustrates one way to prove it.

One good thing about this theorem is that it gives you a straightforward way to describe any link over the phone (or any time you can't draw a picture). Rather than saying, "It's knot number 10_{161} in the standard table," which only helps if you have a standard table and you can find your knot in the table and you are talking to someone who has the same table, or "It goes over and then under, and then back around to where it was before only on the other side, and then …," which is the sort of thing people say and which has some rather obvious drawbacks, you can instead simply say, "It's the knot you get by closing up the 3–string braid $\sigma_1\sigma_2^{-1}\sigma_1\sigma_2^2\sigma_1^2\sigma_2^3$." Much better, right?

Alexander's theorem doesn't give a one-to-one correspondence between links and braids, though. Lots of different braids can close up to give the same link, which is obvious once you think about it for a moment. (Think about it for a moment.) You can even have braids with different numbers of strings that close up to give the same link.

Exercise 41. How many different 2–string braids close up to the unknot? What about 3–string braids?

Exercise 42. Let β and γ be n–string braids. Show that when you close up β, you get the same link as when you close up $\gamma\beta\gamma^{-1}$.

Conjugating a braid, as in the previous exercise, doesn't change the number of strings. Here is a move that does: it is called a Markov move. See Figure 18.18.

Exercise 43. Show that these two braids close up to give the same link.

In fact it is a theorem that any time you have two braids that close up to the same link, you can get from one braid to the other by a sequence of conjugations and Markov moves. You don't need any more operations besides these two. (This theorem is attributed to Markov although he may not have actually proven it.)

In the modern era, the connection between braids and links led to the discovery of the Jones polynomial, which is one of the most successful tools for distinguishing knots and links. To learn more about knots, braids, and knot polynomials including the Jones polynomial, see [4] or [182].

Aside: Reidemeister moves.   If you have studied knots before, you must know about Reidemeister moves. If not, you could either skip the rest of this section or else go look up Reidemeister moves on the internet (for example in Wikipedia). In summary, these are three different types of changes you can make to a knot diagram that don't change the actual knot. Moreover they have the wonderful property that you can get from any diagram of a knot to any other diagram of the same knot by a sequence of Reidemeister moves. Only recently, a bound has been proven on how many moves this could take (as a function of the number of crossings in the diagram). This implies a conceptually simple algorithm for determining whether two diagrams represent the same knot: try all possible sequences of moves on one of the diagrams, and see if you ever produce the other diagram. Once you try all possible sequences with no more moves than the known bound, you know that if you haven't seen the other diagram, you never will. Unfortunately this isn't yet a practical algorithm, as the bound is still tremendously large.[7] Nevertheless, Reidemeister moves are also useful for a lot of theoretical purposes, like establishing many important knot invariants.

For braids, there is an analog of Reidemeister moves; namely, the Reidemeister moves themselves! Except that you don't need the Type I move, because of the condition that each string is monotonic. This means you can never have a loop, so you can never do a move of Type I.

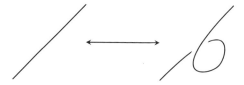

What about Type II? If you see a part of a braid that looks as in Figure 18.5, then you can straighten both strings. This is a Type II move. Algebraically, it corresponds to cancelling $\sigma_i \overline{\sigma}_i$ (or $\overline{\sigma}_i \sigma_i$) out of a word, which of course you can do. You can also insert such a cancelling pair any time you want to without changing the braid.

Finally, what about Type III? Well, look closely at the picture you drew for Exercise 6 (or Figure 18.6). It's a Type III Reidemeister move on braids!

The point is that using Type II and III Reidemeister moves, you can get from any braid diagram to any other diagram of an equivalent braid. (Notice that these moves don't change the *parity* of the number of crossings in the braid. This is related to Exercise 13.)

---

[7] For details see the articles [92, 161, 190].

18.6 CONNECTION: ROBOTICS

Let's turn to a connection that has concrete real-world applications. The subject is again configuration spaces, but this time we will study particles moving around on a graph. These spaces are useful models for network traffic and various automated processes involving many moving parts. We will refer to the particles as *robots*. If you are in a context where it is important that the robots (or cars, or packets, or whatever) not crash into each other, then what you really want to study is the configuration space. To safely coordinate the motions of several robots, you really ought to understand the geometry, i.e., the paths and loops and so on, in a configuration space.

Let G be a finite graph, made of vertices and edges. We have already defined the configuration spaces, both ordered and unordered, of n robots on G:

$$C_n(G) = \{(p_1, \ldots, p_n) \in G \times \cdots \times G \mid p_i \neq p_j \text{ if } i \neq j\},$$

and $UC_n(G)$ is the unordered version of n-point subsets of G^n. There are, correspondingly, pure graph braid groups $PB_n(G)$ and graph braid groups $B_n(G)$ defined as the fundamental groups of these spaces, or, in other words, the equivalence classes of based loops of (labeled or unlabeled) configurations.

What does a nontrivial element of $PB_n(G)$ look like? Such an element is a loop in the configuration space $C_n(G)$, but more concretely it is a "nontrivial" motion of n robots on G. How do you picture what nontrivial means?

Exercise 44. Let Y be the graph with one vertex O connected by edges to each of the three vertices A, B, and C, and with no other edges or vertices. Act out a nontrivial braid in $PB_2(Y)$.

Discretization.   Visualizing these configuration spaces can be tricky, even when $n = 2$. But there is a technique called *discretization* that helps a lot. It works like this: instead of allowing the robots to move anywhere they want on the graph (subject to the rule that they don't collide), we will restrict their motions somewhat. Imagine the vertices of the graph as "stations." Each station has a manager who is responsible for any robot that is either at the station or on (the interior of) an edge that meets the station. Thus a robot at a station occupies just one station manager, but a robot between two stations (meaning, on an edge) occupies the managers at both end stations of the edge.

Now consider all (ordered) configurations of robots where no station manager is ever occupied by more than one robot. This is a subset of $C_n(G)$, and it is called the *discretized configuration space* $D_n(G)$. Similarly, if you start with the unlabeled configurations $UC_n(G)$ and follow the same procedure, you get the *unordered discretized configuration space* $UD_n(G)$.

Let's consider an example.

Example 18.6.1. *Let G be the graph Y from Exercise 44, and consider $D_2(Y)$. Then $D_2(Y)$ has 12 vertices, corresponding to the possible locations of the two robots. From the vertex AB, there are edges leading to OB (corresponding to moving*

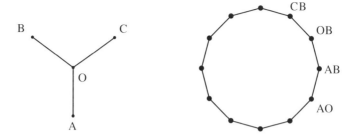

Figure 18.19 The space $D_2(Y)$ is a 12–gon. Exercise: finish labeling the vertices.

the first robot) and AO (corresponding to moving the second robot). Continuing in this manner, we construct the space shown in Figure 18.19.

Exercise 45. (*a*) Fill in the details of this example.
(*b*) Describe the loop in $D_2(Y)$ that you followed when you did Exercise 44.

You can now see why this is called discretizing: the motions of the particles are being modeled as discrete motions. There are several other equivalent ways to define $D_n(G)$ which further explain the term. These are based on the idea that the graph G is a one-dimensional cell complex (with vertices and edges), and so the product G^n is an n–dimensional cell complex. In fact, each (closed) cell of G^n is of the form $c_1 \times \cdots \times c_n$, where c_i is either a vertex or an edge of G. In particular, each cell of G^n has the combinatorial structure of a cube (of some dimension $\leq n$). (Think of the edges as being "closed," meaning they include their endpoints.)

Exercise 46. (*a*) Show that $D_n(G)$ is the largest subcomplex of G^n (that is, the largest union of closed cubes in G^n) that is contained in $C_n(G)$.
(*b*) Show that a cell $c_1 \times \cdots \times c_n$ of G^n is in $D_n(G)$ if and only if the (closed) cells c_i are pairwise disjoint in G.
(*c*) Show that an i–dimensional cell of $D_n(G)$ is a product of i (closed) edges and $n - i$ vertices of G that are pairwise disjoint.
(*d*) Show that any loop in $D_n(G)$ can be moved continuously until it only uses the edges of $D_n(G)$. This means that, although it is legal for several robots to be moving at the same time, it is always possible to exchange such a motion for an equivalent one where only one robot is moving at a time.

You can now see why the space $D_2(Y)$ is one-dimensional: the graph doesn't contain any pairs of disjoint edges, which is what you would need to get two-dimensional cells in D_2.

Exercise 47. (*a*) In terms of the graph G, how many vertices does $D_2(G)$ have? How many edges?
(*b*) Draw $D_2(G)$ for various graphs G: the letter X, a triangle, a square, a pentagon, the complete graphs K_4 and K_5, and any others you'd like to try. In each case, can you predict how many squares $D_2(G)$ will have?

Drawing $D_2(K_5)$ is a bit tricky: it can't be drawn in the plane, but it can be drawn in 3–space. It is hard to resist including a picture, but you should do it yourself!

Figure 18.20 The utilities graph. Look up the "utility problem" to learn the origin of the
name.

Exercise 48. If you gave up on drawing $D_2(K_5)$, try again. What does a neighborhood of a point look like? *Hint: The space is a closed surface. Which surface is it?
(Use the Euler characteristic, introduced in Office Hour 17.)*

Exercise 49. The "utilities graph" $K_{3,3}$ is shown in Figure 18.20. Draw $D_2(K_{3,3})$.
*Hint: This one is also a closed surface! Which surface is it? (Use the Euler
characteristic again.)*

No graph other than K_5 and $K_{3,3}$ yields a closed surface when you construct D_2.
Not only that, but for $n > 2$, it is never the case that $D_n(G)$ is a closed n–manifold.
So these two examples are really quite special! For more on these and various other
examples, see [3] or [2].

At the beginning of this section we mentioned that the process of discretization
provides a helpful way to visualize the configuration spaces $C_n(G)$. You may have
noticed that we never really justified this: what is the relationship supposed to be,
other than that D_n is a subset of C_n? The answer is provided by the following
theorem.

THEOREM 18.11. *If the graph G is subdivided enough, then $D_n(G)$ is a deformation retract of $C_n(G)$.*

This means that $C_n(G)$ is basically a "thickened up" version of $D_n(G)$. In particular, they have the same fundamental group, so our implicit identification of the
graph braid group $B_n(G)$ with the fundamental group of $D_n(G)$ (rather than $C_n(G)$)
is justified, provided that the hypothesis of the theorem holds. What does "subdivided enough" mean? Insert enough valence 2 vertices along each edge so that
any motion you can imagine in $C_n(G)$ can be approximated by a motion in $D_n(G)$.
This is subdivided enough. In particular, if there are two robots, it is enough to have
a "simple graph," which is a graph that has no loops consisting of just one or two
edges. In general, what you actually need is enough stations so that all the robots
can fit on any given edge at the same time, plus one extra on each loop so that there
is room to move around.

18.7 CONNECTION: HYPERPLANE ARRANGEMENTS

We will now describe one more geometric connection with braid groups. Do you
know what a hyperplane in a vector space is? It is a linear subspace of codimension
1, i.e., of dimension 1 less than the dimension of the vector space. In a real vector
space $\mathbb{R}^d$, every hyperplane is isomorphic to $\mathbb{R}^{d-1}$. If you remove a hyperplane
from $\mathbb{R}^d$, you are left with two connected components, each of which is an open
half-space. What happens if you remove more than one hyperplane?

Exercise 50. *(a)* If you remove k hyperplanes from $\mathbb{R}^2$, how many components are left?

(b) If you remove two hyperplanes from $\mathbb{R}^3$, how many components are left?
(c) If you remove three hyperplanes from $\mathbb{R}^3$, how many components are left?
(d) If you remove two hyperplanes from $\mathbb{R}^4$, how many components are left?

Actually, this exercise is a bit of a trick question. The number of components left over when you remove k hyperplanes from $\mathbb{R}^d$ doesn't just depend on k and d. It also depends on how the hyperplanes intersect each other. One part of Exercise 50 has multiple possible answers. Do you know which part?

Anyway, again we ask the question: what does this have to do with braids? The idea is to think about complex vector spaces instead of real ones. The linear algebra doesn't change: a hyperplane H in $\mathbb{C}^d$ is isomorphic as a vector space to $\mathbb{C}^{d-1}$, and it has dimension $d - 1$ as a vector space over $\mathbb{C}$. But as a topological space, the dimension of $\mathbb{C}^d$ is $2d$, and the dimension of H is $2d - 2$. It has "complex codimension 1" but "real codimension 2." What this means is that if you delete a hyperplane from $\mathbb{C}^d$, the result remains connected! It is like deleting a line from $\mathbb{R}^3$: instead of losing connectivity, the space loses simple-connectivity (as in Office Hour 8, a path-connected space is *simply connected* if its fundamental group is trivial, or if every loop is homotopic to a point). A hyperplane H in $\mathbb{C}^d$ has a "linking circle" that wraps around H exactly once; this circle generates the fundamental group of $\mathbb{C}^d - H$, which is isomorphic to $\mathbb{Z}$.

Exercise 51. *(a)* Convince yourself that if you start with the space $\mathbb{C}^2$, with complex coordinates z and w, and you delete the hyperplane $z = 0$, you are left with a space X that is connected.

(b) Convince yourself that the unit circle $|z| = 1$ in the hyperplane $w = 0$ is *(i)* contained in X and *(ii)* a *linking circle* for the deleted hyperplane $z = 0$. One way to see *(ii)* is to show that the circle doesn't bound any disk that is disjoint from the hyperplane $z = 0$.

That's the situation with one hyperplane. If you delete a bunch of hyperplanes from $\mathbb{C}^d$, you always get a connected space but the fundamental group can be quite complicated.

So now we see how this relates to braids. As we learned in Section 18.3, the pure braid group PB_n is the fundamental group of the configuration space $C_n(\mathbb{R}^2)$. If we identify $\mathbb{R}^2$ with the complex numbers $\mathbb{C}$, then we can write PB_n as the fundamental group of the space

$$C_n(\mathbb{C}) = \{(z_1, \ldots, z_n) \in \mathbb{C}^n \mid z_i \neq z_j \text{ if } i \neq j\}$$
$$\subset \mathbb{C}^n.$$

This is the space of pairwise distinct n–tuples of complex numbers. It is a subset of the vector space $\mathbb{C}^n$. What pairwise distinct means is that z_1 is not allowed to be equal to z_2 or z_3 or any of the others, and so on. But look: the (linear) equation $z_1 = z_2$ describes a hyperplane in $\mathbb{C}^n$. So $C_n(\mathbb{C})$ is exactly what you get if you start with $\mathbb{C}^n$ and delete the $\binom{n}{2}$ hyperplanes $z_i = z_j$.

Figure 18.21 A braid in the wild.

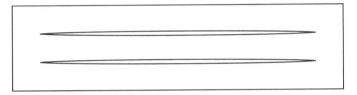

Figure 18.22 Slit your paper.

Thus the pure braid group PB_n can be viewed as the fundamental group of the complement of a collection of hyperplanes in $\mathbb{C}^n$. It turns out that lots of other interesting groups also arise as fundamental groups of complements of "hyperplane arrangements." To learn more about this subject, see [215] (or Wikipedia to get started).

18.8 A STYLISH AND PRACTICAL FINALE

Have you ever seen a braided belt, like the ones in Figure 18.21?

Did you ever wonder how they're made? Here's the basic idea. Take a strip of paper and cut two slits in it as in Figure 18.22.

Now, by braiding part of the strands and then passing one end through the slits, possibly several times, see if you can get a braid into the strip (without any further cutting). You want the individual strands to lie flat, i.e., face up, the whole time. In practice it is easier if the slits are long and thin and numerous. Try it with three, four, five, six, or more.

To carry out a similar experiment with your hair, the situation is slightly different, because there is no restriction that the strands "lie flat." Imagine braiding your hair by first making a ponytail and then doing the braiding. Which braids can you make this way? See Project 8.

FURTHER READING

We have seen several different ways to think about braid groups, and several connections with other areas of mathematics, but there are still many more! For further reading you can look up braids in conjunction with topics such as Artin groups, spaces of polynomials (this arises in Project 6), planar algebras, representation theory (this arises in Project 10), hypergeometric functions, quantum computing, homotopy groups of spheres, and surely others as well.

For more on braids themselves, and especially their role in the world of topology, check out *The Knot Book* by Adams [4] (an undergraduate-level knot theory book), the books by Rolfsen and by Birman [41, 228] (classical graduate-level geometric topology books), the survey paper by Birman and Brendle [42] (more advanced but still accessible), and the book by Farb and Margalit [127] (a more modern and thorough treatment of some of the ideas introduced in this office hour).

PROJECTS

Project 1. Read about the presentations of the pure braid group in the book by Farb and Margalit [127]. Learn the details about the presentation of the pure braid group by Margalit and McCammond [200].

Project 2. *(a)* Learn about as many different solutions to the word problem in B_n as you can. Compare their complexities: which solutions are most efficient? Some places to start: [38, 99, 107]. Note that several solutions are discussed in this office hour!

(b) Likewise, the *conjugacy problem* is also solvable in the braid groups. This means that it is possible to tell (with an algorithm) whether two given braids are conjugate. Learn about some solutions to the conjugacy problem, and how fast they are. Can you do better?

Project 3. In Section 18.3 we realized the braid group as the fundamental group of a reasonably nice space. From this perspective it is possible to use some basic tools from algebraic topology to derive the presentation of the braid group given in Theorem 18.1. The basic idea is to decompose $UC_n(\mathbb{R}^2)$ into cells using walls that correspond to configuraions in which different points have the same x–coordinate. Investigate the procedure for deriving the fundamental group of a space from a cell decomposition and use this to derive the presentation of Theorem 18.1. Use Chapter 2 of the book by Fathi, Laudenbach, and Poénaru [129] as a guide.

Project 4. This project introduces a property of groups called *orderability*.

(a) A group G is *left orderable* if there is a concept of "less than" for group elements that is compatible with the group operation. Precisely, G is left orderable

if there is a total order[8] $<$ such that for all $g, h, k \in G$, if $g < h$ then $kg < kh$. Show that the group of real numbers under addition is left orderable, but the group of nonzero complex numbers under multiplication is not.

(b) An alternative definition of left orderability is that there exists a subset P of G, called the *positive* elements, such that P is closed under multiplication and for every $g \neq 1$, either g or g^{-1} is in P. Show that the two definitions are equivalent.

(c) Let P be the set of right veering braids. Show that P satisfies the conditions given in part *(b)*. Thus the braid groups are left orderable.

(d) Use left orderability to give a different and very quick proof of Theorem 18.10.

(e) What do you think *right orderable* means? Show that a group is left orderable if and only if it is right orderable. There is also a notion of *bi-orderability*, where the order has to be compatible with multiplication on both the left and the right. Perhaps surprisingly, there are groups that are left orderable (hence right gorderable) but not bi-orderable. In fact, the braid groups have this property! Show that in a bi-orderable group, if $g > 1$, then $hgh^{-1} > 1$ for all h. Thus, by Exercise 9, the braid groups B_n are not bi-orderable for $n \geq 3$. The pure braid groups PB_n, on the other hand, are bi-orderable!

(f) Put together the various parts of this project to conclude that there must be braids α and β such that $\alpha\beta$ is right veering but $\beta\alpha$ is left veering. (This is a bit subtle.) Then find examples of such braids.

(g) This is not the original proof that braids are left orderable. In the original, due to Dehornoy, braid words are put into a certain form from which their positivity can be determined. This process is algorithmic, which means it gives yet another solution to the word problem. (This is because a word represents the identity braid if and only if neither it nor its inverse is positive.) Learn Dehornoy's algorithm from [107] or [109]; it involves a process called "handle reduction" (not a diet plan!) that is also described in the video [99]. This is one of the fastest known solutions to the word problem in braid groups.

(h) There is a lot more to be said about orderings of braid groups. Learn about the types of orderings, how they relate to each other, and many different ways to think about them from [109, 110] or another source.

Project 5. This project uses an idea that is similar to that of curve diagrams, but slightly different. Historically, this was probably the first known solution to the word problem in braid groups.

Pick a base point on the boundary of the disk D_n and draw n loops $x_1, \ldots, x_n$ from the base point, one going counterclockwise around each puncture. The fundamental group of D_n is a free group F_n generated by these loops, i.e., any loop in D_n can be decomposed into a product of these loops and their inverses, and there are no relations among these loops.

It helps to have some familiarity with fundamental groups for this project, but it is not too important. The goal of the project is to study the *action* of the braid group B_n on the fundamental group F_n. Figure 18.23 depicts the action of a particular braid on a particular loop. These images are taken from [98].

---

[8] A *total order* is simply a partial order such that for any two distinct elements a and b, either $a < b$ or $b < a$.

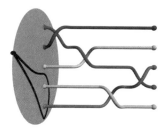 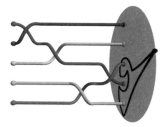

Figure 18.23 This braid acts on the loop x_1x_2 (shown in black), producing the loop x_1x_3.

(a) Show that σ_1 changes the loop x_1 into the loop x_2, and that σ_1 changes the loop x_2 into the loop $x_2^{-1}x_1x_2$.

(b) In general, show that σ_i moves x_i to x_{i+1} and moves x_{i+1} to $x_{i+1}^{-1}x_ix_{i+1}$ and leaves the other x_j fixed.

(c) In the language of Office Hour 6, each braid σ_i produces an automorphism $\alpha(\sigma_i)$ of the free group F_n generated by the x_i. Read that sentence again, slowly. The automorphism $\alpha(\sigma_i)$ is a map from F_n to F_n, and it takes x_i to x_{i+1}, x_{i+1} to $x_{i+1}^{-1}x_ix_{i+1}$, and x_j to x_j for all $j \neq i, i+1$. Make sure you understand how this map on the x_i's extends to a homomorphism from F_n to itself.

(d) Show that $\alpha(\sigma_i)$ is indeed an automorphism of F_n. *Hint: Find the inverse homomorphism!*

(e) It is not just σ_i that gives an automorphism of F_n but any braid in B_n. Given a word in the σ_i, the word will act on F_n one letter at a time. In other words, this circle of ideas is saying that we have defined a map α from B_n to $\mathrm{Aut}(F_n)$. Show that this map is a homomorphism, and compute $\alpha(\sigma_1\sigma_2)$.

(f) In fact, α is injective, so the braid group B_n is a subgroup of $\mathrm{Aut}(F_n)$. So here is how to solve the word problem: given a word in the letters $\sigma_i^{\pm 1}$, you compute the automorphism it induces on F_n. This means you compute where x_1 goes, where x_2 goes, and so on, using the rules we've outlined. At the end, if each x_i goes to itself, then the automorphism is the identity map, so by injectivity of α the braid must be the identity braid. If any x_i fails to go to itself, then the braid can't be the identity braid.

(g) Compute $\alpha(\sigma_1\sigma_2\sigma_1\sigma_2^{-1}\sigma_1^{-1}\sigma_2^{-1})$. What should the answer be? Note that you don't actually need to draw any punctured disks for this!

(h) How fast is this algorithm?

By the way, the map α is far from being surjective: there are lots of elements of $\mathrm{Aut}(F_n)$ that do not arise from braids. The group $\mathrm{Aut}(F_n)$ was introduced in Office Hour 6, and is currently a popular object of study for geometric group theorists.

Project 6. This project uses some algebraic topology.

(a) If you are familiar with knot theory, and in particular with fundamental groups of knot complements, compute the fundamental group Π of the complement of the trefoil knot and show that $B_3 \cong \Pi$. (See Exercise 7.)

(b) There is an explanation of this apparent coincidence. Try to understand the isomorphism $B_3 \cong \Pi$ by filling in the (many) details in the following chain of reasoning:

- B_3 is the fundamental group of the configuration space $UC_3(\mathbb{R}^2)$;
- by considering the particles as locations of roots, this configuration space is the same as the space of monic cubic polynomials (over $\mathbb{C}$) with no repeated roots;
- by keeping the "center of mass" of the roots fixed at 0, such a cubic can be put into the normal form $z^3 + az + b$;
- once in the form $z^3 + az + b$, the distinctness of roots is equivalent to $4a^3 + 27b^2 \neq 0$;
- the space of pairs $(a, b) \in \mathbb{C}^2$ satisfying $4a^3 + 27b^2 \neq 0$ can be squished onto the unit sphere in $\mathbb{C}^2$, also known as S^3, in which $4a^3 + 27b^2 = 0$ is the equation of a trefoil knot (hint: this requires a "twisted" radial projection!);
- therefore B_3 is isomorphic to the fundamental group of the trefoil knot complement.

Project 7. This project introduces the *edge machine*, which is a way to realize graph braid groups as subgroups of right-angled Artin groups.

(a) If you haven't done so already, see Office Hour 14 on right-angled Artin groups and learn the definition.

(b) Given a graph G, make a new graph G' as follows: G' has a vertex for each edge of G, and two vertices of G' are adjacent in G' if and only if the corresponding edges of G are disjoint, i.e., do not share a vertex. Draw a few examples of G and G'.

(c) Consider each edge of G to be oriented. Starting with a given configuration of robots on vertices, a loop in the space $UD_n(G)$ can be described by listing a sequence ϵ of (oriented) edges of G, namely the edges traversed by the robots as they describe the loop. Convince yourself that this sequence of oriented edges can be interpreted as an element of the right-angled Artin group $A_{G'}$ determined by the graph G'.

(d) Show that the function $\epsilon : B_n(G) \to A_{G'}$ is well defined and a homomorphism.

(e) Show that ϵ is injective. Thus every graph braid group embeds in a right-angled Artin group.

Project 8. *(a)* Make some belts as described in Section 18.8.

(b) Which braids can you get this way? Note that a "beltable braid" must be pure. Is the set of beltable braids a subgroup of PB_n? Is it all of PB_n?

(c) Which braids can you put in your hair after first making a ponytail? See [240].

Project 9. Are you familiar with the old-school fold-up sunshades that people put on their car windshields? When they fold up, they fold over themselves three times, as in Figure 18.24.

Why three? Turns out it is not by chance—it is a physical and mathematical necessity! The relevant mathematical fact is that it is impossible to fold one of

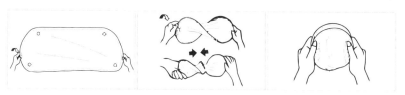

Figure 18.24 Folding a sunshade.

these in half without causing the metal frame to twist. This phenomenon also occurs, for example, with certain (flat, sufficiently loose) rubber bands, and the same phenomenon provides some insight into the geometric conformations produced by strands of DNA. Learn about this phenomenon and give a talk to your math club about it. A reference is [130].

Project 10. This project uses both linear algebra and algebraic topology (compare this project with Project 7 in Office Hour 17).

A group is called *linear* (as in linear algebra) if it is isomorphic to a (multiplicative) group of $n \times n$ matrices (with, say, complex entries), for some n. This is the same as saying that there is a faithful representation of the group into a matrix group. ("Faithful" means injective and "representation" means homomorphism.) Since we know how to do matrix multiplication, it is easy to do computations in a linear group if you've successfully represented the group as a bunch of matrices (for instance, it is very easy to solve the word problem in a linear group). However, not all groups are linear. For many decades it was unknown whether the braid groups are linear. But they are!

(a) Learn something about linear groups. Show that all of the most basic groups you can think of (finite cyclic groups, symmetric groups, the integers, free groups, etc.) are linear.

(b) It is not so easy to show that a group is not linear. For an in-depth reading project, learn how to do this. The easiest way I know has two steps, each substantial: one, there is a theorem due to Malcev that every finitely generated linear group is *residually finite*; and two, some groups are not residually finite, such as $\langle a, b \mid ba^2b^{-1} = a^3 \rangle$ (this is an example of a *Baumslag–Solitar group*). Another possibility: if you like $\mathrm{Aut}(F_n)$, you may enjoy learning why it is linear for $n = 2$ but not linear for $n \geq 3$.

(c) Show that the group B_3 is linear, as follows. Let $\omega = e^{\frac{2\pi i}{12}}$, and let a and b be any two complex numbers. Show that the map sending σ_1 to the matrix $\begin{bmatrix} \omega & a \\ 0 & \omega^{-1} \end{bmatrix}$ and σ_2 to the matrix $\begin{bmatrix} \omega & b \\ 0 & \omega^{-1} \end{bmatrix}$ extends to a well-defined homomorphism from B_3 to the group $SL_2(\mathbb{C})$ of 2×2 matrices over $\mathbb{C}$ with determinant 1. Show that this map is not injective if $a = 0$ (consider σ_1^{12}), if $b = 0$ (consider σ_2^{12}), if $a = b$ (consider $\sigma_1\sigma_2^{-1}$), or if $a = -b$ (consider $(\sigma_1\sigma_2)^6$). Then show that there are values of a and b that make the map injective, which proves B_3 is linear.

(d) For a long time there were various known procedures for representing braids as matrices, the most famous of which was due to Burau, and yet it was unknown whether these representations were faithful. Learn how the Burau representation

works. In particular the Burau representation is faithful for $n = 3$ but not for $n \geq 5$; learn how to prove this (see [33]). As of now, nobody knows whether the Burau representation of B_4 is faithful!

(e) This is the part that uses algebraic topology. The braid groups can also be represented as matrices using certain vector spaces called "homology groups of configuration spaces." If you have encountered homology before, you can learn what this means. These representations are faithful for all n, showing that the braid groups are linear. See [35].

Bibliography

[1] J. M. Aarts and T. Nishiura. *Dimension and Extensions*, volume 48 of *North-Holland Mathematical Library*. North-Holland Publishing Co., 1993.

[2] Aaron Abrams. Configuration spaces of colored graphs. *Geometriae Dedicata*, 92:185–194, 2002.

[3] Aaron Abrams and Robert Ghrist. Finding topology in a factory: Configuration spaces. *The American Mathematical Monthly*, 109(2):140–150, 2002.

[4] Colin C. Adams. *The Knot Book*. American Mathematical Society, Providence, RI, 2004. An elementary introduction to the mathematical theory of knots, Revised reprint of the 1994 original.

[5] Jared Adams, Eric Freden, and Marni Mishna. From indexed grammars to generating functions. *RAIRO Theor. Inform. Appl.*, 47(4):325–350, 2013.

[6] Kazushi Ahara. Teruaki. http://www51.atwiki.jp/kazushiahara/pages/32.html, November 2011.

[7] Juan M. Alonso. Inégalités isopérimétriques et quasi-isométries. *C. R. Acad. Sci. Paris Sér. I Math.*, 311(12):761–764, 1990.

[8] M. Amchislavska and T. Riley. Lamplighters, metabelian groups, and horocyclic products of trees, 2014. arXiv:1405.1660.

[9] A. V. Anīsīmov. The group languages. *Kibernetika (Kiev)*, (4):18–24, 1971.

[10] Yago Antolín and Laura Ciobanu. Geodesic growth in right-angled and even Coxeter groups. *European J. Combin.*, 34(5):859–874, 2013.

[11] E. Artin. Theory of braids. *Ann. of Math. (2)*, 48:101–126, 1947.

[12] Emil Artin. Theorie der Zöpfe. *Abh. Math. Sem. Univ. Hamburg*, 4(1):47–72, 1925.

[13] Thomas Kwok-Keung Au, Feng Luo, and Tian Yang. Lectures on the mapping class group of a surface. In *Transformation groups and moduli spaces of curves*, volume 16 of *Adv. Lect. Math. (ALM)*, pages 21–61. Int. Press, Somerville, MA, 2011.

[14] Patrick Bahls. *The isomorphism problem in Coxeter groups*. Imperial College Press, London, 2005.

[15] Laurent Bartholdi, Markus Neuhauser, and Wolfgang Woess. Horocyclic products of trees. *J. Eur. Math. Soc. (JEMS)*, 10(3):771–816, 2008.

[16] A. Baudisch. Subgroups of semifree groups. *Acta Math. Acad. Sci. Hungar.*, 38(1–4):19–28, 1981.

[17] G. Baumslag, C. F. Miller, III, and H. Short. Isoperimetric inequalities and the homology of groups. *Invent. Math.*, 113(3):531–560, 1993.

[18] Gilbert Baumslag. Wreath products and finitely presented groups. *Math. Z.*, 75:22–28, 1960/1961.

[19] Gilbert Baumslag. A non-cyclic one-relator group all of whose finite quotients are cyclic. *J. Austral. Math. Soc.*, 10:497–498, 1969.

[20] Jason Behrstock, Bruce Kleiner, Yair Minsky, and Lee Mosher. Geometry and rigidity of mapping class groups. *Geom. Topol.*, 16(2):781–888, 2012.

[21] James Belk and Kai-Uwe Bux. Thompson's group F is maximally non-convex. In *Geometric Methods in Group Theory*, volume 372 of *Contemp. Math.*, pages 131–146. Amer. Math. Soc., Providence, RI, 2005.

[22] James M. Belk and Kenneth S. Brown. Forest diagrams for elements of Thompson's group F. *Internat. J. Algebra Comput.*, 15(5–6):815–850, 2005.

[23] G. Bell and A. Dranishnikov. Asymptotic dimension. *Topology Appl.*, 155(12):1265–1296, 2008.

[24] Gregory C. Bell and Koji Fujiwara. The asymptotic dimension of a curve graph is finite. *J. Lond. Math. Soc. (2)*, 77(1):33–50, 2008.

[25] Aldo A. Bernasconi. *On HNN-extensions and the complexity of the word problem for one-relator groups*. ProQuest LLC, Ann Arbor, MI, 1994. Thesis (Ph.D.)–The University of Utah.

[26] J. Berstel and L. Boasson. Context-free languages. In *Handbook of Theoretical Computer Science, Vol. B*, pages 59–102. Elsevier, Amsterdam, 1990.

[27] Mladen Bestvina. A Bers-like proof of the existence of train tracks for free group automorphisms. *Fund. Math.*, 214(1):1–12, 2011.

[28] Mladen Bestvina. Geometric group theory and 3-manifolds hand in hand: The fulfillment of Thurston's vision. *Bull. Amer. Math. Soc. (N.S.)*, 51(1): 53–70, 2014.

[29] Mladen Bestvina and Noel Brady. Morse theory and finiteness properties of groups. *Invent. Math.*, 129(3):445–470, 1997.

[30] Mladen Bestvina, Ken Bromberg, and Koji Fujiwara. Constructing group actions on quasi-trees and applications to mapping class groups. *Publ. Math. Inst. Hautes Études Sci.*, 122:1–64, 2015.

[31] Mladen Bestvina and Michael Handel. Train tracks and automorphisms of free groups. *Ann. of Math. (2)*, 135(1):1–51, 1992.

[32] Robert Bieri. Normal subgroups in duality groups and in groups of cohomological dimension 2. *J. Pure Appl. Algebra*, 7(1):35–51, 1976.

[33] Stephen Bigelow. The Burau representation is not faithful for $n = 5$. *Geom. Topol.*, 3:397–404, 1999.

[34] Stephen Bigelow and Ryan Budney. The mapping class group of a genus two surface is linear. *Algebr. Geom. Topol.*, 1(34):699–708, 2001.

[35] Stephen J. Bigelow. Braid groups are linear. *J. Amer. Math. Soc.*, 14(2): 471–486 (electronic), 2001.

[36] Nicholas Billington. Growth of groups and graded algebras. *Comm. Algebra*, 12(19-20):2579–2588, 1984.

[37] J.-C. Birget, A. Yu. Ol'shanskii, E. Rips, and M. V. Sapir. Isoperimetric functions of groups and computational complexity of the word problem. *Ann. of Math. (2)*, 156(2):467–518, 2002.

[38] Joan Birman, Ki Hyoung Ko, and Sang Jin Lee. A new approach to the word and conjugacy problems in the braid groups. *Adv. Math.*, 139(2):322–353, 1998.

[39] Joan Birman, Dan Margalit, and William Menasco. Efficient geodesics and an effective algorithm for distance in the complex of curves. *Math. Annalen (to appear)*, 2016.

[40] Joan S. Birman. On Siegel's modular group. *Math. Ann.*, 191:59–68, 1971.

[41] Joan S. Birman. *Braids, Links, and Mapping Class Groups*. Princeton University Press, Princeton, N.J.; University of Tokyo Press, Tokyo, 1974. Annals of Mathematics Studies, No. 82.

[42] Joan S. Birman and Tara E. Brendle. Braids: A survey. In *Handbook of Knot Theory*, pages 19–103. Elsevier B. V., Amsterdam, 2005.

[43] Oleg Bogopolski. *Introduction to Group Theory*. EMS Textbooks in Mathematics. European Mathematical Society (EMS), Zürich, 2008. Translated, revised and expanded from the 2002 Russian original.

[44] William W. Boone. Certain simple, unsolvable problems of group theory. I. *Nederl. Akad. Wetensch. Proc. Ser. A.*, 57: 231–237 = Indag. Math. 16, 231–237 (1954), 1954.

[45] B. H. Bowditch. A short proof that a subquadratic isoperimetric inequality implies a linear one. *Michigan Math. J.*, 42(1):103–107, 1995.

[46] Noel Brady, Martin R. Bridson, Max Forester, and Krishnan Shankar. Snowflake groups, Perron-Frobenius eigenvalues and isoperimetric spectra. *Geom. Topol.*, 13(1):141–187, 2009.

[47] Noel Brady, Tim Riley, and Hamish Short. *The Geometry of the Word Problem for Finitely Generated Groups*. Advanced Courses in Mathematics. CRM Barcelona. Birkhäuser Verlag, Basel, 2007. Papers from the Advanced Course held in Barcelona, July 5–15, 2005.

[48] Marcus Brazil. Growth functions for some nonautomatic Baumslag-Solitar groups. *Trans. Amer. Math. Soc.*, 342(1):137–154, 1994.

[49] Tara E. Brendle and Benson Farb. Every mapping class group is generated by 6 involutions. *Journal of Algebra*, 278(1):187–198, 2004.

[50] Martin Bridson, José Burillo, Murray Elder, and Zoran Šunić. Groups with polynomial geodesic growth, 2006. arXiv:1009.5051v1.

[51] Martin R. Bridson. Fractional isoperimetric inequalities and subgroup distortion. *J. Amer. Math. Soc.*, 12(4):1103–1118, 1999.

[52] Martin R. Bridson. The geometry of the word problem. In *Invitations to geometry and topology*, volume 7 of *Oxf. Grad. Texts Math.*, pages 29–91. Oxford Univ. Press, Oxford, 2002.

[53] Martin R. Bridson, José Burillo, Murray Elder, and Zoran Šunić. On groups whose geodesic growth is polynomial. *Internat. J. Algebra Comput.*, 22(5):1250048, 13, 2012.

[54] Martin R. Bridson and André Haefliger. *Metric Spaces of Non-positive Curvature*, volume 319 of *Grundlehren der Mathematischen Wissenschaften [Fundamental Principles of Mathematical Sciences]*. Springer-Verlag, Berlin, 1999.

[55] Matthew G. Brin. Higher dimensional Thompson groups. *Geom. Dedicata*, 108:163–192, 2004.

[56] Matthew G. Brin. The algebra of strand splitting. I. A braided version of Thompson's group V. *J. Group Theory*, 10(6):757–788, 2007.

[57] Matthew G. Brin and Craig C. Squier. Groups of piecewise linear homeomorphisms of the real line. *Invent. Math.*, 79(3):485–498, 1985.

[58] Brigitte Brink and Robert B. Howlett. A finiteness property and an automatic structure for Coxeter groups. *Math. Ann.*, 296(1):179–190, 1993.

[59] Peter Brinkmann. Xtrain. http://math.sci.ccny.cuny.edu/pages?name= XTrain.

[60] J. L. Britton. The word problem for groups. *Proc. London Math. Soc. (3)*, 8:493–506, 1958.

[61] John L. Britton. The word problem. *Ann. of Math. (2)*, 77:16–32, 1963.

[62] Kenneth S. Brown. Presentations for groups acting on simply-connected complexes. *J. Pure Appl. Algebra*, 32(1):1–10, 1984.

[63] Kenneth S. Brown and Ross Geoghegan. An infinite-dimensional torsion-free FP_∞ group. *Invent. Math.*, 77(2):367–381, 1984.

[64] Maxim Bruckheimer and Abraham Arcavi. Farey series and Pick's area theorem. *Math. Intelligencer*, 17(4):64–67, 1995.

[65] Dmitri Burago, Yuri Burago, and Sergei Ivanov. *A Course in Metric Geometry*, volume 33 of *Graduate Studies in Mathematics*. American Mathematical Society, Providence, RI, 2001.

[66] J. Burillo, S. Cleary, and M. I. Stein. Metrics and embeddings of generalizations of Thompson's group *F*. *Trans. Amer. Math. Soc.*, 353(4):1677–1689, 2001.

[67] José Burillo and Sean Cleary. Metric properties of braided Thompson's groups. *Indiana Univ. Math. J.*, 58(2):605–615, 2009.

[68] José Burillo and Sean Cleary. Metric properties of higher-dimensional Thompson's groups. *Pacific J. Math.*, 248(1):49–62, 2010.

[69] José Burillo and Jennifer Taback. Equivalence of geometric and combinatorial Dehn functions. *New York J. Math.*, 8:169–179 (electronic), 2002.

[70] J. W. Cannon. Geometric group theory. In *Handbook of Geometric Topology*, pages 261–305. North-Holland, Amsterdam, 2002.

[71] J. W. Cannon and W. J. Floyd. What is . . . Thompson's group? *Notices Amer. Math. Soc.*, 58(8):1112–1113, 2011.

[72] J. W. Cannon, W. J. Floyd, and W. R. Parry. Introductory notes on Richard Thompson's groups. *Enseign. Math. (2)*, 42(3-4):215–256, 1996.

[73] James W. Cannon. The combinatorial structure of cocompact discrete hyperbolic groups. *Geom. Dedicata*, 16(2):123–148, 1984.

[74] James W. Cannon, William J. Floyd, Richard Kenyon, and Walter R. Parry. Hyperbolic geometry. In *Flavors of geometry*, volume 31 of *Math. Sci. Res. Inst. Publ.*, pages 59–115. Cambridge Univ. Press, Cambridge, 1997.

[75] David Carter and Gordon Keller. Bounded elementary generation of $SL_n(\mathcal{O})$. *Amer. J. Math.*, 105(3):673–687, 1983.

[76] Ruth Charney. An introduction to right-angled Artin groups. *Geom. Dedicata*, 125:141–158, 2007.

[77] Ruth Charney and John Meier. The language of geodesics for Garside groups. *Math. Z.*, 248(3):495–509, 2004.

[78] N. Chomsky and M. P. Schützenberger. The algebraic theory of context-free languages. In *Computer programming and formal systems*, pages 118–161. North-Holland, Amsterdam, 1963.

[79] Noam Chomsky. On certain formal properties of grammars. *Information and Control*, 2:137–167, 1959.

[80] Sean Cleary, Murray Elder, and Jennifer Taback. Cone types and geodesic languages for lamplighter groups and Thompson's group *F*. *J. Algebra*, 303(2):476–500, 2006.

[81] Sean Cleary and Jennifer Taback. Thompson's group *F* is not almost convex. *J. Algebra*, 270(1):133–149, 2003.

[82] Sean Cleary and Jennifer Taback. Combinatorial properties of Thompson's group *F*. *Trans. Amer. Math. Soc.*, 356(7):2825–2849 (electronic), 2004.

[83] Sean Cleary and Jennifer Taback. Dead end words in lamplighter groups and other wreath products. *Q. J. Math.*, 56(2):165–178, 2005.

[84] Sean Cleary and Jennifer Taback. Metric properties of the lamplighter group as an automata group. In *Geometric Methods in Group Theory*, volume 372 of *Contemp. Math.*, pages 207–218. Amer. Math. Soc., Providence, RI, 2005.

[85] Daniel E. Cohen. The mathematician who had little wisdom: a story and some mathematics. In *Combinatorial and Geometric Group Theory (Edinburgh, 1993)*, volume 204 of *London Math. Soc. Lecture Note Ser.*, pages 56–62. Cambridge Univ. Press, Cambridge, 1995.

[86] Daniel E. Cohen, Klaus Madlener, and Friedrich Otto. Separating the intrinsic complexity and the derivational complexity of the word problem for finitely presented groups. *Math. Logic Quart.*, 39(2):143–157, 1993.

[87] M. Cohen, Wolfgang Metzler, and A. Zimmermann. What does a basis of $F(a, b)$ look like? *Math. Ann.*, 257(4):435–445, 1981.

[88] D. J. Collins, M. Edjvet, and C. P. Gill. Growth series for the group $\langle x, y \mid x^{-1}yx = y^l \rangle$. *Arch. Math. (Basel)*, 62(1):1–11, 1994.

[89] Donald J. Collins. Relations among the squares of the generators of the braid group. *Invent. Math.*, 117(3):525–529, 1994.

[90] Briana Cook, Eric M. Freden, and Alisha McCann. A simple proof of a theorem of Whyte. *Geom. Dedicata*, 108:153–162, 2004.

[91] Thierry Coulbois. Train-tracks for sage. http://www.latp.univ-mrs.fr/~coulbois/train-track/.

[92] Alexander Coward and Marc Lackenby. An upper bound on Reidemeister moves. *Amer. J. Math.*, 136(4):1023–1066, 2014.

[93] H. S. M. Coxeter. Discrete groups generated by reflections. *Ann. of Math. (2)*, 35(3):588–621, 1934.

[94] H. S. M. Coxeter. The complete enumeration of finite groups $r_i^2 = (r_i r_j)^{k_{i,j}} = 1$. *J. London Math. Soc*, s1-10(1):21–25, 1935.

[95] John Crisp and Benson Farb. The prevalence of surface groups in mapping class groups. Preprint.

[96] John Crisp, Eddy Godelle, and Bert Wiest. The conjugacy problem in subgroups of right-angled Artin groups. *J. Topol.*, 2(3):442–460, 2009.

[97] John Crisp and Bert Wiest. Embeddings of graph braid and surface groups in right-angled Artin groups and braid groups. *Algebr. Geom. Topol.*, 4:439–472, 2004.

[98] Ester Dalvit. *New proposals for the popularization of braid theory*. PhD thesis, University of Trento, 2011. http://eprints-phd.biblio.unitn.it/663/.

[99] Ester Dalvit. Braids. Chapter 2: The word problem, 2013. https://www.youtube.com/watch?v=VhryJaoJT1Q.

[100] Ester Dalvit. Braids. Chapter 3: The world of knots, 2013. https://www.youtube.com/watch?v=1goWirL46qo.

[101] Ester Dalvit. Braids. Chapter 4: Hilden dances, 2013. https://www.youtube.com/watch?v=WxKs3GtxUbk.

[102] Michael W. Davis. *The Geometry and Topology of Coxeter Groups*, volume 32 of *London Mathematical Society Monographs Series*. Princeton University Press, Princeton, NJ, 2008.

[103] Michael W. Davis and Tadeusz Januszkiewicz. Right-angled Artin groups are commensurable with right-angled Coxeter groups. *J. Pure Appl. Algebra*, 153(3):229–235, 2000.

[104] Pierre de la Harpe. *Topics in Geometric Group Theory*. Chicago Lectures in Mathematics. University of Chicago Press, Chicago, IL, 2000.

[105] M. Dehn. Über unendliche diskontinuierliche Gruppen. *Math. Ann.*, 71(1):116–144, 1911.

[106] Max Dehn. *Papers on Group Theory and Topology*. Springer-Verlag, New York, 1987. Translated from the German and with introductions and an appendix by John Stillwell, With an appendix by Otto Schreier.

[107] Patrick Dehornoy. A fast method for comparing braids. *Adv. Math.*, 125(2):200–235, 1997.

[108] Patrick Dehornoy. The group of parenthesized braids. *Adv. Math.*, 205(2):354–409, 2006.

[109] Patrick Dehornoy, Ivan Dynnikov, Dale Rolfsen, and Bert Wiest. *Why are Braids Orderable?*, volume 14 of *Panoramas et Synthèses [Panoramas and Syntheses]*. Société Mathématique de France, Paris, 2002.

[110] Patrick Dehornoy, Ivan Dynnikov, Dale Rolfsen, and Bert Wiest. *Ordering Braids*, volume 148 of *Mathematical Surveys and Monographs*. American Mathematical Society, Providence, RI, 2008.

[111] Warren Dicks. Simplified Mineyev, 2012. http://mat.uab.es/~dicks/SimplifiedMineyev.pdf.

[112] Volker Diekert and Jürn Laun. On computing geodesics in Baumslag-Solitar groups. *Internat. J. Algebra Comput.*, 21(1–2):119–145, 2011.

[113] W. Dison, E. Einstein, and T. R. Riley. Taming the hydra: The word problem and extreme integer compression, 2015. arXiv:1509.02557.

[114] Will Dison, Murray Elder, Timothy R. Riley, and Robert Young. The Dehn function of Stallings' group. *Geom. Funct. Anal.*, 19(2):406–422, 2009.

[115] Will Dison and Timothy R. Riley. Hydra groups. *Comment. Math. Helv.*, 88(3):507–540, 2013.

[116] A. Dranishnikov. On asymptotic inductive dimension. *JP J. Geom. Topol.*, 1(3):239–247, 2001.

[117] A. Dranishnikov and T. Januszkiewicz. Every Coxeter group acts amenably on a compact space. *Topology Proc.*, 24(Spring):135–141, 1999.

[118] Alexander Dranishnikov. On asymptotic dimension of amalgamated products and right-angled Coxeter groups. *Algebr. Geom. Topol.*, 8(3):1281–1293, 2008.

[119] Cornelia Drutu and Michael Kapovich Preface. Lectures on geometric group theory.

[120] Anna Dyubina. Instability of the virtual solvability and the property of being virtually torsion-free for quasi-isometric groups. *Internat. Math. Res. Notices*, (21):1097–1101, 2000.

[121] M. Edjvet and D. L. Johnson. The growth of certain amalgamated free products and HNN-extensions. *J. Austral. Math. Soc. Ser. A*, 52(3):285–298, 1992.

[122] Murray Elder, Éric Fusy, and Andrew Rechnitzer. Counting elements and geodesics in Thompson's group *F*. *J. Algebra*, 324(1):102–121, 2010.

[123] Moritz Epple. Geometric aspects in the development of knot theory. In *History of topology*, pages 301–357. North-Holland, Amsterdam, 1999.

[124] David B. A. Epstein, James W. Cannon, Derek F. Holt, Silvio V. F. Levy, Michael S. Paterson, and William P. Thurston. *Word Processing in Groups*. Jones and Bartlett Publishers, Boston, MA, 1992.

[125] Alex Eskin, David Fisher, and Kevin Whyte. Coarse differentiation of quasi-isometries I: Spaces not quasi-isometric to Cayley graphs. *Ann. of Math. (2)*, 176(1):221–260, 2012.

[126] Benson Farb. *Problems on Mapping Class Groups and Related Topics*, volume 74 of *Proc. Sympos. Pure Math.* Amer. Math. Soc., Providence, RI, 2006.

[127] Benson Farb and Dan Margalit. *A Primer on Mapping Class Groups*, volume 49 of *Princeton Mathematical Series*. Princeton University Press, Princeton, NJ, 2012.

[128] Daniel S. Farley. Finiteness and CAT(0) properties of diagram groups. *Topology*, 42(5):1065–1082, 2003.

[129] Albert Fathi, François Laudenbach, and Valentin Poénaru. *Thurston's Work on Surfaces*, volume 48 of *Mathematical Notes*. Princeton University Press, Princeton, NJ, 2012. Translated from the 1979 French original by Djun M. Kim and Dan Margalit.

[130] Curtis Feist and Ramin Naimi. Topology explains why automobile sunshades fold oddly. *College Math. J.*, 40(2):93–98, 2009.

[131] Philippe Flajolet. Analytic models and ambiguity of context-free languages. *Theoret. Comput. Sci.*, 49(2-3):283–309, 1987. Twelfth international colloquium on automata, languages and programming (Nafplion, 1985).

[132] Philippe Flajolet and Robert Sedgewick. *Analytic Combinatorics*. Cambridge University Press, Cambridge, 2009.

[133] S. Blake Fordham. Minimal length elements of Thompson's group *F*. *Geom. Dedicata*, 99:179–220, 2003.

[134] Eric M. Freden and Teresa Knudson. Recent growth results. In *Groups St. Andrews 2005. Vol. 1*, volume 339 of *London Math. Soc. Lecture Note Ser.*, pages 341–355. Cambridge Univ. Press, Cambridge, 2007.

[135] Eric M. Freden, Teresa Knudson, and Jennifer Schofield. Growth in Baumslag-Solitar groups I: Subgroups and rationality. *LMS J. Comput. Math.*, 14:34–71, 2011.

[136] Eric M. Freden and Jennifer Schofield. The growth series for Higman 3. *J. Group Theory*, 11(2):277–298, 2008.

[137] Peter Freyd and Alex Heller. Splitting homotopy idempotents. II. *J. Pure Appl. Algebra*, 89(1–2):93–106, 1993.

[138] Thanos Gentimis. Asymptotic dimension of finitely presented groups. *Proc. Amer. Math. Soc.*, 136(12):4103–4110, 2008.

[139] Ross Geoghegan. *Topological Methods in Group Theory*, volume 243 of *Graduate Texts in Mathematics*. Springer, New York, 2008.

[140] S. M. Gersten. Dehn functions and l_1-norms of finite presentations. In *Algorithms and Classification in Combinatorial Group Theory (Berkeley, CA, 1989)*, volume 23 of *Math. Sci. Res. Inst. Publ.*, pages 195–224. Springer, New York, 1992.

[141] S. M. Gersten. Introduction to hyperbolic and automatic groups. In *Summer School in Group Theory in Banff, 1996*, volume 17 of *CRM Proc. Lecture Notes*, pages 45–70. Amer. Math. Soc., Providence, RI, 1999.

[142] Steve M. Gersten. Isoperimetric and isodiametric functions of finite presentations. In *Geometric Group Theory, Vol. 1 (Sussex, 1991)*, volume 181 of *London Math. Soc. Lecture Note Ser.*, pages 79–96. Cambridge Univ. Press, Cambridge, 1993.

[143] Étienne Ghys and Pierre de la Harpe. Infinite groups as geometric objects (after Gromov). In *Ergodic Theory, Symbolic Dynamics, and Hyperbolic Spaces (Trieste, 1989)*, Oxford Sci. Publ., pages 299–314. Oxford Univ. Press, New York, 1991.

[144] Étienne Ghys and Vlad Sergiescu. Sur un groupe remarquable de difféomorphismes du cercle. *Comment. Math. Helv.*, 62(2):185–239, 1987.

[145] Robert H. Gilman. Formal languages and their application to combinatorial group theory. In *Groups, Languages, Algorithms*, volume 378 of *Contemp. Math.*, pages 1–36. Amer. Math. Soc., Providence, RI, 2005.

[146] Paul Glenn, William W. Menasco, Kayla Morrell, and Matthew Morse. Metric in the curve complex, 2014. micc.github.io.

[147] Constantin Gonciulea. Infinite Coxeter groups virtually surject onto **Z**. *Comment. Math. Helv.*, 72(2):257–265, 1997.

[148] E. Green. *Graph Products of Groups*. PhD thesis, The University of Leeds, 1990.

[149] Rostislav Grigorchuk and Tatiana Nagnibeda. Complete growth functions of hyperbolic groups. *Invent. Math.*, 130(1):159–188, 1997.

[150] Rostislav Grigorchuk and Igor Pak. Groups of intermediate growth: An introduction. *Enseign. Math. (2)*, 54(3–4):251–272, 2008.

[151] Rostislav I. Grigorchuk and Sergei V. Ivanov. On Dehn functions of infinite presentations of groups. *Geom. Funct. Anal.*, 18(6):1841–1874, 2009.

[152] Rostislav I. Grigorchuk and Andrzej Żuk. The lamplighter group as a group generated by a 2-state automaton, and its spectrum. *Geom. Dedicata*, 87(1–3):209–244, 2001.

[153] R. I. Grigorčuk. On Burnside's problem on periodic groups. *Funktsional. Anal. i Prilozhen.*, 14(1):53–54, 1980.

[154] M. Gromov. Hyperbolic groups. In *Essays in Group Theory*, volume 8 of *Math. Sci. Res. Inst. Publ.*, pages 75–263. Springer, New York, 1987.

[155] M. Gromov. Asymptotic invariants of infinite groups. In *Geometric Group Theory, Vol. 2 (Sussex, 1991)*, volume 182 of *London Math. Soc. Lecture Note Ser.*, pages 1–295. Cambridge Univ. Press, Cambridge, 1993.

[156] Mikhael Gromov. Groups of polynomial growth and expanding maps. *Inst. Hautes Études Sci. Publ. Math.*, (53):53–73, 1981.

[157] V. S. Guba. On the properties of the Cayley graph of Richard Thompson's group *F*. *Internat. J. Algebra Comput.*, 14(5–6):677–702, 2004. International Conference on Semigroups and Groups in honor of the 65th birthday of Prof. John Rhodes.

[158] V. S. Guba. The Dehn function of Richard Thompson's group *F* is quadratic. *Invent. Math.*, 163(2):313–342, 2006.

[159] Victor Guba and Mark Sapir. Diagram groups. *Mem. Amer. Math. Soc.*, 130(620):viii+117, 1997.

[160] Marshall Hall, Jr. Coset representations in free groups. *Trans. Amer. Math. Soc.*, 67:421–432, 1949.

[161] Joel Hass and Jeffrey C. Lagarias. The number of Reidemeister moves needed for unknotting. *J. Amer. Math. Soc.*, 14(2):399–428 (electronic), 2001.

[162] A. Hatcher and W. Thurston. A presentation for the mapping class group of a closed orientable surface. *Topology*, 19(3):221–237, 1980.

[163] Allen Hatcher. Topology of numbers. http://www.math.cornell.edu/~hatcher/TN/TNpage.html.

[164] Allen Hatcher. *Algebraic Topology*. Cambridge University Press, Cambridge, 2002.

[165] Susan Hermiller and John Meier. Algorithms and geometry for graph products of groups. *J. Algebra*, 171(1):230–257, 1995.

[166] Derek F. Holt, Sarah Rees, Claas E. Röver, and Richard M. Thomas. Groups with context-free co-word problem. *J. London Math. Soc. (2)*, 71(3):643–657, 2005.

[167] Derek F. Holt and Claas E. Röver. Groups with indexed co-word problem. *Internat. J. Algebra Comput.*, 16(5):985–1014, 2006.

[168] John E. Hopcroft and Jeffrey D. Ullman. *Introduction to Automata Theory, Languages, and Computation.* Addison-Wesley Publishing Co., Reading, Mass., 1979. Addison-Wesley Series in Computer Science.

[169] Heinz Hopf. Enden offener Räume und unendliche diskontinuierliche Gruppen. *Comment. Math. Helv.*, 16:81–100, 1944.

[170] James E. Humphreys. *Reflection Groups and Coxeter Groups*, volume 29 of *Cambridge Studies in Advanced Mathematics.* Cambridge University Press, Cambridge, 1990.

[171] Stephen P. Humphries. Generators for the mapping class group. In *Topology of Low-Dimensional Manifolds (Proc. Second Sussex Conf., Chelwood Gate, 1977)*, volume 722 of *Lecture Notes in Math.*, pages 44–47. Springer, Berlin, 1979.

[172] Stephen P. Humphries. On representations of Artin groups and the Tits conjecture. *J. Algebra*, 169(3):847–862, 1994.

[173] Witold Hurewicz and Henry Wallman. *Dimension Theory.* Princeton Mathematical Series, v. 4. Princeton University Press, 1941.

[174] A. Hurwitz. Ueber Riemann'sche Flächen mit gegebenen Verzweigungspunkten. *Math. Ann.*, 39(1):1–60, 1891.

[175] Gerald Janusz and Joseph Rotman. Outer automorphisms of S_6. *Amer. Math. Monthly*, 89(6):407–410, 1982.

[176] Tadeusz Januszkiewicz and Jacek Świątkowski. Filling invariants of systolic complexes and groups. *Geom. Topol.*, 11:727–758, 2007.

[177] D. L. Johnson. Growth of groups. *Arab. J. Sci. Eng. Sect. C Theme Issues*, 25(2):53–68, 2000.

[178] Dennis Johnson. The structure of the Torelli group. I. A finite set of generators for $\mathcal{I}$. *Ann. of Math. (2)*, 118(3):423–442, 1983.

[179] Dennis Johnson. A survey of the Torelli group. In *Low-Dimensional Topology (San Francisco, Calif., 1981)*, volume 20 of *Contemp. Math.*, pages 165–179. Amer. Math. Soc., Providence, RI, 1983.

[180] Egbert R. Van Kampen. On Some Lemmas in the Theory of Groups. *Amer. J. Math.*, 55(1–4):268–273, 1933.

[181] Svetlana Katok. *Fuchsian Groups.* Chicago Lectures in Mathematics. University of Chicago Press, Chicago, IL, 1992.

[182] Louis H. Kauffman. *On Knots*, volume 115 of *Annals of Mathematics Studies*. Princeton University Press, Princeton, NJ, 1987.

[183] O. Kharlampovich, A. Myasnikov, and M. Sapir. Algorithmically complex residually finite groups, 2012. arXiv:1204.6506.

[184] Sang-hyun Kim. Co-contractions of graphs and right-angled Artin groups. *Algebr. Geom. Topol.*, 8(2):849–868, 2008.

[185] Sang-hyun Kim and Thomas Koberda. Embedability between right-angled Artin groups. *Geom. Topol.*, 17(1):493–530, 2013.

[186] Entry: Stephen Kleene. Free online dictionary of computing. http://foldoc.org.

[187] F. Klein. Neue beiträge zur riemann'schen functionentheorie. *Math. Ann.*, 21(2):141–218, 1883.

[188] Thomas Koberda. Right-angled artin groups and a generalized isomorphism problem for finitely generated subgroups of mapping class groups. *Geom. Funct. Anal.*, 22(6):1541–1590, 2012.

[189] Catherine Labruère and Luis Paris. Presentations for the punctured mapping class groups in terms of Artin groups. *Algebr. Geom. Topol.*, 1:73–114 (electronic), 2001.

[190] Marc Lackenby. A polynomial upper bound on Reidemeister moves. *Ann. of Math.*, to appear.

[191] Amy N. Langville and Carl D. Meyer. *Google's PageRank and Beyond: The Science of Search Engine Rankings*. Princeton University Press, Princeton, NJ, 2006.

[192] J. Lehnert and P. Schweitzer. The co-word problem for the Higman-Thompson group is context-free. *Bull. Lond. Math. Soc.*, 39(2):235–241, 2007.

[193] Yash Lodha and Justin Tatch Moore. A finitely presented group of piecewise projective homeomorphisms, 2013. arXiv:1308.4250.

[194] Roger C. Lyndon and Paul E. Schupp. *Combinatorial Group Theory*. Classics in Mathematics. Springer-Verlag, Berlin, 2001. Reprint of the 1977 edition.

[195] Klaus Madlener and Friedrich Otto. Pseudonatural algorithms for the word problem for finitely presented monoids and groups. *J. Symbolic Comput.*, 1(4):383–418, 1985.

[196] W. Magnus. Das Identitätsproblem für Gruppen mit einer definierenden Relation. *Math. Ann.*, 106(1):295–307, 1932.

[197] Wilhelm Magnus, Abraham Karrass, and Donald Solitar. *Combinatorial Group Theory: Presentations of Groups in Terms of Generators and Relations*. Interscience Publishers [John Wiley & Sons, Inc.], New York-London-Sydney, 1966.

[198] Jean Mairesse and Frédéric Mathéus. Growth series for Artin groups of dihedral type. *Internat. J. Algebra Comput.*, 16(6):1087–1107, 2006.

[199] Dan Margalit. Automorphisms of the pants complex. *Duke Math. J.*, 121(3):457–479, 2004.

[200] Dan Margalit and Jon McCammond. Geometric presentations for the pure braid group. *J. Knot Theory Ramifications*, 18(1):1–20, 2009.

[201] A. Markoff. On the impossibility of certain algorithms in the theory of associative systems. *C. R. (Doklady) Acad. Sci. URSS (N.S.)*, 55:583–586, 1947.

[202] John Meier. *Groups, Graphs and Trees*, volume 73 of *London Mathematical Society Student Texts*. Cambridge University Press, Cambridge, 2008. An introduction to the geometry of infinite groups.

[203] Carl Meyer. *Matrix Analysis and Applied Linear Algebra*. Society for Industrial and Applied Mathematics (SIAM), Philadelphia, PA, 2000. With 1 CD-ROM (Windows, Macintosh and UNIX) and a solutions manual (iv+171 pp.).

[204] J. Milnor. A note on curvature and fundamental group. *J. Differential Geometry*, 2:1–7, 1968.

[205] Igor Mineyev. Groups, graphs, and the Hanna Neumann conjecture. *J. Topol. Anal.*, 4(1):1–12, 2012.

[206] David E. Muller and Paul E. Schupp. Groups, the theory of ends, and context-free languages. *J. Comput. System Sci.*, 26(3):295–310, 1983.

[207] David Mumford, Caroline Series, and David Wright. *Indra's Pearls*. Cambridge University Press, New York, 2002. The vision of Felix Klein.

[208] James R. Munkres. *Topology*. Prentice-Hall Inc., 2nd edition edition, 2000.

[209] Alexei Myasnikov, Alexander Ushakov, and Dong Wook Won. The word problem in the Baumslag group with a non-elementary Dehn function is polynomial time decidable. *J. Algebra*, 345:324–342, 2011.

[210] Hanna Neumann. On the intersection of finitely generated free groups. *Publ. Math. Debrecen*, 4:186–189, 1956.

[211] P. S. Novikov. *Ob algoritmičeskoĭ nerazrešimosti problemy toždestva slov v teorii grupp*. Trudy Mat. Inst. im. Steklov. no. 44. Izdat. Akad. Nauk SSSR, Moscow, 1955.

[212] Piotr W. Nowak and Guoliang Yu. *Large Scale Geometry*. EMS Textbooks in Mathematics. European Mathematical Society (EMS), Zürich, 2012.

[213] A. Yu. Olʹshanskiĭ. Hyperbolicity of groups with subquadratic isoperimetric inequality. *Internat. J. Algebra Comput.*, 1(3):281–289, 1991.

[214] A. Yu. Olʹshanskii. Groups with undecidable word problem and almost quadratic Dehn function. *J. Topol.*, 5(4):785–886, 2012. With an appendix by M. Sapir.

[215] Peter Orlik and Hiroaki Terao. *Arrangements of Hyperplanes*, volume 300 of *Grundlehren der Mathematischen Wissenschaften [Fundamental Principles of Mathematical Sciences]*. Springer-Verlag, Berlin, 1992.

[216] Pierre Pansu. Croissance des boules et des géodésiques fermées dans les nilvariétés. *Ergodic Theory Dynam. Systems*, 3(3):415–445, 1983.

[217] Panagiotis Papasoglu. On the sub-quadratic isoperimetric inequality. In *Geometric Group Theory (Columbus, OH, 1992)*, volume 3 of *Ohio State Univ. Math. Res. Inst. Publ.*, pages 149–157. de Gruyter, Berlin, 1995.

[218] Walter Parry. Growth series of some wreath products. *Trans. Amer. Math. Soc.*, 331(2):751–759, 1992.

[219] David Peifer. Max Dehn and the origins of topology and infinite group theory. *Amer. Math. Monthly*, 122(3):217–233, 2015.

[220] Ch. Pittet. Isoperimetric inequalities for homogeneous nilpotent groups. In *Geometric Group Theory (Columbus, OH, 1992)*, volume 3 of *Ohio State Univ. Math. Res. Inst. Publ.*, pages 159–164. de Gruyter, Berlin, 1995.

[221] A. N. Platonov. An isoparametric function of the Baumslag-Gersten group. *Vestnik Moskov. Univ. Ser. I Mat. Mekh.*, (3):12–17, 70, 2004.

[222] Burkard Polster. *The Mathematics of Juggling*. Springer-Verlag, New York, 2003.

[223] Emil L. Post. Recursive unsolvability of a problem of Thue. *J. Symbolic Logic*, 12:1–11, 1947.

[224] Jerome Powell. Two theorems on the mapping class group of a surface. *Proc. Amer. Math. Soc.*, 68(3):347–350, 1978.

[225] David G. Radcliffe. Rigidity of graph products of groups. *Algebr. Geom. Topol.*, 3:1079–1088, 2003.

[226] R. W. Richardson. Conjugacy classes of involutions in Coxeter groups. *Bull. Austral. Math. Soc.*, 26(1):1–15, 1982.

[227] John Roe. *Lectures on Coarse Geometry*, volume 31 of *University Lecture Series*. American Mathematical Society, Providence, RI, 2003.

[228] Dale Rolfsen. Tutorial on the braid groups. In *Braids*, volume 19 of *Lect. Notes Ser. Inst. Math. Sci. Natl. Univ. Singap.*, pages 1–30. World Sci. Publ., Hackensack, NJ, 2010.

[229] Joseph J. Rotman. *An Introduction to the Theory of Groups*, volume 148 of *Graduate Texts in Mathematics*. Springer-Verlag, New York, fourth edition, 1995.

[230] Lucas Sabalka. Geodesics in the braid group on three strands. In *Group Theory, Statistics, and Cryptography*, volume 360 of *Contemp. Math.*, pages 133–150. Amer. Math. Soc., Providence, RI, 2004.

[231] Andrew P. Sánchez and Michael Shapiro. Growth of $\mathbb{Z}^m$ in cubing extensions, 2016. arXiv/1605.01131.

[232] Mark Sapir. Asymptotic invariants, complexity of groups and related problems. *Bull. Math. Sci.*, 1(2):277–364, 2011.

[233] Mark V. Sapir, Jean-Camille Birget, and Eliyahu Rips. Isoperimetric and isodiametric functions of groups. *Ann. of Math. (2)*, 156(2):345–466, 2002.

[234] Caroline Series. The geometry of Markoff numbers. *Math. Intelligencer*, 7(3):20–29, 1985.

[235] Caroline Series. The modular surface and continued fractions. *J. London Math. Soc. (2)*, 31(1):69–80, 1985.

[236] Jean-Pierre Serre. Amalgames et points fixes. In *Proceedings of the Second International Conference on the Theory of Groups (Australian Nat. Univ., Canberra, 1973)*, pages 633–640. Lecture Notes in Math., Vol. 372, Berlin, 1974. Springer.

[237] Jean-Pierre Serre. *Trees*. Springer Monographs in Mathematics. Springer-Verlag, Berlin, 2003. Translated from the French original by John Stillwell, Corrected 2nd printing of the 1980 English translation.

[238] Herman Servatius. Automorphisms of graph groups. *J. Algebra*, 126(1): 34–60, 1989.

[239] Herman Servatius, Carl Droms, and Brigitte Servatius. Surface subgroups of graph groups. *Proc. Amer. Math. Soc.*, 106(3):573–578, 1989.

[240] J. A. H. Shepperd. Braids which can be plaited with their threads tied together at each end. *Proc. Roy. Soc. Ser. A*, 265:229–244, 1961/1962.

[241] Daniel S. Silver. Knots in the nursery: *(Cats) Cradle Song* of James Clerk Maxwell. *Notices Amer. Math. Soc.*, 61(10):1186–1194, 2014.

[242] John Stallings. *Group Theory and Three-Dimensional Manifolds*. Yale University Press, New Haven, Conn.-London, 1971. A James K. Whittemore Lecture in Mathematics given at Yale University, 1969, Yale Mathematical Monographs, 4.

[243] John R. Stallings. On torsion-free groups with infinitely many ends. *Ann. of Math. (2)*, 88:312–334, 1968.

[244] John R. Stallings. Topology of finite graphs. *Invent. Math.*, 71(3):551–565, 1983.

[245] John R. Stallings. Foldings of *G*-trees. In *Arboreal Group Theory (Berkeley, CA, 1988)*, volume 19 of *Math. Sci. Res. Inst. Publ.*, pages 355–368. Springer, New York, 1991.

[246] Melanie Stein and Jennifer Taback. Metric properties of Diestel-Leader groups. *Michigan Math. J.*, 62(2):365–386, 2013.

[247] John Stillwell. *Classical Topology and Combinatorial Group Theory*, volume 72 of *Graduate Texts in Mathematics*. Springer-Verlag, New York, second edition, 1993.

[248] Michael Stoll. Rational and transcendental growth series for the higher Heisenberg groups. *Invent. Math.*, 126(1):85–109, 1996.

[249] Michał Stukow. Dehn twists on nonorientable surfaces. *Fund. Math.*, 189(2):117–147, 2006.

[250] Michał Stukow. A finite presentation for the mapping class group of a nonorientable surface with Dehn twists and one crosscap slide as generators. *J. Pure Appl. Algebra*, 218(12):2226–2239, 2014.

[251] Dennis Sullivan. On the ergodic theory at infinity of an arbitrary discrete group of hyperbolic motions. In *Riemann Surfaces and Related Topics: Proceedings of the 1978 Stony Brook Conference (State Univ. New York, Stony Brook, N.Y., 1978)*, volume 97 of *Ann. of Math. Stud.*, pages 465–496. Princeton Univ. Press, Princeton, N.J., 1981.

[252] Shin'ichi Suzuki. On homeomorphisms of a 3-dimensional handlebody. *Canad. J. Math.*, 29(1):111–124, 1977.

[253] A. S. Švarc. A volume invariant of coverings. *Dokl. Akad. Nauk SSSR (N.S.)*, 105:32–34, 1955.

[254] Richard J. Thompson. Embeddings into finitely generated simple groups which preserve the word problem. In *Word Problems, II (Conf. on Decision Problems in Algebra, Oxford, 1976)*, volume 95 of *Stud. Logic Foundations Math.*, pages 401–441. North-Holland, Amsterdam, 1980.

[255] William P. Thurston. *Three-Dimensional Geometry and Topology. Vol. 1*, volume 35 of *Princeton Mathematical Series*. Princeton University Press, Princeton, NJ, 1997. Edited by Silvio Levy.

[256] J. Tits. Free subgroups in linear groups. *J. Algebra*, 20:250–270, 1972.

[257] Pekka Tukia. On quasiconformal groups. *J. Analyse Math.*, 46:318–346, 1986.

[258] Richard D. Wade. Folding free-group automorphisms. *Q. J. Math.*, 65(1):291–304, 2014.

[259] Bronislaw Wajnryb. Mapping class group of a surface is generated by two elements. *Topology*, 35(2):377–383, 1996.

[260] Bronisław Wajnryb. Mapping class group of a handlebody. *Fund. Math.*, 158(3):195–228, 1998.

[261] Herbert S. Wilf. *generatingfunctionology*. A K Peters, Ltd., Wellesley, MA, third edition, 2006.

[262] Daniel T. Wise. *From Riches to Raags: 3-Manifolds, Right-Angled Artin groups, and Cubical Geometry*, volume 117 of *CBMS Regional Conference Series in Mathematics*. Published for the Conference Board of the Mathematical Sciences, Washington, DC; by the American Mathematical Society, Providence, RI, 2012.

[263] W. Woess. What is a horocyclic product, and how is it related to lamplighters?, 2014. arXiv:1401.1976.

[264] Wolfgang Woess. Lamplighters, Diestel-Leader graphs, random walks, and harmonic functions. *Combin. Probab. Comput.*, 14(3):415–433, 2005.

[265] Robert Young. The Dehn function of $SL(n; \mathbb{Z})$. *Ann. of Math. (2)*, 177(3):969–1027, 2013.

[266] Robert Young. Filling inequalities for nilpotent groups through approximations. *Groups Geom. Dyn.*, 7(4):977–1011, 2013.

[267] Guoliang Yu. The Novikov conjecture for groups with finite asymptotic dimension. *Ann. of Math. (2)*, 147(2):325–355, 1998.

Index